Omics Studies of Medicinal Plants

Herbal drugs play a pivotal role in modern medicine and pharmaceutical care; however, only limited biotechnology applications have been seen in medicinal plants. Revolutions in high-throughput approaches emphasize omics approaches, such as genomics, transcriptomics, proteomics, and metabolomics. A volume in the Exploring Medicinal Plants series, this book provides a comprehensive and in-depth analysis of breakthroughs in high-throughput approaches for the research of medicinal plants.

Exploring the principles and applications of omics technologies, this book is essential for those working on or are involved in the modern research of medicinal and aromatic plants. There is also a strong focus on practical implications of these technologies through exploring the safety aspects and conservation strategies of various plants.

From informative discussions on the latest research to a holistic evaluation of their potential applications, this book appeals to students, researchers and professionals working with medicinal and aromatic plants, as well as healthcare professionals interested in the area.

EXPLORING MEDICINAL PLANTS

Series Editor Azamal Husen
Wolaita Sodo University, Ethiopia

Medicinal plants render a rich source of bioactive compounds used in drug formulation and development; they play a key role in traditional or indigenous health systems. As the demand for herbal medicines increases worldwide, supply is declining as most of the harvest is derived from naturally growing vegetation. Considering global interests and covering several important aspects associated with medicinal plants, the Exploring Medicinal Plants series comprises volumes valuable to academia, practitioners, and researchers interested in medicinal plants. Topics provide information on a range of subjects including diversity, conservation, propagation, cultivation, physiology, molecular biology, growth response under extreme environment, handling, storage, bioactive compounds, secondary metabolites, extraction, therapeutics, mode of action, and healthcare practices.

Led by Azamal Husen, Ph.D., this series is directed to a broad range of researchers and professionals consisting of topical books exploring information related to medicinal plants. It includes edited volumes, references, and textbooks available for individual print and electronic purchases.

Omics Studies of Medicinal Plants, Altaf Ahmad and Ambreen Asif

Traditional Herbal Therapy for the Human Immune System, Azamal Husen

Environmental Pollution and Medicinal Plants, Azamal Husen

Herbs, Shrubs and Trees of Potential Medicinal Benefits, Azamal Husen

Phytopharmaceuticals and Biotechnology of Herbal Plants, Sachidanand Singh, Rahul Datta, Parul Johri, and Mala Trivedi

Exploring Poisonous Plants: Medicinal Values, Toxicity Responses, and Therapeutic Uses, Azamal Husen

Plants as Medicine and Aromatics: Conservation, Ecology, and Pharmacognosy, Mohd Kafeel Ahmad Ansari, Bengu Turkyilmaz Unal, Munir Ozturk and Gary Owens

Omics Studies of Medicinal Plants

Edited by
Altaf Ahmad and Ambreen Asif

CRC Press
Taylor & Francis Group
Boca Raton London New York

CRC Press is an imprint of the
Taylor & Francis Group, an **informa** business

First edition published 2023
by CRC Press
6000 Broken Sound Parkway NW, Suite 300, Boca Raton, FL 33487-2742

and by CRC Press
4 Park Square, Milton Park, Abingdon, Oxon, OX14 4RN

CRC Press is an imprint of Taylor & Francis Group, LLC

ISBN: 9781032014791 (hbk)
ISBN: 9781032015675 (pbk)
ISBN: 9781003179139 (ebk)

DOI: 10.1201/9781003179139

Typeset in Times
by Newgen Publishing UK

Contents

Preface

Medicinal plants (MPs) deliver some of the most imperative sources of life-preserving medications to the entire world. A majority of the world's inhabitants chiefly rely on herbal remedies. Medicinal plants are also a good source of income. The herbal medicine market, over the last decade, has shown immense growth and is projected to reach $107 billion by the end of 2020. The emerging demand for herbal medicine has also burdened their original stocks. A cumulative worldwide urge for herbal drugs is accompanied by a dwindling reserve of medicinal plants due to over-harvesting. Approximately 90 percent of the harvest comes from wild and naturally growing resources, which are not explicitly cultivated for commercial usage. This overutilization of medicinal plants from their natural resources has directed these plant species to the danger of extinction. To cope with this rising demand for medicinal plants, their genetic improvement is required for having higher yields and more active ingredients. These ameliorated high-yielding MPs can be utilized for commercial cultivation purposes.

Biotechnological intrusions have likewise played an essential role in upgrading the yield and quality of crops. Crops have also been engineered to produce valuable enzymes, heterologous proteins, and antibodies. Today, natural-product drugs have a very pivotal and definite role in modern medicine and pharmaceutical care. Despite great progress in plant molecular biology for the improvement of crop plants, only limited biotechnology applications have been seen in medicinal plants. Recent revolutions in high-throughput approaches have emphasized the research intentions toward omics approaches, such as genomics, transcriptomics, proteomics and metabolomics. In recent years, extensive attention has been paid to the up-gradation of medicinal and aromatic plants using "omics" approaches. Various groups of scientists are using these approaches to characterize medicinal plants at genes, proteins, and metabolite levels. The first chapter of the book deals with the role of medicinal plants in the treatment of various health ailments. The second chapter focuses on the *in vitro* biosynthesis, elicitation, and upregulation of secondary metabolites, the role of integrative omics approaches, and the prospects of herbal-based clinical practices. Chapter 3 provides valuable insights into how biotechnology approaches have helped address various challenges faced by medical plant research and illustrate future strategies that can be adopted. Recent interventions that have the potential to revolutionize medicinal plant research for pharmaceutical industry application are also discussed. The current scenario of omics data in medicinal plant research is described in Chapter 4 with a brief discussion on issues that need to be addressed while analyzing these data. There have been quite a favorable number of reports available for omics analyses in several medicinal plants. The authors have also attempted to document selected research works in this chapter. An overview of the genomics, transcriptomics and proteomics approaches for the improvement of medicinal plants is given in Chapter 5. Information about genomic profiling of medicinal plants has increased greatly in the past decade. Efforts have been undertaken to characterize medicinal plants at genome levels with the help of molecular markers for justifiable application of the gene pool, its conservation and future research programs. Chapter 6 documents the genome profiling of medicinal plants using the DNA fingerprinting approach. Adulteration has become prevalent in the herbal healthcare industry and is becoming a threat to the safety of consumers. This is further exacerbated by the fact that plants and their products sold in the herbal market are devoid of apparent phenotypic features that mask their authenticity due to being either in a dried or powdered state. Chapter 7 explains that the challenge of misidentification can be overcome by the molecular technique of DNA barcoding. This method uniquely identifies every living organism, just like every product in the supermarket is barcoded and labeled. This will aid consumers to easily and efficiently identify medicinal plants and their products irrespective of their physical state. Customs officers and other related agencies could use these generated barcodes for checking illegal trade and thus carry out conservation efforts. DNA barcoding can also help in the conservation of endangered plant species and protect biodiversity. With species-specific barcode databases, chances of adulterations, accidents related to herbal medication, and over-exploitation of flora will be reduced. The author has further stressed that, en masse, this will be a big step toward the conservation of resources. Transcriptomics analysis of genes is another advancement in the analysis of gene expression at mRNA transcript levels. In the case of medicinal plants, the information on the use of this technique is meager because genomic sequences of limited medicinal plants are available. The author of Chapter 8 has discussed the relevance of transcriptomic approaches in the analysis of the expression of the gene of metabolic pathways associated with the biosynthesis of the secondary metabolites. While genomics and transcriptomics technologies illustrate the expression profiling of the genes in an organism and help understand the structures and functions of expressed proteins, the profiling of metabolites of an organism is provided by metabolomics at various life levels. Several techniques, such as LC-MS, GC-MS, NMR-MS, and so on, are associated to illustrate the metabolic pattern of medicinal plants, qualitatively and quantitatively. The LC-MS is one of the growing and highly advanced techniques used for metabolomic analysis due to its high specificity, selectivity and precision and accuracy in an overall assessment of data selectivity and integrity. Chapter 9 describes the role of LC-MS in metabolomic and the various steps to reduce the associated problems such as experimental design, selection of method, sample preparation, and method optimization. Metabolomics is also an important tool in quality control and in standardizing the production of medicinally important natural products. It has also been used in identifying and characterizing rare or endangered species of medicinal plants. It has provided useful information regarding the biochemical evolutionary relationships among different medicinal plants. Chapter 10 provides information on the metabolomic approaches used to study some common medicinal plants. Since the genetic variability in medicinal plants is very narrow, induced mutagenesis is witnessed as a practical tool for broadening the genetic variation to circumvent these bottleneck conditions. Plant mutagenesis allows the deliberate induction of viable mutations to generate the novel mutants with upgraded agronomic characters and increased yield. With the rapid understanding of different types of mutagens and other mutation technologies, plant mutagenesis has become a right-hand tool in mutation breeding programs that embrace its increased use in related research fields. Chapter 11 discusses a valuable vision of mutagenesis on account of ample knowledge of biological, morphological, yield, and cytological parameters. Also, the expanding research horizons in terms of phytochemical, molecular, and proteomics allow the significant possibility to assess induced mutagenesis in these medicinal plants. Chapter 12 summarizes the fabrication of nanoparticles from various parts of herbal plants, characterization techniques, their potential application in antimicrobial activities and future perspectives. Adulteration with medicinal plants always remains a major concern of users for reasons of safety and effectiveness. Correct authentication and identification of genuine medicinal herbs has always been a great challenge for the people using them. With an increase in the consumption of herbal medicine, the authentication and characterization of these medicinal plants are becoming a critical issue. Further, over-exploitation of the medicinal plant in the preparation of herbal formulations without any prior steps for their conservation has led to their becoming endangered. Even many plant species are on the verge of extinction. Chapter 13 provides an insight into the safety aspects and conservation strategy of the medicinal plants. In this last chapter, the phytochemical evaluation of some medicinal plants has been provided by the author, and the biotechnological approach for the conservation of these medicinal plants is discussed. The present book, therefore, provides a comprehensive treatise of the chapters to provide an in-depth analysis of the recent breakthroughs in high-throughput approaches for the research of medicinal plants. We hope that this book will cater to the needs of all those who are working on or have an interest in the modern research of medicinal plants.

Altaf Ahmad
Ambreen Asif
Editors

Editor Biographies

Altaf Ahmad, Ph.D., is Professor of Plant Biotechnology in the Department of Botany, Aligarh Muslim University, Aligarh, India. Dr. Ahmad has 22 years of research experience and more than 125 significant publications to his credit. A regular reviewer and on the panel of editorial boards of national and international journals, Dr. Ahmad is also a member of prestigious national and international scientific societies. He has guided the thesis research of 30 Ph.D. students (12 as supervisor and 18 as co-supervisor) and over 30 M.Sc. students. He has organized several national and international seminars/symposia. Dr. Ahmad handled several research projects funded by various agencies such as UKRI-GCRF, BBSRS-UK, DST, DBT, CSIR, UGC, CCRUM, ICAR, and INSA. He was Visiting Scientist at Virginia Polytechnic and State University, USA, University of Freiburg, University of Aberdeen, and the University of Cologne. Based on his research contribution, Dr. Ahmad has been honored with the INSA Medal for Young Scientist (2003) and Professor LSS Kumar Memorial Award (2003) of Indian National Science Academy, Young Scientist Award (2005) of Council of Science & Technology, Govt. of Uttar Pradesh, and was elected an Associate of National Academy of Agricultural Sciences.

Ambreen Asif, Ph.D., is a postdoctoral fellow in the Department of Botany, Aligarh Muslim University, Aligarh, India. She earned her M.Sc. (Botany) degree in 2013 from Aligarh Muslim University and secured first position in order of merit, and thereafter was awarded her Ph.D. degree on the topic "Characterization of genetic variability and mutagenecity induced by mutagens in *Nigella sativa* L. – an important medicinal herb of India" in 2019. She has published several significant research papers in journals of national and international repute such as *Industrial Crop Products*, *Scientific Reports* and *Genes and Plants*. She is a frequent participant in scientific conferences and seminars in India.

Dr. Asif is a thorough and meritorious student, evident by awards such as the "University Gold Medal" for securing first position in her M.Sc. (Botany), "Anjalina Khan Award" for securing first position in her B.Sc. (Honors), and "Mrs. M.J. Haider Academic Memorial Award" for securing first position at her school in Aligarh Muslim University, Aligarh. She is a recipient of the Innovation in Science Pursuit for Inspired Research (INSPIRE) fellowship, a flagship program of the Department of Science and Technology, Government of India for her Ph.D. work. Based on her research contribution, she received the Junior Scientist of the Year award (2018) from the National Environmental Science Academy and the Emerging Scientist award from the Indo-US A.P.J. Abdul Kalam STEM ER Centre of The Ohio State University (USA) and Aligarh Muslim University (India), 2018. She was also awarded with the prestigious "NEWS India-UK Fellowship" of BBSRC, UK to work as a visiting scientist at the University of Aberdeen, Scotland, UK (2017).

Contributors

Limenew Abate, Nanotechnology Centre of Excellence, Addis Ababa Sciences and Technology University, Addis Ababa, Ethiopia; Department of Industrial Chemistry, Addis Ababa Sciences and Technology University, Addis Ababa, Ethiopia

Sahrish Aftab, Department of Botany, Aligarh Muslim University, Aligarh, U.P., India

Lokesh Kumar Agarwal, Department of Chemistry, Mohanlal Sukhadia University, Udaipur, Rajasthan, India

Altaf Ahmad, Department of Botany, Aligarh Muslim University Aligarh, Uttar Pradesh, India

Sayeed Ahmad, Bioactive Natural Product Laboratory, Department of Pharmacognosy and Phytochemistry, School of Pharmaceutical Education and Research, Jamia Hamdard, New Delhi, India

Neha Arora, Department of Integrative Biology and Pharmacology, McGovern Medical School, University of Texas Health Science Center at Houston, TX, USA

Mohd Ashraf Ashfaq, Department of Botany, School of Chemical and Life sciences, Jamia Hamdard, New Delhi, India

Fekadu Ashine C., Chemical and Construction Inputs Industry Development Institute, Addis Ababa, Ethiopia; Department of Chemical Engineering, Addis Ababa Science and Technology University, Ethiopia

Ambreen Asif, Department of Botany, Aligarh Muslim University, Aligarh, U.P., India

Archana Bachheti, Department of Environment Science, Graphic Era University, Dehradun, Uttarakhand, India

Rakesh Kumar Bachheti, Nanotechnology Centre of Excellence, Addis Ababa Sciences and Technology University, Addis Ababa, Ethiopia; Department of Industrial Chemistry, Addis Ababa Sciences and Technology University, Addis Ababa, Ethiopia

Faizia Bano, Department of Biotechnology, Jamia Hamdard, New Delhi, India

Geetanjali Baruah, Department of Biotechnology, Assam Kaziranga University, Jorhat, Assam, India

Sana Basri, Department of Botany, Aligarh Muslim University, Aligarh, U.P., India

Mohd Yasin Bhat, Department of Plant Sciences, School of Life Sciences, University of Hyderabad, Prof. C. R. Rao Road, Gachibowli, Hyderabad, Telangana, India

Ankita Chatterjee, National Chemical Laboratory, Dr. Homi Bhabha Road, Pune, Maharashtra, India

Neha Chaurasiya, Department of Botany, Aligarh Muslim University, Aligarh, U.P., India

Siva P. K. Chetri, Department of Botany, Dimoria College, Kamroop Khetri, Assam, India

Gaurav, Bioactive Natural Products Laboratory, Department of Pharmacognosy and Phytochemistry, School of Pharmaceutical Education and Research, Jamia Hamdard, New Delhi, India

Irfan A. Ghazi, Department of Plant Sciences, School of Life Sciences, University of Hyderabad, Prof. C. R. Rao Road, Gachibowli, Hyderabad, Telangana, India

Mir Zahoor Gul, Department of Plant Sciences, School of Life Sciences, University of Hyderabad, Prof. C. R. Rao Road, Gachibowli, Hyderabad, Telangana, India, and Department of Biochemistry, University College of Sciences, Osmania University, Hyderabad, Telangana, India

Amit Kumar Gupta, Department of Botany, Mohanlal Sukhadia University, Udaipur, Rajasthan, India

Azamal Husen, Wolaita Sodo University, Wolaita, Ethiopia

Shipra Rani Jha, K.L. Mehta Dayanand College for Woman, Fariadabad, Haryana, India

Samia Khan, King Abdul Aziz University, Jeddah, Saudi Arabia

Ratan Lal, Department of Botany, Mohanlal Sukhadia University, Udaipur, Rajasthan, India

Mohd Mughees, Department of Biotechnology, School of Chemical and Life Sciences, Jamia Hamdard, New Delhi, India; Department of Breast Medical Oncology, the University of Texas MD Anderson Cancer Center, Houston, Texas, USA

Md. Salik Noorani, Department of Botany, School of Chemical and Life sciences, Jamia Hamdard, New Delhi, India

Renu, Department of Botany, Aligarh Muslim University, Aligarh, U.P., India

Rabea Parveen, Bioactive Natural Product Laboratory, Department of Pharmacognosy and Phytochemistry, School of Pharmaceutical Education and Research, Jamia Hamdard, New Delhi, India

Elizabeth P. Ryan, Department of Environmental and Radiological Health Sciences, Colorado State University and Colorado School of Public Health, Fort Collins, CO, USA

Jagajjit Sahu, GyanArras Academy, Gothapatna, Malipada, Bhubaneswar, Odisha, India

Anindita Sengupta, Department of Biotechnology, Jamia Hamdard, New Delhi, India

Kuldeep Sharma, Department of Botany, Mohanlal Sukhadia University, Udaipur, Rajasthan, India

Kritika Shukla, Department of Botany, School of Chemical and Life Sciences, Jamia Hamdard, New Delhi, India

Vibhuti Singh, Department of Botany, Swami Shraddhanand College, University of Delhi, Delhi, Delhi, India

Akanksha Vashishtha, Department of Botany, CCS University, Meerut, U.P., India

Saima Wajid, Department of Biotechnology, School of Chemical and Life Sciences, Jamia Hamdard, New Delhi, India

1 Medicinal Plants for Health Care

Neha Chaurasiya and Altaf Ahmad

CONTENTS

1.1 INTRODUCTION

Natural food has successfully formed the basis of medicines for the treatment of various health ailments. Every plant is similar to an industrial unit and competent in synthesizing an infinite quantity of extremely multifaceted and unusual chemical substances (Kinghorn et al. 2011). It is believed that at least 120 distinctive chemical substances derived from plants are well thought out as vital drugs now in use in the world, while a number of other drugs are uncomplicated artificial modifications of the natural products (Farooqi and Sreeramu 2001). Regardless of the long times past of accomplishment in inventions of drugs from natural sources, natural products have fallen out of favor in the present elevated throughput selection. People of the developing world (around 80% of the world's population) primarily rely on herbal products for the treatment of most health ailments. Many developed countries have also included herbal remedies. Complementary and alternative medicines (CAMs) are widespread in the United Kingdom, Europe, North America and Australia (Braun 2010) because of the belief that herbal medicine will promote healthier

DOI: 10.1201/9781003179139-1

living. Another reason for the increasing demand for these medicines in developed countries is a balanced and moderate approach to healing. Therefore, there has been a big boom in sales of medicines of plant origin. This represents a considerable proportion of the world's drug market. Approximately 21,000 plants have been considered by the World Health Organization as having medicinal value (WHO 2005). The WHO has considered 252 drugs as basic and essential. Interestingly, the contribution of plant-based drugs is 11% in this list (Akinyemi, Oyewole and Jimoh 2018). In addition to this, a large number of synthetic drugs are obtained from medicinal plants. Digoxin is obtained from *Digitalis* spp., atropine from *Atropa belladonna*, quinine and quinidine from *Cinchona* spp., vincristine and vinblastine from *Catharanthus roseus*. Many anti-tumor and anti-infectious drugs are available on the market. Ninety species of plants provide around 120 chemicals in pure forms. These chemicals are used in medicines. The number of plants known for their medicinal utility is consistently increasing with new ethnomedicinal claims coming up fast. One of the priority programs of WHO is to examine the drugs that are used in traditional medicine in various countries. Medicinal plants represent a priceless resource with much use and potential value to humanity.

The last century witnessed a return to nature, particularly in matters related to medicine. Rapid development in omics technologies is contributing immensely to the exploration of biodiversity for new medicines. The advances in these omics' technologies have further redefined the overall scope and utility of the exploration of medicinal plants for commercially valuable resources. Detailed studies of medicinal plants at genome, transcriptome, proteome and metabolome levels are being carried out to identify biomarkers and targets for therapies for various diseases (Nicora et al. 2020). A search for new therapeutic resources led to the re-examination of nature that has given remarkable results. Many ancient medicinal herbs like *Podophyllum peltatum, Rauwolfia serpentina, Catharanthus roseus* and *Stephania cepharantha* are some classic examples. *Lonchocarpus cyanescens, Costus after, Terminalia, Dianthus barbatus, Aspidosperma quebracho-blanco, Brucea antidysenterica, Jatropha gossypiifolia, Maytenus serrata, Camptotheca acuminata, Sanguinaria canadensis, Silybum marianum, Lithospermum ruderale, Dioscorea composita, Myrsine africana* and *Oldenlandia affinis* have been found to be active against a variety of human ailments when examined through pharmacological trials. *Podophyllum hexandrum*, which is used in India in the Ayurvedic systems of medicine extensively, is now proving useful in the treatment of cancer, Hodgkin's disease, monocytic leukemia and non-Hodgkin's lymphoma. For the treatment of *Condyloma acuminatum* and *Tinea capitis*, this drug is a well-thought-out drug of choice (Goel, Prasad and Sharma 1998). With the increased experimental evidence of the potential of medicinal plants to treat various health ailments, research interests in different aspects of medicinal plants have developed tremendously. Increased involvement of drug companies in manufacturing herbal products for herbal healthcare services has further intensified this interest in the research of medicinal plants. However, the growing demand for herbal drugs has led to an uncontrolled wild collection that, in turn, has become one of the potential threats to medicinal plants. Being an invaluable natural resource, there is a need to conserve and sustainably utilize medicinal plants for the benefit of the present and future human generations. In the present chapter, an update of medicinal plants, plant-derived drugs and their use in various therapies is provided.

1.2 MEDICINAL PLANTS FOR THE TREATMENT OF THE GASTRO-INTESTINAL SYSTEM

The human body has a number of organ systems; among them, the gastro-intestinal system is one of the most important organs, which is highly prone to several disorders. The medicines available on the market have less potential and result in several complexities. Prasad, Shyma and Raghavendra (2013) recorded and analyzed various medicinal plants that are orally transmitted for the treatment of digestive system diseases by the tribal population in Wayanad district. The parts of the plant are used solely or in union to treat various diseases such as piles, ulcers, abdominal pain, diarrhea, dysentery, vomiting, acidity, jaundice and stomach pain. Based on results obtained from the review, several plants have been suggested in certain studies to get rid of digestive disorders and gastrointestinal cancers, including *Glycyrrhiza glabra*, *Ribes nigrum*, *Phyllanthus niruri*, *Lantana camara*, *Streblus asper*, *Cordia myxa*, *Coriandrum sativum*, *Cyperus rotundus*, *Azadirachta indica*, *Aloe vera*, *Plantago asiatica*, *Passiflora* spp. In several investigations, there are some principle bioactive compounds: kobusone and isokobusone, rotundone, auxins and phenolic compounds: isoflavone glabrene, glabridin, phytoestrogens, glycosides, flavonoids and saponins which are isolated from the medicinal plants results in inhibition in sufferance of patients by reducing pain (Anbari et al. 2019). Out of 482 vascular plants were recorded, *Phyllanthus emblica*, *Centella asiatica* and *Aegle marmelos*, *Carica papaya*, *Portulaca oleracea* and *Psidium guajava* are present in the Kuki-Chin region whose boundaries are shared by Bangladesh, India and Myanmar are used in the treatment of GAS-Ds like abdominal pain epigastric, gastroenteritis presumed infections and diarrhea.

1.3 MEDICINAL PLANTS FOR THE TREATMENT OF THE CARDIOVASCULAR SYSTEM

The disorders of the cardiovascular system affect the pumping organs: heart, arteries, veins and capillaries. In today's world, it is the leading cause of death. The diseases related to this system constitute 31% of annual global mortality, according to the World Health Organization (WHO). The major factors responsible for such problems are smoking, high consumption of saturated fat and cholesterol and physical inactivity (Adegbola et al. 2017). Some of the common medicinal plants, with their functional bioactive compound, have the potential for the treatment of cardiovascular diseases Table 1.1. *Allium sativum*, commonly called garlic, is used in the treatment of atherosclerosis and hyperlipidemia (Sun, Wang and Qin 2018). *Crocus sativus*, or Saffron, belongs to the family Iridaceae and is helpful in vasodilation (Razavi et al. 2016). *Hibiscus sabdariffa*, also known as roselle, has the potential in lowering blood pressure (Ojeda et al. 2010). The extracts of *Camellia sinensis* (Tea) can lower high blood pressure by elevating brachial artery flow-mediated dilation (FMD) (Ras, Zock and Draijer 2011). *Commiphora wightii* (guggal) and *Terminalia arjuna* (Arjuna) have an effect on cardiovascular diseases, including ischemic heart disease, congestive heart failure, arrhythmias and high blood pressure (Rastogi, Pandey and Rawat 2016).

1.4 NEURO-PROTECTIVE PLANTS

The protection of the nervous system refers to the strategies and associated mechanisms which are competent to shield the central nervous system (CNS) against injury of the nervous system caused

TABLE 1.1
Medicinal Plants and Their Functional Bioactive Compound Having Potential of the Treatment of Cardiovascular Diseases

Medicinal plants	Functional bioactive compounds
Salix alba	Aspirin
Digitalis purpurea	digoxin
Ephedra sinica	ephedrine
Monascus purpureus	lovastatin
Taxus brevifolia	taxol
Astragalus membranaceus	Astragaloside IV
Rauvolfia serpentina	reserpine

due to both short-term (e.g., stroke or trauma) and long-term diseases (e.g., Alzheimer's disease (AD) and Parkinson's disease (PD)). Some medicinal plants like *Ginkgo biloba* L., *Panax ginseng* C.A. Mayer and *Scutellaria baicalensis* Georgi help in the treatment of the above-mentioned diseases (Iriti et al. 2010). Ginseng has been found to help in memory improvement, better learning performance and motor activity and is extract-rich in bioactive compounds from its roots (called aphrodisiac) that help in the treatment of several different types of (neuro) inflammatory diseases (Oliynyk and Oh 2013). There is an abundance of clinically important EOs and their constituents from *Nigella sativa*, *Acorus gramineus*, *Lavandula angustifolia*, *Eucalyptus globulus*, *Mentha piperita*, *Rosmarinus officinalis*, *Jasminum sambac*, *Piper nigrum* and so many other plants, have the potential to treat neurodegenerative disorders (Ayaz et al. 2017).

1.5 LIVER-PROTECTIVE PLANTS

The liver is one of the vital organs which is accountable for catabolism, anabolism, elimination of toxic waste and homeostasis in the body of a human (Chiang 2014). The damaging effects of toxicants and pathogens on the liver and its cells include inflammatory (i.e., hepatitis) or non-inflammatory (i.e., hepatosis), liver diseases, liver fibrosis (i.e., cirrhosis), jaundice and alcoholic liver disease. The disease related to the liver affects the majority of the population around the world, with about 20 lakhs dying annually irrespective of age, sex, region and race (Asrani et al. 2013). Various research has proven the liver-protective role of different herbs and their formulations (Table 1.2). The term "hepatoprotective" means to protect or prevent the liver from severe damage. *Aloe vera*, *Andrographis paniculata*, *Cassia fistula*, *Curcuma longa* and *Boerhaavia diffusa*, with their bioactive components, namely, andrographolide, aloe-emodin, curcumin, karaviloside III, catechin, chebulic acid and gallic acid, are found to be effective in hepatoprotective drug discovery development. The activities of hepatoprotection involve intensifying liver cell regeneration or reducing the programmed cell death of liver cells and later reducing the leakage as well as inhibiting lipid peroxidation and elevating activities of antioxidant enzymes of liver.

1.6 ANTICOAGULANT PLANTS

Anticoagulants are also called as blood thinners, antiplatelet drugs and anti-thrombosis. They prevent the formation of blood clots. The treatment of blood clots is essential as clots in the circulatory

TABLE 1.2
Medicinal Plants Having Hepatoprotective Properties

Plants	References
Andrographis lineata	Sangameswaran et al. 2008
Andrographis paniculate	Handa and Sharma 1990
Azadirachta indica	Chattopadhyay et al. 1992
Careya arborea	Senthilkumar et al. 2009
Cassia fistula	Bhakta et al. 2001
Cleome viscosa	Gupta and Dixit 2009
Phyllanthus amarus	Naaz et al. 2007
Morinda citrifolia	Wang et al. 2008
Polygala arvensis	Dhanabal et al. 2006
Pterocarpus santalinus	Manjunatha 2006
Phyllanthus reticulates	Das et al. 2008
Picrorhiza kurroa	Ansari et al. 2009
Solanum nigrum	Sultana et al. 1995

system (heart and blood vessels) can cause heart attacks, strokes and blockages. They allow the natural clot-dissolving activity of the body to sweep away at deep venous thrombosis (DVT) or pulmonary embolism (PE) that has already formed and put a stop to the formation of additional clots (Goldhaber and Grasso-Correnti 2002). Warfarin, heparins (low molecular weight heparins, LMWH), direct oral anticoagulants (DOACs), vitamin K antagonists and their derivatives are the major anticoagulant drugs that have played a critical role in the clinical setting for more than 50 years (Stone 2007, Bounameaux 2009). In the United States, each year 500,000 persons go through chronic anticoagulant therapy to get rid of venous and arterial thromboembolism (Stults, Dere and Caine 1989). Due to the biological activities of plants, they may play an important role as substitutes for the development of new anticoagulant agents (Matsubara et al. 2001). In the Eastern Cape Province of South Africa, nine plans were tested for their antithrombotic activity. On several plant parts, organic (methanol) and aqueous (distilled water) extractions were performed and utilization of the thrombin assay and clotting time assays (thrombin-induced and $CaCl_2$-induced) was possible (Kee et al. 2008).

In contrast to standard warfarin, it is found that the medicinal plant, *Careya arborea*, possesses anticoagulant activities. In Ayurveda, it is commonly known as "Kumbhi". Its main constituents are terpenoids, flavonoids, alkaloids, saponins and tannins (Wadkar et al. 2008). The methanolic bark extract of *C. arborea* (Subhadra et al. 2010) and a new dicoumarin, named dimeresculetin, isolated from *Viola yedoensis* Makino (Zhou et al. 2009) extended the time taken for blood clotting, and there was a substantial increase in the activated Partial Thromboplastin Time, Prothrombin Time and Thrombin Time and the aqueous and ethanol extract of *Synclisia scabrida* increases the prothrombin time of normal plasma (Afonne et al. 2000).

The aqueous leaf extract of *Melastoma malabathricum* L. exhibited anticoagulant activity, which influences the intrinsic pathway of the coagulation cascade by causing clotting factor(s) deficiency (Manicam et al. 2013). Almost the same effect is shown by the methanol extract from the leaf of Eichhornia crassipes, commonly known as "Common Water Hyacinth": this coagulating activity is due to the presence of polysaccharides (Gomes et al. 2009). Extracts from the leaves of *Gloriosa superba* Linn., also known as "Glory lily", exhibited the property of anticoagulation by hindering thrombin-induced clotting (Kee et al. 2008). *Bauhinia forficata* is commonly known as "Pata de vaca" in Brazil. The aqueous extract from aerial parts of *Bauhinia forficata* is a source of natural inhibitors of serine-protease involved in blood clotting disturbances induced by snake venom (Oliveira et al. 2005). A soft wooded shrub, *Jatropha curcas* L., is commonly found in rural areas in India. The latex of *J. curcas* decreased substantially the time of coagulation of human blood. Its latex also shows procoagulant activities (Osoniyi and Onajobi 2003). Procoagulants are the substances that promote the conversion of the inactive protein prothrombin to the clotting enzyme thrombin in the blood. The highest coagulant activity is shown by the polysaccharides (Galactose, galacturonic acid and mannose) extracted from *Porana volubilis*. Its activity is arbitrated by the increment of thrombin inhibition which in turn is mediated by heparin cofactor II but not by antithrombin (Yoon et al. 2002). The polyphenolic polysaccharide preparation of "Horseweed" (*Erigeron canadensis*) also shows the same effect as that *Porana volubilis* (Pawlaczyk et al. 2010).

1.7 ANTICANCER PLANTS

After cardiovascular diseases, cancer is the second most deadly disease. The treatment of cancer includes therapies (radio, chemo, etc.) that have their own pros and cons, but nowadays, for cancer treatment, plants are found and proved to be outstanding therapeutic agents (Khazir et al. 2014). The drugs extracted from plants are very much essential because they are not synthetic and can be obtained easily. They play a major role in the dietary intake of patients as they can be taken orally (Cornblatt et al. 2007, Amin et al. 2009). The plants are in high demand as they do not cause any toxicity in normal cells of humans, they produce cytotoxic effects on cancerous cells and the body's tolerance power is remarkable as these compounds are naturally procured (Unnati et al. 2013). They also prove to be the potential inhibitors of cancer cell lines (Sivaraj et al. 2014). In 2007, the estimation by World Health Organization was that the worth of drugs derived from plants was US$ 100 billion, and in 2050 the expected trade would be US$5 trillion (Rajeswara Rao 2007).

The anticancer agent sulforaphane, isolated from *Brassica*, induces phase 2 detoxification enzyme in breast cancers. It prevents the growth of the tumor and also shows antiproliferative effects (Heiss 2001, Cornblatt et al. 2007, Pledgie-Tracy, Sobolewski and Davidson 2007),.Paclitaxel, sold under the brand name Taxol, is classified as a plant alkaloid, which is isolated from *Taxus brevifolia*; it causes disruption of microtubule and spindle fibers, stops division of cells, causes programmed cell death and stops the mechanism of protein synthesis (Pezzuto 1997, Jordan and Wilson 2004, Cragg and Newman 2005, Pawar, Vinugala and Bothiraja 2014, Che et al. 2015). The interference in the cell cycle and the effects in pro-apoptosis are shown by Epipodophyllotoxin derived from *Podophyllum peltatum* (Acharya 2013, Solowey et al. 2014). Madagascar periwinkle (*Catharanthus roseus*) contains around 70 alkaloids that are found to be effective in the treatment of various types of cancers. The isolated agents are vincristine, vincristine, vinorelbine, vindesine, Vinflunine, and so on, have many clinical approaches in cancer treatment. The major functions are in the stoppage of cell division, binding to the plus end of microtubules, causing stabilization and destabilization of microtubules, promoting arrest of the cell cycle and showing pro-apoptotic and antitumoral activities (Amin et al. 2009, Acharya 2013, Solowey et al. 2014). Combretastatin A-4 phosphate, isolated from *Combretum caffrum,* is an anti-angiogenic. It causes necrosis of tumor. Roscovitine from *Raphanus sativus* proved to be helpful in hindering cyclin-dependent kinases and decreasing the pace of cell cycle progression and flavopiridol from *Dysoxylum binectariferum* has actions of anti-inflammation, tyrosine kinase activity and has roles in the inhibition of growth (Acharya 2013). Green tea extracts Epigallacotechin-3-gallate have antioxidant activity, prevent DNA damage from oxidative stress, stop the functioning of important kinases and prevent carcinogenesis chemically or lower wavelength radiations (Amin et al. 2009).

1.8 ANTIDIABETIC PLANTS

Diabetes is a syndrome or a group of diseases and a long-lasting health condition that results from too much sugar in the blood. This condition generally arises either when the cells of the pancreas are unable to produce enough insulin or when the cells are not able to use the available insulin. According to World Health Organization, between 2000 and 2016, there was a 5% rise in premature death rates from diabetes, and in 2019, diabetes was the ninth leading cause of death, with an estimated 1.5 million deaths directly caused by diabetes. By 2030, the number of diabetics is estimated to reach over 366 million (Wild et al. 2004). The alarming rise in the cases of diabetes is a matter of serious issue for the health of human beings throughout the world. Many studies have shown the effect of medicinal plants in the treatment of diabetes mellitus by their hypoglycemic effects (Malviya, Jain and Malviya 2010). The leaves of the herb *Anacardium occidentale* have been found to prevent diabetes-induced changes in regard to functional and histopathology. It also prevents the growth of mucopolysaccharides in the kidneys (Tedong et al. 2006). *Annona squamosa* L. (custard apple) contains both hypoglycemic and antidiabetic activity. It acts by stimulating pancreatic cells by increasing insulin levels and glucose utilization in muscles and also prevents glucose moving from the liver. So the leaves of this plant are found to be useful in the maintenance of blood sugar and cholesterol levels (Kumar et al. 2005). The roots of *Boerhaavia diffusa* L. are used in the treatment of diabetes (Nalamolu, Boini and Nammi 2004). This herb makes the body more prone to insulin and decreases blood glucose levels (Rao, Boini and Nammi 2004). An ornamental plant, *Bougainvillea spectabilis* Linn. is commonly found in the gardens of India. It is an aggressive climbing vine or shrub which is native to Brazil. Its antihyperglycemic activity is shown by the ethanolic extract of its leaves in increasing the breakdown of glycogen in the liver, which enhances glucose uptake and also possesses the role in insulin sensitivity (Purohit and Sharma 2006). *Canavalia ensiformis* DC, commonly called horse bean, has its origin in Central America and West Indies. Recent research has shown that the seeds of this medicinal plant have hypoglycemic activities (Enyikwola et al. 1991). The aqueous extract of its seeds can be taken orally, which decreases the levels of urinary and blood glucose and increases the levels of triacylglycerol, ketone bodies and cholesterol in regard to pancreatic diabetes (John, John and Lawrence 2003). The concentrated liquor obtained from the boiling of leaves of *Catharanthus roseus* L. is found to be useful in diabetic treatment (Don 1999). The increased secretion of insulin is possessed by the dichloromethane-methanol extract of the leaves. In some research, it is also found that it enhances glucose utilization and is helpful in the inhibition of damage caused by reactive oxygen species (Singh et al. 2001). In Ayurveda, a herbal preparation named Triphala obtained from *Terminalia chebula*, has been used in the treatment of diabetes for over 1000 years, according to the writing of great physician Charak (Rao and Nammi 2006, Peterson, Denniston and Chopra 2017). *Bridelia ndellensis*, *Terminalia catappa*, *Panax ginseng* and *Syzygium cumini* are also some potent plants that show antidiabetic properties.

1.9 LIPID-LOWERING PLANTS

Elevated levels of blood lipids, largely cholesterol and triglycerides, in the human body cause hyperlipidemia. Hyperlipidemia is the key risk factor for cardiovascular disease. There is also an increase in the level of cholesterol esters and triglycerides in the blood. The excess quantity of the lipid circulates in the blood and binds to the protein. This condition is called hyperlipoproteinemia. Abnormalities in plasma lipids are the most common reason for mortality in developing and developed countries (Bauman 2004). Awareness of the link between reduction of low-density lipoprotein and decreased cardiovascular disease mortality is increasing. Several medicinal plants, having antihyperlipidemic activity, play an imperative role in the lessening of cardiovascular diseases. *Vitis vinifera* (grapevine) contains a lot of active components such as flavonoids, polyphenols, anthocyanins, proanthocyanidins, procyanidins and the stilbene derivative resveratrol, which are extracted from its seeds and skin (Nassiri-Asl and Hosseinzadeh 2009). The grape seed extract functions in gene regulation involved in cholesterol, bile acid and lipid metabolism in liver and adipose tissues, which reduces very-low-density lipoprotein (VLDL), low-density lipoprotein (LDL) and plasma concentrations of total cholesterol (Kim et al. 2014). *Syzygium cumini* is an evergreen tree. In India, it is commonly called Jamun or Indian blackberry (Indian blackberry) and is widely cultivated throughout the Indian sub-continent, Southeast Asian countries and Eastern Africa. Its fruit contains a variety of components such as tannins, alkaloids, steroids, flavonoids, terpenoids, fatty acids, phenols, minerals, carbohydrates and vitamins (Srivastava and Chandra 2013). A lipid-lowering agent, quercetin, is present in Indian blackberry (Seo et al. 2015). *Apium graveolens* (Celery), belonging to the Apiaceae family, is an annual or perennial plant and is distributed in tropical and subtropical regions of Europe and Africa and Asia (Rubatzky and Yamaguchi 2012). From the seeds of celery, n-butanol is isolated, which inhibits the synthesis of lipid formation (Al-Sa'aidi, Alrodhan and Ismael 2012). *Camellia sinensis* (tea) is consumed by the people of India, China, Japan and Thailand (Namita, Mukesh and Vijay 2012). It contains polyphenols that constitute one-third of 4000 bioactive components (Mahmood, Akhtar and Khan 2010). Epigallocatechin gallate (EGCG) binds directly to the enzyme, modifying its conformation and catalytic activity, resulting in the inhibition of uptake lipid cellularly (Shishikura, Khokhar and Murray 2006).

Trigonella foenum-graecum (Fenugreek) belongs to Fabaceae and is abundantly grown throughout the world. It contains some bioactive compounds such as alkaloids, flavonoids, steroids, saponins, and so on (Snehlata and Payal 2012). The seeds of this plant have been found to show a lipid-lowering effect (Vinayagam and Xu 2015). *Zingiber officinale* (Ginger) belongs to the family Zingiberaceae and is used by the inhabitants of China and in Ayurvedic and Tibb-Unani medicines for the treatment of several disorders (Tapsell et al. 2006). Ginger reduces the triolein hydrolysis, which in turn inhibits the triacylglycerol level in plasma (Wang and Wang 2005).

1.10 ANTIOXIDANT PLANTS

The main reason for several diseases is oxidative stress. Oxidative stress is generated because of an imbalance between oxidants and antioxidants, where the level of oxidants become higher than antioxidants. For maintaining health, a proper balance of oxidants and antioxidants is needed. Though our body synthesizes antioxidants, an exogenous supply of the antioxidants or improving the antioxidant defense system of the body is a promising way of fighting oxidative stress. Plants have been an excellent source of exogenous antioxidants for a long time. The discovery and further isolation of ascorbic acid from plants have developed for the first time an interest in the exogenous plant antioxidants (Szent-Giörgyi 1963). A list of potential antioxidants of plant origin is given in Table 1.3.

1.11 IMMUNOMODULATORY PLANTS

A system of the body that stands against various diseases and infections is our immune system. It works as an armor for our body. Suppression/stimulation of the immune system causes many disorders. For balancing immunity, immunomodulators are used according to therapeutic requirements. Natural products have been used since ancient times to modulate the immune system. In Iran, there are five medicinal plants – *Silybum marianum*, *Matricaria chamomilla*, *Calendula officinalis*, *Cichorium intybus* and *Dracocephalum kotschyi* – that are found to have immunomodulatory potential (Amirghofran, Azadbakht and Karimi 2000). An extract of *Silybum marianum* called a milk thistle extract, or silymarin, contains an abundant number of flavonolignans, including silybin, isosilybin, silydianin, silychristi and dehydrosilybin, dehydrosilychristin, neosilyhermin, silyhermin and silybinome (Banaee et al. 2011). *Matricaria chamomilla*, commonly called chamomile, belongs to the family Asteraceae and has its origin in southern and Eastern Europe. It was brought to India during the Mughal period and in present times it is cultivated in Punjab, Uttar Pradesh, Maharashtra, and Jammu and Kashmir. They are widely grown in North Africa, Asia, North and South America, Australia and New Zealand (Flax et al. 1979). Its flower contains more than 120 secondary metabolites (Pirzad et al. 2006). A genus of the Labiatae family, *Dracocephalum kotschyi*, has a compound, calycopterin. This compound has been reported to have an inhibitory effect on the proliferation of

TABLE 1.3
List of Potential Antioxidants of Plant Origin

Antifibrotic compounds	Plant and its parts	References
Curcumin	stem or roots of *Curcuma longa*	Zheng et al. 2006
Silymarin	Seeds/fruits of *Curcuma longa*	Jia et al. 2001
Silibinin or silybin	seeds of *Silybum marianum*	Ezhilarasan et al. 2012
Resveratrol	roots of *Veratrum grandiflorum* and *Polygonum cuspidatum*	Kawada et al. 1998
Salvianolic acids A & B	*Salvia miltiorrhiza*	Tao et al. 2013, Tsai et al . 2010
Berberine	fruits of *Berberis aristate* and *Berberis aquifolium*	Zhang et al. 2008
Epigallocatechin-3-gallate (EGCG),	*Camellia sinensis*	Latief et al. 2018
Osthole Hesperidin	fruits of *Cnidium monnieri* L. Cusson Citrus species	Basnet et al. 2001 Ghafar et al. 2010

lymphocytes (Faham et al. 2008). From the stem of *Tinospora cordifolia* (Guduchi), belonging to the family *Menispermaceae*, arabinogalactan polysaccharide (G1-4A) has been reported that has potential of suppression of induced suppression of immune system (Desai et al. 2007). The ethanol extract of rhizome of *Acorus calamus* was found to have immunosuppressive potential, which prevents the rapid increase of mitogen (phytohaemagglutinin) and antigen, encouraging human peripheral blood mononuclear cells (PBMCs) (Mehrotra et al. 2003). A parasitic plant *Cuscuta europea* has a C3 binding glycoprotein which has an immunomodulatory effect (Stanilova, Zhelev and Dobreva 2000). Coumarinolignoids, extracted from the *Cleome viscosa* seeds, have the potential to modify humoral and cell-mediated immune responses (Khanuja et al. 2007). An Indian Ayurvedic formulation, Triphala, which is prepared by the combination of three plants, *Terminalia chebula*, *Terminalia belerica* and *Emblica officinalis*, has the potential of immunomodulation (Srikumar, Parthasarathy and Devi 2005).

1.12 ANTITUSSIVE PLANTS

Approximately eight liters of air are inhaled by a person per minute. A large number of germs and impurities are present in the air. A sudden and noisy expulsion of air from the lungs is called a cough. It is a reflex action against irritation of the air passages to protect the respiratory apparatus from harmful germs and impurities. The cough can be mild, tormenting and noisy on the basis of its intensity. Chronic or persistent cough needs proper treatment. For its treatment, medicinal plant-based remedies derived are available. A genus of Meliaceae family, *Trichilia emetica*, is an evergreen tree. The powder from the roots of *T. emetica* with some additives (honey, milk) has antitussive potential (Diallo et al. 2003). *Crossopteryx febrifuga* is a plant of the family Rubiaceae. It is widely cultivated in West Africa. It comes under the expectorants (medicine which induces sputum release by air passages, used to treat coughs) as it contains Syrup Balembo, which is traditionally commonly used medicine containing its fruit extracts (Occhiuto et al. 1999, Maiga et al. 2006). Several plants have antitussive potential when induced by some agents like *Nigella sativa* and *Linum usitatissimum* by sulfur dioxide, *Cuminum cyminum* and *Glycyrrhiza glabra* by citric acid and by ammonia on *Artemisia absinthium* (Saadat, Shakeri and Boskabady 2018). An herb called ma huang, formed from Chinese ephedra, *Ephedra sinica*, is the origin of some agents like d-pseudoephedrine, l-ephedrine and other sympathomimetic ones (Rotblatt and Ziment 2002). These are generally used to get rid of bronchospasm and mucosal congestion. Ephedrine and pseudoephedrine, constituents of ma huang, are used in the treatment of several disorders related to pulmonary and nasal airways (Rotblatt and Ziment 2002). The extracts of cocaine and marijuana from *Erythroxylum coca* and *Cannabis sativa*, respectively, are effective in lower airways, having central effects (Rotblatt and Ziment 2002). There are several plants with parts containing polysaccharides that were tested and found to have cough-reducing potential. These are the following plant *Althaea officinalis*, *Malva mauritiana*, *Salvia officinalis*, *Verbascum thapsiforme*, *Mahonia aquifolium* and *Rudbeckia fulgida*, and a wild plant *Arctium lappa* (Nosál'ová et al. 2006). The tribal people of North American Indian use a roots decoction as a cough remedy from *Rosa damascena*. It belongs to the family Rosaceae, also known as Gole Mohammadi (Boskabady et al. 2011). In Iran it is called as Damask rose (Kaul, Singh and Singh 2000). *Mahonia aquifolium* is commonly called an Oregon grape. Acid polysaccharides (GX, MO25-K1) extracted from its stem were found to have cough-reducing potential (Nosalova et al. 2005).

1.13 ANTI-INFLAMMATORY PLANTS

Our body responds defensively to various hazardous stimuli in the form of inflammation. Many disorders like cancer, cardiovascular dysfunctions, metabolic syndrome, autoimmune diseases and allergies are the result of uncontrolled inflammatory responses. Various medicines are available for the control of inflammatory diseases. Many drugs such as immunosuppressants and steroids are used to treat this disease, but these have adverse effects on health. Herbal drugs are showing promising results in treating this disease with the least unwanted side effects. *Eryngium creticum* Lam., also known as field eryngo, is a spiny perennial plant and has its origin in Greece, Israel, Jordan, Lebanon, Palestine and Syria. It is found to possess anti-inflammatory effects (Strzelecka et al. 2005). *Nigella sativa* (black caraway, cumin, kalonji) is an annual plant. It is used traditionally as an anti-inflammatory in Eastern Europe and South and Southwest Asia (Hajhashemi, Ghannadi and Jafarabadi 2004). *Teucrium polium* (felty germander) is a wild-growing herb/flowering plant commonly found in the western Mediterranean region. It is found to be effective in inflammations (Tariq et al. 1989). There are some members of the Asteraceae family that have the potential to be good anti-inflammatory agents. *Arnica montana* and *Calendula officinalis* are used by Europeans for several inflammation types (Akihisa et al. 1996). In China, for curing the long-term diseases caused by inflammation, *Sigesbeckia glabrescens* has been used. Some of its species are commonly called St. Paul's work. The mechanism of its bioactive substances is still unclear (Zhong et al. 2019). In Brazil, *Peltastes peltatus* is found to be effective in anti-inflammation (Alice et al. 1991). *Bouchea fluminensis* leaves contain iridoid and steroid glycosides that show anti-inflammatory activities (Costa et al. 2003). In Brazil, *Sida cordifolia* is commonly called as "malva-branca" or "malva-branca-sedosa". The aqueous extract of leaves of *S. cordifolia* possesses potential for anti-inflammation (Franzotti et al. 2000).

1.14 ANTIMUTAGENIC PLANTS

Changes in the environment due to industrialization and the population explosion have led to the inevitable exposure to mutagens in our life. Exposure to mutagens cause toxic effects on human health and is a serious threat to human being. It has been reported that many medicinal plants have a potent antimutagenic effect. Evidences show that mutation-related problems can be treated by the use of medicinal plants. *Bauhinia galpinii* is a shrub native to eastern and western Africa where it is commonly called as "Pride of De Kaap". In other parts of the world, it is also known as red bauhinia, orchid tree, and so on. Extracts of this plant have shown antimutagenic properties (Verschaeve and Van Staden 2008). *Rotheca myricoides*, or blue butterfly bush, is a flowering plant, having its origin in Africa. It is almost found throughout the world and has shown to be effective as an antimutagenic (Verschaeve and Van Staden 2008). *Pteris multifida* is an herbaceous folk herb, which is commonly called Spider brake (Wang et al. 2007). Against picrolonic acid-induced alterations, this plant is found to be the most effective antimutagenic (Lee and Lin 1988). In some research, it is found that there are four medicinal plants being used traditionally in antimutagenicity. The methanol extract was tested and the rhizome of *Acorus calamus*, stem of *Hemidesmus indicus*, bark of *Holarrhena antidysenterica* and root of *Plumbago zeylanica* have antimutagenic potential against direct-acting mutagens, methyl methanesulphonate (MMS) and sodium azide (NaN_3). The highest potential is shown by *A. calamus* and the lowest by *P. zeylanica* (Aqil, Zahin and Ahmad 2008). In South Africa, the herbal teas *Aspalathus linearis* (rooibos), the species of *Cyclopia*, *C. subternata*, *C. genistoides* and *C. sessiliflora*, which is commonly called honeybush and *Camellia sinensis*, commonly called tea tree, all have antimutagenic potential. The compounds isolated from the above listed plants are flavonoid dihydrochalcone aspalathin from *Aspalathus linearis*, mangiferin, a xanthone isolated from species of *Cyclopia* and the ample amount of catechin, (−)epigallocatechin gallate (EGCG) from *Camellia sinensis*. These compounds have antimutagenic activity. Two folk medicinal plants, *Maytenus ilicifolia* and *Peltastes peltatus*, show antimutagenicity due to the presence of a rich number of flavonoids and tannins (Horn and Vargas 2003). The native range of *M. ilicifolia* hails from Bolivia to Brazil. *Peltasts peltatus* is extensively grown in areas of subtropical forest. Aqueous extract of the stem is used in the antimutagenic treatment (Alice et al. 1991).

TABLE 1.4
Some Antiprotozoan Medicinal Plants

Name of plant	Plant part/compound used for the treatment	References
Pentaclethra macrophylla	Stem bark	Yamssi et al. 2020
Psidium guajava	Leaves	Yamssi et al. 2020.
Morinda lucida	Molucidin, ML-2-3 and ML-F52 from leaves	Kwofie et al. 2016
Alnus japonica	Oregonin	Choi et al. 2009
Harungana madagascariensis	Ethanolic stem bark extract	Iwalewa et al. 2008
Acacia nilotica	Tannins, alkaloids, saponins of stem bark	Ohashi et al. 2018
Acanthospermum hispidum	Essential oil and alkaloids	
Alstonia boonei	Indole alkaloids, triterpenoids, tannins isolated from stem bark and leaves	
Dichroea febnifuga	Alkaloids febrifugine (J3-dichroine), isofebrifugine (a-dichroine)	Chang et al. 1987
Aspidosperma megalocarpon	Fendlerine, aspidoalbine and aspidolimidine from the stem bark	Mitaine et al. 1998
Flaveria trinervia	Methanol-aqueous extract of contains coumarins, flavonoids, terpenoids and tannins	Rizk et al. 1986
Boerhavia coccinea	Methanol-aqueous extract of the stems and the hexane crude extract	Tapia-Pérez et al. 2003
Tournefortia densiflora	Hexane and methanol-aqueous extracts of the stems	Tapia-Pérez et al. 2003

1.15 ANTIPROTOZOAL PLANTS

Protozoa are the causal organisms for a number of major diseases. These diseases spread huge misery and death in the tropical world. Many dreadful protozoal diseases are malaria, trypanosomiasis, amoebiasis, giardiasis and leishmaniasis. Control of these protozoal diseases in the allopathy is limited due to drug toxicity, lengthy parenteral treatment and affordability. On the other hand, medicinal plants are being used for the treatment of these diseases. Some important medicinal plants are given in Table 1.4.

1.16 ANTIVIRAL PLANTS

One of the serious challenges to mankind's survival is the control of viral infections. Infections that are associated with human immunodeficiency virus and newly emerging infectious viruses are of main concern. Promising results have been found when several medicinal plants were used to treat a number of viral problems. Some important medicinal plants showing antiviral activities are given in Table 1.5.

1.17 ANTIFIBROTIC PLANTS

Excess growth of fibrous connective tissue in an organ causes abnormal function and disturbed architecture. This is called fibrosis. Originating from the mesenchymal tissue, fibrosis can extend to any organ. Fibrosis in any tissue and organ is fatal with high mortality. No specific drug is available to cure fibrosis. Available drugs can only reduce the symptoms. Some compounds of medicinal plants have been reported to show potential for the treatment of fibrosis. These are curcumin extracted from the stem or roots of *Curcuma longa* (Fu et al. 2008), silymarin extracted from seeds/fruits of *Curcuma longa* (Shafik et al. 2011), silibinin or silybin from seeds of *Silybum marianum* (Ezhilarasan, Karthikeyan and Vivekanandan 2012), baicalein from roots of *Scutellariae radix* (Sun

TABLE 1.5
Antiviral Medicinal Plants

Plant name	Plant parts/active compounds	References
Aframomum melegueta	Tannins, saponin, flavonoids, steroid extracted from seeds	Ohashi et al. 2018
Cinnamomum zeylanicum	essential oils, alkaloids, tannins, triterpenoids, coumarins	Ohashi et al. 2018
Euphorbia denticulata L.	Bioactive compound, botulin, of the aerial parts	Shamsabadipour et al. 2013
Quercus brantii	Chloroform extract of fruit	Karimi et al. 2017
Prunus dulcis	Methanol/HCl extract of almond seed skin	Bisignano et al. 2017
Peganum harmala	Methanol extract of seed	Benzekri et al. 2018
Vigna radiata	Methanol/HCl extract of sprouts	Hafidh et al. 2015
Embelia ribes	Ethyl acetate extract of fruit	Hossan et al. 2018
Ribes nigrum folium	Water extract of leaves	Haasbach et al. 2014
Aesculus parviflora	Methanol extract of leaves	Yousaf et al. 2018

et al. 2010), resveratrol from roots of *Veratrum grandiflorum* and *Polygonum cuspidatum* (Chávez et al. 2008), salvianolic acids A & B from roots of *Salvia miltiorrhiza* (Li et al. 2012), tetrandrine from roots of *Stephania tetrandra*, berberine from fruits of *Berberis aristate* and *Berberis aquifolium* and epigallocatechin-3-gallate (EGCG) from *Camellia sinensis* (Latief et al. 2018).

1.18 ANTIEMETIC PLANTS

Forceful expulsion of stomach content via the mouth is called emesis. There are some plants that show antiemetic activity. Methanolic extract of some plants of Leguminosae family, *Adenanthera pavonina* L., *Peltophorum roxburghii* L., *Prosopis cineraria* L. and *Prosopis juliflora* DC., and so on, show antiemetic potential (Hasan et al. 2012). *Adenanthera pavonina* (red-bead tree) contains flavonoids, glycosides, saponins and steroids. Bergenin isolated from leaves of *Peltophorum roxburghii* L. has antiemetic activity (Tasleem, Ahmed and Azhar 2015). *Prosopis juliflora* contains a number of alkaloids, juliflorine, julifloricine and julifloridine (Micheloud et al. 2019). *Grewia asiatica* contains phenols, saponins, flavonoids and tannins. The methanolic extracts of leaves of this plant are found to be effective in antiemetic effects (Zia-Ul-Haq et al. 2012). *Spilanthes paniculata* is a short-lived perennial herb or annual plant which is widely used in toothache and throat infections, so it is also called a toothache plant. It is commonly called "Shormoni" (Biplab Kumar et al. 2014). The leaf extract of this plant shows the presence of tannins, saponin and lots of flavonoids. In Bangladesh, for the first time, the ethanolic extract, n-hexane and ethyl acetate soluble fractions of leaves of *S. paniculata* have antiemetic potential (Hossain et al. 2012). A strong-smelling weed of *Capparidaceae* family, *Cleome viscosa*, is widely distributed in the tropical world. It is commonly known as wild mustard, dog mustard and sticky cleome (Mali 2010). The fixed oil extracted from the seeds of *Cleome viscosa* has antiemetic potential (Ahmed et al. 2011).

1.19 ANTINOCICEPTIVE PLANTS

Papaver somniferum (poppy) contains narcotic analgesics, codeine and morphine. Morphine has an important feature in modifying pain perception, which results in an elevation in harmful stimuli threshold. The role of antinociception promoted by morphine is due to opioid receptors activation; that's why they are called antagonists of opioid receptors (Calixto et al. 2000). Several studies have shown that methanol extract of *Saxifraga paniculata* has antinociceptive potential (Kumar et al. 2014). *Cannabis sativa* is a multipurpose plant due to the presence of fibers, oil and molecules. For a very long period of time, several substances like marijuana, hashish and other preparations have been used for pain relief. The major active ingredient in the marijuana plant is the presence of THC (delta-9-tetrahydrocannabinol), which is present in the leaves and in some flowering parts. This active constituent, THC, is found to be effective in (Hirst, Lambert and Notcutt 1998). *Capsicum*, a genus in the family Solanaceae, contains capsaicin which accounts for 70% of the total pungent acid amides present. This compound has the potential for antinociception. It can also be used in the treatment of atypical odontalgia and "burning mouth" syndrome (Vickers et al. 1998). *Salix* sp., or willows, is an economically-important plant as it produces timber, paper, shelter and many more things. It also produces major biologically active compounds. Almost 332 metabolites which are secondary in nature are produced from this plant (Tawfeek et al. 2021). Acetylsalicylic acid or other salicylates are most potent in relieving pain (Calixto et al. 2000).

1.20 ANTIFUNGAL PLANTS

The diseases caused by invading fungus are the leading reasons for the increasing rate of disease and death rates in the persons who have a weak immune system as in HIV AIDS patients and cancer patients undergoing chemotherapy (Lemke, Kiderlen and Kayser 2005). The earliest and most efficient antimycotic agent is a broad-spectrum drug Amphotericin B, but recent research gathered information about the repetition in infections caused by the fungus and, thus, makes it less prone to the drug (Arif et al. 2009). Some antifungal compounds isolated from various medicinal plants are given in Table 1.6.

TABLE 1.6
Some Antifungal Compounds Isolated from Various Medicinal Plants

Antifungal Compounds	Name of Plants	References
Tannins and salicylic acid	*Gaullher procumbens, Rhammus purshiand and Anacardum pulsatilla*	Schultes 1978
3-hydroxy-4-geranyl-5-methoxybiphenyl	*Garcinia mangostana*	Dharmaratne et al. 2005
Polyisoprenylated benzophenone	*Cuban propolis*	Rubio et al. 1999
Prenylindoles	*Monodora angolensis* and *Isolona cauliflora*	Nkunya et al. 2004
Amentoflavone	*Selaginella tamariscina*	Jung et al. 2006
Eupomatenoid-3, eupomatenoid-5, conocarpan and orientin	*Piper solmsianum*	De Campos et al. 2005
2-hydroxy maackiain	*Hildegardia barteri*	Meragelman et al. 2005
Hispidulin and belamcanidin	*Artemisia giraldii*	Wang et al. 1999
Clausenidin, dentatin, nor-dentatin and carbazole derivatives and clauszoline	*Clausena excavata*	Sunthitikawinsakul et al. 2003
Dichloromethane	*Baccharis pedunculata*	Rahalison et al. 1995
Prenylated coumarins	*Fsafetida foetida*	Houghton et al. 2006
Naphthoquinones kigelinone, isopinnatal, dihydro-a-lapachone, lapachol	*Kigelia pinnata*	Binutu et al. 1996
11-hydroxy-16-hentriacontanone	*Annona squamosa*	Shanker et al. 2007
Dimethoxy-p-benzoquinone	*Rhizophora apiculata*	Kokpol et al. 1993
1,3-dihydroxy-2-methyl-5,6-dimethoxyanthraquinone	*Prismatomeris fragrans*	Kanokmedhakul et al. 2005
Smilagenin 3-O-b-Dglucopyranoside, disporoside A and Spirostanol steroidal saponins	*Smilax medica*	Sautour et al. 2005
Ypsilandroside B, ypsilandroside A, isoypsilandroside A, isoypsilandroside B and isoypsilandrogaine	*Ypsilandra thebetica*	Xie et al. 2006

(continued)

TABLE 1.6 (Continued)
Some Antifungal Compounds Isolated from Various Medicinal Plants

Antifungal Compounds	Name of Plants	References
Caledonixanthone E	*Calophyllum caledonicum*	Larcher et al. 2004
1,3,6-Trihydroxy-2,5-dimethoxyxanthone	*Monnina obtusifolia*	Pinto et al. 1994
2-(3,4-dimethyl2,5-dihydro-1H-pyrrol-2-yl)-1-methylethyl pentanoate	*Datura metel*	Dabur et al. 2005
Cinnamodial and cinnamosmolide	*Pleodendron costaricense*	Amiguet et al. 2006

TABLE 1.7
Some Antivenom Medicinal Plants

Medicinal plants	Active compounds	References
Ageratum houstonianum Mill	Precocene-II, precocene-I and caryophyllene from leaves and inflorescence	Muhstaq et al. 2006, Kurade et al. 2010
Albizia lebbeck	Alkaloid, tannins, carbohydrates, flavonoids, proteins, echinocystic acid and amino acids	Chulet et al. 2010
Althaea officinalis Linn	Fatty oil, butyric acid, phytosterin, mucilage and starch from leaves and roots	Husain et al. 2008
Bauhinia variegate	Gum, tannins and fatty oil from roots	Baquar 1989
Calotropis procera (Wild.)	Calotropin, calotropagenin, uscharine, calotoxin, calactin and voruscharine from milky latex	Khan 2002
Cassia occidentalis	Emodin, oxymethylanthraquinones, toxalbumin, tannic acid, mucilage and fatty oil from roots	Ahmad 2007
Coccinia grandis Linn	Triterpenoids, carotenoids, flavonoids, alkaloids and fatty acids	Singh et al. 2007, Farrukh et al. 2008
Chenopodium album	Essential oil, carotene and vitamin C from fruits and roots	Shinwari 2000
Nerium indicum Mill	Cardiotoxins, neriodorin, nerioderin, odorin and Karabin from leaves, roots and flowers	Naqvi et al. 1994).
Verbena officinalis Linn	Glucoside, verbenalin and verbenin	Khan et al. 2006
Vicia sativa	Hydrocyanic acid, vicianine and glucosides	Baquar 1989
Euphorbia caducifolia Haines	Euphol, tirucallol, cycloartenol and cyclocaducinol from latex	Khan 2009
Andrographis paniculata and *Aristolochia indica*	Methanol extract	Meenatchisundaram et al. 2009
Aristolochia radix	Aristolochic acid	Vishwanath et al. 1987
Hemidesmus indicus R. Br	2-hydroxy-4-methoxy benzoic acid and lepuol acetate	Alam et al. 1998
Azadirachta indica	AIPLI purified from methanolic leave	Mukherjee et al. 2008
Cardiospermum halicacabum	Isoquinoline	Chandra et al. 2011

1.21 ANTIVENOM PLANTS

A major health hazard, leading to high mortality, is the snake bite. The only specific treatment for this is antivenom immunotherapy. Antivenom sera have various side effects. The World Health Organization has declared snakebite a "Neglected Tropical Disease". Several medicinal plants have antivenom properties. Some antivenom medicinal plants are listed in Table 1.7.

1.22 CONCLUSIONS AND FUTURE PROSPECT

Undoubtedly, the demand for medicinal plants for the treatment of various health ailments has increased throughout the world. All the medicinal plants reviewed in this chapter showed potent activity and confirmed their ability for the treatment of various diseases. There is, therefore, huge scope for the use of medicinal plants for the treatment of various diseases. The importance of research in the field of medicinal plants is felt more than ever. While the medicines of plant origin potentially contribute to the progress of healthcare, there is need to evaluate the safety, side effects, accurate dose, duration of treatment, acute and chronic toxicities as well as the standardization of these medicines. Further, the scientific studies, active compound identification, and effects of these compounds can lead to the finding of new therapeutic aids and the manufacture of herbal products in the future. It is therefore necessary to utilize advance omics technologies to resolve these issues. Integrated omics approaches can help in improving these plants so that these plants can be used as a safe, effective and affordable form of health care.

REFERENCES

Acharya, N. 2013. Novel anticancer agents from plant sources. *Chinese J. nat. med.* 11(1): 16–23.

Adegbola, P., Aderibigbe, I., Hammed, W. and Omotayo, T. 2017. Antioxidant and anti-inflammatory medicinal plants have potential role in the treatment of cardiovascular disease: a review. *Am. J. Cardiovascular Disease* 7(2): 19–32.

Afonne, O. J., Orisakwe, O. E., Obi, E., Orish, C. and Akumka, D. D. 2000. Some pharmacological properties of *Synclisia scabrida* III. *Indian J. of Pharmacol.* 32(3): 239–241.

Ahmad, S. S. 2007. Medicinal wild plants from Lahore-Islamabad motorway (M-2). *Pakistan J. Bot.* 39(2): 355.

Ahmed, S., Sultana, M., Mohtasheem, M., Hasan, U. and Azhar, I. 2011. Analgesic and antiemetic activity of *Cleome viscosa* L. *Pakistan J. Bot.* 43: 119–122.

Akihisa, T., Yasukawa, K., Oinuma, H., Kasahara, Y., Yamanouchi, S., Takido, M. ... and Tamura, T. 1996. Triterpene alcohols from the flowers of Compositae and their anti-inflammatory effects. *Phytochem.* 43(6): 1255–1260.

Akinyemi, O., Oyewole, S. O. and Jimoh, K. A. 2018. Medicinal plants and sustainable human health: a review. *Horticult. Int J.* 2(4): 194–195.

Alam, M. I. and Gomes, A. 1998. Adjuvant effects and antiserum action potentiation by a (herbal) compound 2-hydroxy-4-methoxy benzoic acid isolated from the root extract of the Indian medicinal plantsarsaparilla (*Hemidesmus indicus* R. Br.). *Toxicon*, 36(10): 1423–1431.

Alice, C. B., Vargas, V. M. F., Silva, G. A. A. B., De Siqueira, N. C. S., Schapoval, E. E. S., Gleye, J. ... and Henriques, A. T. 1991. Screening of plants used in south Brazilian folk medicine. *J. Ethnopharmacol.* 35(2): 165–171.

Al-Sa'aidi, J. A., Alrodhan, M. N. and Ismael, A. K. 2012. Antioxidant activity of n-butanol extract of celery (*Apium graveolens*) seed in streptozotocin-induced diabetic male rats. *Res. Pharm. Biotech.* 4(2): 24–29.

Amiguet, V., Petit, P., Ta, C. A., Nuñez, R., Sánchez-Vindas, P., Alvarez, L. P. ... and Durst, T. 2006. Phytochemistry and antifungal properties of the newly discovered tree *Pleodendron c ostaricense*. *J. Nat. Prod.* 69(7):1005–1009.

Amin, A., Gali-Muhtasib, H., Ocker, M. and Schneider-Stock, R. 2009. Overview of major classes of plant-derived anticancer drugs. *Int. J. Biomed. Sci. IJBS* 5(1): 1.

Amirghofran Z., Azadbakht M. and Karimi, M. H. 2000 Evaluation of the immunomodulatory effects of five herbal plants. *J. Ethnopharmacol.* 72(1–2): 167–172.

Anbari, K., Hasanvand, A., Andevari, A.N. and Abbaszadeh, S. 2019. Concise overview: A review on natural antioxidants and important herbal plants on the gastro-intestinal system. *Res. J. Pharmacy Tech.* 12(2): 841–847.

Ansari, R. A., Tripathi, S. C., Patnaik, G. K. and Dhawan, B. N. 1991. Antihepatotoxic properties of picroliv: An active fraction from rhizomes of *Picrorhiza kurrooa*. *Journal of Ethnopharmacology* 34: 61–68.

Aqil, F., Zahin, M. and Ahmad, I. 2008. Antimutagenic activity of methanolic extracts of four ayurvedic medicinal plants. *Indian J. Exp. Biol.* 46(9): 668–672.

Arif, T., Bhosale, J. D., Kumar, N., Mandal, T. K., Bendre, R. S., Lavekar, G. S. and Dabur, R. 2009. Natural products–antifungal agents derived from plants. *J. Asian Nat. Prod. Res.* 11(7): 621–638.

Asrani, S. K., Larson, J. J., Yawn, B., Therneau, T. M. and Kim, W. R. 2013. Underestimation of liver-related mortality in the United States. *Gastroenterology* 145(2): 375–382.

Ayaz, M., Sadiq, A., Junaid, M., Ullah, F., Subhan, F. and Ahmed, J. 2017. Neuroprotective and anti-aging potentials of essential oils from aromatic and medicinal plants. *Front. Aging Neurosci.* 9: 168.

Banaee M., Sureda A., Mirvaghefi A. R. and Rafei, G. R. 2011. Effects of long-term silymarin oral supplementation on the blood biochemical profile of rainbow trout (Oncorhynchus mykiss). Fish *Physiol. Biochem.* 37(4): 885–896.

Baquar, S. R. 1989. *Medicinal and Poisonous Plants of Pakistan.* Karachi, Pakistan: Printas.

Basnet, P., Yasuda, I., Kumagai, N., Tohda, C., Nojima, H., Kuraishi, Y. and Komatsu, K. 2001. Inhibition of itch-scratch response by fruits of *Cnidium monnieri* in mice. *Biol. Pharm. Bull.* 24(9): 1012–1015.

Bauman, A. E. 2004. Updating the evidence that physical activity is good for health: an epidemiological review 2000–2003. *J. Sci. Med. Spor*t 7(1): 6–19.

Benzekri, R., Bouslama, L., Papetti, A., Hammami, M., Smaoui, A. and Limam, F. 2018. Anti HSV-2 activity of *Peganum harmala* (L.) and isolation of the active compound. *Microb. Pathog.* 114: 291–298.

Bhakta, T., Banerjee, S., Subhash, C. M., Tapan, K. M., Saha, B. P. and Pal, M. 2001. Hepatoprotective activity of *Cassia fistula* leaf extract. *Phytomedicine* 8(3): 220–224.

Binutu, O. A., Adesogan, K. E. and Okogun, J. I. 1996. Antibacterial and antifungal compounds from *Kigelia pinnata. Planta Med.* 62(04): 352–353.

Biplab Kumar, D. A. S., Ahmed, K., Uddin, A., Bhattacharjee, R. and Al-Amin, M. M. 2014. Antinociceptive activity of methanol extract of *Spilanthes paniculata* Linn. *Turkish J. Pharm. Sci.* 11(2): 137–144.

Bisignano, C., Mandalari, G., Smeriglio, A., Trombetta, D., Pizzo, M. M., Pennisi, R. and Sciortino, M. T. 2017. Almond skin extracts abrogate HSV-1 replication by blocking virus binding to the cell. *Viruses* 9(7): 178.

Boskabady, M. H., Shafei, M. N., Saberi, Z. and Amini, S. 2011. Pharmacological effects of *Rosa damascena. Iranian J.basic Med. Sci.*, 14(4): 295.

Bounameaux, H. 2009. The novel anticoagulants: Entering a new era. *Swiss med. Wkly.* 139(5–6): 60–64.

Bourgou, S., Ksouri, R., Bellila, A., Skandrani, I., Falleh, H. and Marzouk, B. 2008. Phenolic composition and biological activities of Tunisian Nigella sativa L. shoots and roots. *C. R. Biol.* 331(1): 48–55.

Braun, L. A., Tiralongo, E., Wilkinson, J. M., Spitzer, O., Bailey, M., Poole, S. et al. 2010. Perceptions, use and attitudes of pharmacy customers on complementary medicines and pharmacy practice. *BMC Complement. Altern. Med.* 10: 38.

Calixto, J. B., Beirith, A., Ferreira, J., Santos, A. R., Filho, V. C. and Yunes, R. A. 2000. Naturally occurring antinociceptive substances from plants. *Phytother. Res.: Int. J. Devot. Pharmacol. Toxicol. Eval. Nat. Prod. Deriv.* 14(6): 401–418.

Chandra, D. N., Prasanth, G. K., Singh, N., Kumar, S., Jithesh, O., Sadasivan, C. ... and Haridas, M. 2011. Identification of a novel and potent inhibitor of phospholipase A2 in a medicinal plant: crystal structure at 1.93 Å and Surface Plasmon Resonance analysis of phospholipase A2 complexed with berberine. *Biochimica et Biophysica Acta (BBA)-Proteins and Proteom.* 1814(5): 657–663.

Chang, H. M. and But, P. P. H. 1987. Dangshen. In *Pharmacol. Apple. Chinese Materia Medica:* (Volume II) (Eds. Paul Pui-hay But and Hson-mou Chang). World Scientific Publishing, Singapore, pp. 989–994).

Chattopadhyay, R. R., Sarkar, S. K., Ganguly, S., Banerjee, R. N., Basu, T. K. and Mukherjee, A. Hepatoprotective activity of *Azadirachta indica* leaves on paracetamol-induced hepatic damage in rats. *Indian J. Exp. Biol.* 30(8): 738–740.

Chávez, E., Reyes-Gordillo, K., Segovia, J., Shibayama, M., Tsutsumi, V., Vergara, P. ... and Muriel, P. 2008. Resveratrol prevents fibrosis, NF-κB activation and TGF-β increases induced by chronic CCl4 treatment in rats. *J. Appl. Toxicology: Int. J.* 28(1): 35–43.

Che, E., Gao, Y., Wan, L., Zhang, Y., Han, N., Bai, J. ... and Wang, S. 2015. Paclitaxel/gelatin-coated magnetic mesoporous silica nanoparticles: Preparation and anti-tumor efficacy in vivo. *Micropor. Mesopor. Mat.* 204: 226–234.

Chiang, J. 2014. Liver physiology: metabolism and detoxification. In Pathophysiology of Human Diseases (Eds. L. M. McManus and R. N. Mitchell), New York: Elsevier, pp. 1770–1782.

Choi, S. E., Jeong, M. S., Kang, M. J., Lee, D. I., Joo, S. S., Lee, C. S. ... and Lee, M. W. 2009. Effect of topical application and intraperitoneal injection of oregonin on atopic dermatitis in NC/Nga mice. *Exp. Dermatol.* 19(8): e37–e43.

Chulet, R., Lincy, J., Methew, G. and Pankaj, P. 2010. Pharmacognostic standardization and phytochemical screening of *Albizzia lebbeck. J. Chem. Pharm. Res.* 2(1): 432–443.

Costa, V. B., Coube, C. S., Marinho, B. G., Matheus, M. E., Leitão, S. G. and Fernandes, P. D. 2003. Anti-inflammatory and analgesic activity of *Bouchea fluminensis*. *Fitoterapia*, 74(4): 364–371.

Cragg, G. M. and Newman, D. J. 2005. Plants as a source of anticancer agents. *J. Ethnopharmacol.* 100(1–2): 72–79.

Dabur, R., Chhillar, A. K., Yadav, V., Kamal, P. K., Gupta, J. and Sharma, G. L. 2005. In vitro antifungal activity of 2-(3, 4-dimethyl-2, 5-dihydro-1H-pyrrol-2-yl)-1-methylethyl pentanoate, a dihydropyrrole derivative. *J. med. microbio.* 54(6): 549–552.

Das, B. K., Bepary, S., Datta, B. K., Chowdhury, A. A., Ali, M. S. and Rouf, A. S. 2008. Hepatoprotective activity of *Phyllanthus reticulates. Pak. J. Pharmaceu. Sci.* 21(4): 333–337.

De Campos, M. P., Cechinel Filho, V., Da Silva, R. Z., Yunes, R. A., Zacchino, S., Juarez, S. ... and Cruz, A. B. 2005. Evaluation of antifungal activity of *Piper solmsianum* C. DC. var. solmsianum (Piperaceae). *Biol. Pharm. Bull.* 28(8): 1527–1530.

Desai, V. R., Ramkrishnan, R., Chintalwar, G. J. and Sainis, K. B. 2007. G1–4A, an immunomodulatory polysaccharide from *Tinospora cordifolia*, modulates macrophage responses and protects mice against lipopolysaccharide-induced endotoxic shock. *Int. Immunopharmacol.* 7(10): 1375–1386.

Dhanabal, S. P., Syamala, G., Satish, K. M. N. and Suresh, B. 2006. Hepatoprotective activity of the Indian medicinal plant *Polygala arvensis* on D-galactosamine-induced hepatic injury in rats. *Fitoterapia* 77(6): 472–474.

Dharmaratne, H. R. W., Piyasena, K. G. N. P. and Tennakoon, S. B. 2005. A geranylated biphenyl derivative from *Garcinia mangostana. Nat. Prod. Res.* 19(3): 239–243.

Diallo, D., Paulsen, B. S., Liljebäck, T. H. and Michaelsen, T. E. 2003. The malian medicinal plant *Trichilia emetica*; studies on polysaccharides with complement-fixing ability. *J. Ethnopharmacol.* 84(2–3): 279–287.

Don, G. 1999. *Catharanthus roseus*. In Medicinal Plants of the World (Ed. I. A. Ross). Totowa, NJ: Human Press, pp. 109–118.

Enyikwola, O., Addy, E. O. and Adoga, G. I. 1991. Hypoglycaemic effect of *Canavalia ensiformis* L. (Leguminosae) in albino rats. *Discov. Innov.* 3(3): 61–63.

Faham, N., Javidnia, K., Bahmani, M. and Amirghofran, Z. 2008. Calycopterin, an immunoinhibitory compound from the extract of *Dracocephalum kotschyi. Phytother. Res.* 22(9): 1154–1158.

Farooqi, A. A. and Sreeramu, B. S. 2001. *Cultivation of Medicinal and Aromatic Crops*. p. 518. Hyderabad, India: Universities Press (India) Ltd.

Farrukh, U., Shareef, H., Mahmud, S., Ali, S. A. and Rizwani, G. H. 2008. Antibacterial activities of *Coccinia grandis* L. *Pak. J. Bot.* 40(3): 1259–1262.

Flax, J. W., Ivens, R. E., Wagenfeld, M. O. and Weiss, R. 1979. Mental health and rural America: An overview. *Community Ment. Health Rev.* 3(5–6): 1–15.

Franzotti, E. M., Santos, C. V. F., Rodrigues, H. M. S. L., Mourão, R. H. V., Andrade, M. R. and Antoniolli, A. R. 2000. Anti-inflammatory, analgesic activity and acute toxicity of *Sida cordifolia* L. (*Malva-branca*). *J. Ethnopharmacol.* 72(1–2): 273–277.

Fu, Y., Zheng, S., Lin, J., Ryerse, J. and Chen, A. 2008. Curcumin protects the rat liver from CCl4-caused injury and fibrogenesis by attenuating oxidative stress and suppressing inflammation. *Mol. Pharmacol.* 73(2): 399–409.

Ghafar, M. F., Prasad, K. N., Weng, K. K. and Ismail, A. 2010. Flavonoid, hesperidine, total phenolic contents and antioxidant activities from Citrus species. *African J. Biotech.* 9(3): 326–330.

Goel, H. C., Prasad, J. and Sharma, A. 1998. Antitumor and radioprotective action of Podophyllum hexandrum. *Indian J. Exp. Bio.* 36: 583–587.

Goldhaber, S. Z. and Grasso-Correnti, N. 2002. Treatment of blood clots. *Circ.* 106(20): e138–e140.

Gomes, D. L., Silva, J. M. C., Santo's dos ND, R. H. and Leite, E. L. 2009. *Eichhornia crassipes* is a source of anticoagulant compounds. *Pub.* 5: 42–50.

Gupta, N. K. and Dixit, V. K. 2009. Evaluation of hepatoprotective activity of *Cleome viscosa* L. extract. *Indian J. Pharmacol.* 41: 36–40.

Haasbach, E., Hartmayer, C., Hettler, A., Sarnecka, A., Wulle, U., Ehrhardt, C. ... and Planz, O. 2014. Antiviral activity of Ladania067, an extract from wild black currant leaves against influenza A virus in vitro and in vivo. *Front. Microbiol.* 5: 171.

Hafidh, R. R., Abdulamir, A. S., Abu Bakar, F., Sekawi, Z., Jahansheri, F. and Jalilian, F. A. 2015. Novel antiviral activity of mung bean sprouts against respiratory syncytial virus and herpes simplex virus – 1: an in vitro study on virally infected Vero and MRC-5 cell lines. *BMC Complement Altern. Med.* 15(1): 1–16.

Hajhashemi, V., Ghannadi, A. and Jafarabadi, H. 2004. Black cumin seed essential oil, as a potent analgesic and antiinflammatory drug. *Phytother. Res.: Int J. Devot. Pharmacol. Toxicol. Eval. Nat. Prod. Deriv.* 18(3): 195–199.

Handa, S.S. and Sharma, A. 1990. Hepatoprotective activity of andrographolide from *Andrographis paniculata* against carbon tetrachloride. *Indian J. Med. Res.* 92: 276–283.

Hasan, M. M. U., Azhar, I., Muzammil, S., Ahmed, S. and Ahmed, S. W. 2012. Antiemetic activity of some leguminous plants. *Pakistan J. Bot.* 44(1): 389–391.

Heiss, E., Herhaus, C., Klimo, K., Bartsch, H. and Gerhäuser, C. 2001. Nuclear factor κB is a molecular target for sulforaphane-mediated anti-inflammatory mechanisms. *J. Biol. Chem.* 276(34): 32008–32015.

Hirst, R. A., Lambert, D. G. and Notcutt, W. G. 1998. Pharmacology and potential therapeutic uses of *Cannabis. British J. Anaesth.* 81(1): 77–84.

Horn, R. C. and Vargas, V. M. F. 2003. Antimutagenic activity of extracts of natural substances in the *Salmonella/* microsome assay. *Mutagen.*18(2): 113–118.

Hossain, H., Shahid-Ud-Daula, A. F. M., Hasan, K., Mansur, A. A. and Haq, M. M. 2012. Anti-inflammatory activity, total flavonoids and tannins content from the ethanolic extract of *Spilanthes paniculata* leaf growing in Bangladesh. *Int, J. Pharm*, 2(2): 271–277.

Hossan, M. S., Fatima, A., Rahmatullah, M., Khoo, T. J., Nissapatorn, V., Galochkina, A. V. ... and Wiart, C. 2018. Antiviral activity of *Embelia ribes* Burm. f. against influenza virus in vitro. *Arch. Virol.* 163(8): 2121–2131.

Houghton, P. J., Ismail, K. M., Maxia, L. and Appendino, G. 2006. Antidermatophytic prenylated coumarins from asafetida. *Planta Med.* 72(11): S_008.

Husain, S. Z., Malik, R. N., Javaid, M. and Bibi, S. 2008. Ethnobotanical properties and uses of medicinal plants of Morgah biodiversity park, Rawalpindi. *Pakistan J. Bot*, 40(5): 1897–1911.

Iriti, M., Vitalini, S., Fico, G. and Faoro, F. 2010. Neuroprotective herbs and foods from different traditional medicines and diets. *Molecules* 15(5): 3517–3555.

Iwalewa, E. O., Omisore, N. O., Adewunmi, C. O., Gbolade, A. A., Ademowo, O. G., Nneji, C. ... and Daniyan, O. M. 2008. Anti-protozoan activities of *Harungana madagascariensis* stem bark extract on trichomonads and malaria. *J. Ethnopharmacol.* 117(3): 507–511.

Jia, J. D., Bauer, M., Cho, J. J., Ruehl, M., Milani, S., Boigk, G. ... and Schuppan, D. 2001. Antifibrotic effect of silymarin in rat secondary biliary fibrosis is mediated by downregulation of procollagen α1 (I) and TIMP-1. *J. Hepatol.* 35(3): 392–398.

John, L. S., John, T. A. and Lawrence, A. L. 2003. Antidiabetic plants. *J Am Coll Nutr.* 22: 524.

Jordan, M. A. and Wilson, L. 2004. Microtubules as a target for anticancer drugs. *Nat. Rev. Cancer* 4(4): 253–265.

Jung, H. J., Sung, W. S., Yeo, S. H., Kim, H. S., Lee, I. S., Woo, E. R. and Lee, D. G. 2006. Antifungal effect of amentoflavone derived *from Selaginella tamariscina Arch. Pharm. Res.*, 29(9): 746–751.

Kanokmedhakul, K., Kanokmedhakul, S. and Phatchana, R. 2005. Biological activity of anthraquinones and triterpenoids from *Prismatomeris fragrans. Journal of Ethnopharmacology* 100: 284–288.

Karimi, A., Rafieian-Kopaei, M., Moradi, M. T. and Alidadi, S. 2017. Anti-herpes simplex virus type-1 activity and phenolic content of crude ethanol extract and four corresponding fractions of *Quercus brantii* L Acorn. *Evid. Based Complementary Altern. Med.* 22(3): 455–461.

Kaul, V. K., Singh, V. and Singh, B. 2000. Damask rose and marigold: prospective industrial crops. *J. Med. Arom. Plant Sci.* 22(1B): 313–318.

Kawada, N., Seki, S., Inoue, M. and Kuroki, T. 1998. Effect of antioxidants, resveratrol, quercetin, and N-acetylcysteine, on the functions of cultured rat hepatic stellate cells and Kupffer cells. *Hepatol.* 27(5): 1265–1274.

Kee, N. L. A., Mnonopi, N., Davids, H., Naude, R. J. and Frost, C. L. 2008. Antithrombotic/anticoagulant and anticancer activities of selected medicinal plants from South Africa. *African J. Biotech.* 7(3).

Khan, M. S. 2002. *A Guide to the Snakes of Pakistan* (Vol. 16). Frankfurt am Main: Edition Chimaira.

Khan, F. M. 2009. Ethno-veterinary medicinal usage of flora of Greater Cholistan desert (Pakistan). *Pakistan Vet. J.* 29(2).

Khan, M. A. and Qaiser, M. 2006. Halophytes of Pakistan: Characteristics, distribution and potential economic usages. In *Sabkha ecosystems* (Eds. M. A. Khan, B. Böer, G. S. Kust and H. J. Barth) (pp. 129–153). Dordrecht: Springer.

Khanuja, S. P. S., Pal, A., Chattopadhyay, S. K., Darokar, M. P., Patel, R. P., Gupta, A. K., Negi, A. S., Kaur, T., Tandon, S., Kahol, A. P. and Garg, A. 2007. Immunomodulatory pharmaceutical composition and a process for preparation thereof. US 2007/0258989 A1.

Khazir, J., Mir, B. A., Pilcher, L. and Riley, D. L. 2014. Role of plants in anticancer drug discovery. *Phytochem. Lett.* 7: 173–181.

Kim, H., Bartley, G. E., Arvik, T., Lipson, R., Nah, S. Y., Seo, K. and Yokoyama, W. 2014. Dietary supplementation of chardonnay grape seed flour reduces plasma cholesterol concentration, hepatic steatosis, and abdominal fat content in high-fat diet-induced obese hamsters. *J. Agricul. food Chem.* 62(8): 1919–1925.

Kinghorn, A. D., Pan, L., Fletcher, J. N. and Chai, H 2011. The relevance of higher plants in lead compound discovery programs. *J. Nat. Prod.* 74(6): 1539–1555.

Kokpol, U., Chavasiri, W., Chittawong, V., Bruce, M., Cunningham, G. N. and Miles, D. H. 1993. Long chain aliphatic alcohols and saturated carboxylic acids from heartwood of Rhizophora apiculata. *Phytochemistry* 33: 1129–1131.

Kumar, C. D., Rajendran, K., Lobo, R. and Shirwaikar, A. 2005. An identity based pharmacognostical profile of folium *Annona squamosa. Nat. Prod. Sci.* 11(4): 213–219.

Kumar, B., D. A. S., Ahmed, K., Uddin, A., Bhattacharjee, R. and Al-Amin, M. M. 2014. Antinociceptive activity of methanol extract of *Spilanthes paniculata* Linn. *Turkish J. Pharm. Sci* 11(2): 137–144.

Kurade, N. P., Jaitak, V., Kaul, V. K. and Sharma, O. P. 2010. Chemical composition and antibacterial activity of essential oils of *Lantana camara, Ageratum houstonianum and Eupatorium adenophorum. Pharm. Biol.* 48(5): 539–544.

Kwofie, K. D., Tung, N. H., Suzuki-Ohashi, M., Amoa-Bosompem, M., Adegle, R., Sakyiamah, M. M. ... and Ayi, I. 2016. Antitrypanosomal activities and mechanisms of action of novel tetracyclic iridoids from *Morinda lucida* Benth. *Antimicrob. Agents Chemother.* 60(6): 3283–3290.

Larcher, G., Morel, C., Tronchin, G., Landreau, A., Séraphin, D., Richomme, P. and Bouchara, J. P. 2004. Investigation of the antifungal activity of caledonixanthone E and other xanthones against *Aspergillus fumigatus. Planta Med.* 70(06): 569–571.

Latief, U. and Ahmad, R. 2018. Herbal remedies for liver fibrosis: A review on the mode of action of fifty herbs. *J. Tradit. Complement. Med.* 8(3): 352–360.

Lee, H. and Lin, J. Y. 1988. Antimutagenic activity of extracts from anticancer drugs in Chinese medicine. *Mutat. Res. Genet. Toxicol. Environ. Mutagen* 204(2): 229–234.

Lemke, A., Kiderlen, A. F. and Kayser, O. 2005. Amphotericin b. *Appl. Microbiol. Biotechnol.* 68(2): 151–162.

Li, S., Wang, L., Yan, X., Wang, Q., Tao, Y., Li, J., ... and Liu, C. 2012. Salvianolic acid B attenuates rat hepatic fibrosis via downregulating angiotensin II signaling. *Evid.-Based Complementary Altern. Med.* 2012: 160726.

Mahmood, T., Akhtar, N. and Khan, B. A. 2010. The morphology, characteristics, and medicinal properties of *Camellia sinensis* tea. *J. Med. Plants Res.* 4(19): 2028–2033.

Maiga, A., Malterud, K. E., Diallo, D. and Paulsen, B. S. 2006. Antioxidant and 15-lipoxygenase inhibitory activities of the Malian medicinal plants *Diospyros abyssinica* (Hiern) F. White (Ebenaceae), *Lannea velutina* A. Rich (Anacardiaceae) and *Crossopteryx febrifuga* (Afzel) Benth. (Rubiaceae). *J. Ethnopharmacol.* 104(1–2): 132–137.

Mali, R. G. 2010. *Cleome viscosa* (wild mustard): A review on ethnobotany, phytochemistry, and pharmacology. *Pharm. l Bio.* 48(1): 105–112.

Malviya, N., Jain, S. and Malviya, S. A. P. N. A. 2010. Antidiabetic potential of medicinal plants. *Acta. pol. pharm.* 67(2): 113–118.

Manicam, C., Abdullah, J. O., Tohit, E. R. M., Seman, Z., Chin, S. C. and Hamid, M. 2013. In vitro anticoagulant activities of *Melastoma malabathricum* Linn. aqueous leaf extract: A preliminary novel finding. *J. Med. Plant Res.* 4(14): 1464–1472.

Manjunatha, B. K. 2006. Hepatoprotective activity of *Pterocarpus santalinus* L.f., an endangered medicinal plant. *Indian J. Pharmacol.* 38: 25–28.

Matsubara, K., Matsuura, Y., Bacic, A., Liao, M. L., Hori, K. and Miyazawa, K. 2001. Anticoagulant properties of a sulfated galactan preparation from a marine green alga, *Codium cylindricum*. *Int. J. Biol. Macromol.* 28(5): 395–399.

Meenatchisundaram, S., Parameswari, G. and Michael, A. 2009. Studies on antivenom activity of *Andrographis paniculata* and *Aristolochia indica* plant extracts against *Daboia russelli* venom by in vivo and in vitro methods. *Indian J. Sci. Tech.* 2(4): 76–79.

Mehrotra, S., Mishra, K. P., Maurya, R., Srimal, R. C., Yadav, V. S., Pandey, R. and Singh, V. K. 2003. Anticellular and immunosuppressive properties of ethanolic extract of *Acorus calamus* rhizome. *Int. Immunopharmacol.* 3(1): 53–61.

Meragelman, T. L., Tucker, K. D., McCloud, T. G., Cardellina, J. H. and Shoemaker, R. H. 2005. Antifungal flavonoids from *Hildegardia barteri*. *J. Nat. Prod.* 68(12): 1790–1792.

Micheloud, J. F., Caro, L. A. C., Cholich, L. A., Martínez, O. G. and Gimeno, E. J. 2019. Suspected poisoning in beef cattle from ingestion of *Prosopis nigra* pods in North-western Argentina. *Toxicon* 157: 80–83.

Mitaine, A. C., Weniger, B., Sauvain, M., Lucumi, E., Aragon, R. and Zeches-Hanrot, M. 1998. Indole alkaloids from the trunk bark of *Aspidosperma megalocarpon*. *Planta Med.* 64(05): 487–487.

Muhstaq, A., Khan, M. A., Muhammad, Z. and Shazia, S. 2006. Ethnomedicinal demography and ecological diversification of some important weeds from District Attock-Pakistan. *Pak. J. Weed Sci. Res.* 12(1/2): 37–46.

Mukherjee, A. K., Doley, R. and Saikia, D. 2008. Isolation of a snake venom phospholipase A2 (PLA2) inhibitor (AIPLAI) from leaves of *Azadirachta indica* (Neem): Mechanism of PLA2 inhibition by AIPLAI in vitro condition. *Toxicon* 51(8): 1548–1553.

Naaz, F., Javed, S. and Abdin, M. Z. 2007. Hepatoprotective effect of ethanolic extract of *Phyllanthus amarus* Schum. et Thonn. on aflatoxin B1-induced liver damage in mice. *J. Ethnopharmacol.* 113(3): 503–509.

Nalamolu, R. K., Boini, K. M. and Nammi, S. 2004. Effect of chronic administration of *Boerhaavia diffusa* Linn. leaf extract on experimental diabetes in rats. *Trop. J. Pharm. Res.* 3(1): 305–309.

Namita, P., Mukesh, R. and Vijay, K. J. 2012. *Camellia sinensis* (green tea): A review. *Glob. J. Pharmacol.* 6(2): 52–59.

Naqvi, B. S., Shaikh, M. R., Maleka, F. A. and Shaikh, D. 1994. Studies of antibacterial activity of ethanolic extracts from *Nerium indicum* and *Hibiscus rosa sinensis*. *J. Islam. Stud.* 7(3): 167–168.

Nassiri-Asl, M. and Hosseinzadeh, H. 2009. Review of the pharmacological effects of *Vitis vinifera* (Grape) and its bioactive compounds. *Phytother. Res.: An Int. J. Devot. Pharmacol. Toxicol. Eval. Nat. Prod. Deriv.* 23(9): 1197–1204.

Nicora, G., Vitali, F., Dagliati, A., Geifman, N. and Bellazzi, R. 2020. Integrated multi-omics analyses in oncology: a review of machine learning methods and tools. *Front. Oncol.* 10: 1030.

Nkunya, M. H., Makangara, J. J. and Jonker, S. A. 2004. Prenylindoles from Tanzanian *Monodora* and *Isolona* species. *Nat. Prod. Res.* 18(3): 253–258.

Nosál'ová, G., Capek, P., Šutovská, M., Fraňová, S. and Matulova, M. 2006. Antitussive active polysaccharides from ornamental-medicinal plants. *Flor. Ornam. Plant Biotech.* 4: 472–481.

Nosalova, G., Šutovská, M., Mokry, J., Kardosova, A., Capek, P. and Khan, M. T. H. 2005. Efficacy of herbal substances according to cough reflex. *Minerva Biotechnol.* 17(3): 141.

Occhiuto, F., Sanogo, R., Germano, M. P., Keita, A., D'angelo, V. and De Pasquale, R. 1999. Effects of some Malian medicinal plants on the respiratory tract of guinea-pigs. *J. Pharm. Pharmacol.* 51(11): 1299–1303.

Ohashi, M., Amoa-Bosompem, M., Kwofie, K. D., Agyapong, J., Adele, R., Sakyiamah, M. M. ... and Ohta, N. 2018. In vitro antiprotozoal activity and mechanisms of action of selected Ghanaian medicinal plants against Trypanosoma, leishmania, and Plasmodium parasites. *Phytother. Res.* 32(8): 1617–1630.

Ojeda, D., Jimenez-Ferrer, E., Zamilpa, A., Herrera-Arellano, A., Tortoriello, J. and Alvarez, L. 2010. Inhibition of angiotensin-converting enzyme (ACE) activity by the anthocyanins delphinidin- and cyanidin-3-O-sambubiosides from *Hibiscus sabdariffa*. *J. Ethnopharmacol.* 127(1): 7–10.

Oliveira, C. Z., Maiorano, V. A., Marcussi, S., Sant'Ana, C. D., Januário, A. H., Lourenço, M. V. ... and Soares, A. M. 2005. Anticoagulant and antifibrinogenolytic properties of the aqueous extract from *Bauhinia forficata* against snake venoms. *J. Ethnopharmacol.* 98(1–2): 213–216.

Oliynyk, S. and Oh, S. 2013. Actoprotective effect of ginseng: Improving mental and physical performance. *J. Ginseng Res.* 37(2): 144–166.

Osoniyi, O. and Onajobi, F. 2003. Coagulant and anticoagulant activities in *Jatropha curcas* latex. *J. Ethnopharmacol.* 89(1): 101–105.

Pawar, A. P., Vinugala, D. and Bothiraja, C. 2014. Nanocochleates derived from nanoliposomes for paclitaxel oral use: preparation, characterization, in vitro anticancer testing, bioavailability and biodistribution study in rats. *Biomed. Pharmacother.* 3502: 1–9.

Pawlaczyk, I., Czerchawski, L., Kuliczkowski, W., Karolko, B., Pilecki, W., Witkiewicz, W. and Gancarz, R. 2010. Anticoagulant and antiplatelet activity of polyphenolic-polysaccharide preparation isolated from the medicinal plant *Erigeron canadensis* L. *Thrombo. Res.* 127(4): 328–340.

Peterson, C. T., Denniston, K. and Chopra, D. 2017. Therapeutic uses of Triphala in ayurvedic medicine. *J Altern Complement Med.* 23(8): 607–614.

Pezzuto, J. M. 1997. Plant-derived anticancer agents. *Biochem. Pharmacol.* 53(2): 121–133.

Pinto, D. C., Fuzzati, N., Pazmino, X. C. and Hostettmann, K. 1994. Xanthone and antifungal constituents from *Monnina obtusifolia. Phytochem.* 37(3): 875–878.

Pirzad, A., Alyari, H., Shakiba, M. R., Zehtab-Salmasi, S. and Mohammadi, A. 2006. Essential oil content and composition of German chamomile (*Matricaria chamomilla* L.) at different irrigation regimes. *J. Agron.* 5(3): 451–455.

Pledgie-Tracy, A., Sobolewski, M. D. and Davidson, N. E. 2007. Sulforaphane induces cell type-specific apoptosis in human breast cancer cell lines. *Mol. Cancer Ther.* 6(3): 1013–1021.

Prasad, A. D., Shyma, T. B. and Raghavendra, M. P. 2013. Plants used by the tribes for the treatment of digestive system disorders in Wayanad district, Kerala. *J. Appl. Pharmaceu. Sci.* 3: 171.

Purohit, A. and Sharma, A. 2006. Blood Glucose Lowering Potential of *Bougainvillea Spectabilis* Willd. Leaf Extract in Streptozotocin-Induced Type-I Diabetic Albino Rats. *Indian Drugs-Bombay* 43(7): 538.

Rahalison, L., Benathan, M., Monod, M., Frenk, E., Gupta, M. P., Solis, P. N. ... and Hostettmann, K. 1995. Antifungal principles of *Baccharis pedunculata. Planta Med.* 61(04): 360–362.

Rajeswara Rao, B. R., Singh, K., Sastry, K. P., Singh, C. P., Kothari, S. K. and Rajput, D. K. 2007. Bhattacharya AK. Cultivation technology for economically important medicinal plants. In: Janardhan Reddy K. Bahadur B. Bhadraiah B, Rao MLN (eds) *Advan. Med. Plants*, Universities Press India Pvt Ltd: Hyderabad, 112–122.

Rao, N. K., Boini, K. M. and Nammi, S. 2004. Effect of chronic administration of Boerhavia diffusa Linn. Leaf extract on experimental diabetes in rats experimental diabetes rats. *Tropical Journal of Pharmaceutical Research* 3(1): 305–309.

Rao, N. K. and Nammi, S. 2006. Antidiabetic and renoprotective effects of the chloroform extract of *Terminalia chebula* Retz. seeds in streptozotocin-induced diabetic rats. *BMC Complement Altern. Med.* 6(1): 1–6.

Ras, R. T., Zock, P. L. and Draijer, R. 2011. Tea consumption enhances endothelial-dependent vasodilation: a meta-analysis. *PloS One* 6(3): e16974.

Rastogi, S., Pandey, M. M. and Rawat, A. K. S. 2016. Traditional herbs: a remedy for cardiovascular disorders. *Phytomedicine* 23(11): 1082–1089.

Razavi, B. M., Amanloo, M. A., Imenshahidi, M. and Hosseinzadeh, H. 2016. The relaxant activity of safranal in isolated rat aortas is mediated predominantly via an endothelium-independent mechanism: -vasodilatory mechanism of safranal. *J. Pharmacopuncture* 19(4): 329–335.

Rizk, A. M., Heiba, H. I., Ma Ayergi, H. A. and Batanouny, K. H. 1986. Constituents of plants growing in Qatar. *Fitoterapia* 1: 3–9.

Rotblatt, M. and Ziment, I. 2002. *Evid.-based Herbal Med.* Philadelphia: Hanley & Belfus Inc.

Rubatzky, V. E. and Yamaguchi, M. 2012. *World Vegetables: Principles, Production, and Nutritive Values.* Dordrecht: Springer Science and Business Media.

Rubio, O. C., Cuellar, A. C., Rojas, N., Castro, H. V., Rastrelli, L. and Aquino, R. 1999. A polyisoprenylated benzophenone from *Cuban propolis. J. Nat. Prod.* 62(7): 1013–1015.

Saadat, S., Shakeri, F. and Boskabady, M. H. 2018. Comparative antitussive effects of medicinal plants and their constituents. *Alternative Therapies* 24(4): 36–49.

Sangameswaran B., Reddy, T. C. and Jayakar, B. 2008. Hepatoprotective effect of leaf extracts *of Andrographis lineata* Nees on liver damage caused by carbon tetrachloride in rats. *Phytother. Res.* 22(1): 124–126.

Sautour, M., Miyamoto, T. and Lacaille-Dubois, M. A. 2005. Steroidal saponins from *Smilax medica* and their antifungal activity. *J.nat. prod.* 68(10): 1489–1493.

Schultes, R. E. 1978. The kingdom of plants. In *Medicines from the earth* (Ed. W. A. R. Thomson). New York: McGraw-Hill Book Co., p. 208.

Senthilkumar, N., Badami, S., Dongre, S. H. and Bhojraj, S. 2009. Antioxidant and hepatoprotective activity of the methanol extract of *Careya arborea* bark in Ehrlich ascites carcinoma-bearing mice. *J. Ethnopharmacol.* 116(1): 1–6.

Seo, Y. S., Kang, O. H., Kim, S. B., Mun, S. H., Kang, D. H., Yang, D. W. ... and Kwon, D. Y. 2015. Quercetin prevents adipogenesis by regulation of transcriptional factors and lipases in OP9 cells. *Int. J. mol. med.* 5(6): 1779–1785.

Shafik, A. N., Khodeir, M. M., Gouda, N. A. and Mahmoud, M. E. 2011. Improved antifibrotic effect of a combination of verapamil and silymarin in rat-induced liver fibrosis. *Arab J. Gastroenterol.* 12(3): 143–149.

Shamsabadipour, S., Ghanadian, M., Saeedi, H., Rahimnejad, M. R., Mohammadi-Kamalabadi, M., Ayatollahi, S. M. and Salimzadeh, L. 2013. Triterpenes and steroids from *Euphorbia denticulata* Lam. with anti-Herpes simplex virus activity. *Iranian J. pharmaceutical Res.: IJPR*, 12(4): 759.

Shanker, K. S., Kanjilal, S., Rao, B. V. S. K., Kishore, K. H., Misra, S. and Prasad, R. B. N. 2007. Isolation and antimicrobial evaluation of isomeric hydroxy ketones in leaf cuticular waxes of *Annona squamosa. Phytochemical Analysis: Int. J. Plant Chem. Bioch. Tech.* 18(1): 7–12.

Shinwari, M. I. and Khan, M. A. 2000. Folk use of medicinal herbs of Margalla hills national park, Islamabad. *J. Ethnopharmacol.* 69(1): 45–56.

Shishikura, Y., Khokhar, S. and Murray, B. S. 2006. Effects of tea polyphenols on emulsification of olive oil in a small intestine model system. *J. Agricul. food Chem.* 54(5): 1906–1913.

Singh, S. N., Vats, P., Suri, S., Shyam, R., Kumria, M. M. L., Ranganathan, S. and Sridharan, K. 2001. Effect of an antidiabetic extract of *Catharanthus roseus* on enzymic activities in streptozotocin-induced diabetic rats. *J. Ethnopharmacol.* 76(3): 269–277.

Singh, G., Gupta, P., Rawat, P., Puri, A., Bhatia, G. and Maurya, R. 2007. Antidyslipidemic activity of polyprenol from *Coccinia grandis* in high-fat diet-fed hamster model. *Phytomed.* 14(12): 792–798.

Sivaraj, R., Rahman, P. K., Rajiv, P., Narendra, S. and Venkatesh, R. 2014. Biosynthesis and characterization of *Acalypha indica* mediated copper oxide nanoparticles and evaluation of its antimicrobial and anticancer activity. *Spectrochimica Acta Part A: Mol. Biomol. Spectrosc.* 129: 255–258.

Snehlata, H. S. and Payal, D. R. 2012. Fenugreek (*Trigonella foenum-graecum* L.): An overview. *Int. J. Curr. Pharm. Rev. Res.* 2(4): 169–87.

Solowey, E., Lichtenstein, M., Sallon, S., Paavilainen, H., Solowey, E. and Lorberboum-Galski, H. 2014. Evaluating medicinal plants for anticancer activity. *Sci. World J.* 2014: 72140.

Srikumar, R., Parthasarathy, N. J. and Devi, R. S. 2005. Immunomodulatory activity of Triphala on neutrophil functions. *Bio. Pharm. Bull.* 28(8): 1398–1403.

Srivastava, S. and Chandra, D. 2013. Pharmacological potentials of Syzygium cumini: a review. *J. Sci. Food Agricul.* 93(9): 2084–2093.

Sultana, S., Perwaiz, S., Iqbal, M. and Athar, M. 1995. Crude extracts of hepatoprotective plants, Solanum nigrum and Cichorium intybus inhibit free radical-mediated DNA damage. *J Ethnopharmacol.* 45(3): 189–192.

Stanilova, S. A., Zhelev, Z. D. and Dobreva, Z. G. 2000. Preliminary studies on the immunomodulatory effect of the C3 binding glycoprotein isolated from *Cuscuta europea. Int. J. Immunopharmacol.* 22(1): 15–24.

Stone, W. M., Tonnessen, B. H. and Money, S. R. 2007. The new anticoagulants. *Perspect. Vasc. Surg. Endovasc. Ther.* 19(3): 332–335.

Strzelecka, M., Bzowska, M., Koziel, J., Szuba, B., Dubiel, O., Rivera Nuñez, D. ... and Bereta, J. 2005. Anti-inflammatory effects of extracts from some traditional Mediterranean diet plants. *J. Physiol. Pharmacol. Supplement*, 56(1): 139–156.

Stults, B. M., Dere, W. H. and Caine, T. H. 1989. Long-term anticoagulation. Indications and management. *West. J. Med.* 151(4): 414.

Subhadra, V. D., Christy, J., Kumar, K. A. and Maheswari, M. U. 2010. Anticoagulant activity of methanolic bark extract of *Careya arborea* Roxb. *Inter J Pharma. Sci. Biotech.* 1: 93–95.

Sun, H., Che, Q. M., Zhao, X. and Pu, X. P. 2010. Antifibrotic effects of chronic baicalein administration in a CCl4 liver fibrosis model in rats. *European J. Pharmacol.* 631(1–3): 53–60.

Sun, Y. E., Wang, W. and Qin, J. 2018. Anti-hyperlipidemia of garlic by reducing the level of total cholesterol and low-density lipoprotein: A meta-analysis. *Medicine* (Baltimore) 97(18): e0255.

Sunthitikawinsakul, A., Kongkathip, N., Kongkathip, B., Phonnakhu, S., Daly, J. W., Spande, T. F. ... and Rochanaruangrai, S. 2003. Coumarins and carbazoles from *Clausena excavata* exhibited antimycobacterial and antifungal activities. *Planta Med.* 69(02): 155–157.

Szent-Giörgyi, A. 1963. Lost in the twentieth century. *Annu. Rev. Biochem.* 36: 1–15.

Tao, Y. Y., Wang, Q. L., Shen, L., Fu, W. W. and Liu, C. H. 2013. Salvianolic acid B inhibits hepatic stellate cell activation through transforming growth factor beta-1 signal transduction pathway in vivo and in vitro. *Exp. Biol.med.* 238(11): 1284–1296.

Tapia-Pérez, M. E., Tapia-Contreras, A., Cedillo-Rivera, R., Osuna, L. and Meckes, M. 2003. Screening of Mexican medicinal plants for antiprotozoal activity–Part II. *Pharmaceutical Biol.* 41(3): 180–183.

Tapsell, L. C., Hemphill, I., Cobiac, L., Sullivan, D. R., Fenech, M., Patch, C. S. ... and Inge, K. E. 2006. Health benefits of herbs and spices: the past, the present, the future. *Med. J. Aust.* 185(S4): S1–S24.

Tariq, M., Ageel, A. M., Al-Yahya, M. A., Mossa, J. S. and Al-Said, M. S. 1989. Anti-inflammatory activity of *Teucrium polium*. *Int. J. tissue react.* 11(4): 185–188.

Tasleem, F., Ahmed, S. and Azhar, I. 2015. Antiemetic activity of bergenin from *Peltophorum roxburghii* L. *Indo American J. Pharm Res.* 5(02): 760–764.

Tawfeek, N., Mahmoud, M. F., Hamdan, D. I., Sobeh, M., Farrag, N., Wink, M. and El-Shazly, A. M. 2021. Phytochemistry, pharmacology and medicinal uses of plants of the genus *Salix*: an updated review. *Front. Pharmacol.* 12: 593856.

Tedong, L., Dimo, T., Dzeufiet, P. D. D., Asongalem, A. E., Sokeng, D. S., Callard, P. and Pierre, K. J. F. F. 2006. Antihyperglycemic and renal protective activities of *Anacardium occidentale* (Anacardiaceae) leaves in streptozotocin-induced diabetic rats. *Afr. J. Trad. Complement. Altern. Med.* 3(1): 23–35.

Tsai, M. K., Lin, Y. L. and Huang, Y.T. 2010. Effects of salvianolic acids on oxidative stress and hepatic fibrosis in rats. *Toxicol. Appl. Pharmacol.* 242(2): 155–164.

Unnati, S., Ripal, S., Sanjeev, A. and Niyati, A. 2013. Novel anticancer agents from plant sources. *Chinese J. Nat. Med.* 11(1): 0016–0023.

Verschaeve, L. and Van Staden, J. 2008. Mutagenic and antimutagenic properties of extracts from South African traditional medicinal plants. *J. Ethnopharmacol.* 119(3): 575–587.

Vickers, E. R., Cousins, M. J., Walker, S. and Chisholm, K. 1998. Analysis of 50 patients with atypical odontalgia: a preliminary report on pharmacological procedures for diagnosis and treatment. *Oral Surg Oral Med Oral Pathol Oral Radiol* 85(1): 24–32.

Vinayagam, R. and Xu, B. 2015. Antidiabetic properties of dietary flavonoids: A cellular mechanism review. *Nutr. metabol.* 12(1): 1–20.

Vishwanath, B. S., Kini, R. M. and Gowda, T. V. 1987. Characterization of three edema-inducing phospholipase A2 enzymes from habu (*Trimeresurus flavoviridis)* venom and their interaction with the alkaloid aristolochic acid. *Toxicon*, 25(5): 501–515.

Wadkar, K. A., Magdum, C. S. and Kondawar, M. S. (2008). Use of Careya arborea Roxb. Leaf Extract as an Indicator in Acid-Base Titrations. *Res. J. Pharm. Technol.* 1(4): 535–536.

Wang, Q., Terreaux, C., Marston, A., Tan, R. X., Stoeckli-Evans, H. and Hostettmann, K. 1999. A new 2-hydroxyflavanone from *Mosla soochouensis*. *Planta Med.* 65(08): 729–731.

Wang, W. H. and Wang, Z. M. 2005. Studies of commonly used traditional medicine-ginger. *Zhongguo Zhong Yao za Zhi Zhongguo Zhongyao Zazhi China J. Chinese Materia Med.* 30(20): 1569–1573.

Wang, T. C., Su, Y. P., Hsu, T. Y., Yang, C. C. and Lin, C. C. 2007. 28-Day oral toxicity study of the aqueous extract from spider brake (*Pteris multifida* Poiret) in rats. *Food Chem. Toxicol.* 45(9): 1757–1763.

Wang M. Y., Anderson, G., Nowicki, D. and Jensen, J. 2008. Hepatic protection by noni fruit juice against CCl_4-induced chronic liver damage in female SD rats. *Plant Foods Hum. Nutr.* 63(3): 141–145.

WHO 2005. WHO global atlas of traditional, complementary and alternative medicine. In Map Volume (Eds. C. K. Ong, G. Bodeker, C. Grundy, G. Burford and K. Shein). Geneva, Switzerland: World Health Organization, p. 216.

Wild, S., Roglic, G., Green, A., Sicree, R. and King, H. 2004. Global prevalence of diabetes: estimates for the year 2000 and projections for 2030. *Diabetes care* 27(5): 1047–1053.

Xie, B. B., Liu, H. Y., Ni, W., Chen, C. X., Lue, Y., Wu, L. and Zheng, Q. T. 2006. Five new steroidal compounds from *Ypsilandra thibetica*. *Chem. biodivers.* 3(11): 1211–1218.

Yamssi, C., Payne, V. K., Noumedem Anangmo, C. N., Tateng Ngouateu, A., Megwi, L. and Kuiate, J. R. 2020. Acute Toxicity of *Pentaclethra macrophylla* and *Psidium guajava* Use as Antiprotozoan Medicinal Plants. *J. Drug Discov. Develop. Deliv.* 6(1): 1037.

Yoon, S. J., Pereira, M. S., Pavão, M. S., Hwang, J. K., Pyun, Y. R. and Mourão, P. A. 2002. The medicinal plant *Porana volubilis* contains polysaccharides with anticoagulant activity mediated by heparin cofactor II. *Thromb. Res.* 106(1): 51–58.

Yousaf, T., Rafique, S., Wahid, F., Rehman, S., Nazir, A., Rafique, J. ... and Shah, S. M. 2018. Phytochemical profiling and antiviral activity of *Ajuga bracteosa, Ajuga parviflora, Berberis lycium and Citrus lemon* against Hepatitis C Virus. *Microb. Pathog.* 118: 154–158.

Zhang, B. J., Xu, D., Guo, Y., Ping, J., Chen, L. B. and Wang, H. 2008. Protection by an antioxidant mechanism of berberine against rat liver fibrosis induced by multiple hepatotoxic factors. *Clin. Exp. Pharmacol. Physiol.* 35(3): 303–309.

Zheng, S. and Chen, A. 2006. Curcumin suppresses the expression of extracellular matrix genes in activated hepatic stellate cells by inhibiting gene expression of connective tissue growth factor. *American J. Physiol. Gastrointest. Liver Physiol.* 290(5): G883–G893.

Zhong, Z., Zhang, Q., Tao, H., Sang, W., Cui, L., Qiang, W. ... and Wang, Y. 2019. Anti-inflammatory activities of *Sigesbeckia glabrescens* Makino: combined in vitro and in silico investigations. *Chinese Med.* 14(1): 1–12.

Zhou, H. Y., Hong, J. L., Shu, P., Ni, Y. J. and Qin, M. J. 2009. A new coumarin and anticoagulant activity from *Viola yedoensis* Makino. *Fitoterapia* 80(5): 283–285.

Zia-Ul-Haq, M., Shahid, S. A., Muhammed, S., Qayum, M., Khan, I. and Ahmad, S. 2012. Antimalarial, antiemetic and antidiabetic potential of *Grewia aslatica* L. leaves. *J. Med. Plants Res.* 6(16): 3087–3092.

Ziment, I. 2002. Herbal antitussives. *Pulm. Pharmacol. Ther.* 15(3): 327–333.

2 Translational Research for Medicinal Plants

Integrative Approaches, Trends and Perspectives

Siva P. K. Chetri, Lokesh Kumar Agarwal, Akanksha Vashishtha, Vibhuti Singh, Ratan Lal, Amit Kumar Gupta and Kuldeep Sharma

CONTENTS

2.1 INTRODUCTION

The COVID-19 pandemic has reoriented global opinion and compelled the world to look toward plants and plant-based products for its food security, therapeutics, pharmaceuticals and nutraceuticals all over again. A new wave of interest has arisen in plant-based products, functional foods and herbal formulations (Assocham 2017, Joseph 2021). Globally, the functional food and herbal formulations-based market is at its torrid pace and expecting a big boom in the time ahead, and the plant products-based companies are gearing up their capital investments worldwide (Assocham 2017, Joseph 2021). Phytohormones, polyamines, vitamins, alkaloids, flavonoids, phenolics, terpenoids and other primary and secondary metabolites are distributed differentially among different plant taxonomic groups that have no recognized role in plant life but are of immense therapeutic and clinical potential in serving humanity (Hussein et al. 2019). Bioactive compounds from medicinal plants are of particular interest and nowadays, there is an excellent focus on drug discovery and therapeutically active formulations based on natural products. Traditional healthcare knowledge is now increasingly being applied to the usage of such herbal formulations to treat a variety of health issues all around the world (Kunwar et al. 2013, Ekor 2014, Atanasov et al. 2015, Lim et al. 2021, Shah et al. 2021). Many plant metabolites are usually isolated either from wild or cultivated plants, and their content depends on the plant's physiological and developmental stage (Oksman-Caldentey and Inze

DOI: 10.1201/9781003179139-2

2004). The availability of raw materials, their identification and verification in terms of quality control standards are some of the challenging considerations for medicinal plants (Balekundri and Mannur 2020).

The detection of the elite germplasm of a medicinal plant with the help of advanced molecular techniques such as NGS, DNA barcoding, molecular markers, and so on, would lead to proper identification, authentication, enhanced cultivation and conservation practices for the germplasm of interest, and further scaling up of the bioactive metabolites for commerce (Kumar and Gupta 2008, Chetri et al. 2014, Chetri, Sharma and Agrawal 2013, Techen et al. 2014, Sarwat and Yamdagni 2016, Lo and Shaw 2019). Molecular technologies enable for the identification of elite species, genotypes, chemotypes and variations, as well as sex determination, and are cost-effective, trustworthy, and rapid approaches for conserving and harnessing plant genetic resources (Agrawal et al. 2007, Heikrujam et al. 2015, Sarwat and Yamdagni 2016, Kumar et al. 2019, Lo and Shaw 2019, Zhang, Qingdu and Li 2019a, Zhang et al. 2019b).

Advances in plant research *in vitro* have also given a plethora of information on the general genetics and biochemistry involved in the biosynthesis of plant bioactive constituents (Parast et al. 2011, Chetri, Kapoor and Agrawal 2016, Niazian 2019, Babich et al. 2020, Marchev et al. 2020, Khan et al. 2021). A variety of genetic engineering techniques have shown the potential to boost the quantity of bioactive constituents in such therapeutically important plants (Oksman-Caldentey and Inzé 2004, Kowalczyk et al. 2020); however, for a large-scale improvement of bioactive constituents, more comprehensive and accurate information on the regulatory and biosynthetic pathways, as well as their intermediate byproducts, are of prime importance. The recent advancement of new "omics" approaches such as genomics, transcriptomics, proteomics and metabolomics, has been a boon to understanding plant system biology and elucidating the multiple levels of information involved in biosynthesis of plant metabolites (Rai, Saito and Yamazaki 2017, Pandita et al. 2021).

In the present article, *in vitro* biosynthesis of secondary metabolites, the elicitation and upregulation of secondary metabolites, the essential enzyme genes involved in metabolic pathways, omic tools and techniques, the therapeutic use of herbal products, and the prospects and perspectives for herbal-based clinical practices have been reviewed briefly.

2.2 *IN VITRO* BIOSYNTHESIS AND ELICITATION OF SECONDARY METABOLITES

Plant biotechnology, especially plants tissue culture offers possibilities for mass propagation of elite clones for a regular supply of raw materials to the industries as well as for the conservation of the elite and rare germplasm (Ahmad et al. 2019, Chetri, Kapoor and Agrawal 2016, 2014, Parast et al. 2011, Rout et al. 2000).

Plant cell tissue culture is an intriguing approach because it may provide ideal conditions for the biosynthesis of such bioactive molecules that are scarcely available from natural sources (rare, endangered, or overexploited species), from difficult-to-cultivate sources, or that can only be produced under *in vitro* conditions (Isah et al. 2018). The steady advancements in plant biotechnology and metabolic engineering have opened several prospects for the *in vitro* synthesis of biomolecules of interest (Goncalves and Romano 2018).

The pace at which intermediate substances from core metabolic pathways are re-routed to secondary biosynthesis pathways determines the formation of secondary metabolites. Therefore, cell growth and quantity of bioactive compounds can be modified and enhanced by changing the media composition and culture conditions, elicitation, and so on (Espinosa-Leal, Puente-Garza and García-Lara 2018, Halder, Sarkar and Jha 2019).

Different strategies can also be applied to improve the production of bioactive compounds to overcome obstacles such as accurate identification and speciation, elicitation and upregulation, hairy root culture, metabolic engineering, and so on (Halder, Sarkar and Jha 2019, Chetri, Kapoor and Agrawal 2016, Baenas, García-Viguera and Moreno 2014).

One of the most successful ways for increasing the synthesis of such metabolites has been the elicitation procedure for induced *in vitro* upregulation of plant bioactive compounds (Halder et al. 2019, Chetri, Kapoor and Agrawal 2016, Baenas, García-Viguera and Moreno 2014, Roberts and Shuler 1997). Elicitors are classed abiotically and biotically based on their source. Abiotic elicitors are non-biological sources like salts or physical factors and biotic elicitors are derived from biological sources and can be either exogenous or endogenous. The overall ability of such elicitors is distinctive and multidimensional and have the ability to regulate a variety of cellular functions at both the biochemical and molecular levels (Isah 2019, Baenas, García-Viguera and Moreno 2014, Zhao et al. 2005). The elicitors act as signaling compounds leading to enhanced production of natural plant products such as flavonoids, alkaloids, terpenoids, phenylpropanoids, and so on (Wang and Lei 2018, Giri and Zaheer 2016, van der Fits and Memelink 2000). Plant parts such as leaves, shoots, roots and other plant parts have been used to accomplish *in vitro* production of bioactive compounds of high medicinal importance via callus, cell suspension, immobilized cells and differentiated culture (Ahmad et al. 2019, Goncalves and Romano 2018).

The genetic and physiological status of the plant regulates the effect of such stimulus on plants (Ramirez-Estrada et al. 2016) and the augmented synthesis of plant bioactive substances is determined by specific phases of plant development and differentiation (Brukhin and Morozova 2011). Several researchers have employed plant cell biosynthetic capacities *in vitro* using a range of elicitors to scale up useful enhanced metabolites from a variety of medicinal plants (Table 2.1).

2.3 CHARACTERIZATION OF KEY ENZYME GENE(S) FROM METABOLIC PATHWAYS

Plant bioactive compounds with unique pharmaceutical activity have led to increased interest in exploring their biosynthetic pathways. The plant genomes encode high metabolic complexity. Therefore, deciphering this data and translating anticipated sequences into enzymes, and metabolites is a daunting task (Medema and Osbourn 2016). However, the metabolic pathways are sequential chemical processes mediated by a series of different enzymes taking place within plant cells producing several intermediate compounds and culminating in the formation of a metabolite regulated by genes. Therefore, it is possible to control the key enzyme gene(s) of a plant's biosynthesis pathway and to upregulate/downregulate the biosynthesis of a bioactive substance of interest. Recent advances in metabolic engineering for the characterization of key enzyme gene(s) and respective enzymes of the metabolic pathways have opened up new horizons (Chetri, Kapoor and Agrawal 2016, Parast et al. 2011, Kumar and Gupta 2008). Several researchers have reported the cloning, characterization and functional analysis of different key enzymes and related genes involved in the production of a range of bioactive substances in medicinal as well as commercially important plants in the recent past (Table 2.1).

2.4 INTEGRATIVE APPROACHES

2.4.1 Omics Technology: Genomics, Transcriptomics and Proteomics

The complex plant metabolite research gradually gained huge momentum in the last decade with newer and improved tools and techniques. The advancements in the emerging "omics" technologies, such as genomics, transcriptomics and proteomics, have provided extensive information on biological system in plants and assisted in the know-how of desired plant metabolite biosynthesis (Pinu et al. 2019, Rai, Saito and Yamazaki 2017, Lagunin et al. 2014).

TABLE 2.1
Biotechnological Interventions for Elicitation and *in Vitro* Enhancement of Bioactive Constituents, Identification of Key Enzyme Genes and Bioefficacy of Herbal Extracts From Some Medicinal Plants

S. No.	Medicinal plant	Elicitors	Tissue/ organ/ callus	Bioactive compounds	Reference
Elicitation and *in vitro* enhancement of bioactive secondary metabolites					
1	*Securinega suffruticosa*	9 auxins [2,4-dichlorophenoxyacetic acid (2,4- D), 2,4,5- trichlorophenoxyacetic acid, indole-3-acetic acid (IAA), 4-chlorophenoxyacetic acid, indole-3-butyric acid, 4-amino, 3,5,6- trichloropicolinic acid, b-naphthoxyacetic acid, 1-naphthaleneacetic acid, indole-3-propionic acid, one polar auxin transport inhibitor (2,3,5- triiodobenzoic acid)] and 7 cytokinins [2-isopentenyladenine, 1-phenyl-3-(1,2,3-thiadiazol-5-yl)urea, N-(2-chloro-4-pyridinyl)-N0-phenylurea, diphenylurea, zeatin, kinetin (Kn), 6-benzyl aminopurine]	Callus cultures	Indolizidine alkaloids	Raj et al. 2015
2	*Schisandra chinensis (Turcz.)*	Different spectral properties of monochromatic light (far-red, red, blue, UVA), and in darkness and under white light (control conditions)	Shoot-differentiating callus cultures	lignans and phenolic acids	Szopa and Ekiert 2016
3	*Scutellaria alpina*	Bacterial nanocellulose supported liquid culture system	Shoot tips	Baicalin, wogonoside, luteolin, luteolin-7 glucoside and verbascoside	Grzegorczyk-Karolak et al. 2017
4	*Decalepi ssalicifolia* (Bedd. Ex Hook.f.) Venter	Chitosan and yeast extract	Root tuber derived callus suspension culture	2-hydroxy-4 methoxybenzaldehyde	Ahmad et al. 2019
5	*Rhodiola rosea*	Phenylalanine, *trans*-cinnamic acid, and cinnamyl alcohol	In vitro leaf derived callus	Cinnamyl alcohol glycosides viz., rosin, rosavin, rosarin	Mirmazloum et al. 2019
6	*Ocimum basilicum*	Methyl jasmonate	Cambial meristematic cells	Triterpenoids oleanolic acid and ursolic acid	Mehring et al. 2020

7	*Eryngium alpinum* L.	Stationary and agitated culture systems on solidified and in liquid media	Shoot cultures	Phenolic acids: 3,4-dihydroxyphenylacetic, caftaric, caffeic, neochlorogenic, chlorogenic, isochlorogenic, and rosmarinicacids, and Flavonoids – isoquercetin, quercitrin and robinin	Kikowska et al. 2020

Characterization of key enzymes involved in biosynthesis of bioactive secondary metabolites

S. No.	Plant source	Key enzyme gene (s)	Biosynthesis pathway	Usefulness of the bioactive compounds	Reference
1	*Punica granatum* L.	Chalcone synthase, chalcone isomerase, flavanone-3 hydroxylase, dihydroflavonol 4-reductase, anthocyanidin synthase (ANS), UDP-glucose-flavonoid 3-*O*-glucosyltransferase	Anthocyanin	Polyphenolic compounds involved in attracting pollinators and protection against stress (biotic and abiotic)	Zhao et al. 2015
2	*Cassia angustifolia* Vahl	Isochorismate synthase	Sennosides	Anticancer activity	Chetri et al. 2016
3	*Salvia miltiorrhiza* Bunge	Squalene synthase	Squalene	Inhibits chemically induced colon, lung, and skin cancers, has bactericidal and antifungal properties, markedly increases both cellular and non-specific immune functions, and reduces serum cholesterol levels	Rong et al. 2016
4	*Betula platyphylla*	BpMYC4 and BpbHLH9	Triterpenoids	Possess anti-tumor pharmacological activities	Yin et al. 2017
5	*Salvia guaranitica*	Terpenoid synthase	Terpenes	Possess antimicrobial, antimutagenic, anticancer, antioxidant, anti-inflammatory, choleretic and antimicrobial activities	Ali et al. 2018
6	*Carthamus tinctorius*	Chalcone isomerase	Flavonoid	Protect plants from different biotic and abiotic stresses and function as signal molecules, detoxifying agents and antimicrobial defensive compounds, etc.	Liu et al. 2019

(*continued*)

TABLE 2.1 (Continued)
Biotechnological Interventions for Elicitation and *in Vitro* Enhancement of Bioactive Constituents, Identification of Key Enzyme Genes and Bioefficacy of Herbal Extracts From Some Medicinal Plants

Characterization of key enzymes involved in biosynthesis of bioactive secondary metabolites

S. No.	Plant source	Key enzyme gene (s)	Biosynthesis pathway	Usefulness of the bioactive compounds	Reference
7	*Arachis hypogaea*	Flavonol synthase	Flavonol	Possess antioxidant, antimicrobial, hepatoprotective and anti-inflammatory properties	Hou et al. 2020
8	*Andrographis paniculata* (Burm. f) Nees	1-Deoxy-D-xylulose 5-Phosphate Synthase (DXS)	1-Deoxy-D-xylulose 5-phosphate (DXP)	A branch point metabolite in isoprenoid biosynthesis, and is also used in the biosynthesis of thiamin (vitamin B1) and pyridoxal (vitamin B6)	Srinath et al. 2021

Bioefficacy and therapeutic potential of phytochemical constituents

S. No.	Plant name	Chemical constituents	Activity tested	Type of disease/ cancer cell lines studied	Reference
1	*Sansevieria liberica* Gerome and Labroy	Hydroethanolic root extract	Anticancer	Sarcoma-180 (S-180) ascites and solid tumor, and L1210 lymphoid leukemia	Akindele et al. 2015
2	*Plectranthus stocksii* Hook. f.	Powdered plant parts in petroleum ether, chloroform, ethyl acetate, acetone and methanol	Antioxidant and anticancer	MCF-7, Caco-2 and RAW264.7 cancerous cell line	Muniyandi et al. 2017
3	*Trigonella foenum-graecum, Cassia acutifolia Rhazya stricta*	Leaf extract in 70% ethanol	Antioxidant and anticancer	Human hepatoma (HepG2) and HUVEC cancer cell lines	Al-Dabbagh et al. 2018
4	*Cardiospermum halicacabum L., Gomphrena celosioides* Mart. *Scoparia dulcis L.*	Ethanolic extracts	Anticancer	Cholangiocarcinoma (CCA) cell lines KKU-100 (JCRB 1568) and KKU-213 (JCRB 1557)	Promraksa et al. 2019
5	*Azadirachta indica* A. Juss *Melia azedarach* Linn.	Dried aerial parts extracted with ether, petroleum, methanol, hexane and water	Anticancer	MCF cell lines	Malar et al. 2020
6	*Adenosma bracteosum* (Bonati)	Ethanol extract and its derived fractions (chloroform, ethyl acetate, butanol, and aqueous) of aerial parts	Anticancer	Human large cell lung carcinoma (NCI-H460) and hepatocellular carcinoma (HepG2)	Nguyen et al. 2020

Genomics encompasses the study of the genome that includes the sum total of nuclear DNA (nr- genome), mitochondrial DNA and chloroplast DNA of a cell or an organism. The genome of an organism is accountable for several phenotypic expressions and their inheritance, expression and regulation of genes in an organism. With the advent of sequencing techniques in the 1970s, several quick and cost-effective, high-throughput, comprehensive DNA sequencing methods have emerged in the recent past (Levy and Mayers 2016, Mardis 2011, 2008, Pease et al. 1994, Maxam and Gilbert 1977, Sanger and Coulson 1975). So far, 236 plant genomes from 31 taxonomic orders have been sequenced either partially or fully (Chen et al. 2018). The nuclear, mitochondrial and chloroplast genomes pertaining to herbal genomics have also been successfully sequenced for certain medicinal plants, viz. *Azadirachta indica*, *Camptotheca acuminata*, *Catharanthus roseus*, *Glycyrrhiza uralensis*, *Ocimum sanctum*, *Panax* spp., *Salvia miltiorrhiza*, *Sophora* spp., and so on (Wei et al. 2020, Duan et al. 2019, Zhang, Qingdu and Li 2019a, Zhang et al. 2019b, Chen et al. 2018, Mochida et al. 2017, Xu et al. 2017, Kellner et al. 2015). The diverse genomics approaches for herbal plants paved way for vital information on gene analysis in terms of structure and function and their interactions for a better understanding of the total biosynthetic potential (Chakraborty 2018). The genomic resources of medicinal plants are also assisting in barcoding and assessment of molecular phylogeny and phylogenomics to assure the quality and safety of raw herbal materials and also to predict and discover chemo-diversity and newer plants of medicinal significance (Yu et al. 2021, Ji et al. 2009).

Besides tools to sequence DNA or RNA partially and determination of the genetic constitution, a variety of markers of any vital genome have also been explored over the past few decades (Manzoni et al. 2018, Heikrujam et al. 2015). The genomic information of the medicinal herbs can boost development of vital genomic resources, for example, multiple homologous genes of metabolic pathways, evolution of secondary metabolites, enhanced capacity to knock out genes to determine their role, and so on (Rai, Saito and Yamazaki 2017). Furthermore, the genome information delivers broader information on the origin and evolution of plants, the transmissible traits as well as the basis for chemo-diversity and promotes the selection of cultivars with superior features and increased metabolite content of pharmaceutical importance (Dhanapal and Govindaraj 2015, Hao and Xiao 2015).

Interestingly, recent findings have revealed the appearance of similar co-regulated metabolic gene aggregation in certain medicinal plants and prokaryotic organisms (Mochida et al. 2017, Kellner et al. 2015). Therefore, further efforts are required to scan the genome and/or genomic sequences of medicinal herbs for the predictions and validation of such gene clusters responsible for synthesizing various metabolites of different classes on a broader scale. Such analysis would be helpful for structural and functional analysis of the new genes and also in understanding the evolutionary patterns for a diverse class of natural products (Hernández-Domínguez et al. 2019, Chakraborty 2018, Chavali and Rhee 2018). An alternative approach, the genome-wide association studies (GWAS) or association mapping (AM) approaches, based on a comparison of frequency and distribution of SNPs across the genome or between the different genomes of plants, is needed to dig out among medicinal herbs for such analysis (Abe et al. 2012, Bergelson and Roux 2010, Huang et al. 2010, Lu et al. 2003). In addition to this, the application of advanced genome mining and editing tools such as CENH3 (centromere histone H3), CRISPR/Cas (clustered regularly interspaced short palindromic repeats/CRISPR-associated), ZFNs zinc-finger nucleases, transcription activator-like effector nucleases, targeting-induced local lesions in genomes, in association with bioinformatics with reference to the secondary metabolites networks of diversified biosynthetic pathways among medicinal plants are being explored to look for an efficient and a faster combinatorial biosynthesis approach for tailor-made, tuned, custom-designed and user-friendly bioactive molecules in recent years (Shabir 2021, Niazian 2019, Malzahn, Lowder and Qi 2017, Yamazaki et al. 2018, Weeks, Spalding and Yang 2016).

Transcriptomics is the survey of transcriptomes that represents the total of all transcripts in a cell for a specific phase that provides a genome-wide understanding of the complex mechanisms in the expression, function and regulation of genes (Tripathi et al. 2016, Wang et al. 2009). The recent developments in economical and faster transcriptome profiling (RNA-seq-based) and de novo transcriptome assembly, co-expression as well as comparative transcriptome analysis have generated tissue-specific genetic resources for several plants of biomedical importance (Kalmankar et al. 2020, Stander et al. 2020, Thakur et al. 2020, Wu et al. 2020, Kim et al. 2019, Liu et al. 2018, Wang et al. 2016). Today, with the advances in the RNA sequencing techniques, the differential expression analysis for cell or tissue-specific functional characterization and regulation of enzymes and their genetic counterparts (mRNAs and regulatory RNAs) of plant secondary metabolite biosynthesis, and to distinguish the unknown homoeologous and paralogous gene copies and expression analysis of enzymes and their genes to trace the secondary metabolite biosynthesis pathway, is quite apparent (Wang et al. 2020, Niazian 2019, Rai, Saito and Yamazaki 2017).

Proteomics involves analysis of a subset of differentially expressed functional and regulatory cellular proteins at a given time in an organism, organ, tissue, or a cell and has a vital role in elucidating plant stress responses under different conditions (Pedrete, Hauser-Davis and Moreira 2019, Gong, Hu and Wang 2015, Ahmad et al. 2016, Kosová et al. 2018). Akin to this, recent technological advances have widely expanded proteomics to identify the candidate proteins and enzymes in the tissues involved in active plant metabolite biosynthesis, the transporter and carrier proteins, and also the transcriptional factors and all the regulatory proteins have also assisted in a wider understanding of the secondary metabolite biosynthesis and different aspects of protein-protein interactions in several medicinal plants (Pedrete et al. 2019, Hashiguchi, Tian and Komatsu 2017, Paz et al. 2017, Ma et al. 2016, Martínez-Esteso et al. 2015, Bhattacharyya et al. 2012, Desgagné -Penix et al. 2010).

Hence, for the broader role and identification of genes, proteins and enzymes of plant metabolite synthesis, Omics studies through the integrative approaches of genomics, transcriptomics and proteomics have been extensively utilized. The application of integrative omics technologies in medicinal plant research is being proved to be of immense significance in tracing the respective secondary metabolite biosynthesis pathways, and further, their correlation with nutraceuticals, pharmaceuticals, bio-efficacies and therapeutic outcomes has opened up new avenues bridging the certain knowledge gaps and leading to the future research endeavors.

2.4.2 Computational Tools: Bioinformatics and Cheminformatics in Herbal Drug Discovery

Bioinformatics, an interdisciplinary approach, covers genomics, transcriptomics, proteomics, population genetics and molecular phylogenetics. Bioinformatics possesses essential techniques for analyzing and interpreting high-throughput molecular data by developing and implementing computational algorithms and software tools (Rhee, Dickerson and Xu 2006). The bioinformatics technique is crucial in drug development that provides biological databases and computational tools for identifying the associative gene(s) and metabolic pathways of phytochemicals of clinical and therapeutic significance (Saito and Matsuda 2010). The bioinformatics approach helps to reduce the time and expense of innovative medication development by applying *in silico* tools for possible disease cures by utilizing phytochemicals of vital interest. Systems biology is a subfield of bioinformatics that focuses on researching the physiological properties of diseases at the molecular, cellular, tissue, organ and organism levels (Csermely et al. 2013). Hence, bioinformatics approaches help to forecast the likely genes or proteins implicated in the manifestation of a disease (Saito 2013, Schenone et al. 2013, Ziegler et al. 2013).

On the other hand, chemoinformatics uses appropriate databases to maintain an enormous amount of chemical data generated globally by chemists. The cheminformatics approach in the drug development process includes different steps like target identification, lead optimization, drug likeliness, efficacy, validation and safety, molecular docking, molecular simulations, pre-clinical trials, subsequent clinical trials and approval from the drug authority (Hazam, Goyal and Ramakrishnan 2019,

Huang et al. 2019, Malathi and Ramaiah 2018, Durrant and McCammon 2011, He et al. 2010, Yadav et al. 2010, Guha and Jurs 2004).

Computer-aided drug design (CADD) is a constructive approach for developing innovative therapeutic compounds from known natural molecules with medicinal significance. For a decade, the development of computational tools has prompted the application of secondary plant metabolites by recognizing secondary metabolite biosynthetic gene clusters (BGCs). These BGCs consist of organized information for biosynthetic pathways producing plant metabolites, which is the key to discovering new drugs. Blin et al. (2017) summarized the computational tools to study the mechanism of BGCs for secondary metabolites, emphasizing genome mining and compound identification. Recently used computational tools, with their URL and main functions, are shown in Table 2.2. The *in silico* techniques have been utilized for the identification of potent secondary metabolites to be used for novel medicine design (Xuhua 2017, Lagunin et al. 2014). The *in silico* approaches for drug design focus on the different strategies that include (1) virtual high-throughput screening (vHTS) to filter vast chemical libraries into smaller groups of a drug like potential compounds for further *in vitro* verification; (2) molecular simulation of lead compounds for absorption, distribution, metabolism, excretion and toxicity (ADMET) with drug metabolism

TABLE 2.2
URLs for Computational Tools With Their Core Functions

S. No.	Tool	Functions	URL	Reference
1.	antiSMASH 4	Genome mining BGC analysis Domain analysis	http://antismash.secondarymetabolites.org	Blin et al. 2017
2.	ARTS	Genome mining	http://arts.ziemertlab.com	Alanjary et al. 2017
3.	BAGEL 3	Genome mining	http://bagel.molgenrug.nl	vanHeel et al. 2013
4.	CASSIS	BGC-boundary prediction	https://sbi.hki-jena.de/cassis/cassis.php	Wolf et al. 2016
5.	CRISPy-web	sgRNA design	http://crispy.secondarymetabolites.org	Blin et al. 2016
6.	eSNaPD v2	Genome mining	http://esnapd2.rockefeller.edu	Reddy et al. 2014
7.	FunGeneClusterS	BGC-boundary prediction	https://fungiminions.shinyapps.io/FunGeneClusterS	Vesth et al. 2016
8.	fungiSMASH	Genome mining BGC analysis Domain analysis	http://fungismash.secondarymetabolites.org	Blin et al. 2016
9.	GRAPE/GARLIC	Genome mining	https://magarveylab.ca/gast	Dejong et al. 2016, Johnston et al. 2016
10.	NaPDoS	Genome mining	http://napdos.ucsd.edu	Ziemert et al. 2012
11.	NP searcher	Genome mining	http://dna.sherman.lsi.umich.edu	Li et al. 2009
12.	plantiSMASH	Genome mining, BGC analysis	http://plantismash.secondarymetabolites.org	Kautsar et al. 2017
13.	PRISM 3	Genome mining BGC analysis Domain analysis	http://magarveylab.ca/prism	Skinnider et al. 2017
14.	RODEO	Genome mining RiPP analysis	www.ripprodeo.org	Tietz et al. 2017
15.	Smiles2Monomers	Retro-biosynthetic monomer prediction	http://bioinfo.lifl.fr/norine/smiles2monomers.jsp	Dufresne et al. 2015
16.	SMURF	Genome mining	www.jcvi.org/smurf	Khaldi et al. 2010

and pharmacokinetics (DMPK) properties; and (3) design and synthesis of novel compounds. The keen interest of structure-based CADD is screening and designing of pharmaceutically active compounds with large binding free energy, that bind tightly to the target, significant DMPK/ADMET properties, and target species are the keen interest of structure-based CADD (Jorgensen 2010, 2004). Structure-based and ligand-based are the two main approaches for CADD. Structure-based CADD uses the available cheminformatics database of the target molecular structure to calculate interaction energies for all lead molecules. This approach is appropriate for obtaining high-resolution structural data on receptor molecules. CADD based on ligands, on the other hand, depends on quantitative structure-activity relation (QSAR) models and the knowledge of pharmaceutically active and inactive molecules through chemical resemblance (Kalyaanamoorthy and Chen 2011). The methods of quantitative or qualitative structure-activity relationships (QSAR), molecular modeling, and simulated screening are broadly used in the drug design and development process on the notion that the activity of a compound is connected to its structure (Lagunin et al. 2014). As a result, CADD employs a significantly more focused search than standard HTS and combinatorial chemistry, resulting in a better success rate for new therapeutic molecules. So, if the structures of phytochemicals are available, these methods may be used to explore their biological activity. Molecular simulation studies of these plant molecules having certain pharmacophores predict potential molecules for new drug design. These studies are not limited to synthesized or naturally isolated molecules but toward exploration of the pharmacokinetic activity of possible derivatives that might reveal improved activity.

In the drug discovery process, both the bioinformatics and cheminformatics approaches play an important role and work in coordination with the other "-omics-" tools, narrowing the phenomenon to achieve a target to cure a specific disease. Further, options for personalized medicine employing both approaches are possible as they can identify the exact molecular mechanism involved in a particular condition in a particular patient (Whirl-Carrillo et al. 2012). Hence, advanced computational tools encourage the experimental organic chemists to synthesize structurally resembled molecules and derivatives based on their predicted values. Biological screening of these synthesized molecules needs to identify potential drug discovery research molecules. A vast amount of data is usually generated annually on the biological activity of herbal compounds along with the disease target macromolecules related to their therapeutic effects, and *in silico* methods to evaluate the biological activity of the medicinal plants are advantageous. In the future, it would be useful to use computational tools to enhance lead to cover a larger chemical space while reducing the number of compounds that must be created and assessed *in vitro*. Researchers in the pharmaceutical industry would pay greater consideration to CADD because of its lower cost compared to chemical synthesis and *in vitro* characterization of substances.

2.5 BIOEFFICACY AND THERAPEUTIC USEFULNESS OF HERBAL PRODUCTS

Researchers have extensively explored drug development using bioactive compounds, which are used in various bioefficacy research. Bioefficacy and therapeutic utility research in medicinal plants have resulted in a variety of bio-activity tests, lead identification, recognition of target sites and tracing of the probable mechanisms of action for biologically active phytochemicals. Further, plant extracts are being screened for the potent bioactive compounds responsible for anticancer, antioxidant, insecticidal activity, and so on.

Nowadays, the abnormal proliferation of cells known as cancer is a growing health problem globally. The wider prevalence of cancer with 18.1 million new diagnoses and 9.6 million deaths in 2018 has primarily afflicted males with colorectal, liver, lung, prostate and stomach cancer, and most women with breast, cervical, colorectal, lung and thyroid cancer (Khan et al. 2019, Bray et al. 2018). Various epidemiological and pharmacological research has indicated that consuming specific phytochemicals on a regular basis might reduce the occurrence of several forms of cancer (Russo et al. 2010, Vauzour et al. 2010). Furthermore, inflammation and oxidative stress are also involved in the development of a variety of diseases (Pandey et al. 2020). Plant extracts also contain many active compounds, which are fruitful against oxidative damage. Since numerous synthetic agents used to cure diseases display toxicity, so research interests are more inclined toward plant-derived chemotherapeutic agents.

Nowadays, the therapeutic use of bioactive herbal products is gaining increasing weightage over chemotherapy to cure cancer because of their ease of access, availability, dietary nature, and fewer side effects. Chemoprevention using dietary phytochemicals has proven to be a potential technique for reducing cancer growth. Besides, synergistic use of herbal compounds along with pharmaceuticals has been suggested with an added advantage of increased efficacy and lowered daily intake, and suppressed drug resistance of synthetic drugs to cure cancer (Liu 2004). Moreover, phytochemicals also interfere with multiple cell-signaling pathways (Aggarwal et al. 2006). The bioefficacy, cytotoxicity, chemopreventive and therapeutic properties of herbal compounds such as flavonoids, terpenoids and alkaloids, and so on, derived from different plant parts, have been described by the researchers (Table 2.1).

2.6 PROSPECTS AND PERSPECTIVES FOR HERBAL-BASED CLINICAL PRACTICES – A NEW LEAD FOR MODERN MEDICINE

Plants possess several bioactive compounds of immense medicinal and therapeutic importance crucial for human health and thus have been endowed as valuable, noble and reliable sources of phytomedicines. Increasing the production, quality and quantity of these natural phytochemicals to meet the needs of the human population and pharmaceutical companies, however, remains a major concern. Besides, the identification of plant species, prediction of plant metabolite content and dose of candidate metabolites are crucial quality control factors for pharmaceutically significant plants (Kumar and Gupta 2008). To develop proper herbal medicines, it is necessary to elucidate the unique physiology, identify the active principles and understand the knowledge of the intricate genetics and biochemistry involved in the biogenesis and regulation of the metabolite pathway of bioactive compounds of medicinal value (Isah 2019). It is, therefore, imperative to develop approaches for enhancing the bioactive compounds and for isolation of the phytochemical constituents and developing methods for its large-scale *in vitro* production (Espinosa-Leal, Puente-Garza and García-Lara 2018).

The contemporary surge in rejuvenation and awareness for herbal medications among common people may be attributed mainly to dissatisfaction with the results of commercial synthetic pharmaceuticals, their high cost and their side effects (Ekor 2014). The herbal treatments containing a variety of bioactive compounds are believed to function together in a synergistic manner, efficiently, effectively and safely (Ekor 2014, Firenzuoli and Gori 2007). However, a piece of appropriate knowledge and information based on disease-specific herbal treatments and the duration of the treatment is of prime importance. As a result, the relevant regulatory agencies must take adequate steps to protect public health in order to guarantee that all herbal medications are safe and of sufficient quality (Ekor 2014). Every nation, through its drug regulatory framework, should ensure that plant-based herbal medicines conform to the requisite standards of safety and quality, and efficacy.

2.7 CONCLUSION

This chapter briefly describes a comprehensive review of herbal plants translational research and their biotechnological interventions. The *in vitro* biosynthesis of secondary metabolites, elicitation, upregulation of secondary metabolites and characterization of key enzyme genes from biosynthetic pathways have been discussed. The developments in the field of biological research have led to broader integrative approaches involving tools and techniques of genomics, proteomics and transcriptomics. Various researchers globally have reported their positive research findings on those integrative and synergistic approaches with enhanced *in vitro* accumulation of requisite plant metabolites. The elicitors have their individual and synergistic effects *in vitro* on improved production of bioactive substances by activating signal molecules for up/down-regulation of key enzymes in biosynthesis pathways. A variety of plants having significant anticancerous, antioxidant and insecticidal activities have already been reported earlier. The gradual development in ethnopharmacological research has led to improved and broader prospects for its usage; however, a proper drug regulatory framework can better ensure the adequate safety, quality and efficacy of herbal medicines. Further research updates on *in vitro* elicitation of plant metabolites involving wider integrative approaches with a better understanding of their regulatory interactions at the molecular level will result in enhanced biosynthesis of plant metabolites. Finally, improved research on disease-specific herbal medicines in terms of better standardization of disease-specific herbal drugs, adequate dose and duration of treatment will ensure a larger utility of these plant-based herbal medicines globally.

ACKNOWLEDGMENTS

SPKC is grateful to DST-SERB-TARE (TAR/2020/000184). KDS, AKG and LKA acknowledge the MLS University, Udaipur, Rajasthan, India, MHRD- RUSA, Govt. of India. KDS and AKG also grateful to University Grants Commission (UGC), New Delhi, India [FRPS-SRG: F.30-476/2019 (BSR)].

REFERENCES

Abe, A., Kosugi, S., Yoshida, K., Natsume, S., Takagi, H., Kanzaki, H., Matsumura, H., Yoshida, K., Mitsuoka, C., Tamiru, M. and Innan, H. 2012. Genome sequencing reveals agronomically important loci in rice using MutMap. *Nat. Biotech.* 30: 174–178.

Aggarwal, S., Ichikawa, H., Takada, Y., Sandur, S.K., Shishodia, S. and Aggarwal, B.B. 2006. Curcumin (diferuloylmethane) down-regulates expression of cell proliferation and antiapoptotic and metastatic gene products through suppression of IκBα kinase and Akt activation. *Mol. Pharma.* 69: 195–206.

Agrawal, V., Sharma, K., Gupta, S., Kumar, R. and Prasad, M. 2007. Identification of sex using RAPD marker in male and female plants of jojoba. *Plant Biotech. Rep.* 1: 207–210.

Ahmad, P., Abdel Latef, A.A., Rasool, S., Akram, N.A., Ashraf, M. and Gucel, S. 2016. Role of proteomics in crop stress tolerance. *Front. Plant Sci.* 7: 1336.

Ahmad, Z., Shahzad, A. and Sharma, S. 2019. Chitosan versus yeast extract driven elicitation for enhanced production of fragrant compound 2-hydroxy-4-methoxybenzaldehyde (2H4MB) in root tuber derived callus of *Decalepis salicifolia* (Bedd. Ex Hook. F.) Venter. *Plant Cell Tiss. Org. Cult.* 136: 29–40.

Akindele, A.J., Wani, Z.A., Sharma, S., Mahajan, G., Satti, N.K., Adeyemi, O.O., Mondhe, D.M. and Saxena, A.K. 2015. *In vitro* and *in vivo* anticancer activity of root extracts of *Sansevieria liberica* Gerome and Labor (Agavaceae). *Evidence-Based Complementary and Alternative Medicine*, 2015: 1–11.

Alanjary, M., Kronmiller, B., Adamek, M., Blin, K., Weber, T., Huson, D., Philmus, B. and Ziemert, N. 2017. The Antibiotic Resistant Target Seeker (ARTS), an exploration engine for antibiotic cluster prioritization and novel drug target discovery. *Nucleic Acids Res.* 45: W42–W48.

Al-Dabbagh, B., Elhaty, I.A., Al Sakkaf, R., El-Awady, R., Ashraf, S.S. and Amin, A. 2018. Antioxidant and anticancer activities of Trigonella foenum-graecum, *Cassia acutifolia* and *Rhazya stricta. BMC Comp. Alt. Med.* 18: 1–12.

Ali, M., Hussain, R.M., Rehman, N.U., She, G., Li, P., Wan, X., Guo, L. and Zhao, J. 2018. De novo transcriptome sequencing and metabolite profiling analyses reveal the complex metabolic genes involved in the terpenoid biosynthesis in Blue Anise Sage (*Salvia guaranitica* L.). *DNA Res.* 25: 597–617.

Assocham 2017. www.assocham.org/newsdetail.php?id=6259

Atanasov, A.G., Waltenberger, B., Pferschy-Wenzig, E.M., Linder, T., Wawrosch, C., Uhrin, P., Temml V., Wang, L., Schwaiger, S., Heiss, E.H. and Rollinger, J.M. 2015. Discovery and resupply of pharmacologically active plant-derived natural products: A review. *Biotechnol. Adv.* 33: 1582–1614.

Babich, O., Sukhikh, S., Pungin, A., Ivanova, S., Asyakina, L. and Prosekov, A. 2020. Modern trends in the *in vitro* production and use of callus, suspension cells and root cultures of medicinal plants. *Molecules* 25: 5805.

Baenas, N., García-Viguera, C. and Moreno, D.A. 2014. Elicitation: A tool for enriching the bioactive composition of foods. *Molecules*, 19: 13541–13563.

Balekundri, A. and Mannur, V. 2020. Quality control of the traditional herbs and herbal products: a review. *Future J. Pharma. Sci.* 6: 1–9.

Bergelson, J. and Roux, F. 2010. Towards identifying genes underlying ecologically relevant traits in *Arabidopsis thaliana*. *Nat. Rev. Genet.* 11: 867–879.

Bhattacharyya, D., Sinha, R., Ghanta, S., Chakraborty, A., Hazra, S. and Chattopadhyay, S. 2012. Proteins differentially expressed in elicited cell suspension culture of *Podophyllum hexandrum* with enhanced podophyllotoxin content. *Proteome Sci.* 10: 1–12.

Blin, K., Pedersen, L.E., Weber, T. and Lee, S.Y. 2016. CRISPy-web: An online resource to design sgRNAs for CRISPR applications. *Synth. Systems Biotech.* 1: 118–121.

Blin, K., Wolf, T., Chevrette, M.G., Lu, X., Schwalen, C.J., Kautsar, S.A., Suarez Duran, H.G., De Los Santos, E.L., Kim, H.U., Nave, M. and Dickschat, J.S. 2017. antiSMASH 4.0 – improvements in chemistry prediction and gene cluster boundary identification. *Nuc. Acids Res.* 45: W36–W41.

Bray, F., Ferlay, J., Soerjomataram, I., Siegel, R.L., Torre, L.A. and Jemal, A. 2018. Global cancer statistics 2018: GLOBOCAN estimates of incidence and mortality worldwide for 36 cancers in 185 countries. *CA: A Cancer J. Clinic*. 68: 394–424.

Brukhin, V. and Morozova, N. 2011. Plant growth and development – Basic knowledge and current views. *Math. Model. Nat. Phenom.* 6: 1–53.

Chakraborty, P. 2018. Herbal genomics as tools for dissecting new metabolic pathways of unexplored medicinal plants and drug discovery. *Biochimie Open* 6: 9e16.

Chavali, A.K. and Rhee, S.Y. 2018. Bioinformatics tools for the identification of gene clusters that biosynthesize specialized metabolites. *Brief. Bioinfo.* 19: 1022–1034.

Chen, F., Dong, W., Zhang, J., Guo, X., Chen, J., Wang, Z., Lin, Z., Tang, H. and Zhang, L. 2018. The sequenced angiosperm genomes and genome databases. *Front. Plant Sci.* 9: 418.

Chetri S.P.K., Sharma K., Subhan S. and Agrawal V. 2014. Efficient protocol for micropropagation of high psoralen yielding chemotype of *Cullen corylifolium* (L.) Medik.- a traditional herb of immense medicinal value. *Int. J. Pharma BioSci.* 5: (B) 567–579.

Chetri S.P.K., Kapoor, H. and Agrawal V. 2016. Marked enhancement of sennoside bioactive compounds through precursor feeding in *Cassia angustifolia* Vahl and cloning of isochorismate synthase gene involved in its biosynthesis. *Plant Cell Tiss. Org. Cult.* 124: 431–446.

Chetri S.P.K., Sharma K. and Agrawal V. 2013. Genetic diversity analysis and screening of high psoralen yielding chemotype of *Psoralea corylifolia* from different regions of India employing HPLC and RAPD marker. *Vegetos* 26: 88–95.

Csermely, P., Korcsmáros, T., Kiss, H.J., London, G. and Nussinov, R. 2013. Structure and dynamics of molecular networks: a novel paradigm of drug discovery: a comprehensive review. *Pharmaco. Therapy.* 138: 333–408.

Dejong, C.A., Chen, G.M., Li, H., Johnston, C.W., Edwards, M.R., Rees, P.N., Skinnider, M.A., Webster, A.L. and Magarvey, N.A. 2016. Polyketide and nonribosomal peptide retro-biosynthesis and global gene cluster matching. *Nat. Chem. Bio.* 12: 1007–1014.

Desgagné-Penix, I., Khan, M.F., Schriemer, D.C., Cram, D., Nowak, J. and Facchini, P.J. 2010. Integration of deep transcriptome and proteome analyses reveals the components of alkaloid metabolism in opium poppy cell cultures. *BMC Plant Bio.* 10: 1–17.

Dhanapal, A.P. and Govindaraj, M. 2015. Unlimited thirst for genome sequencing, data interpretation, and database usage in genomic era: the road towards fast-track crop plant improvement. *Gen. Res. Internat.* 2015: 1–15.

Duan, N., Deng, Y., Liu, Y., Zhang, Y., Zhang, L. G., Wang, C.Y. and Liu, B.B. 2019. The complete chloroplast genome of *Sophora alopecuroides* (Fabaceae). Mitochondria. DNA Part B 4: 1336–1337.

Dufresne, Y., Noé, L., Leclère, V. and Pupin, M. 2015. Smiles2Monomers: a link between chemical and biological structures for polymers. *J. Cheminfo.* 7: 1–11.

Durrant, J.D. and McCammon, J.A. 2011. Molecular dynamics simulations and drug discovery. *BMC Bio.* 9: 1–9.

Ekor, M. 2014. The growing use of herbal medicines: issues relating to adverse reactions and challenges in monitoring safety. *Front. Pharma.* 4: 177.

Espinosa-Leal, C.A., Puente-Garza, C.A. and García-Lara, S. 2018. *In vitro* plant tissue culture: means for production of biological active compounds. *Planta* 248: 1–18.

Firenzuoli, F. and Gori, L. 2007. Herbal medicine today: clinical and research issues. *Evidence-Based. Comp. Altern. Med.* 4(S1): 37–40.

Giri, C.C. and Zaheer, M., 2016. Chemical elicitors versus secondary metabolite production in vitro using plant cell, tissue and organ cultures: recent trends and a sky eye view appraisal. *Plant Cell Tiss. Org. Cult.* 126: 1–18.

Goncalves, S. and Romano, A. 2018. Production of plant secondary metabolites by using biotechnological tools. Secondary metabolites, sources and applications. In Vijaykumar R. and Raja S.S.S. (Eds.), Secondary Metabolites: Sources and Applications, pp. 81–99 London: IntechOpen.

Gong, F., Hu, X. and Wang, W. 2015. Proteomic analysis of crop plants under abiotic stress conditions: where to focus our research?. *Front. Plant Sci.* 6: 418.

Grzegorczyk-Karolak, I., Rytczak, P., Bielecki, S. and Wysokinska, H. 2017. The influence of liquid systems for shoot multiplication, secondary metabolite production and plant regeneration of *Scutellaria alpina. Plant Cell Tiss Organ Cult* 128: 479–486.

Han-Karolak, I., Rytczak, P., Bielecki, S. and Wysokińska, H. 2017. The influence of liquid systems for shoot multiplication, secondary metabolite production and plant regeneration of *Scutellariaalpina. Plant Cell Tiss. Org. Cult.* 128: 479–486.

Guha, R. and Jurs, P.C. 2004. Development of QSAR models to predict and interpret the biological activity of artemisinin analogues. *J. Chem. Info. Comp. Sc.* 44: 1440–1449.

Halder, M., Sarkar, S. and Jha, S. 2019. Elicitation: A biotechnological tool for enhanced production of secondary metabolites in hairy root cultures. *Eng. Life Sc.* 19: 880–895.

Hao, D.C. and Xiao, P.G. 2015. Genomics and evolution in traditional medicinal plants: road to a healthier life. *Evolut. Bioinfo.* 11: EBO-S31326.

Hashiguchi, A., Tian, J. and Komatsu, S. 2017. Proteomic contributions to medicinal plant research: from plant metabolism to pharmacological action. *Proteomes* 5: 35.

Hazam, P.K., Goyal, R. and Ramakrishnan, V. 2019. Peptide based antimicrobials: Design strategies and therapeutic potential. *Prog. Biophys. Mol. Bio.* 142: 10–22.

He, Z., Zhang, J., Shi, X.H., Hu, L.L., Kong, X., Cai, Y.D. and Chou, K.C. 2010. Predicting drug-target interaction networks based on functional groups and biological features. *PloS One* 5: e9603.

Heikrujam M., Sharma K., Kumar J. and Agrawal V. 2015. Review on different mechanisms of sex determination and sex- linked molecular markers in dioecious crops- A current update. *Euphytica* 201: 161–194.

Hernández-Domínguez, E.M., Castillo-Ortega, L.S., García-Esquivel, Y., Mandujano-González, V., Díaz-Godínez, G. and Álvarez-Cervantes, J. 2019. Bioinformatics as a tool for the structural and evolutionary analysis of proteins. In Behzadi, P. and Bernabò N. (Eds.), Computational Biology and Chemistry. London: Intech Open, pp. 1–27.

Hou, M., Zhang, Y., Mu, G., Cui, S., Yang, X. and Liu, L. 2020. Molecular cloning and expression characterization of flavonol synthase genes in peanut (*Arachis hypogaea*). *Sci. Rep.* 10: 1–11.

Huang, S., Feng, K. and Ren, Y. 2019. Molecular modelling studies of quinazolinone derivatives as MMP-13 inhibitors by QSAR, molecular docking and molecular dynamics simulations techniques. *Med. Chem. Comm.* 10: 101–115.

Huang, X., Sang, T., Zhao, Q., Feng, Q., Zhao, Y., Li, C., Zhu, C., Lu, T., Zhang, Z., Li, M. and Fan, D. 2010. Genome-wide association studies of 14 agronomic traits in rice landraces. *Nat. Genet.* 42: 961–967.

Hussein, R.A. and El-Anssary, A.A. 2019. Plants secondary metabolites: The key drivers of the pharmacological actions of medicinal plants. *Herbal Medicine* 1: 13.

Isah, T. 2019. Stress and defense responses in plant secondary metabolites production. *Biol. Res.* 52: 39.

Isah, T., Umar, S., Mujib, A., Sharma, M.P., Rajasekharan, P.E., Zafar, N. and Frukh, A. 2018. Secondary metabolism of pharmaceuticals in the plant in vitro cultures: Strategies, approaches, and limitations to achieving higher yield. *Plant Cell Tiss. Org. Cult.* 132: 239–265.

Ji, H.F., Li, X.J. and Zhang, H.Y. 2009. Natural products and drug discovery: Can thousands of years of ancient medical knowledge lead us to new and powerful drug combinations in the fight against cancer and dementia? *EMBO Reports* 10: 194–200.

Johnston, C.W., Skinnider, M.A., Dejong, C.A., Rees, P.N., Chen, G.M., Walker, C.G., French, S., Brown, E.D., Bérdy, J., Liu, D.Y. and Magarvey, N.A. 2016. Assembly and clustering of natural antibiotics guides target identification. *Nat. Chem. Bio.* 12: 233–239.

Jorgensen, W.L. 2004. The many roles of computation in drug discovery. *Science* 303: 1813–1818.

Jorgensen, W.L. 2010. Pulled from a protein's embrace. *Nature* 466: 42–43.

Joseph, J. 2021. Food ingredients trends, market & growth forecast by 2023. www.fnbnews.com/Top-News/food-ingredients-trends-market--growth-forecast-by-2023–62061

Kalmankar, N.V., Venkatesan, R., Balaram, P. and Sowdhamini, R. 2020. Transcriptomic profiling of the medicinal plant *Clitoriaternatea*: Identification of potential genes in cyclotide biosynthesis. *Sci. Rep.* 10: 1–20.

Kalyaanamoorthy, S. and Chen, Y.P. 2011. Structure-based drug design to augment hit discovery. *Drug. Discov. Today* 16: 831–839.

Kautsar, S.A., Suarez Duran, H.G., Blin, K., Osbourn, A. and Medema, M.H. 2017. plantiSMASH: automated identification, annotation and expression analysis of plant biosynthetic gene clusters. *Nuc. Acids Res.* 45: W55–63.

Kellner, F., Kim, J., Clavijo, B.J., Hamilton, J.P., Childs, K.L., Vaillancourt, B., Cepela, J., Habermann, M., Steuernagel, B., Clissold, L. and McLay, K. 2015. Genome-guided investigation of plant natural product biosynthesis. *Plant J.* 82: 680–692.

Khaldi, N., Seifuddin, F.T., Turner, G., Haft, D., Nierman, W.C., Wolfe, K.H. and Fedorova, N.D. 2010. SMURF: Genomic mapping of fungal secondary metabolite clusters. *Fungal. Genet. Biol.* 47: 736–41.

Khan, T., Khan, M. A., Karam, K., Ullah, N., Mashwani, Z. U. R. and Nadhman, A. 2021. Plant in vitro culture technologies: A promise into factories of secondary metabolites against COVID-19. *Front. Plant Sci.* 12: 356.

Khan, T., Ali, M., Khan, A., Nisar, P., Jan, S.A., Afridi, S. and Shinwari, Z.K. 2019. Anticancer plants: A review of the active phytochemicals, applications in animal models, and regulatory aspects. *Biomolecules* 10: 47.

Kikowska, M., Thiem, B., Szopa, A. and Ekiert, H. 2020. Accumulation of valuable secondary metabolites: Phenolic acids and flavonoids in different *in vitro* systems of shoot cultures of the endangered plant species – *Eryngium alpinum* L. *Plant Cell Tiss. Org. Cult.* 17: 1–11.

Kim, J.A., Roy, N.S., Lee, I.H., Choi, A.Y., Choi, B.S., Yu, Y.S., Park, N.I., Park, K.C., Kim, S., Yang, H.S. and Choi, I.Y. 2019. Genome-wide transcriptome profiling of the medicinal plant *Zanthoxylum planispinum* using a single-molecule direct RNA sequencing approach. *Genomics* 111: 973–979.

Kosová, K., Vítámvás, P., Urban, M.O., Prášil, I.T. and Renaut, J. 2018. Plant abiotic stress proteomics: The major factors determining alterations in cellular proteome. *Front. Plant Sci.* 9: 122.

Kowalczyk, T., Wieczfinska, J., Skała, E., Śliwiński, T. and Sitarek, P. 2020. Transgenesis as a tool for the efficient production of selected secondary metabolites from plant in vitro cultures. *Plants* 9: 132.

Kumar J., Heikrujam M., Sharma K. and Agrawal V. 2019. SRAP and SSR marker-assisted genetic diversity, population structure analysis and sex identification in Jojoba (*Simmondsia chinensis*). *Indust. Crops Prod.* 133: 118–132.

Kumar, J. and Gupta, P.K. 2008. Molecular approaches for improvement of medicinal and aromatic plants. *Plant Biotech. Rep.* 2: 93–112.

Kunwar, R.M., Mahat, L., Acharya, R.P. and Bussmann, R.W. 2013. Medicinal plants, traditional medicine, markets and management in far-west Nepal. *J. Ethnobio. Ethnomed.* 9: 1–10.

Lagunin, A.A., Goel, R.K., Gawande, D.Y., Pahwa, P., Gloriozova, T.A., Dmitriev, A.V., Ivanov, S.M., Rudik, A.V., Konova, V.I., Pogodin, P.V., Druzhilovsky, D.S. and Poroikov, V.V. 2014. Chemo- and bioinformatics resources for in silico drug discovery from medicinal plants beyond their traditional use: a critical review. *Nat. Prod. Rep.* 31: 1585–611.

Levy, S.E. and Myers, R.M. 2016. Advancements in next-generation sequencing. *Annu. Rev. Genomics Hum. Genet.* 17: 95–115.

Li, M.H., Ung, P.M., Zajkowski, J., Garneau-Tsodikova, S. and Sherman, D.H. 2009. Automated genome mining for natural products. *BMC Bioinf.* 10: 1–10.

Lim, X. Y., Teh, B. P. and Tan, T. Y. C. 2021. Medicinal plants in COVID-19: Potential and limitations. *Front. Pharmacol.* 12: 355.

Liu, J., Xu, C., Zhang, H., Liu, F., Ma, D. and Liu, Z. 2018. Comparative transcriptomics analysis for gene mining and identification of a cinnamyl alcohol dehydrogenase involved in methyleugenol biosynthesis from *Asarum sieboldii* Miq. *Molecules* 23: 3184.

Liu, R.H. 2004. Potential synergy of phytochemicals in cancer prevention: Mechanism of action. *J. Nut.* 134: 3479S–3485S.

Liu, X., Ahmad, N., Yang, L., Fu, T., Kong, J., Yao, N., Dong, Y., Wang, N., Li, X., Wang, F. and Liu, X. 2019. Molecular cloning and functional characterization of chalcone isomerase from *Carthamus tinctorius. AMB Express* 9: 1–12.

Lo, Y. T. and Shaw, P. C. 2019. Application of next-generation sequencing for the identification of herbal products. *Biotechnol. Adv.* 37: 107450.

Lu, H., Rate, D.N., Song, J.T. and Greenberg, J.T. 2003. ACD6, a novel ankyrin protein, is a regulator and an effector of salicylic acid signaling in the *Arabidopsis* defense response. *Plant Cell* 15: 2408–2420.

Ma, R., Sun, L., Chen, X., Mei, B., Chang, G., Wang, M. and Zhao, D. 2016. Proteomic analyses provide novel insights into plant growth and ginsenoside biosynthesis in forest cultivated *Panax ginseng* (F. Ginseng). *Front. Plant Sci.* 7: 1.

Malar, T.J., Antonyswamy, J., Vijayaraghavan, P., Kim, Y.O., Al-Ghamdi, A.A., Elshikh, M.S., Hatamleh, A.A., Al-Dosary, M.A., Na, S.W. and Kim, H.J. 2020. *In-vitro* phytochemical and pharmacological bio-efficacy studies on *Azadirachta indica* A. Juss and Melia azedarach Linn for anticancer activity. *Saudi J. Biol. Sci.* 27: 682–688.

Malathi, K. and Ramaiah, S. 2018. Bioinformatics approaches for new drug discovery: A review. *Biotech. Genet. Eng. Rev.* 34: 243–260.

Malzahn, A., Lowder, L. and Qi, Y. 2017. Plant genome editing with TALEN and CRISPR. *Cell Biosci. 7*: 1–18.

Manzoni, C., Kia, D.A., Vandrovcova, J., Hardy, J., Wood, N.W., Lewis, P.A. and Ferrari, R. 2018. Genome, transcriptome and proteome: The rise of omics data and their integration in biomedical sciences. *Brief. Bioinfo.* 19: 286–302.

Marchev, A., Yordanova, Z.P. and Georgiev, M. 2020. Green (cell) factories for advanced production of plant secondary metabolites. *Critical Reviews in Biotechnology* 40: 443-458.

Mardis, E.R. 2008. Next-generation DNA sequencing methods. *Annu. Rev. Genom. Hum. Genet.* 9: 387–402.

Mardis, E.R. 2011. A decade's perspective on DNA sequencing technology. *Nature* 470: 198–203.

Martínez-Esteso, M.J., Martínez-Márquez, A., Sellés-Marchart, S., Morante-Carriel, J.A. and Bru-Martínez, R. 2015. The role of proteomics in progressing insights into plant secondary metabolism. *Front. Plant Sc.* 6: 504.

Maxam, A.M. and Gilbert, W. 1977. A new method for sequencing DNA. *P.N.A.S.* 74: 560–564.

Medema, M.H. and Osbourn, A. 2016. Computational genomic identification and functional reconstitution of plant natural product biosynthetic pathways. *Nat. Prod. Rep.* 33: 951–962.

Mehring, A., Haffelder, J., Chodorski, J., Stiefelmaier, J., Strieth, D. and Ulber, R. 2020. Establishment and triterpenoid production of *Ocimum basilicum* cambial meristematic cells. *Plant Cell Tiss. Org. Cult.* 143: 573–581.

Mirmazloum, I., Kiss, A., Ladányi, M. and György, Z. 2019. Production of cinnamyl alcohol glycosides by biotransformation in roseroot callus cells. *Plant Cell Tiss. Org. Cult.* 139: 29–37.

Mochida, K., Sakurai, T., Seki, H., Yoshida, T., Takahagi, K., Sawai, S., Uchiyama, H., Muranaka, T. and Saito, K. 2017. Draft genome assembly and annotation of *Glycyrrhiza uralensis*, a medicinal legume. *Plant J.* 89: 181–194.

Muniyandi, K., George, E., Mudili, V., Kalagatur, N.K., Anthuvan, A.J., Krishna, K., Thangaraj, P. and Natarajan, G. 2017. Antioxidant and anticancer activities of *Plectranthusstocksii* Hook. f. leaf and stem extracts. *Agri. Nat. Resour.* 51: 63–73.

Nguyen, N.H., Ta, Q.T.H., Pham, Q.T., Luong, T.N.H., Phung, V.T., Duong, T.H. and Vo, V.G. 2020. Anticancer activity of novel plant extracts and compounds from *Adenosmabracteosum* (bonati) in human lung and liver cancer cells. *Molecules* 25: 2912.

Niazian, M. 2019. Application of genetics and biotechnology for improving medicinal plants. *Planta* 249: 953–973.

Oksman-Caldentey, K.M. and Inzé, D. 2004. Plant cell factories in the post-genomic era: New ways to produce designer secondary metabolites. *Trends Plant Sci.* 9: 433–440.

Pandey, B.P., Adhikari, K., Pradhan, S.P., Shin, H.J., Lee, E.K. and Jung, H.J. 2020. *In-vitro* antioxidant, anti-cancer, and anti-inflammatory activities of selected medicinal plants from western Nepal. *Fut. J. Pharma. Sci.* 6: 1–12.

Pandita, D., Pandita, A., Wani, S.H., Abdelmohsen, S.A., Alyousef, H.A., Abdelbacki, A.M. ... and Elansary, H.O. (2021). Crosstalk of multi-omics platforms with plants of therapeutic importance. *Cells* 10: 1296.

Parast, B.M., Chetri, S.K., Sharma, K. and Agrawal, V. 2011. *In vitro* isolation, elicitation of psoralen in callus cultures of *Psoralea corylifolia* and cloning of psoralen synthase gene. *Plant Physiol. Biochem.* 49: 1138–1146.

Paz, T.A., dos Santos, V.A., Inácio, M.C., Dias, N.B., Palma, M.S., Pereira, A.M.S. and Furlan, M. 2017. Proteome profiling reveals insights into secondary metabolism in *Maytenus ilicifolia* (Celastraceae) cell cultures producing quinonemethide triterpenes. *Plant Cell Tissue Org. Cult.* 130: 405–416.

Pease, A.C., Solas, D., Sullivan, E.J., Cronin, M.T., Holmes, C.P. and Fodor, S.P. 1994. Light-generated oligonucleotide arrays for rapid DNA sequence analysis. *P.N.A.S.* 91: 5022–5026.

Pedrete, T.A., Hauser-Davis, R.A. and Moreira, J.C. 2019. Proteomic characterization of medicinal plants used in the treatment of diabetes. *Internat. J. Bio. Macromol.* 140: 294–302.

Pinu, F.R., Beale, D.J., Paten, A.M., Kouremenos, K., Swarup, S., Schirra, H.J. and Wishart, D. 2019. Systems biology and multi-omics integration: Viewpoints from the metabolomics research community. *Metabolites* 9: 76.

Promraksa, B., Phetcharaburanin, J., Namwat, N., Techasen, A., Boonsiri, P. and Loilome, W. 2019. Evaluation of anticancer potential of Thai medicinal herb extracts against cholangiocarcinoma cell lines. *PloS One* 14: e0216721.

Rai, A., Saito, K. and Yamazaki, M. 2017. Integrated omics analysis of specialized metabolism in medicinal plants. *Plant J.* 90: 764–787.

Raj, D., Kokotkiewicz, A., Drys, A. and Luczkiewicz, M. 2015. Effect of plant growth regulators on the accumulation of indolizidine alkaloids in *Securinega suffruticosa* callus cultures. *Plant Cell Tiss. Org. Cult.* 123: 39–45.

Ramirez-Estrada, K., Vidal-Limon, H., Hidalgo, D., Moyano, E., Golenioswki, M., Cusidó, R.M. and Palazon, J. 2016. Elicitation, an effective strategy for the biotechnological production of bioactive high-added value compounds in plant cell factories. *Molecules* 21: 182.

Reddy, B.V.B., Milshteyn, A., Charlop-Powers, Z. and Brady, S.F. 2014. eSNaPD: a versatile, web-based bioinformatics platform for surveying and mining natural product biosynthetic diversity from metagenomes. *Chem. Biol.* 21: 1023–1033.

Rhee, S.Y., Dickerson, J. and Xu, D. 2006. Bioinformatics and its applications in plant biology. *Annu. Rev. Plant Biol.* 57:335–360.

Roberts, S.C. and Shuler, M.L. 1997. Large-scale plant cell culture. *Curr. Opin. Biotech.* 8: 154–159.

Rong, Q., Jiang, D., Chen, Y., Shen, Y., Yuan, Q., Lin, H., Zha, L., Zhang, Y. and Huang, L. 2016. Molecular cloning and functional analysis of squalene synthase 2 (SQS2) in *Salvia miltiorrhiza* Bunge. *Front. Plant Sc.* 7: 1274.

Rout, G.R., Samantaray, S. and Das, P. 2000. *In vitro* manipulation and propagation of medicinal plants. *Biotech. Adv.* 18: 91–120.

Russo, M., Spagnuolo, C., Tedesco, I. and Russo, G.L. 2010. Phytochemicals in cancer prevention and therapy: truth or dare?. *Toxins* 2: 517–551.

Saito, K. and Matsuda, F. 2010. Metabolomics for functional genomics, systems biology, and biotechnology. *Annu. Rev. Plant Biol.* 61: 463–89.

Saito, K. 2013. Phytochemical genomics – A new trend. *Curr. Opin. Plant Bio.* 16: 373–380.

Sanger, F. and Coulson, A.R. 1975. A rapid method for determining sequences in DNA by primed synthesis with DNA polymerase. *Journal Mol. Bio.* 94: 441–448.

Sarwat, M. and Yamdagni, M.M. 2016. DNA barcoding, microarrays and next generation sequencing: Recent tools for genetic diversity estimation and authentication of medicinal plants. *Crit. Rev. Biotechnol.* 36: 191–203.

Schenone, M., Dančík, V., Wagner, B.K. and Clemons, P.A. 2013. Target identification and mechanism of action in chemical biology and drug discovery. *Nat. Chem. Biol.* 9: 232–240.

Shabir, P.A. 2021. CRISPR/Cas9-mediated genome editing in medicinal and aromatic plants: developments and applications. In Aftab, T and Hakeem, K.R. (Eds.), *Medicinal and Aromatic Plants* pp. 209–221, USA: Academic Press.

Shah, S. M. A., Nisar, Z., Nisar, J., Akram, M., Ghotekar, S. and Oza, R. (2021). Nanobiomedicine: A new approach of medicinal plants and their therapeutic modalities. *J. Mater. Environ. Sci.* 12: 1–14.

Skinnider, M.A., Merwin, N.J., Johnston, C.W. and Magarvey, N.A. 2017. PRISM 3: Expanded prediction of natural product chemical structures from microbial genomes. *Nucl. Acids Res.* 45: W49–54.

Srinath, M., Shailaja, A., Bindu, B.B.V. and Giri, C.C. 2021. Molecular cloning and differential gene expression analysis of 1-deoxy-D-xylulose 5-phosphate synthase (DXS) in *Andrographis paniculata* (Burm. f) Nees. *Mol. Biotech.* 63: 109–124.

Stander, E.A., Williams, W., Mgwatyu, Y., Heusden, P.V., Rautenbach, F., Marnewick, J., Roes-Hill, M.L. and Hesse, U. 2020. Transcriptomics of the Rooibos (*Aspalathus linearis*) species complex. *BioTech* 9: 19.

Szopa, A. and Ekiert, H. 2016. The importance of applied light quality on the production of lignans and phenolic acids in *Schisandra chinensis* (Turcz.) Baill. cultures in vitro. *Plant Cell Tiss. Org. Cult.* 127: 115–121.

Techen, N., Parveen, I., Pan, Z. and Khan, I. A. 2014. DNA barcoding of medicinal plant material for identification. *Curr. Opin. Biotechnol.* 25: 103–110.

Thakur, V., Bains, S., Pathania, S., Sharma, S., Kaur, R. and Singh, K. 2020. Comparative transcriptomics reveals candidate transcription factors involved in costunolide biosynthesis in medicinal plant- *Saussurea lappa*. *Int. J. Biol. Macromolecules* 150: 52–67.

Tietz, J.I., Schwalen, C.J., Patel, P.S., Maxson, T., Blair, P.M., Tai, H.C., Zakai, U.I. and Mitchell, D.A. 2017. A new genome-mining tool redefines the lasso peptide biosynthetic landscape. *Nat. Chem. Biol.* 13: 470–478.

Tripathi, S., Jadaun, J.S., Chandra, M. and Sangwan, N.S. 2016. Medicinal plant transcriptomes: The new gateways for accelerated understanding of plant secondary metabolism. *Plant Genet. Resour.* 14: 256–269.

van der Fits, L. and Memelink, J. 2000. ORCA3, a jasmonate-responsive transcriptional regulator of plant primary and secondary metabolism. *Science* 289: 295–297.

van Heel, A.J., de Jong, A., Montalban-Lopez, M., Kok, J. and Kuipers, O.P. 2013. BAGEL3: Automated identification of genes encoding bacteriocins and (non-)bactericidal posttranslationally modified peptides. *Nucl. Acids Res.* 41: W448–53.

Vauzour, D., Rodriguez-Mateos, A., Corona, G., Oruna-Concha, M.J. and Spencer, J.P. 2010. Polyphenols and human health: Prevention of disease and mechanisms of action. *Nutrients* 2: 1106–1131.

Vesth, T.C., Brandl, J. and Andersen, M.R. 2016. FunGeneClusterS: Predicting fungal gene clusters from genome and transcriptome data. *Synth. Syst. Biotech.* 1: 122–129.

Wang, Y.N., Tang, L., Hou, Y., Wang, P., Yang, H. and Wei, C.L. 2016. Differential transcriptome analysis of leaves of tea plant (*Camellia sinensis*) provides comprehensive insights into the defense responses to *Ectropis oblique* attack using RNA-Seq. *Funct. Integ. Genomics* 16: 383–398.

Wang, Y.P. and Lei, Q.Y. 2018. Metabolite sensing and signaling in cell metabolism. *Sig. Transduct. Target. Ther.* 3: 1–9.

Wang, Z., Gerstein, M. and Snyder, M. 2009. RNA-Seq: a revolutionary tool for transcriptomics. *Nat. Rev. Genet.* 10: 57–63.

Wang, Z., Zhao, M., Cui, H., Li, J. and Wang, M. 2020. The transcriptomic landscape of medicinal *Dendrobium* reveals genes associated with the biosynthesis of bioactive components. *Front. Plant Sc.* 11: 391.

Weeks, D.P., Spalding, M.H. and Yang, B. 2016. Use of designer nucleases for targeted gene and genome editing in plants. *Plant Biotechn. J.* 14: 483–495.

Wei, F., Tang, D., Wei, K., Qin, F., Li, L., Lin, Y., Zhu, Y., Khan, A., Kashif, M.H. and Miao, J. 2020. The complete chloroplast genome sequence of the medicinal plant *Sophora tonkinensis*. *Sci. Rep.* 10: 1–13.

Whirl-Carrillo, M., McDonagh, E., Hebert, J., Gong, L., Sangkuhl, K., Thorn, C., Altman, R. and Klein, T.E. 2012. Pharmacogenomics knowledge for personalized medicine. *Clin. Pharmacol. Ther.* 92: 414–417.

Wolf, T., Shelest, V., Nath, N. and Shelest, E. 2016. CASSIS and SMIPS: promoterbased prediction of secondary metabolite gene clusters in eukaryotic genomes. *Bioinformatics* 32: 1138–1143.

Wu, X., Li, X., Wang, W., Shan, Y., Wang, C., Zhu, M., Li, Q., Zhong, Y., Xu, Y., Nan, P. and Li, X. 2020. Integrated metabolomics and transcriptomics study of traditional herb *Astragalus membranaceus*Bge. var. mongolicus (Bge.) Hsiao reveals global metabolic profile and novel phytochemical ingredients. *BMC Genomics* 21: 1–16.

Xu, J., Chu, Y., Liao, B., Xiao, S., Yin, Q., Bai, R., Su, H., Dong, L., Li, X., Qian, J., Zhang, J., Zhang, Y., Zhang, X., Wu, M., Zhang, J., Li, G., Zhang, L., Chang, Z., Zhang, Y., Jia, Z., Liu, Z., Afresh, D., Nahurira, R., Zhang, L., Cheng, R., Zhu, Y., Zhu, G., Rao, W., Zhou, C., Qiao, L., Huang, Z., Cheng, Y.C. and Chen, S. 2017. *Panax ginseng* genome examination for ginsenoside biosynthesis. *Gigascience* 6: 1–15.

Xuhua, X. 2017. Bioinformatics and drug discovery. *Curr. Topics Med. Chem.* 17: 1709–1726.

Yadav, D.K., Meena, A., Srivastava, A., Chanda, D., Khan, F. and Chattopadhyay, S.K. 2010. Development of QSAR model for immunomodulatory activity of natural coumarinolignoids. *Drug Des. Dev. Ther.* 4: 173–186.

Yamazaki, M., Rai, A., Yoshimoto, N. and Saito, K. 2018. Perspective: Functional genomics towards new biotechnology in medicinal plants. *Plant Biotech. Rep.* 12: 69–75.

Yin, J., Li, X., Zhan, Y., Li, Y., Qu, Z., Sun, L., Wang, S., Yang, J. and Xiao, J. 2017. Cloning and expression of BpMYC4 and BpbHLH9 genes and the role of BpbHLH9 in triterpenoid synthesis in birch. *BMC Plant Bio.* 17: 1–14.

Yu, J., Wu, X., Liu, C., Newmaster, S., Ragupathy, S. and Kress, W.J. 2021. Progress in the use of DNA barcodes in the identification and classification of medicinal plants. *Ecotoxic. Environ. Safety* 208: 1–7.

Zhang, W., Qingdu, F. and Li, G. 2019a. Characterization of the complete mitochondrial genome of shrubby sophora (*Sophora flavescens* Ait.). *Mitochon. DNA Part B* 4: 696–697.

Zhang, C., Liu, T., Yuan, X., Huang, H., Yao, G., Mo, X. ... and Yan, H. 2019b. The plastid genome and its implications in barcoding specific-chemotypes of the medicinal herb Pogostemon cablin in China. *PloS one* 14: 1–14.

Zhao, J., Davis, L.C. and Verpoorte, R. 2005. Elicitor signal transduction leading to the production of plant secondary metabolites. *Biotech. Adv.* 23: 283–333.

Zhao, X., Yuan, Z., Feng, L. and Fang, Y. 2015. Cloning and expression of anthocyanin biosynthetic genes in red and white pomegranate. *J. Plant Res.* 128: 687–696.

Ziegler, S., Pries, V., Hedberg, C. and Waldmann, H. 2013. Target identification for small bioactive molecules: Finding the needle in the haystack. *Angew Chem. Int. Ed.* 52: 2744–2792.

Ziemert, N., Podell, S., Penn, K., Badger, J.H., Allen, E. and Jensen, P.R. 2012. The natural product domain seeker NaPDoS: a phylogeny based bioinformatic tool to classify secondary metabolite gene diversity. *PloS One* 7: 1–0.

3 Unraveling Medicinal Plant Chemical Diversity for Novel Drug Discovery Through Biotechnological Interventions

Mir Zahoor Gul, Mohd Yasin Bhat, Elizabeth P. Ryan and Irfan A. Ghazi

CONTENTS

3.1 INTRODUCTION

The plant kingdom is one of the earth's prime life-supporting systems. In addition to the apparent nutritional benefits, the therapeutic attributes of plants have been established since the aboriginal times, as they are "green chemical factories" with extensive chemical diversity (Nogueira et al. 2018). Recent advances in modern therapeutics have encouraged the use of medicinal plants as natural therapies with dual roles in the prevention and treatment of various ailments. The biologically active phytochemicals found in medicinal plants and herbs merit individual and collective attention for their contributions in the management of human and animal diseases. For traditional herbal remedies, the plant material, powders and tinctures/potions have been used by a variety of cultures and tribal communities with varying levels of success to promote vitality and to mitigate and treat different ailments (Qiu 2007). Archaeological evidence indicates the use of plant-based systems by humankind for therapeutic reasons through the paleolithic age, with the earliest documented records dating back to the Sumerians, almost 5000 years ago (Hardy 2018, Petrovska 2012, Qiu 2007). Indigenous communities within Native American, South American, Asian and African cultures have used herbal products in their healing ritual practices, whereas, other traditional healthcare systems

DOI: 10.1201/9781003179139-3

were developed in India (Ayurveda, Unani, Siddha), China (Traditional Chinese Medicine-TCM) and Egypt, where herbal remedies are being used effectively (Petrovska 2012, Qiu 2007). Traditional medicine knowledge and practice have been shared and passed down across centuries, mostly through selected community healers and tradition. The first systematic attempt to isolate, purify and chemically analyze the therapeutically active constituents from herbal medicines happened during the 19th century. Quinine (antimalarial), salicylic acid (analgesic) and morphine (pain killer) were the first such therapeutic molecules. Even though contemporary synthetic drugs are primarily used in the western world for allopathic medicine, the usage of herbal medicine has also been well embraced and is becoming integrated into practice in many developed countries.

Even today plants are not only indispensable in health care but form the best hope of source for safe future medicines (Hamburger and Hostettmann 1991). It can be asserted that several years of bioprospecting of medicinal plants – *the search for biological gold* – have led to the discovery of the most important and reasonably abundant bioactive molecules of significant therapeutic-pharmaceutical importance. Modern medicine today utilizes active compounds isolated from higher plants, and a significant percentage of these active ingredients indicate a positive correlation between their modern therapeutic use and the traditional uses. This is validated by the fact that between 1983 and 1994, more than 60% of Food and Drug Administration (FDA) -approved and pre-New Drug Application (NDA) stage chemotherapeutic drugs were of biological origin. Likewise, 78% of all the recently certified antimicrobial drugs were also of biological origin (Concepcion, Lazaro and Hyde 2001). Among the 1562 therapeutic drugs the FDA approved between 1981 to 2014, 64 (4%) were unmanipulated, 141 (9%) were herbal medications (mixtures), 320 (21%) were derivatives of natural products and 61 (4%) were synthesized drugs, albeit with natural product pharmacophore (Newman and Cragg 2016). Of the 252 drugs that the World Health Organization (WHO) deems fundamental and appropriate, 11% of those have been derived from various plant sources (Rates 2001).

Furthermore, about 40% of the pharmacological lead molecules used for the synthesis of modern drugs came from natural resources (Chin et al. 2006, Katiyar et al. 2012). Various other references highlight the prominence and value of novel products from natural plant sources in modern societies. A plethora of therapeutic agents that have proved to be nature's boon in treating and managing various diseases were derived from medicinal plants and herbs (Table 3.1) (Sharma and Arora 2006).

It is not only the individual bioactive molecules but the tinctures and even crude plant concoctions, are often used in several parts of the world. Furthermore, new antibiotics with specific targets are needed expeditiously to combat the existing and emerging problems of antimicrobial resistance (Aslam et al. 2018). Thus, bioprospecting for antibiotic-like molecules and other pharmacologically active molecules especially from plants, need to be globally prioritized. This quest of scientific research is prompted by the crises of resistance in pathogenic organisms like *Staphyloccocus*, *Streptococcus*, *Mycobacterium*, etc., towards the available antimicrobial drugs and the frightening existence of naturally resistant disease-causing pathogens. The main emphasis of medicinal plant research is on discovering and developing novel medicines to effectively fight new infections and diseases such as COVID-19, severe acute respiratory syndrome (SARS), Ebola, HIV/AIDS, etc. These diseases not only require targeted drugs, but they also need new therapies to combat auxiliary infections in immunocompromised patients like candidiasis, encephalopathy, pneumonia, caused by an impaired immune system. Novel and improved drug therapies are also needed to combat various life-threatening diseases/infections like pneumonia, malaria, cholera, trypanosomiasis, filariasis, leishmaniases. Although various antibiotics have been effective for the treatment of these diseases, the global problem of microbial resistance to the current antibiotics and other antimicrobial medicines has led to the treatment failure of infections. These antibiotics are becoming increasingly ineffective as incidents of epidemics due to such drug-resistant microorganisms are now a common global problem, posing enormous public health concerns (Aslam et al. 2018). Notably, malaria took

TABLE 3.1
Selected List of Pharmaceuticals Derived From Medicinal Plants That Have Therapeutic Applications for Diverse Chronic and Infectious Diseases

S. No.	Compound/Drug	Class	Therapeutic Use	Plant Origin
1	Aconitine	Alkaloid	Rheumatic pains	*Aconitum* spp
2	Ajmaline	Alkaloid	Antiarrhythmic	*Rauwolfia serpentina*
3	Artemisinin	Sesquiterpene lactone	Antimalarial	*Artemisia annua*
4	Atropine	Alkaloid	Parasympatholytic agent	*Atropa belladonna, Datura innoxia, D. metel, D. stramonium*
5	Atropine, hyoscyamine, scopolamine	Tropane alkaloid	Anticholinergic	*Solanaceous* spp.
6	Berberine	Alkaloid	Cardiovascular disease, Glucose-lipid metabolism	*Berberis aquifolium, B. vulgaris, B. aristata, Hydrastis canadensis, Mahonia aquifolium, Xanthorhiza simplicissima, Phellodendron amurens* etc.
7	Caffeine	Alkaloid	CNS stimulant	*Tea, Coffee beans, Cocoa*
8	Camptothecin	Indole alkaloid	Antineoplastic/ Cancer therapy drug	*Camptotheca acuminata*
9	Capsaicin	Phenylalkyl-amine alkaloid	Rheumatic pains	*Capsicum* spp. (Chili pepper)
10	Cocaine	Tropane alkaloi	Local anesthetic, CNS stimulant	*Erythroxylon coca*
11	Codeine, morphine	Opium alkaloid	Analgesic, antitussive	*Papaver somniferum*
12	Colchicine	Isoquinoline alkaloid	Antigout	*Colchium autumnale*
13	Digoxin, Digitoxin	Steroidal glycosides	Cardiotonic	*Digitalis* spp.
14	Diosgenin	Steroids	Oral contraceptives, corticosteroids, Sexual hormones	*Dioscorea* spp.
15	Emetine	Isoquinoline alkaloid	Antiamoebic	*Cephaelis ipecacuanha*
16	Galantamine	Isoquinoline alkaloid	Acetylcholine esterase inhibitor/Alzheimer's disease	*Galanthus nivalis, Galanthus caucasicus, Galanthus woronowii,*
17	Genistein	Isoflavones	Angiogenesis inhibitor, Phytoestrogen	*Genista tinctoria*
18	Daidzein	Isoflavones	Menopausal relief, Osteoporosis, Blood cholesterol	Soybeans and other Legumes
19	Hypericin	Anthraquinone derivative	Antibiotic, Antiviral, Non-specific kinase inhibitor	*Hypericum perforatum*
20	Hyperoside	Flavonol glycoside	Anti-inflammatory, Antithrombotic, Antidiabetic, Hepatoprotective	*Hypericum* spp., *Crataegus* spp.
21	Ipecac	Mixture of ipecac alkaloids and other components	Emetic	*Cephaelis ipecacuanha*

(continued)

TABLE 3.1 (Continued)
Selected List of Pharmaceuticals Derived From Medicinal Plants That Have Therapeutic Applications for Diverse Chronic and Infectious Diseases

S. No.	Compound/Drug	Class	Therapeutic Use	Plant Origin
22	Morphine	Alkaloid	Pain reliever, Narcotic/	*Papaver somniferum*
23	Nicotine	Pyrrolidine alkaloid	Smoking cessation therapy	*Nicotiana* spp.
24	Paclitaxel	Alkaloid	Cancer/chemo therapeutic	*Taxus brevifolia*
25	Papaverine	Opium alkaloid	Phosphodiesterase inhibitor, Antispasmodic drug	*Papaver somniferum*
26	Physostigmine	Indole alkaloid	Dementia–Alzheimer's Disease	*Physostigma venenosum*
27	Pilocarpin	Imidazole alkaloid	Cholinergic, Glaucoma	*Pilocarpus microphyllus*
28	Podophyllotoxin	Alkaloid	Cancer therapy	*Podophyllum hexandrum, P. peltatum*
29	Quinidine	Quinoline alkaloid	Cardiac depressant	*Cinchona* sp.
30	Quinine	Quinoline alkaloid	Antimalaria	*Cinchona ledgeriana, C. succirubra*
31	Reserpine	Alkaloid	Antihypertensive drug	*Rauvolfia serpentina*
32	Sennosides A and B	Hydroxy anthracene glycosides	Laxatine	*Cassia angustifolia*
33	Silibinin	Phenol	Hepatoprotective	*Silybum marianum*
34	Strychnine	Alkaloid	CNS stimulant	*Strychnos nux-vomica*
35	Theophylline	Alkaloid	Antitussive	*Camellia sinensis, Coffea arabica*
36	Tubocurarine	Bisbenzyl isoquinolone alkaloid	Skeletal muscle relaxant	*Chondodendron tomentosum/ Strychnos toxifera*
37	Vinblastine and vincristine	Alkaloid	Cancer therapy	*Catharanthus roseus*
38	Yohimbine	Indole alkaloid	Aphrodisiac	*Pausinystalia johimbe*

around 405,000 lives worldwide in 2018, compared to any other infectious agent other than TB and AIDS (World Health Organization 2019). Traditional plant medicines still enjoy a significant position in the modern-day drug industries due to the minor side effects as well as the synergistic action of the combination of compounds. The WHO endorses and promotes the addition of herbal drugs in national health care programs because they are easily accessible at a price within reach of the common man and are time tested and thus considered to be much safer than the modern synthetic drugs (Choudhary and Sekhon 2011). Therefore, research teams across the globe are turning their attention to medicinal plants with effective antimicrobial properties. Currently, artemisinin-based combination therapies (ACTs) are by far the most commonly used antimalarial therapies, formulated using a pure artemisinin, extracted from the herbaceous aromatic plant, *Artemisia annua*. At present, almost all global malarial strains can be treated with at least one of the ACTs recommended by WHO (WHO 2019). In addition, the investigators in the United States of America have shown that *A. annua* extracts inhibit the replication of extreme acute respiratory coronavirus (SARS-CoV-2), an entity responsible for the ongoing 2019 pandemic of coronavirus disease (COVID-19) (Nair et al. 2021).

Even with the human experience of hundreds of years and the transition into a new era of science and technology, it is evident that bioactive compounds of natural origin have a tremendous impact on modern drug regimes. Most natural products are compounds derived from primary metabolites such as amino acids, carbohydrates and fatty acids and are generally categorized as secondary metabolites. These secondary metabolites offer excellent prototypes for the design and subsequent development of chemotherapeutics. Besides the pharmacological significance of medicinal plants, there is an evolving trend toward enhancing life by integrating plant materials and other functional foods and dietary additives into the daily diet. Secondary metabolites, viz., alkaloids, flavonoids, essential oils, phenolics, terpenes, tannins, sterols, lignins, etc. are considered the end-product of primary metabolism and are generally not involved in metabolic activity. During the first 50 years of the 20th century, the scientific focus was on identifying these compounds and their role in the prevention of various dietary deficiency diseases. Scientists also began to identify physiologically active components in foods from plants that potentially could reduce the risk for a variety of chronic diseases. Many food varieties are formulated with medicinal plants to develop functional foods. Medicinal plant products can be used as nutraceutical materials for many potential industrial uses like growth promoters, antioxidants, antimicrobials, pigments. The daily use of certain medicinal plants/herbs has been shown to reduce the risk of chronic diseases like obesity, diabetes mellitus, cardiovascular diseases, cancer, by serving as natural remedies (Martin et al. 2013). Based on their recognized safety, prospective plant-based nutraceuticals/bioactive compounds have received a great deal of attention (Nasri et al. 2014). In the last 20 years, there has been a noticeable decrease in the number of new natural product-based drugs entering the clinical arena, despite the varying degree of success in drug discovery stages.

Some reasons for the decline in the number of new natural products entering the clinic with drug classifications include, but may not be limited to:

- During the late 20th century, scientific innovations in drug discovery approaches saw a shift to automated testing, which allowed high throughput screening of hundreds to thousands of potential drug targets. This transition mandated fair access to huge compound libraries that could be established with modern synthetic approaches, viz., combinatorial chemistry, not with conventional plant extracts.
- Another drawback with natural products is the repetitive and tedious purification process to identify and isolate the active constituents, often in fairly low quantities from complex mixtures.
- There have been some very expensive pharmaceuticals on the market since they are found only in rare plants and often in quite low quantities.
- Arguably the main challenge to using natural products during the last few decades is the fact that there have been more than 200,000 reports of such compounds in the scientific literature (Walsh and Tang 2017). This implies that in using traditional approaches, a large number of natural product compounds have been identified, and there is a need to eliminate the deduplication, which is very complicated and time-consuming.
- More than 100 databases list natural products and chemical structures, which have been described in scientific literature as bioactive during the last two decades. However, the advancement in research depends upon precise information about molecular structures, including the stereochemistry which affects the bioactivity of these natural products. It is pertinent to mention that nearly 12% of the molecules in these databases have stereocenters that do not contain information on stereochemistry. Another drawback is the inaccessibility of many of these databases, which causes a significant loss of the natural product data (Sorokina and Steinbeck 2020).

In summary, there are many diverse hurdles that have diminished the promise of natural products, especially from medicinal plants in drug discovery. Traditionally, natural products have been *pathfinder* products, with great structural heterogeneity unmatched by even the combinatorial repositories. These products can serve as lead molecules, the functions of which could be improved by structural alterations through synthetic and combinatorial chemistry approaches. Given the significant interest in the recent trends and emerging perspectives in medicinal plant biotechnology, in this chapter we will highlight some of the recent biotechnological approaches in medicinal plant research and for the specific purpose of discovering pharmacologically active compounds.

3.2 BIOTECHNOLOGY – MULTIFACETED APPROACHES IN PHYTOMEDICINE

The term "biotechnology" was first given in 1919 by Hungarian engineer Karl Ereky for the branch of science that involves methods and applications which facilitate the manufacture of products from raw materials with the help of living organisms. It is a multidisciplinary science that has proven to be a boon in the lives of human beings and is rightly considered a "technology of hope" (Gupta et al. 2016). With the broad acceptance of phytotherapy and the optimism for beliefs around the safety and efficacy of herbal compounds, various technologies have been adopted for enhancing bioactive molecules in medicinal plants. Biotechnological tools are important for the multiplication and genetic enhancement of medicinal plants by adopting techniques such as *in vitro* regeneration and genetic transformation. It could also be harnessed to produce secondary metabolites using plant bioreactors. Advances in the tissue culture, combined with improvement in genetic engineering techniques, specifically transformation technology, have opened new avenues for the high-volume production of pharmaceuticals, nutraceuticals and other beneficial substances. The myriad of biotechnological tools is important for the multiplication and genetic enhancement of medicinal plants by adopting techniques such as *in vitro* regeneration and genetic transformations (Figure 3.1). Recent trends in biotechnology indicate that there is a need for multi-pronged strategies, and an assemblage of new and innovative biotechnological approaches (bioinformatics, transcriptomics, nanotechnology, etc.) coupled with leads from other conventional approaches is likely to make beneficial strides in novel drug discovery in the years to come. The effective use of these biotechnological tools makes it possible to develop next-generation drugs to address both contemporary and long-term health challenges.

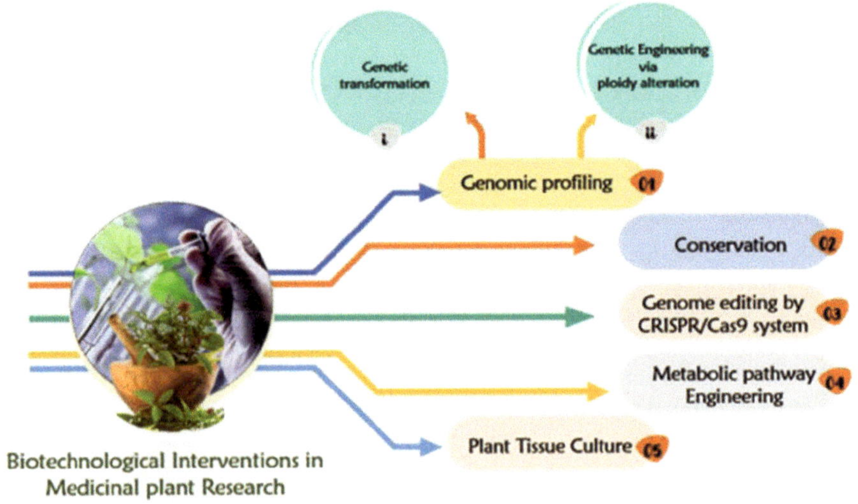

FIGURE 3.1 Biotechnological interventions in medicinal plant and natural products chemistry research.

3.2.1 Challenges and Goals of Medicinal Plants in the Product Development for Pharmaceuticals

Biotechnology approaches described herein help overcome the many challenges and barriers to medicinal plant research. Some of these include the collection of medicinal plants from natural habitats or *wildcrafted*, which has resulted in excessive exploitation of plants, as was realized in the case of *Arctostaphylos uvaursa*, *Cimicifuga racemose*, *Glycyrrhiza glabra*, *Podophyllum* and *Taxus* species and *Piper methysticum*, etc., whereby both habitat devastation and loss of genetic diversity culminated (Vines 2004). At present, just 10% of the medicinally important plants that are used are grown, while the greater proportion of them are wild (Julsing, Quax and Kayser 2008).

Because of this abuse, several medicinally important plants are on the verge of extinction and have been notified as well as listed as an endangered and threatened species on the International Union for Conservation of Nature List (IUCN) and Convention on International Trade in Endangered Species of Wild Fauna and Flora (CITES). Moreover, the mining of these bioactive molecules is often arduous, due to their relatively low concentrations, sluggish plant growth rates, complex build-up patterns and extreme vulnerability to environmental and geographical conditions.

To overcome these hurdles, extensive scientific investigations have been performed worldwide to develop alternative approaches to cultivate these plants and produce the desired constituents. Therefore, the domestic cultivation of these medicinally important plants is widely recognized for cultivating such plants on a large scale in favorable conditions. Also, such approaches aim to resolve other inherent flaws in herbal extracts, including standardization of extracts, plant variability, reduced toxic ingredients/contaminants, elevated quantities of preferred constituents, and breeding as per the Internationally recognized "Good Agricultural Practice" (GAP) guidelines. Biotechnology has shown great promise in the selection, multiplication and conservation of critical genotypes of medicinal plants.

3.2.2 Plant Tissue Culture

Recent advances in biotechnology and molecular biology, especially in biotransformation technology, have brought a paradigm shift for large-scale cultivation of rare plants as well as the formation and accumulation of desirable pharmacologically important bioactive molecules/compounds that they synthesize is certainly one of the best solutions for the above-described hurdles. In addition to preserving rare and endangered medicinal plant species and offering a frequent, steady supply of secondary metabolites, the integrated approaches of tissue culture pave the way for the development of standardized therapeutic agents, the large-scale cultivation of desired plant material for biochemical characterization and evaluation of bioactive molecules/compounds under highly controlled facilities (Shahzad et al. 2013, Chen et al. 2016, Espinosa-Leal, Puente-Garza and García-Lara 2018). The principal advantage of this technology compared to traditional agricultural growth is that it may provide a continuous, reliable source of plant medicinal compounds and could also be used for the large-scale culture of plant cells from which these metabolites can be extracted under highly controlled conditions. The commercial exploitation of various genetically modified plants and the increasing development of a broad range of therapeutically important active molecules with complex structural arrangements represents the core objectives of plant genetic engineering of medicinal plants. Over the past several years, plant biologists and microbiologists across the globe have undertaken numerous investigations on the biosynthetic capabilities of various cell cultures. A combinatorial biosynthesis is a modern approach for developing genetically modified plant species and producing new bioactive molecules that could be identified or have been synthesized chemically earlier. The principal idea of this approach is to integrate metabolic processes in diverse species at the genetic levels. The application of genetically engineered plant cell cultures, like hairy root cultures, provides a systematic approach to upregulation of genes that encode biosynthetic enzymes and circumvent the rate-limiting phases of biosynthesis. In contrast, other strategies, for example bioreactor design, precursor feeding, elicitation, cell immobilization, and *in situ* product elimination, have led to improved production. Several research groups have reported the exceptional potential of plant cell and tissue culture and have also illustrated the inconsistency of biosynthetic potential. This is due to high throughput cultures for widespread commercialization (Scragg 1992, Chen et al. 2016, Lalaleo et al. 2018). Artemisinin, berberine, paclitaxel, rosmarinic acid and shikonin, which are present in plants, albeit in lesser concentrations, are prominent examples that can be commercially processed through the tissue culture approach. Tissue culture growth can be accelerated by employing continuous culture strategies, like bioreactors, which allow efficient, controlled and high-level isolation of secondary metabolites (Wijesekera 2017). This is extremely beneficial for the isolation and production of therapeutically active molecules from medicinal plants that have sluggish growth in natural settings and could also minimize concerns related to overharvesting of rare and endangered plant species (Wijesekera 2017).

In the recent past, several research groups associated with combinatorial biosynthesis have focused on designing new interventions for streaming indispensable bioprocesses to generate biologically active products of pharmacological relevance, which include podophyllotoxin from *Podophyllum hexandrum* and *P. peltatum*, vincristine/vinblastine from *Catharanthus roseus*, paclitaxel from *Taxus species*, camptothecin from *Camptotheca acuminata*, ginsenosides derivatives from *Panax ginseng* and caffeic acid from *Echinacea purpurea*. In addition, strategies have been developed for the cloning of several plants, including *Bacopa monnieri, C. roseus*, *Chlorophytum borivilianum*, *Datura metal*, to name a few (Debnath, Malik and Bisen 2006). Different points in each metabolic pathway can be concurrently regulated, either by upregulation and/or suppression of specific enzymes or using transcriptional regulators to control endogenous genes. Therefore, multipoint metabolic engineering provides a unique opportunity to improve the processing of plant-based chemotherapeutics. The overexpression of the main genes associated with a biosynthetic pathway of bioactive compounds, shift in the regulation of involved genes to produce useful, high-quality biologically active products and the transition of agronomically beneficial genes depends on plant tissue culture methods. The overexpression of geranyl (geranyl) diphosphate synthase [G(G)PPS] and geraniol synthase (GES) genes in *C. roseus* has facilitated a substantial rise in vincristine and vinblastine, monoterpene indole alkaloids in genetically modified plants (Kumar, Shilpashree and Nagegowda 2018). Suppression of rosmarinic acid synthase (SmRAS) gene in *Salvia miltiorrhiza* led to a large spike in the levels of 3,4-dihydroxyphenylacetic acid and, therefore, improved the quality of rosmarinic acid (Zhou et al. 2018). Moreover, a genetically modified variant that is resistant to drought stress was developed by the expression of the transferred AtEDT1 transcription factor (Liu et al. 2017).

The plant tissue culture further serves as a platform for the comprehensive strategy of hair root cultures with greater biosynthetic ability to generate pharmacologically active compounds that significantly enhanced the biochemical and genetic stability compared to classical *in vitro* cultures (Roychowdhury, Halder and Jha 2016). The *in vitro* multiplication and cultivation approaches have been formulated for expeditiously producing high-yielding Himalayan plants (*P. peltatum, P. hexandrum*) by growing excised embryo plants (Sultan et al. 2006, Nandi et al. 2016). The regenerated calli of *P. hexandrum* produced through the transformation of an embryo with multiple strains of *Agrobacterium rhizogenes*, namely, K599, A4 and 15834, resulted in a three-fold surge in the active ingredient of podophyllotoxin compared to the control plants (Giri et al. 2001). The *A. rhizogenes* induced hairy roots are the active sites for the extensive production of pharmacologically valuable compounds employing artificial polyploidy, elicitors, bioreactors and CRISPR/Cas9 genetic manipulation procedures (Dehghan et al. 2012, Kastell et al. 2018, Li et al. 2017, Patra and Srivastava 2018, Xing et al. 2018, Zhou et al. 2018). In *Raphanus sativus* L., the hairy roots stimulated by *A. rhizogenes* yielded elevated levels of phenolic and flavonoid content in comparison to auxin-treated roots of non-transformed plants (Balasubramanian et al. 2018). In the recent past, the effectiveness of the hairy root cultures for the improved levels of phenol carboxylic acids, flavonoids and wedelolactone content in *Sphagneticola calendulacea*, *Isatis tinctoria*, activation of tropanes (anisodamine, hyoscine, hyoscyamine in *Scopolia lurida* plants) have also been substantiated (Jiao et al. 2018, Kundu et al. 2018, Lan et al. 2018). A promising strategy to harness the biosynthetic ability of enzymes in plant cell cultures is the introduction of precursors to media to direct the metabolic flux toward the improved output of desirable active compounds. The introduction of β-cyclodextrin complexed coniferyl alcohol to suspension culture of *P. hexandrum* enhanced podophyllotoxin concentration four-fold, which is approximately 0.013% based on the dry weight (Woerdenbag et al. 1990). The addition of L- phenylalanine, an amino acid, to *Linum flavum* cell cultures led to a three to fivefold surge in 6-methoxypodophyllotoxin level (van Uden et al. 1990). In contrast to *P. hexandrum* suspensions grown separately, co-cultured hairy roots of *P. hexandrum* and *L. flavum* contributed to increased podophyllotoxin content by around 240% and 72% in flasks and dual bioreactors, respectively. The coniferin provided by hairy roots of *L. flavum* served as a precursor for the rapid growth of podophyllotoxin in suspension cultures of *P. hexandrum* (Lin et al. 2003). The addition of potential side-chain paclitaxel precursors, like serine, phenylalanine, N-benzoyl glycine and benzoic acid, has shown elevated levels of paclitaxel in both cell suspension and callus cultures in *Taxus cuspidata* (Fett-Neto et al. 1994, Srinivasan et al. 1996). Furthermore, the application of precursors, such as phenylalanine, hippuric acid, leucine and sodium benzoate, to the cell cultures of *T. wallachiana* substantially enriched the yield of baccatin, 10-deacetyl vacation and paclitaxel (Davey 2017). The beneficial influence of co-cultivation of adventitious roots of *Echinacea purpurea* and *Panax ginseng* using a bioreactor has been documented for relatively elevated production of caffeic acid derivatives and ginsenosides (Wu et al. 2008).

3.2.3 Metabolic Pathway Engineering

Metabolic engineering infers the modulation or redirection of one or more enzymatic reactions and metabolism per se to produce either new desired compounds in an organism or bring about a considerable improvement in the production of existing compounds. This process, in other words, pursues the manipulation of cellular processes intrinsic to the organism itself to produce anticipated products in a simple and economically feasible way (García-Granados, Lerma-Escalera and Morones-Ramírez 2019). Plants are a treasure house of countless chemicals that are synthesized by very integrated and multifaceted biosynthetic metabolic pathways. Metabolic engineering in plants often presents fascinating perspectives for improving the efficiency of the plant as a cell factory for several products. During the recent era of advancement in science and technology, there was a wonderful knowledge explosion and a far better comprehension of the complicated and intricate genetic and biochemical machinery involved in the synthesis of secondary metabolites in plants. With the emergence of new tools and techniques in sequencing and bioinformatics, the genes for the enzymes of biosynthetic pathways have become easier to discover, which has stood us in good stead to abundantly harness the wealth of desired compounds and enzymes that plants provide. A good number of genes have been precisely isolated, and the array of steps in the metabolic pathway has been properly outlined (Tatsis and O'Connor 2016). This approach may generate new avenues and possibilities in diverse fields like agriculture, environment and production of chemicals, biofuels and medicines (Lau et al. 2014, Martin and Li 2017). The enhancement in the production of compounds of interest but present in minute quantities can be successfully achieved by either increasing the generation of precursors or stalling a competitive (parallel) pathway having the same precursor/ intermediate or by overcoming rate-limiting steps in the biosynthetic pathway. Another plausible approach for this purpose is enhancing the number of special cells capable of yielding the desired compound. The alternative approaches can be over expressing the key regulatory genes or the concerned transcription factors that elicit the pathway by impeding or restricting catabolic degradation of the molecule. The targeting of metabolic engineering approaches can only succeed when the rate-influencing genes within a metabolic pathway are identified. One major drawback to the metabolic engineering of plant cells with the aim of enhancing the yield of a secondary metabolite is the dearth of entirely defined and demarcated metabolic pathways (Wilson and Roberts 2012). Metabolic engineering approaches and biotechnological interventions can be used as a substitute production system which can potentially help to overcome the inadequate accessibility of biologically active and commercially valuable plant secondary metabolite molecules (Teske et al. 2020, Zhu and Salmeron 2007). A concise and chronological account of the major metabolic engineering schemes and techniques that encompasses such tactics applied in medicinal plant biotechnology leading to advancements in prospective breeding possibilities of medicinal plants is given as follows:

3.2.3.1 Genetic Transformation/Transgenic Approaches

Genetic engineering in molecular biology implies any change or modification of genes within a cell, which is a follow-up of uptake and efficacious incorporation of any exogenous DNA segment in the host cell genome via the cell membrane. The fundamental necessities for genetic engineering are proficient transformation and tissue culture systems that allow for the feasibility of selection and regeneration of transgenic plants. The introduction and transfer of DNA segments from unrelated or evolutionary distant species afford the possibility of selectively developing new or desired characters into crop plants. It is thus very useful in expanding the gene pool far beyond what could have been available in traditional plant breeding systems (Li et al. 1996, Gómez-Galera et al. 2007). It has a definite edge over conventional breeding methods to mitigate or completely overcome a few of the important agronomic and environmental issues (Sharma et al. 2005, Datta 2013). Transgenic technology has been employed on a commercial scale to produce new crops with better insect resistance, herbicide tolerance and enhanced post-harvest quality. In contemporary times, advancements in plant genetics and recombinant DNA technology have paved the way to boost research studies into plant secondary metabolism. The chief focus of such endeavors is the identification of crucial enzymes in a metabolic pathway followed by their manipulation to modulate that pathway in the desired manner (Hain and Grimmig 2000). Although biotechnological approaches have a very limited application in medicinal plants in comparison to other crops, in contemporary decades, the genetic manipulation of agronomic characters in medicinal plants has led to remarkable progress for their development. Their genetic transformation encompasses enhancement in yield of medicinally important active compounds but without being detrimental to them (Alamgir 2018). The transformation tactics can be employed to target or manipulate a rate-limiting step via management of the expression of specific structural genes. The genetic transformation coupled with *in vitro* plant regeneration facilities helps to redesign secondary product metabolic pathways favorably (Gómez-Galera et al. 2007). The principal reason for plant transformation in medicinal plants is to enhance the quantity of naturally occurring bioactive molecules/compounds and the production of therapeutically active compounds (Bae et al. 2005). In the past and to date, there have been a plethora of attempts at the genetic transformation of medicinal plants belonging to several families such as *Apocynaceae*, *Araceae*, *Araliaceae*, *Asphodelaceae*, *Asteraceae*, *Begoniaceae*, *Crassulaceae*, *Fabaceae*, *Lamiaceae*, *Linaceae*, *Papaveraceae*, *Plantaginaceae*, *Scrophulariaceae* and *Solanaceae* (Sharma et al. 2018). Many transformation approaches are employed to yield recombinant pharmaceuticals in the plant system. Apart from being economically feasible, *Agrobacterium*-mediated transformation has an edge over direct approaches such as the prospects to transfer only single or few copies of desired DNA fragments possessing the target gene with high efficiency and the transfer of sizable DNA fragments with marginal rearrangement (Gelvin 2003).

3.2.3.1.1 Agrobacterium-facilitated Gene Transfer

Genetic transformation of medicinal plants is accomplished by employing the strains of *A. rhizogenes* or *A. tumefaciens* to achieve transformed cells. The transformed cells have the option to be maintained as cell cultures or can be regenerated to complete plants. *A. tumefaciens* possessing the tumor-inducing (Ti) mega-plasmid (more than 200 kb) causes gall formation on roots and crowns whereas *A. rhizogenes* harboring the root-inducing (Ri) plasmid elicits abnormal root production in host plants (Nester 2015). These are a type of natural engineers that transform or alter characters in dicot plants, although there are many reports of infection of some monocots as well as gymnosperm plants (Hwang, Yu and Lai 2017). When the infection takes place, the mobile section of the plasmid, T-DNA, is transported to the nucleus of infected cells. It is followed by stable integration into the host genome and transcription, and finally manifesting as the crown gall disease. The oncogenic genes in T-DNA code for enzymes which are important for the synthesis of plant growth regulators auxins and cytokinins promotes cell division leading to tumor formation. The other genes encode for opines that are chief sources of carbon and nitrogen by the resident bacteria, facilitating a progressively increasing resource of nutrients for the bacteria (Gelvin 2003). Despite the challenges of low success rate and recalcitrance of some species to *A. tumefaciens*-mediated transformation, it continues to remain a powerful tool and an efficient approach for the improvement of many crops (Song, Prieto and Orbovic 2019). Similar success could also be attained in the case of medicinal plants for enhancing the content of pharmacologically active secondary metabolites in their respective tissues. Genetic transformation by *A. tumefaciens* has been reported for a plethora of medicinal plants by researchers around the world. The efficiency of genetic transformation in medicinal plants is dependent upon several factors such as strain of *Agrobacterium*, plant species, age of the plant, method(s) adopted, and the regeneration conditions (Hwang, Yu and Lai 2017). The establishment and standardization of transformation protocols, although they are highly time-consuming, but are prerequisite for successful metabolic engineering for pharmacologically important non-model medicinal plant species. Efficient transformation by *A. tumefaciens* LBA4404 strain possessing the bar gene forwarded by the cauliflower mosaic virus (CaMV) 35S promoter for embryogenic callus inoculation in *Veratrum dahuricum* L. has pharmaceutical significance that stems from yielding the anticancer drug, cyclopamine. It has paved the path for the introduction of biosynthetic pathway genes in this plant that could enhance the content of steroid alkaloid accumulation (Ma et al. 2020). Similarly, such successful transformation and regenerations by *A. tumefaciens* have been obtained in many medicinal plants during the last 3 decades, Some of such plants include; *Artemisia annua* (Wang et al. 2016), *Azadirachta indica,* neem (Naina, Gupta and Mascarenhas 1989), *Camptotheca acuminata* (Wang and Zu 2007), *Centaurea montana* (Abou-Alaiwi et al. 2012), *Chelone glabra* (Gao et al. 2015), *Digitalis* spp. (Sales, Segura and Arrillaga 2003, Li et al. 2014), *Dendrobium lasianthera* (Utami, Hariyanto and Manuhara 2018), *Taxus* species (Han et al. 1994), *Salvia miltiorhiza* (Yan and Wang 2007), *Scrophularia buergeriana*, Figwort (Park, Chae and Facchini 2003), *Withania somnifera* (Pandey et al. 2010).

Improvements in the efficiency of transformation as a metabolic engineering tool has the potential to facilitate the introduction of useful genes into medicinal plants and for enhancing the content of medicinally useful metabolites that make them beneficial breeding material. Recently, a genetic transformation system was reported for *Trachyspermum ammi* with *A. tumefaciens* wherein the TPS2 gene (Terpene Synthase 2), which is a crucial enzyme governing thymol synthesis, was integrated and overexpressed leading to a considerable increment in the thymol content in transgenic plants in comparison to non-transgenic plants (Nomani et al. 2019). Hydroxylase enzyme catalyzes the oxidative reaction steps in the biosynthetic pathway of scopolamine, which is a remarkable anticholinergic drug molecule present in several plants of the Solanaceae family. An *Agrobacterium*-mediated transformation system was utilized to transfer the hydroxylase gene derived from *Hyoscyamus niger* under the regulation of CaMV 35S promoter into hyoscyamine-possessing *Atropa belladonna*, resulting in increased alkaloid content (scopolamine) in the preliminary transformant and its transgenic selfed progeny (Yun, Hashimoto and Yamada 1992). Likewise, artemisinic aldehyde reductase enzyme (DBR2) catalyzes artemisinic aldehyde to form dihydroartemisinic aldehyde, acting as a checking point and has direct involvement in the artemisinin biosynthetic pathway. This enzyme was involved in designing the artemisinin biosynthesis pathway in Agrobacterium-mediated transgenic plants of *Artemisia annua* with overexpression controlled by the CaMV 35S promoter. The yield of artemisinin and its direct precursor dihydroartemisinic acid were notably enhanced in the transgenic plants of *A. annua* with overexpression as demonstrated by high-performance liquid chromatography (HPLC) analysis (Yuan et al. 2015). Taxadiene from Taxus is the first committed precursor to paclitaxel which is by far the best anticancer agent against breast and ovarian cancer. Taxadiene is directly produced from the usual precursor of diterpenoids geranylgeranyl diphosphate (GGPP). Genetic transformation succeeded to produce taxadiene in *A. annua* wherein taxadiene synthase gene, from Taxus, was introduced into *A. annua* via pCAMBIA1304. The transgenic plants were confirmed to produce the Taxadiene by GC-MS without compromising plant growth or photosynthesis. It is pertinent to note that that artemisinin and taxadiene accumulates in the leaves and stems of transgenic plants respectively, making this approach efficient and beneficial to produce two important compounds in two different organs of the same transgenic plant (Li et al. 2015). Genetic transformation of *Bacopa monnieri*, using a gene coding for cryptogenic, a proteinaceous elicitor with disarmed *A. tumefaciens* strain, led to the generation of transgenic plants with significant enhancement in growth and bacopa saponin content (Majumdar, Garai and Jha 2012). Induction of shooty teratoma in *C. roseus* by *A. tumefaciens* led to a tenfold increase in content of dimeric indole alkaloid vincristine, extensively used for cancer chemotherapy as against untransformed control cultures (Begum et al. 2009).

Enhanced productivity of pharmacologically important secondary plant metabolites can also be attained by hairy root cultures after infection by *A. rhizogenes*, a gram-negative soil-dwelling bacterium. The infection by this bacterium at a wounded place manifests in the formation and proliferation of a copious number of small roots as hairy roots. These bacteria possess a special Ri plasmid having rol genes (rolA, rolB and rolC) and aux genes for stable integration of genetic material into host cell DNA and root induction by direct auxin synthesis respectively (Gutierrez-Valdes et al. 2020). Transformed hairy roots act in exactly the same biochemical fashion as in the case of normal roots and there are a sufficient number of examples where they exhibit significantly greater yield of product. Hairy roots of numerous medicinal herbs are cultured in suitable bioreactors not only for enhancement in productivity but for *de novo* production of secondary metabolites (Arora, Mathur and Mathur 2010). Hairy root cultures have the ability to produce highly diverse molecules, genetic stability, indefinite growth without the supply of exogenous hormones, and an augmented capability for secondary metabolites formation as well as accumulation, that makes them an ideal transgenic alternative to medicinal plants (Rekha and Thiruvengadam 2017, Kowalczyk et al. 2020). They can synthesize many compounds whose chemical synthesis, otherwise, is difficult. *A. rhizogenes* mediated hairy root culture system procedures have been customized in many medicinal plants during the last several years such as *Althea officinalis* (Tavassoli and Afshar 2018), *Aconitum heterophyllum* (Giri et al. 1997), *Apocynum venetum* (Jia et al. 2008), *Artemisia absinthium* (Nin et al. 1997), *Berberis aristate* (Brijwal and Tamta 2015), *Gentiana* spp. (Tiwari et al. 2007, Hayta et al. 2011), *Peganum hamala* (Berlin et al. 1992), *Plumbago zeylanica* (Verma et al. 2002), *Rosmarinus officinalis* (Shetty 1998), *Salvia miltiorrhiza* (Zhi and Alfermann 1993), *Tylophora indica* (Chaudhuri et al. 2005). Many endeavors have been made till date to induce hairy roots with the ability to produce specialized metabolites based on genetic engineering techniques. New genetic tools available can be employed to enable enhanced production of tailor-made bioactive molecules which have placed hairy root as chief biotechnological tools (Giri et al. 1997, Gutierrez-Valdes et al. 2020, Kowalczyk et al. 2020). Recently, it was observed that induction of a hairy root culture system in *Lycium ruthenicum* resulted in a high yield of phenolic polyamines along with secondary metabolites. The overexpression by the TCP4 gene using construct (pCAMBIA 1307) and subsequent integration with the genome by Ri-facilitated genetic transformation resulted in transgenic hairy root clones possessing elevated quantities of kukoamine and several other secondary metabolite molecules as compared to normal control hairy roots (Chahel et al. 2019). Another recent study showed that overexpression of SmMDS (2-C-methyl-D-erythritol 2, 4-cyclodiphosphate synthase), a key enzyme in the tanshinone diterpenoid synthesis pathway significantly enhanced the tanshinone content in transgenic *Salvia miltiorrhiza* hairy roots

(Wei et al. 2019). In another study, it was demonstrated that overexpression of a positive regulator SmWRKY2 in hairy roots of the same plant considerably augmented the expression of SmDXS2 and SmCPS, leading to increased accumulation of tanshinones (Deng et al. 2019). The genetic transformation of *Artemisia annua* plant with *A. rhizogenes* possessing the cDNA encoding FDS (Farnesyl diphosphate synthase) under the regulation of CaMV 35S promoter bore three to four times higher artemisinin accumulation in comparison to untreated control plants (Chen et al. 1999). A nine-fold enhancement in the yield of scopolamine was observed in hairy root cultures of black henbane (*Hyoscyamus niger*) by simultaneous overexpression of two genes of the rate-limiting downstream and upstream biosynthetic enzymes (Zhang et al. 2004). This higher production of bioactive molecules and rapid proliferation of hairy root offers exhilarating opportunities for production of biomass on grander scale by bioreactors and steadily high formation of metabolites for the therapeutic industry. Since the extraction of active ingredients is less costly, less arduous and eco-friendly, the hairy root cultures make the year-round production of pharmaceuticals a possibility. Hairy roots are highly differentiated and can cause stable and extensive production of secondary metabolites and therefore, prove economical for commercial production. The widespread utilization of hairy root cultures as a source of raw material will mitigate the reliance of the pharmaceutical industry absolutely on plants from natural habitats. It will ensure the availability of unpolluted, microbial contamination-free raw material without being affected by the seasonal difference in the production of medicinally useful compounds.

3.2.3.1.2 Generation of Transgenic Medicinal Plants by Electroporation

Electroporation is an electrical transformation technique in which an external electric field is applied resulting in surpassing the capacitance of the cell membrane and ultimately culminating in the transient and reversible breakdown or poration. It allows the passage and delivery of micro- or macromolecules across cells such as ions, drug molecules, dyes, antibodies, RNA and DNA (Gehl 2003). The essential requirement for an efficient cell transformation is the physical settings of electroporation, especially the voltage of the electric field, the number, and the time duration of pulses. Due to its feasible regulation and efficiency in cell transfection and appropriate delivery, this method has increasing applications in biology, biotechnology and medicine. Several other merits of this method include rapid application, lesser cost, applicability to a range of cell types, appropriateness for a large number of cells and comparatively higher stable transformation (Shi et al. 2018, Ozyigit 2020). A woody medicinal plant, *Solatium dulcamara* cell suspension protoplasts post-exposure to voltages of 250 to 1250 V cm^{-1} for three consecutive pulses elicited the growth of protoplast-derived tissues accompanied by increased morphogenesis. Regenerated shoots were also shown to develop more abundant root systems than those arising from untreated protoplasts (Chand et al. 1988). Electroporation efficiency in protoplasts of Himalayan gentian (*Gentiana kurroo* Royle) embryogenic cells, electro-transformed by a plasmid having nptII and bar genes was demonstrated to improve by long-term culture of callus, regenerated somatic embryos and presence of 50 mg/L kanamycin (Wójcik and Rybczyński 2015). A shortening of the drying times was reported in Genovese basil (*Ocimum basilicum*) leaves by a pre-treatment with irreversible electroporation; however, it deteriorated the quality of dried leaves (Kwao et al. 2016).

3.2.3.1.3 Generation of Transgenic Medicinal Plants by Chloroplast Transformation

During contemporary times, chloroplast genetic transformation and expression modules have attained the research hotspot in plant bioreactor applied for both basic and applied research (Michoux et al. 2013). Higher plant chloroplasts have unique advantages over nuclear genetic engineering for the expression of recombinant proteins due to their distinct structures and inherited features, such as site-specific incorporation, high expression and the maternal inheritance patterns of exogenous genes. The other merits include lack of risk factor of the transgene being transmitted through pollen to other plants, high level of expression, no effect by gene silencing, etc. Many important proteins of pharmacological significance have been efficaciously expressed in chloroplasts (Daniell, Kumar and Dufourmantel 2005). Vaccine antigens, antibodies, and other important recombinant proteins have been shown to successfully express in tobacco plants which is a model of transformation. More than 40 transgenes have been successfully integrated and expressed using the tobacco chloroplast genome in order to produce desirable agronomic features or to express biopharmaceuticals and vaccine antigens at high quantities (Daniell, Kumar and Dufourmantel 2005). Chloroplast genetic transformation also provides new methods and tactics for studying patterns of gene expression and regulation mechanisms in higher plants. Recently, optimization of a stable chloroplast transformation in the medicinal plant, *A. annua* was performed which could prove to be an exciting prospect for expression of the non-glycosylated proteins of the artemisinin biosynthetic pathway to enhance the production artemisinin. Chloroplast transformation for this purpose is significant due to the fact that nuclear genome transformation has not been as efficient as desired (Kaushal, Abdin and Kumar 2020).

3.2.3.1.4 Particle Gun Method/Biolistic Transformation

Gene/particle gun bombardment is a technique for the physical and direct introduction of DNA into the plant cell. Transformation by this method is an alternative that is appropriate and effective for many plant species. The bombardment of the cell wall of plants is accomplished by many DNA coated metallic particles like gold, platinum, palladium, tungsten, iridium and rhodium, and by using compressed helium as the propellant. These tiny metallic bodily particles ranging in size from 0.45 to1.5 μm in diameter create minute holes in the cell wall facilitating their entry and thus the exogenous DNA of the plant cells. The DNA is shed off from the particles, lodged inside the cells and a portion may be stably incorporated into the host chromosomes (Fang and Trewyn 2012). It is a non-target and non-localized method so that the metal particles are present throughout the cell interior including cytoplasm as well as the nucleus. DNA cargo, after diffusion from the surface of the metal particle carrier, has the chance to impact the intracellular genetic programs. With this approach, the exogenous DNA probably makes entry into intact plant cells via simultaneous multiple gene transfers without any significant biological limitations or host issues, since it is governed only by physical parameters (Kikkert, Vidal and Reisch 2005, Mookkan 2018). Particle bombardment allows DNA-free gene editing of the genome, direct delivery of proteins, RNAs and Ribonucleoprotein (RNPs), which makes it a versatile technique for genetic transformation (Ozyigit and Kurtoglu 2020). Transformation of *Hyoscyamus muticus* (Egyptian henbane) was attained by particle bombardment with the aid of plasmid pAB-6 which contained the *bar* and *gus* genes. Total alkaloid yield was found to be 6.05 mg/g dw in transformed plants as compared to 2.95 mg/g dw in untransformed plants (El-Akkad, Serag and Serag 2018). Marigold (*Tagetes erecta*) leaf explants underwent an efficient transformation after low-pressure particle gun bombardment with plasmid pBI426 (Vanegas et al. 2006). The successful biolistic transformation was done for garlic (*Allium sativum* L.), which led to the generation of transgenic plants resistant to chlorsulfuron, a sulfonylurea herbicide, due to the introduction of acetolactate synthase (ALS) gene from a chlorsulfuron-resistant Arabidopsis mutant (Park et al. 2002). A proficient plastid transformation in *Scoparia dulcis* L., a multipurpose medicinal herb, through particle bombardment was reported recently wherein a plastid transformation vector, KNTc, containing aadA and egfp as a selectable marker and reporter gene, respectively, under the regulation of synthetic promoter pNG1014a targeting inverted repeats, trnR/trnN of the plastid genome. It marks an important improvement of the plastid transformation which has the potential to apply alteration in enhancing and controlling certain key metabolic pathways (Muralikrishna et al. 2016).

The introduction of foreign genetic information into the plants symbolizes a path-breaking step in plant biotechnology concerning the high-volume generation of pharmaceuticals and has opened new windows of prospects for the production of several biologically active compounds for the treatment of several human ailments (Ahmad et al. 2013). Apart from increasing the content of valuable secondary metabolites, transgenic can also be generated for agronomic characteristics like resistance to herbicides, pests and diseases which can serve as useful germplasm for future breeding (Canter, Thomas and Ernst 2005). The major negatives of the transgenic approach are the instability in gene expression, loss of capacity of cells over time to produce target molecules and the high cost of bioreactors. In the transgenic *C. roseus*, the activities of two key enzymes, strictosidine synthases (STR) and tryptophan decarboxylase (TDC) were sometimes shown to fall almost to the levels to those of non-transgenic cultures which resulted in gradual loss of the ability of cell lines to accumulate terpenoid indole alkaloids (TIAs) (Whitmer et al. 2003). One more limitation is the ensuing regeneration of transgenic plants, which is by far a tricky and time-consuming process. A fair chance of introduction of undesirable somaclonal variation in the form of gene mutation or epigenetic alterations through tissue culture regeneration in some cases cannot be ruled out (Ngezahayo 2018). In recent years, there has been a phenomenal increase in the products of plant biotechnology ranging from simple to highly complex biological macromolecules for use in the human healthcare system. Transgenic plant cell culture systems and whole plants can produce antibodies, vaccine epitopes, enzymes, hormones, recombinant proteins and a plethora of biologically active secondary metabolites on a commercial scale for use as diagnostic tools or therapeutic agents (Teli and Timko 2004). The marketable practicability as well as feasibility of bringing medicinal plants into domestic cultivation and the capacity-building potential for enhanced use of modern biotechnologies are likely to be affected by the public perception and views of both biotechnology and medicinal plants. Biopharmaceuticals procured from transgenic plants are largely presumed to be comparatively harmless and less costly than those from animal-based-origin, which are prone to contamination with human pathogens (Canter, Thomas and Ernst 2005).

3.2.3.1.5 Upregulation by Overexpression with Transcription Factors

Transcription factors are a sort of regulatory protein that possesses the capability to control a battery of genes of a metabolic pathway via binding of unique DNA sequences or by protein-protein interactions that can lead to its modulation to produce a significant yield of the anticipated product. They are considered as a more viable and feasible substitute to metabolic engineering *vis-à-vis* a single gene as they can rewire the entire metabolic pathway by regulating the expression of multiple genes and thus can be more beneficial for complex secondary metabolites with indeterminate biosynthetic pathways(Grotewold 2008, Wilson and Roberts 2012). The involvement of transcription factors necessitates the assemblage of information from several fields, such as genomics, proteomics, transcriptomics and metabolomics and can arbitrate as gene activators or repressors, manifesting as an increase or a decrease in mRNA content (Hendrawati et al. 2012). The transcription factors, ORCA1, ORCA2 and ORCA3, belonging to AP2/ERF family in *C. roseus*, are involved in the biosynthesis of terpenoid indole alkaloids. A 6.5 fold enhancement in the production of cantharanthine is caused by the overexpression of ORCA3 in cultured cells, which augmented the expression pattern of the terpenoid indole alkaloid biosynthesis genes such as Tryptophan decarboxylase (TDC), Strictosidine synthases (STR), Cytochrome P450 reductase (CPR), Desacetoxyvindoline 4 – hydroxylase (D4H) and Secologanin synthase (SLS) (Wang et al. 2010). This factor regulates two genes that code for enzymes α – subunit of anthranilate synthase (AS α) and D-1-deoxyxylulose 5- phosphate synthase (DXS) in the primary metabolism leading to terpenoid indole alkaloid precursor formation (Memelink and Gantet 2007). Recently, a study to reveal the transcription factors governing the costunolide (sesquiterpenes lactone) biosynthesis in *Saussurea lappa* was performed by relative transcriptome analysis of tissues of leaf and root and annotated transcripts using *in-silico* tools. Certain putative transcription factors were identified, and bioinformatics tools were employed to predict the cis-regulatory elements, leading to the identification of the MYB family of TFs as an interacting partner of the SlCOS1 (*costunolide synthase*) gene, catalyzing the final key step of costunolide biosynthesis (Thakur et al. 2020).

3.2.3.1.6 Role of Sense/Antisense RNA and RNAi Silencing

The formation of undesirable molecules can be obstructed by suppressing genes that cause upregulation of the pathway or somehow enhance their catabolism and are usually done by adopting certain approaches such as antisense and RNA interference (RNAi) methods. RNA interference has been efficiently used in the engineering of plants by the introduction of short sequences of double-stranded RNA that interfere with the expression of the targeted gene. It is a biological strategy that negates the role of targeted mRNAs through homology-dependent gene silencing. It comprises of gene-specific control by small non-coding RNAs (sncRNAs) and has emerged in contemporary times as a promising tool for plant biotechnologists, not only to decode plant gene function but also for the development of the elite plants with value-added and novel traits by gene manipulation. There is a tremendous potential application of RNAi approaching metabolic alteration owing to the plasticity of plant secondary metabolic pathways (Borgio 2009, Sabzehzari and Naghavi 2019). An attempt of metabolic engineering by RNA interference (RNAi) in transgenic opium poppy (*Papaver somniferum*) by involving hairpin chimeric RNA construct to silence all members of the multigene codeinone reductase (COR) family manifested in high content of the non-narcotic alkaloid reticuline and its methylated derivatives instead of narcotic morphine, codeine, etc. (Allen et al. 2004). A knockdown by RNA interference of berberine bridge enzyme (BBE) caused accumulation of the important key intermediate reticuline up to a maximum level of 310 µg/g-fresh weight in the biosynthetic pathway for isoquinoline alkaloids (morphine, codeine and berberine) in California poppy (*Eschscholzia californica*) cells. Apart from a considerable reduction in expected end-products like sanguinarine, a methylated derivative of reticuline, laudanine, was produced by the transgenic cells which otherwise could not be traced in the usual pathway (Fujii et al. 2007). A DNA-encoded hairpin RNA-mediated suppression approach of the gene, which codes for orphinan pathway enzyme salutaridinol 7-O-acetyltransferase (SalAT) in opium poppy, led to the build-up of another alkaloid salutaridine, instead of alkaloids morphine, codeine and thebaine. Salutaridine is not the substrate of SalAT but of the salutaridine reductase which is the preceding enzyme in the biosynthetic pathway (Allen et al. 2008). Virus-induced gene silencing (VIGS) technology has been known to be beneficial for plants that offer constraints in transformation. By this approach, identification of genes involved in benzylisoquinoline alkaloid (BIA) biosynthesis and two reactions involved in the successive O-demethylations of thebaine to codeine and codeine to morphine in opium poppy have been successfully done (Hagel and Facchini 2010). The successful VIGS mediated suppression of 16-methoxy-2, 3-dihydro-3-hydroxytabersonine N-methyltransferase (NMT) in *C. roseus* showed favorable accumulation of NMT substrate 16-methoxy-2,3-dihydro-3-hydroxytabersonine instead of downstream monoterpenoid indole alkaloid vindoline. It is a practical approach for the comprehension of gene function in *C. roseus* and provides valuable insights into the alkaloid biosynthetic pathway (Liscombe and O'Connor 2011).

3.2.3.1.7 Genome Editing (Engineering) of Medicinal Plants by CRISPR/Cas9 System

The system of microbial defense with clustered regularly interspaced short palindromic repeat (CRISPR) sequences can help accomplish genome editing or engineering in eukaryotic cells. It is a research hotspot in contemporary times for genome editing and regulation, which works by genome silencing, knocking in, transcriptional activation and repression. This handy system has two components: a guide RNA (g RNA), which works simply by precise and specific recognition of base pairing of a 20-nucleotide target sequence within its guide RNA, and a Cas9 nuclease. The latter further recruits the Cas9 endonuclease protein to the target locus and creates double-strand breaks in the target DNA, culminating in a mutation (Linlin, Brianna Marie and Yuan-Xiang 2016, Gupta et al. 2019). The CRISPR/Cas9 method has been rapidly developed as a significant method for targeted genetic modification in a wide variety of plants which includes food crops, fruit crops, ornamental and model plants. It has the potential to become a promising genome editing tool capable of highly desired mutations in plants (Corte et al. 2019). The CRISPR-Cas mediated target editing of Rosamarinic acid synthase (SmRAS) gene of lithospermic acid B biosynthetic pathway was mutated in *Salvia miltiorrhiza* to successfully change the quantity of valuable secondary metabolites. In the successfully edited mutants, a decline in content of rosmarinic acid and lithospermic acid and an elevation in the level of rosmarinic acid precursor 3,4- dihydroxy phenyllactic acid was observed (Zhou et al. 2018). The committed diterpene synthase gene (SmCPS1) involved in tanshinone biosynthesis was accurately knocked out using CRISPR/Cas9 in a separate study in the same plant, thereby opening the door for extensive genome editing in miltiorrhiza, which is crucial for elucidating the pathways of secondary metabolites, improving quality and yield enhancement (Li et al. 2017). With the CRISPR/Cas9 system, 3'-hydroxyl-N-methylcoclaurine 4'-O-methyltransferase (4' OMT2) gene was knocked out in opium poppy (*Papaver somniferum* L.), which efficiently regulated the benzylisoquinoline alkaloids metabolism and biosynthesis. The editing resulted in a reduction in the production of morphine, thebaine, etc, in transgenic plants, in addition to the formation of a novel uncharacterized alkaloid (Alagoz et al. 2016). The main hurdle in genome editing by CRISPR/Cas9 is the lack of sufficient genomic and transcriptome data of a vast number of medicinal plants. In recent times, advancements such as next-generation sequencing can help researchers to attain whole-genome and transcriptome information of medicinal plants, which can prove to be a boon for genome editing via the CRISPR-Cas system (Alok et al. 2020). The other prerequisites include *in vitro* regeneration and successful plant transformation. CRISPR-Cas protocols need to be optimized for medicinal plants to make them more user-friendly and accessible (Alok et al. 2020). This technique bears tremendous potential for redesigning the biosynthetic pathway in medicinal plants with an accurately regulated genetic circuit to maximize the production of valuable biopharmaceuticals (Arshid Shabir 2021).

3.2.3.2 Engineering of Medicinal Plants Via Ploidy Alteration

Apart from several important coordinates like environment, topography, etc genetic variation is the key factor that modulates the number of active constituents in plants (Pradhan, Gupta and Goel 2018). It has generally been observed that there is an increase in the content of secondary metabolites post doubling or multiplying the basal chromosome number as polyploidization is associated with large cell size and enhancement of enzyme synthesis. It represents another economically feasible approach for the enhancement of yield of therapeutic secondary metabolites (Srivastava, Mehrotra and Verma 2017). In conventional breeding, polyploid crops are known for some superior traits like vigor and yield as compared to their diploid counterparts (Zhang et al. 2019). It has been considerably established that the biochemical status is altered such that the quantity of secondary metabolites in polyploids of a medicinal plant is higher as compared to diploids. A study on a medicinal plant, *Siegesbeckia orientalis* L., showed that there is a greater accumulation of secondary metabolites such as phenolics and flavonoids in tetraploids which might have some role in their wider distribution as compared to diploid cytotype (Pradhan, Gupta and Goel ,2018). Such an observation was also made in a very recent study in *Solidago canadensis* wherein a greater build-up of all chemical secondary metabolites, such as phenylpropanoids, flavonoids and terpenoids were observed in hexaploid in contrast to diploid *S. canadensis* roots (Wu et al. 2020). A deliberate induction or artificial polyploidy could thus be a prospective tactic to enrich the pharmaceutically valuable secondary metabolite formation in medicinal plants. The polyploidization, along with the generation of hairy roots with infection by *A. rhizogenes*, also represents an incredible strategy for high production of pharmacologically active metabolites in many medicinal plants (Banerjee 2018). Apart from colchicine, which has been widely engaged as an antimitotic agent leading to polyploidization both *in vitro* and *in vivo*, other herbicides like oryzalin and trifluralin can also be useful for this purpose. Polyploidization for plants is species-dependent and is subject to factors like the choice of agent and the concentration used. Although the success ratio is low, various endeavors in deliberately modulating the ploidy level of medicinal plants for increasing metabolites are presented in Table 3.2. This approach may become more widespread in the future with the discovery of new antimitotic agents with the least harmful effects that can be employed for polyploidy induction in the future (Salma, Kundu and Mandal 2017).

TABLE 3.2
Some Examples of Induced Polyploidy in Medicinal Plants Leading to Enhancement in Production of Valuable Bioactive Metabolites

S. No.	Medicinal Plant	Agent/ Concentration	Alteration in Morphological Features	Enhancement in Bioactive Metabolite	Reference
1	*Allium sativum*	Colchicine (0.5% w/v)	Thicker, dark green leaves.	Allicin (30.7%)	(Dixit and Chaudhary 2014)
2	*Artemisia annua*	Colchicine (250 ppm)	Slow growth, dark thicker larger leaves, increased stomatal length.	Artemisinin (17%)	(Lin et al. 2011)
3	*Cannabis sativa*	Colchicine (0.1–0.2%)	Decreased leaf index and height, increase in stomatal size, larger flowers.	Tetrahydrocannabionol	(Mansouri and Bagheri 2017).
4	*Catharanthus roseus*	Colchicine (0.2%)	Larger stoma and more branches and leaves.	Vindoline (130.9%), Catharanthine 188.6% Vinblastine 122.6%	(Tang et al. 2011)
5	*Centella asiatica*	Colchicine (0.025–0.40%)	Increased biomass, longer stomata, higher stomatal index.	Triterpenoid (11%)	(Kaensaksiri et al. 2011)
6	*Echinacea purpurea*	Colchicine (0.25% w/v)	Larger stomata, pollen grain, seed and flower. Wider and thicker leaves.	Cichoric acid (45%) Chlorogenic acid (71%)	(Abdoli et al. 2013)
7	*Linum album*	Colchicine (2.5 mM)	Larger leaves and stomata and lower density stomata.	Podophyllotoxin 1.39-fold (Leaves) 1.23-fold (Stem)	(Javadian et al. 2017)
8	*Nigella sativa*	Colchicine (0.025–0.1% w/v)	Thicker darker-green leaves, increased number of branches.	Thymoquinone (46%)	(Dixit et al. 2015)
9	*Papaver somniferum*	Colchicine (0.25–0.40%)	Increased stomatal length.	Morphine (25–50 %)	(Mishra et al. 2010)
10	*Thymus persicus*	Colchicine (0.05–0.5 %)	Lower plantlet height, shorter roots, thicker stems and darker leaves, larger stomata etc.	Betulinic acid 69.73% Oleanolic acid 42.76% Ursolic acid 140.67 %	(Tavan et al. 2015)
11	*Trachyspermum ammi*	Colchicine (0.025-0.5% w/v)	Increased plant height, length of leaf, inflorescence, peduncle and seed etc.	Thymol (19.53%)	(Sadat Noori et al. 2017)

3.2.4 Conservation of Medicinal Plants

Natural selection over a span of thousands of years has led to the development of medicinal plants with a huge diversity of characters. According to estimates, more than 80,000 species out of the 0.3 million higher plants are reported to possess at least one or the other medicinal attribute, and around 5000 species have one specific pharmacological property. Although a minuscule percentage of them are cultivated in a field or controlled milieu, a huge sector of the plants involved in the pharmacological industry is procured from natural sources. Only a meager number, about 10%–20%, of the plants used for tonic preparations are grown in fields or under controlled milieu (Tasheva and Kosturkov 2013, Hao and Xiao 2015). This diversity of medicinal plants also forms the foundation for the selection of novel and advanced breeds, having the potential to be exploited for the development of new chemical metabolites with novel therapeutic applications (Arora 2010). Among these plants, about 15,000 species are looming under the serious threat of extinction from factors like overharvesting, habitat destruction, large-scale industrialization, deforestation, forest fires, substitution of local plants with hybrid varieties, climate change, etc. Rapid and intense erosion of these bioresources has already occurred to a certain extent which has led to the extinction of many species whose pharmacological qualities had seldom been discovered. The unabated destruction due to these factors has escalated the risk of extinction of medicinal plants in several diversity-rich countries significantly, including China and India (Arora 2010, Moyo et al. 2013, Chen et al. 2016). Though this important and prized bioresource is in urgent need of conservation, the herbal industry and other concerned stakeholders have paid inadequate consideration to its effective utilization and conservation. The distinctive importance of the conservation of medicinal plants also plays a key part in the culture, livelihood, or economy of a considerable chunk of the population throughout the globe (Zahra et al. 2019). In recent years, there have been both *in situ* and *ex situ* interventions for the conservation of various medicinal plants depending on the feasibility (Arora 2010, Siwach et al. 2013). In the future, we are likely to witness a remarkable escalation in the intensity of efforts towards the conservation of medicinal plants because climate change is expected to exercise its manifestations on a large scale, and several pharmacologically useful plant species will be put on the verge of extinction (Kaye et al. 2019). Nevertheless, both *in situ* and *in vitro* methods are associated with certain advantages as well as disadvantages (Amos and Rus Hoelzel 1992). Signs of progress in biotechnology provide some essential and useful techniques for the selection, management and conservation of elite medicinal plant genetic resources involving tissue culture approaches and molecular biology tools (Attia et al. 2017). Tissue culture facilitates the conservation of medicinal plants by rapid multiplication in the absence of pollinators or seeds, regeneration and by producing new plants from seeds that otherwise have very meager possibilities of germination (Tripathi and Tripathi 2003). Tissue culture-based tactics for the *in vitro* conservation of medicinal plants, in addition to rapid and efficient multiplication of disease-free plants, retain the genetic stability of the material. There are minimal growth and storage space requirements for maintaining thousands of genotypes and *in vitro* storage excludes the requisite extensive and annoying quarantine procedures and protocols to be followed during the movement and exchange of germplasm (Canter, Thomas and Ernst 2005). Conservation of germplasm by this approach can be classified based on time duration, such as short-, medium- and long-term conservation. The latter approach encompasses cryopreservation (Canter, Thomas and Ernst 2005, Arora 2010, Moyo et al. 2013). Apart from tissue culture, some other techniques in biotechnology that are handy for the conservation include callus-mediated organogenesis, genetic engineering and regeneration via somatic embryogenesis (Sharmin et al. 2014, Kumar, Moyo and Van Staden 2017). Plant culture systems generally help to mitigate costs and involve fast production and scalability. Nevertheless, a thorough analysis of the cost-benefit ratio of *in vitro* culture techniques is warranted before application (Espinosa-Leal, Puente-Garza and García-Lara 2018). Storage of germplasm under *in vitro* cultures has become costly with time on one hand and vulnerability to microbial contamination and somaclonal variation has also increased. Conservation of medicinal plants through cryopreservation deserves a special mention as a strategy long-term, safe and economically feasible approach to evade extinction by overexploitation (Kundu, Salma and Gantait 2018). In cryopreservation, since cell division and all the metabolic biochemical processes are put to a halt and regeneration can occur from frozen cultures, it provides prospects and opportunities for conservation of endangered medicinal plants like *Atropa belladonna* (Touno, Yoshimatsu and Shimomura 2006), *Digitalis lanata* (Diettrich et al. 1982). *Rauvolfia serpentina* (Ray and Bhattacharya 2008). Some specialties of this approach, such as the halted metabolism of cells and the absence of subculturing, mitigate the apprehensions of genetic and epigenetic alterations considerably. The quantitative productivity of important bioactive molecules, even after regeneration of cryopreserved samples, is usually retained like tropane alkaloids in the case of hairy roots of

A. belladonna (Touno, Yoshimatsu and Shimomura 2006). Cryopreservation has been involved efficaciously in storing a wide range of plant tissue such as anthers, embryos, pollens, meristems, calli and even protoplasts (Kartha and Engelmann 1994). Only a minor portion of medicinal plants have the possibility of cryopreservation for germplasm conservation and continual efforts to standardize cryopreservation protocols and procedures for a number of endangered medicinal plant species are required (Kundu et al. 2018).

3.2.5 Phyto-genomic Profiling

The plant is an environmentally-friendly green industry for the synthesis of plenty of metabolite molecules that can be used for achieving various therapeutic targets. To date, only a minute percentage of the tremendously diverse plant metabolic pathways have been identified for the possibility of the production of new medicines to target various human ailments (De Luca et al. 2012). Plant biologists were not in a position, even up to recent times, to explore the biosynthetic pathways of these metabolites, chiefly due to the paucity of plant genomic information (Rai, Saito and Yamazaki 2017). Researchers, from time to time, have started projects and programs for not only genome sequencing but also post-genomic functional analysis of secondary metabolite biosynthetic pathways of various medicinal plants. These studies can unveil the mysteries about the origin and formation of the huge biochemical diversity of plant metabolites that encompass approximately 200,000 to one million molecules (Saito 2013). The interpretation of the biosynthesis pathways of various pharmacologically significant metabolites has been made possible by the decrease in the cost of generating omics data sets and the availability of a vast array of computer tools for analysis and integration. (Rai, Saito and Yamazaki 2017). Genetic mapping is a game-changer with regard to the isolation of genes and genetic markers linked with phenotypic characters by explicating their genetic positions (Zhu and Salmeron 2007). Because of their complex nature, not many assembled herbal genomes have been released so far. The genome sequencing of some frequently used medicinal plants such as *Salvia miltiorrihiza*, *Ganoderma lucidum* and *C. roseus* have been accomplished, which has made them valued models for the study of genetics and metabolic pathways (Chakraborty 2018). The genetic diversity of medicinal plants has the potential to serve as the foundation of selection for plant improvement, which warrants identification, characterization and documentation of their gene pool. Genetic resources that have proved to be useful for domestication and improvement programs, like plant breeding and mapping include molecular, genome sequencing, expressed sequence tags sequencing, gene transmit, transcript profiling (Gantait, Debnath and Nasim Ali 2014). Apart from such programs, numerous endeavors for a genomic description of several medicinal plants utilizing molecular markers have been undertaken for sustainable exploitation of gene pool, conservation and prospective benefits. Summarizing the complete genome of the well-identified medicinal plant species based on DNA fingerprinting techniques will enable the compilation of a database of commonly used, but economically significant medicinal plant species. Molecular techniques have proved a boon for authentication of medicinal plants based on phylogenic discrepancy signature on nuclear DNA and chloroplast DNA domains. The molecular markers are a powerful and reliable aid for ascertaining the homogeneity in samples. Genomic profiling is an appropriate contrivance for operational quality and regulation of raw material as it can differentiate not only species but also varieties and identify admixtures and contaminants from the field (Teske et al. 2020). They are useful for the recognition of adulterants and thus certify quality control in the area of research, production, clinical use and forensic examination of herbal medicine formulations (Gantait, Debnath and Nasim Ali 2014). With the increase in demand for medicinal compounds concomitant with the global population, a thorough comprehension of the biosynthesis of specialized metabolites has an imperative role in the identification or development of reliable sources in future (Rai, Saito and Yamazaki 2017).

3.3 CONCLUSION AND PROSPECTS

Traditionally, plants and herbs have been used as a source of therapeutic agents and remedies and have also been identified as promising sources for modern biopharmaceuticals, both in terms of safety and economics. Today, natural product drugs have a very pivotal and definite role in modern medicine and pharmaceutical care. Advances in science and technology have helped unravel diverse phyto-bioactive molecules/compounds and their mode(s) of action to improve human health. In this context, the intensified supply of plant-based therapeutic compounds, coupled with minimal yield issues related to habitat destruction, resulted in the large-scale production of these compounds by employing various biotechnological approaches. The recent introduction of cutting-edge biotechnology and molecular biology strategies has brought a paradigm shift for extensive cultivation of threatened, rare and endangered species as well as the development and accumulation of desired secondary metabolites that they synthesize. The advancement in *in vitro* culture, micropropagation and bioreactors have opened new vistas to exploit secondary metabolites that are pharmacologically viable and achieve their optimum production through a cost-effective manner. Therefore, production optimization has evolved from an analytical approach to one of the myriad improvements in efficiency and synergistic effects. Metabolic engineering is an essential facet of producing valuable phytotherapeutic compounds and comprehensive knowledge of biosynthetic pathways as well as responsible genes that have helped to deliver these compounds at higher and commercially suitable production rates. In addition, genomic approaches have offered the requisite means to fully comprehend intricate signaling and regulatory pathways, dynamic phylogenic interactions, and the underlying genetic structure. The genomic innovations had a impact on natural product research and will inevitably continue to lead to discoveries in the future. In this era of genomics, medicinal plants are not the only source of bioactive compounds but also a repository of genes responsible for the development of these active compounds.

There remains a tremendous space for improvement of medicinal plants and drug product research and development despite the outstanding accomplishments in metabolic engineering and rewiring of metabolic pathways of certain medicinally valuable plants. The discovery and subsequent development of phytotherapeutics are being embraced with the advancement of plant biotechnology approaches, and this is being embraced in an open scientific setting, and thus, it is very unlikely to cease expanding. These innovations will be implemented in the future to strengthen the value of these medicinal herbs and plants as a sustainable source of therapeutically important compounds. In conclusion, the discovery of novel phytotherapeutic molecules remains a significant contributing factor in resolving health issues and achieving health goals. It is possible to accomplish successful production of distinctive, highly effective, yet unknown phytochemicals/biologics with the aid of modern advanced biotechnological tools and methodologies.

ACKNOWLEDGMENTS

The author, MZG is thankful to the Council of Scientific and Industrial Research (CSIR-HRDG), New Delhi, India for their financial assistance under CSIR-Research Associateship Scheme (vide File No. 09/132(0883)/2019-EMR-1). The authors acknowledge Prof. B. Sashidhar Rao, Department of Biochemistry, Osmania University, Hyderabad, for his valuable insights in drafting this book chapter. IAG is thankful to DST-SERB, UGC, DBT, UGC-UoH-UPE Phase I and II for all research funding in the lab. IAG is also thankful to Science and Engineering Research Board (SERB), DST, Govt. of India for the financial support given to the lab for the project EMR/2016/006109. The funding provided by UGC sponsored University of Hyderabad-UPE Phase I and II to IAG lab is gratefully acknowledged. IAG is also grateful to University Grant Commission, New Delhi, India for the Indo-US Raman-postdoctoral fellowship 2013–14 with research at Colorado

State University, Fort Collins, USA. The authors are also thankful to DBT-CREBB, DST-FIST Level-I and II support, DST-PURSE Phase I and II and UGC-SAP-CAS for supporting different infra structural facilities to Department of Plant Sciences and School of Life Sciences, University of Hyderabad.

REFERENCES

Abdoli, M., Moieni, A. and Naghdi Badi, H. 2013. Morphological, physiological, cytological and phytochemical studies in diploid and colchicine-induced tetraploid plants of *Echinacea purpurea* (L.). *Acta Physiol. Plant.* 35(7): 2075–2083.

Abou-Alaiwi, W.A., Potlakayala, S.D., Goldman, S.L., Josekutty, P.C., Karelia, D.N. and Rudrabhatla, S.V. 2012. *Agrobacterium*-mediated transformation of the medicinal plant *Centaurea montana*. *Plant Cell Tiss. Org. Cult.* 109(1): 1–8.

Ahmad, S., Garg, M., Tamboli, E., Abdin, M. and Ansari, S. 2013. In vitro production of alkaloids: Factors, approaches, challenges and prospects. *Pharmacog. Rev.* 7(13): 27–33.

Alagoz, Y., Gurkok, T., Zhang, B. and Unver, T. 2016. Manipulating the biosynthesis of bioactive compound alkaloids for next-generation metabolic engineering in opium poppy Using CRISPR-Cas 9 genome editing technology. *Sci. Rep.* 6: 1–9.

Alamgir, A.N.M. 2018. Molecular pharmacognosy – A new borderline discipline between molecular biology and pharmacognosy. In: *Therapeutic Use of Medicinal Plants and their Extracts* (Ed. A.N.M. Alamgir) (Volume 2; pp. 665–720). Progress in Drug Research, vol 74. Springer, Cham.

Allen, R.S., Miller, J.A.C., Chitty, J.A., Fist, A.J., Gerlach, W.L. and Larkin, P.J. 2008. Metabolic engineering of morphinan alkaloids by over-expression and RNAi suppression of salutaridinol 7-O-acetyltransferase in opium poppy. *Plant Biotech. J.* 6(1): 22–30.

Allen, R.S., Millgate, A.G., Chitty, J.A., Thisleton, J., Miller, J.A.C., Fist, A.J., Gerlach, W.L. and Larkin, P.J. 2004. RNAi-mediated replacement of morphine with the nonnarcotic alkaloid reticuline in opium poppy. *Nat. Biotech.* 22(12): 1559–1567.

Alok, A., Jain, P., Kumar, J., Yajnik, K. and Bhalothia, P. 2020. Genome engineering in medicinally important plants using CRISPR/Cas9 tool. In *Genome Engineering via CRISPR-Cas9 System* (Eds. V. Singh and P.K. Dhar) (pp. 155–161). San Diego: Academic Press.

Amos, B. and Rus Hoelzel, A. 1992. Applications of molecular genetic techniques to the conservation of small populations. *Biol. Conser.* 61(2): 133–144.

Arora, R. 2010. *Medicinal Plant Biotechnology.* Wallingford, UK: C.A.B. International.

Arora, R., Mathur, A. and Mathur, A. K. 2010. Emerging trends in medicinal plant biotechnology. In *Medicinal Plant Biotechnology* (pp. 1–12). Wallingford, UK: C.A.B. International.

Arshid Shabir, P. 2021. CRISPR/Cas9-mediated genome editing in medicinal and aromatic plants: developments and applications. In *Medicinal and Aromatic Plants* (Eds. T. Aftab and K.R. Hakeem) (pp. 209–221). San Diego: Academic Press.

Aslam, B., Wang, W., Arshad, M.I., Khurshid, M., Muzammil, S., Rasool, M.H., Nisar, M.A., Alvi, R.F., Aslam, M.A., Qamar, M.U., Salamat, M.K.F. and Baloch, Z. 2018. Antibiotic resistance: A rundown of a global crisis. *Infection Drug Resis.* 11: 1645–1658.

Attia, A.O., Dessoky, E.D.S., Al-Sodany, Y.M. and Ismail, I.A. 2017. Ex situ preservation for some endemic and rare medicinal plants in Taif, KSA. *Biotech. Biotech. Equip.* 31(5): 912–920.

Bae, T.W., Park, H.R., Kwak, Y.S., Lee, H.Y. and Ryu, S.B. 2005. *Agrobacterium tumefaciens*-mediated transformation of a medicinal plant *Taraxacum platycarpum. Plant Cell Tiss. Org. Cult.* 80(1): 51–57.

Balasubramanian, M., Anbumegala, M., Surendran, R., Arun, M. and Shanmugam, G. 2018. Elite hairy roots of *Raphanus sativus* (L.) as a source of antioxidants and flavonoids. 3 *Biotech* 8(2).

Banerjee, S. 2018. Voyaging through chromosomal studies in hairy root cultures towards unravelling their relevance to medicinal plant research: An updated review. *Nucleus (India):* 61(1): 3–18.

Begum, F., Nageswara Rao, S.S.S., Rao, K., Prameela Devi, Y., Giri, A. and Giri, C.C. 2009. Increased vincristine production from Agrobacterium tumefaciens C58 induced shooty teratomas of *Catharanthus roseus* G. Don. *Nat. Prod. Res.* 23(11): 973–981.

Berlin, J., Rügenhagen, C., Fecker, L., Commandeur, U., Wray, V. and Kuzovkina, I.N. 1992. Hairy root cultures of *Peganum harmala. Zeitschrift Fur Naturforschung* – Section C, *J. Bios.* 47(3–4): 222–230.

Borgio, J.F. 2009. RNA interference (RNAi) technology: A promising tool for medicinal plant research. *J. Med. Plants Res.* 3(13): 1176–1183.

Brijwal, L. and Tamta, S. 2015. *Agrobacterium rhizogenes* mediated hairy root induction in endangered *Berberis aristata* DC. *SpringerPlus*, 4(1): 433.

Canter, P.H., Thomas, H. and Ernst, E. 2005. Bringing medicinal plants into cultivation: Opportunities and challenges for biotechnology. *Trends Biotech.* 23(4): 180–185.

Chahel, A. A., Zeng, S., Yousaf, Z., Liao, Y., Yang, Z., Wei, X. and Ying, W. 2019. Plant-specific transcription factor LrTCP4 enhances secondary metabolite biosynthesis in *Lycium ruthenicum* hairy roots. *Plant Cell Tiss. Org. Cult.* 136(2): 323–337.

Chakraborty, P. 2018. Herbal genomics as tools for dissecting new metabolic pathways of unexplored medicinal plants and drug discovery. *Biochimie Open*, 6: 9–16.

Chand, P.K., Ochatt, S.J., Rech, E.L., Power, J.B. and Davey, M.R. 1988. Electroporation stimulates plant regeneration from protoplasts of the woody medicinal species *Solarium dulcamara* L. *Journal of Experimental Botany*, 39(9): 1267–1274.

Chaudhuri, K.N., Ghosh, B., Tepfer, D. and Jha, S. 2005. Genetic transformation of *Tylophora indica* with Agrobacterium rhizogenes A4: Growth and tylophorine productivity in different transformed root clones. *Plant Cell Rep.* 24(1) 25–35.

Chen, D.H., Liu, C.J., Ye, H.C., Li, G.F., Liu, B.Y., Meng, Y.L. and Chen, X.Y. 1999. Ri-mediated transformation of *Artemisia annua* with a recombinant farnesyl diphosphate synthase gene for artemisinin production. *Plant Cell Tiss. Org. Cult.* 57(3): 157–162.

Chen, S.L., Yu, H., Luo, H.M., Wu, Q., Li, C.F. and Steinmetz, A. 2016. Conservation and sustainable use of medicinal plants: Problems, progress and prospects. *Chinese Medicine (United Kingdom)*: 11(1): 37.

Chin, Y.-W., Balunas, M.J., Chai, H.B. and Kinghorn, A.D. 2006. Drug discovery from natural sources. *The AAPS Journal*, 8(2): E239–E253.

Choudhary, N. and Sekhon, B. 2011. An overview of advances in the standardization of herbal drugs. *Journal of Pharmaceutical Education and Research* 2(2): 55.

Concepcion, G.P., Lazaro, J.E. and Hyde, K.D. 2001. Screening for bioactive novel compounds. In *Bio-exploitation of Filamentous Fungi* (Eds. K.D. Pointing and S.B. Hyde) (pp. 93–130). Fungal Diversity Press, Hong Kong.

Corte, L.E.D., Mahmoud, L.M., Moraes, T.S., Mou, Z., Grosser, J.W. and Dutt, M. 2019. Development of improved fruit, vegetable and ornamental crops using the CRISPR/cas9 genome editing technique. *Plants*, 8(12): 601.

Daniell, H., Kumar, S. and Dufourmantel, N. 2005. Breakthrough in chloroplast genetic engineering of agronomically important crops. *Trends Biotech.* 23(5): 238–245.

Datta, A. 2013. Genetic engineering for improving quality and productivity of crops. *Agriculture and Food Security* 2(1): 15.

Davey, M. 2017. Secondary metabolism in plant cell cultures. In *Encyclopedia of Applied Plant Sciences* (Eds. T. Brian, G. M. Brian and J. M. Denis) (Vol. 2, pp. 462–467). Waltham: Academic Press.

De Luca, V., Salim, V., Atsumi, S. M. and Yu, F. 2012. Mining the biodiversity of plants: A revolution in the making. *Science*, 336(6089): 1658–1661.

Debnath, M., Malik, C. and Bisen, P. 2006. Micropropagation: A tool for the production of high quality plant-based medicines. *Current Pharmaceutical Biotechnology*, 7(1): 33–49.

Dehghan, E., Häkkinen, S.T., Oksman-Caldentey, K.M. and Ahmadi, F.S. 2012. Production of tropane alkaloids in diploid and tetraploid plants and in vitro hairy root cultures of *Egyptian henbane* (Hyoscyamus muticus L.) *Plant Cell Tiss. Org. Cult.* 110(1): 35–44.

Deng, C., Hao, X., Shi, M., Fu, R., Wang, Y., Zhang, Y., Zhou, W., Feng, Y., Makunga, N.P. and Kai, G. 2019. Tanshinone production could be increased by the expression of SmWRKY2 in *Salvia miltiorrhiza* hairy roots. *Plant Science* 284: 1–8.

Diettrich, B., Neumann, D., Schindler, U. and Luckner, M. 1982. Cryopreservation of *Digitalis lanata* cells cultivated in vitro. *Planta Medica* 46(2): 82–87.

Dixit, V. and Chaudhary, B.R. 2014. Colchicine-induced tetraploidy in garlic (*Allium sativum* L.) and its effect on allicin concentration. *J. Hort. Sci. Biotech.* 89(5): 585–591.

Dixit, V., Verma, S. and Chaudhary, B.R. 2015. Changes in ploidy and its effect on thymoquinone concentrations in *Nigella sativa* L. seeds. *J. Hort. Sci. Biotech.* 90(5): 537–542.

El-Akkad, T., Serag, A. and Serag, W. 2018. Regeneration and genetic transformation by pAB6 plasmid of *Hyoscyamus muticus* L. (Egyptian Henbane) using particle bombardment. *J. Agric. Chem. Biotech.* 9(10): 235–241.

Espinosa-Leal, C.A., Puente-Garza, C.A. and García-Lara, S. 2018. In vitro plant tissue culture: means for production of biological active compounds. *Planta* 248(1): 1–18.

Fang, I.J. and Trewyn, B.G. 2012. Application of mesoporous silica nanoparticles in intracellular delivery of molecules and proteins. In *Methods in Enzymology* (Ed. N. Düzgüneş) (Vol. 508, pp. 41–59).

Fett-Neto, A.G., Melanson, S.J., Nicholson, S.A., Pennington, J.J. and DiCosmo, F. 1994. Improved taxol yield by aromatic carboxylic acid and amino acid feeding to cell cultures of taxus cuspidata. *Biotech. Bioenginee.* 44(8): 967–971.

Fujii, N., Inui, T., Iwasa, K., Morishige, T. and Sato, F. 2007. Knockdown of berberine bridge enzyme by RNAi accumulates (S)-reticuline and activates a silent pathway in cultured California poppy cells. *Transgenic Res.* 16(3): 363–375.

Gantait, S., Debnath, S. and Nasim Ali, M. 2014. Genomic profile of the plants with pharmaceutical value. *3 Biotech* 4(6): 563–578.

Gao, Z., Li, Y., Chen, J., Chen, Z. and Cui, M. L. 2015. A rapid and stable *Agrobacterium*-mediated transformation method of a medicinal plant *Chelone glabra* L. *Appl. Biochem. Biotech.* 175(5): 2390–2398.

García-Granados, R., Lerma-Escalera, J.A. and Morones-Ramírez, J.R. 2019. Metabolic engineering and synthetic biology: Synergies, future and challenges. *Front. Bioeng. Biotech.* 7: 36.

Gehl, J. 2003. Electroporation: Theory and methods, perspectives for drug delivery, gene therapy and research. *Acta Physiol. Scandinavica* 177(4): 437–447.

Gelvin, S.B. 2003. *Agrobacterium*-mediated plant transformation: the biology behind the "Gene-Jockeying" tool. *Microbio. Mol. Biol. Rev.* 67(1): 16–37.

Giri, A., Banerjee, S., Ahuja, P.S. and Giri, C.C. 1997. Production of hairy roots in *Aconitum heterophyllum* Wall. using *Agrobacterium rhizogenes. In Vitro Cell. Dev. Biol. – Plant* 33(4): 280–284.

Giri, Archana, Giri, C.C., Dhingra, V. and Narasu, M.L. 2001. Enhanced podophyllotoxin production from *Agrobacterium rhizogenes* transformed cultures of *Podophyllum hexandrum. Natural Prod. Lett.* 15(4): 229–235.

Gómez-Galera, S., Pelacho, A.M., Gené, A., Capell, T. and Christou, P. 2007. The genetic manipulation of medicinal and aromatic plants. *Plant Cell Rep.* 26(10): 1689–1715.

Grotewold, E. 2008. Transcription factors for predictive plant metabolic engineering: are we there yet? *Curr. Opin. Biotech.* 19(2): 138–144.

Gupta, D., Bhattacharjee, O., Mandal, D., Sen, M.K., Dey, D., Dasgupta, A., Kazi, T.A., Gupta, R., Sinharoy, S., Acharya, K., Chattopadhyay, D., Ravichandiran, V., Roy, S. and Ghosh, D. 2019. CRISPR-Cas9 system: A new-fangled dawn in gene editing. *Life Sci.* 232: 116636.

Gupta, V., Sengupta, M., Prakash, J. and Tripathy, B.C. 2016. *Basic and Applied Aspects of Biotechnology*. Ist Edition, Springer, Singapore.

Gutierrez-Valdes, N., Häkkinen, S.T., Lemasson, C., Guillet, M., Oksman-Caldentey, K.M., Ritala, A. and Cardon, F. 2020. Hairy root cultures – A versatile tool with multiple applications. *Front. Plant Sci.*, 11: 33.

Hagel, J.M. and Facchini, P.J. 2010. Dioxygenases catalyze the O-demethylation steps of morphine biosynthesis in opium poppy. *Nat. Chem. Biol.* 6(4): 273–275.

Hain, R. and Grimmig, B. 2000. Modification of plant secondary metabolism by genetic engineering. In *Metabolic Engineering of Plant Secondary Metabolism* (Eds. R. Hain and B. Grimmig) (pp. 217–231. Netherlands: Springer.

Hamburger, M. and Hostettmann, K. 1991. Bioactivity in plants: The link between phytochemistry and medicine. *Phytochemistry* 30(12): 3864–3874.

Han, K.H., Fleming, P., Walker, K., Loper, M., Scott Chilton, W., Mocek, U., Gordon, M.P. and Floss, H.G. 1994. Genetic transformation of mature *Taxus*: an approach to genetically control the in vitro production of the anticancer drug, taxol. *Plant Sci.* 95(2): 187–196.

Hao, D.C. and Xiao, P.G. 2015. Genomics and evolution in traditional medicinal plants: Road to a healthier life. *Evol. Bioinform.* 11, 197–212.

Hardy, K. 2018. Plant use in the Lower and Middle Palaeolithic: Food, medicine and raw materials. *Quat. Sci. Rev.* 191: 393–405.

Hayta, S., Gurel, A., Akgun, I. H., Altan, F., Ganzera, M., Tanyolac, B. and Bedir, E. 2011. Induction of *Gentiana cruciata* hairy roots and their secondary metabolites. *Biologia* 66(4): 618–625.

Hendrawati, O., Woerdenbag, H.J., Hille, J. and Kayser, O. 2012. Metabolic engineering of medicinal plants and microorganisms for the production of natural products. In *Pharmaceutical Biotechnology: Drug Discovery and Clinical Applications* (Eds. O. Kayser and H. Warzecha) (2nd ed., pp. 491–526). Weinheim, Germany: Wiley-VCH Verlag: GmbH & Co.

Hwang, H.-H., Yu, M. and Lai, E.-M. 2017. *Agrobacterium*-mediated plant transformation: Biology and applications. *The Arabidopsis Book* 15: e0186.

Javadian, N., Karimzadeh, G., Sharifi, M., Moieni, A. and Behmanesh, M. 2017. In vitro polyploidy induction: Changes in morphology, podophyllotoxin biosynthesis and expression of the related genes in *Linum album* (Linaceae). *Planta* 245(6): 1165–1178.

Jia, H., Zhao, B., Wang, X. and Wang, Y. 2008. *Agrobacterium* rhizogenes-mediated transformation and regeneration of the *Apocynum venetum. Shengwu Gongcheng Xuebao/Chinese J. Biotech.* 24(10): 1723–1728.

Jiao, J., Gai, Q. Y., Wang, W., Zang, Y.P., Niu, L.L., Fu, Y.J. and Wang, X. 2018. Remarkable enhancement of flavonoid production in a co-cultivation system of *Isatis tinctoria* L. hairy root cultures and immobilized *Aspergillus niger. Ind. Crops Prod.* 112: 252–261.

Julsing, M.K., Quax, W.J. and Kayser, O. 2008. The engineering of medicinal plants: Prospects and limitations of medicinal plant biotechnology. In *Medicinal Plant Biotechnology: From Basic Research to Industrial Applications* (Eds. O. Kayser and W.J. Quax) (pp. 1–8). Weinheim, Germany: Wiley-VCH Verlag: GmbH & Co.

Kaensaksiri, T., Soontornchainaksaeng, P., Soonthornchareonnon, N. and Prathanturarug, S. 2011. In vitro induction of polyploidy in *Centella asiatica* (L.) Urban. *Plant Cell Tiss. Org. Cult.* 107(2): 187–194.

Kartha, K.K. and Engelmann, F. 1994. Cryopreservation and germplasm storage. In *Plant Cell and Tissue Culture* (Eds. I.K. Vasil and T.A. Thorpe) (pp. 195–230). Springer, Dordrecht.

Kastell, A., Schreiner, M., Knorr, D., Ulrichs, C. and Mewis, I. 2018. Influence of nutrient supply and elicitors on glucosinolate production in *E. sativa* hairy root cultures. *Plant Cell Tiss. Org. Cult.* 132(3): 561–572.

Katiyar, C., Kanjilal, S., Gupta, A. and Katiyar, S. 2012. Drug discovery from plant sources: An integrated approach. *AYU (An International Quarterly Journal of Research in Ayurveda)* 33(1): 10.

Kaushal, C., Abdin, M.Z. and Kumar, S. 2020. Chloroplast genome transformation of medicinal plant Artemisia annua. *Plant Biotech. J.* 18(11): 2155–2157.

Kaye, T.N., Bahm, M.A., Thorpe, A.S., Gray, E.C., Pfingsten, I. and Waddell, C. 2019. Population extinctions driven by climate change, population size and time since observation may make rare species databases inaccurate. *PLoS One*, 14(10): e0210378.

Kikkert, J.R., Vidal, J.R. and Reisch, B.I. 2005. Stable transformation of plant cells by particle bombardment/biolistics. *Meth. Mol. Biol. (Clifton, N.J.)* 286: 61–78.

Kowalczyk, T., Wieczfinska, J., Skała, E., Śliwiński, T. and Sitarek, P. 2020. Transgenesis as a tool for the efficient production of selected secondary metabolites from plant in vitro cultures. *Plants* 9(2): 132.

Kumar, S.R., Shilpashree, H.B. and Nagegowda, D.A. 2018. Terpene moiety enhancement by overexpression of geranyl (Geranyl) diphosphate synthase and geraniol synthase elevates monomeric and dimeric monoterpene indole alkaloids in transgenic *Catharanthus roseus. Front. Plant Sci.* 9: 942.

Kumar, V., Moyo, M. and Van Staden, J. 2017. Somatic embryogenesis in *Hypoxis hemerocallidea*: An important African medicinal plant. *South Afr. J. Bot.*, 108: 331–336.

Kundu, S., Salma, U., Ali, M.N., Hazra, A.K. and Mandal, N. 2018. Development of transgenic hairy roots and augmentation of secondary metabolites by precursor feeding in *Sphagneticola calendulacea* (L.) Pruski. *Ind. Crops Prod.* 121: 206–215.

Kundu, S., Salma, U. and Gantait, S. 2018. Cryopreservation of medicinal herbs: Major breakthroughs, hurdles and future. In *Biotechnological Approaches for Medicinal and Aromatic Plants: Conservation, Genetic Improvement and Utilization* (Ed. N. Kumar) (pp. 353–381). Springer, Singapore.

Kwao, S., Al-Hamimi, S., Damas, M.E.V., Rasmusson, A.G. and Gómez Galindo, F. 2016. Effect of guard cells electroporation on drying kinetics and aroma compounds of Genovese basil (*Ocimum basilicum* L.) leaves. *Innovative Food Sci. Emerging Techn.* 38: 15–23.

Lalaleo, L., Testillano, P., Risueño, M. C., Cusidó, R. M., Palazon, J., Alcazar, R. and Bonfill, M. 2018. Effect of in vitro morphogenesis on the production of podophyllotoxin derivatives in callus cultures of *Linum album. J. Plant Physiol.* 228: 47–58.

Lan, X., Zeng, J., Liu, K., Zhang, F., Bai, G., Chen, M., Liao, Z. and Huang, L. 2018. Comparison of two hyoscyamine 6β-hydroxylases in engineering scopolamine biosynthesis in root cultures of *Scopolia lurida*. *Biochem. Biophys. Res. Commu.* 497(1): 25–31.

Lau, W., Fischbach, M.A., Osbourn, A. and Sattely, E.S. 2014. Key applications of plant metabolic engineering. *PLoS Biol.* 12(6): 1–5.

Li, B., Cui, G., Shen, G., Zhan, Z., Huang, L., Chen, J. and Qi, X. 2017. Targeted mutagenesis in the medicinal plant *Salvia miltiorrhiza*. *Sci. Rep.* 7: 43320.

Li, H.Q., Sautter, C., Potrykus, I. and Puonti-Kaerlas, J. 1996. Genetic transformation of cassava (manihot esculenta crantz). *Nat. Biotech.* 14(6): 736–740.

Li, M., Jiang, F., Yu, X. and Miao, Z. 2015. Engineering isoprenoid biosynthesis in *Artemisia annua* L. for the production of taxadiene: A key intermediate of taxol. *BioMed Res. Int.* 2015: 504932.

Li, Y., Gao, Z., Piao, C., Lu, K., Wang, Z. and Cui, M.-L. 2014. A stable and efficient *Agrobacterium tumefaciens*-mediated genetic transformation of the medicinal plant *Digitalis purpurea* L. *Applied Biochem. Biotech.* 172: 1807–1817.

Lin, H.W., Kwok, K.H. and Doran, P.M. 2003. Production of podophyllotoxin using cross-species xoculture of *Linum flavum* hairy roots and *Podophyllum hexandrum* cell suspensions. *Biotech. Prog.* 19(5): 1417–1426.

Lin, X., Zhou, Y., Zhang, J., Lu, X., Zhang, F., Shen, Q., Wu, S., Chen, Y., Wang, T. and Tang, K. 2011. Enhancement of artemisinin content in tetraploid *Artemisia annua* plants by modulating the expression of genes in artemisinin biosynthetic pathway. *Biotech. Appl. Biochem.* 58(1): 50–57.

Linlin, S., Brianna Marie, L. and Yuan-Xiang, T. 2016. The CRISPR/Cas9 system for gene editing and its potential application in pain research. *Translational Perioperative and Pain Medicine*, 1(3): 22–33.

Liscombe, D.K. and O'Connor, S.E. 2011. A virus-induced gene silencing approach to understanding alkaloid metabolism in *Catharanthus roseus*. *Phytochemistry* 72(16): 1969–1977.

Liu, Y., Sun, G., Zhong, Z., Ji, L., Zhang, Y., Zhou, J., Zheng, X. and Deng, K. 2017. Overexpression of AtEDT1 promotes root elongation and affects medicinal secondary metabolite biosynthesis in roots of transgenic *Salvia miltiorrhiza*. *Protoplasma* 254(4): 1617–1625.

Ma, R., Yu, Z., Cai, Q., Li, H., Dong, Y., Oksman-Caldentey, K.M. and Rischer, H. 2020. *Agrobacterium*-mediated genetic transformation of the medicinal plant *Veratrum dahuricum*. *Plants*, 9(2).

Majumdar, S., Garai, S. and Jha, S. 2012. Use of the cryptogein gene to stimulate the accumulation of bacopa saponins in transgenic *Bacopa monnieri* plants. *Plant Cell Rep.*, 31(10): 1899–1909.

Mansouri, H. and Bagheri, M. 2017. Induction of polyploidy and its effect on *Cannabis sativa* L. In *Cannabis sativa L. – Botany and Biotechnology* (Eds. S. Chandra, H. Lata and M. ElSohly) (pp. 365–383). Springer, Cham.

Martin, C. and Li, J. 2017. Medicine is not health care, food is health care: Plant metabolic engineering, diet and human health. *New Phytol.* 216(3): 699–719.

Martin, C., Zhang, Y., Tonelli, C. and Petroni, K. 2013. Plants, diet and health. *Annu. Rev. Plant Biol.* 64: 19–46.

Memelink, J. and Gantet, P. 2007. Transcription factors involved in terpenoid indole alkaloid biosynthesis in *Catharanthus roseus*. *Phytochem. Rev.* 6(2–3): 353–362.

Michoux, F., Ahmad, N., Hennig, A., Nixon, P.J. and Warzecha, H. 2013. Production of leafy biomass using temporary immersion bioreactors: An alternative platform to express proteins in transplastomic plants with drastic phenotypes. *Planta* 237(3): 903–908.

Mishra, B.K., Pathak, S., Sharma, A., Trivedi, P.K. and Shukla, S. 2010. Modulated gene expression in newly synthesized auto-tetraploid of *Papaver somniferum* L. *South Afr. J. Bot.*, 76(3): 447–452.

Mookkan, M. 2018. Particle bombardment-mediated gene transfer and GFP transient expression in *Seteria viridis*. *Plant Signaling Behavior* 13(4): e1441657.

Moyo, M., Aremu, A.O., Gruz, J., Šubrtová, M., Szüčová, L., Doležal, K. and Van Staden, J. 2013. Conservation strategy for *Pelargonium sidoides* DC: Phenolic profile and pharmacological activity of acclimatized plants derived from tissue culture. *J. Ethnopharmacol.* 149(2): 557–561.

Muralikrishna, N., Srinivas, K., Kumar, K.B. and Sadanandam, A. 2016. Stable plastid transformation in *Scoparia dulcis* L. *Physiol. Mol. Biol. Plants* 22(4): 575–581.

Naina, N.S., Gupta, P.K. and Mascarenhas, A.F. 1989. Genetic transformation and regeneration of transgenic neem (*Azadirachta indica*) plants using *Agrobacterium tumefaciens*. *Curr. Sci.* 58(4): 184–187.

Nair, M.S., Huang, Y., Fidock, D.A., Polyak, S.J., Wagoner, J., Towler, M.J. and Weathers, P.J. 2021. *Artemisia annua* L. extracts inhibit the in vitro replication of SARS-CoV-2 and two of its variants. *J Ethnopharmacol*, 274:114016.

Nandi, S.K., Palni, L.M.S., Pandey, H., Chandra, B. and Nadeem, M. 2016. Selection of elites and in vitro propagation of selected high-value Himalayan medicinal herbs for sustainable utilization and conservation. In *Plant Tissue Culture: Propagation, Conservation and Crop Improvement* (Eds. M. Anis and N . Ahmad) (pp. 15–44). Springer, Singapore.

Nasri, H., Baradaran, A., Shirzad, H. and Kopaei, M.R. 2014. New concepts in nutraceuticals as alternative for pharmaceuticals. *Int. J. Preventive Med.* 5(12): 1487–1499.

Nester, E.W. 2015. *Agrobacterium*: Nature's Genetic Engineer. *Front. Plant Sci.*, 5: 730.

Newman, D.J. and Cragg, G.M. 2016. Natural products as sources of new drugs from 1981 to 2014. *J. Nat. Prod.* 79(3): 629–661.

Ngezahayo, F. 2018. Somaclonal variations and their applications in medicinal plant improvement. In *Biotechnological Approaches for Medicinal and Aromatic Plants: Conservation, Genetic Improvement and Utilization* (Eds. N. Kumar) (pp. 503–519). Springer, Singapore.

Nin, S., Bennici, A., Roselli, G., Mariotti, D., Schiff, S. and Magherini, R. 1997. Agrobacterium-mediated transformation of *Artemisia absinthium* L. (wormwood) and production of secondary metabolites. *Plant Cell Rep.* 16(10): 725–730.

Nogueira, M., Enfissi, E.M., Almeida, J. and Fraser, P.D. 2018. Creating plant molecular factories for industrial and nutritional isoprenoid production. *Curr. Opin. Biotech.* 49: 80–87.

Nomani, M., Sadat Noori, S.A., Tohidfar, M. and Ramshini, H. 2019. Overexpression of TPS2 gene to increase thymol content using *Agrobacterium tumefaciens*-mediated transformation in *Trachyspermum ammi* (Qom ecotype). *Ind. Crops Prod.* 130: 63–70.

Ozyigit, I.I. 2020. Gene transfer to plants by electroporation: methods and applications. *Mol. Biol. Rep.*, 47(4): 3195–3210.

Ozyigit, I.I. and Yucebilgili Kurtoglu, K. 2020. Particle bombardment technology and its applications in plants. *Mol. Biol. Rep.*. 47(12): 9831–9847.

Pandey, V., Misra, P., Chaturvedi, P., Mishra, M.K., Trivedi, P.K. and Tuli, R. 2010. *Agrobacterium tumefaciens*-mediated transformation of *Withania somnifera* (L.) Dunal: An important medicinal plant. *Plant Cell Rep.* 29(2): 133–141.

Park, M.Y., Yi, N.R., Lee, H.Y., Kim, S.T., Kim, M., Park, J.H., Kim, J.K., Lee, J.S., Cheong, J.J. and Choi, Y. Do. 2002. Generation of chlorsulfuron-resistant transgenic garlic plants (*Allium sativum* L.) by particle bombardment. *Mol. Breed.* 9(3): 171–181.

Park, S.U., Chae, Y.A. and Facchini, P.J. 2003. Genetic transformation of the figwort, *Scrophularia buergeriana* Miq., an oriental medicinal plant. *Plant Cell Rep.* 21(12): 1194–1198.

Patra, N. and Srivastava, A.K. 2018. Mass production of *Artemisinin* using hairy root cultivation of *Artemisia annua* in bioreactor. In *Bioprocessing of Plant in vitro Systems: Reference Series in Phytochemistry* (Eds. A. Pavlov and T. Bley) (pp. 343–359). Springer, Cham.

Petrovska, B.B. 2012. Historical review of medicinal plants' usage. *Pharmacog. Rev.* 6(11): 1–5.

Pradhan, S.K., Gupta, R.C. and Goel, R.K. 2018. Differential content of secondary metabolites in diploid and tetraploid cytotypes of *Siegesbeckia orientalis* L. *Natural Prod. Res.* 32(20): 2476–2482.

Qiu, J. 2007. Traditional medicine: A culture in the balance. *Nature* 448(7150): 126–128.

Rai, A., Saito, K. and Yamazaki, M. 2017. Integrated omics analysis of specialized metabolism in medicinal plants. *Plant J.* 90(4): 764–787.

Rates, S.M.K. 2001. Plants as source of drugs. *Toxicon* 39(5): 603–613.

Ray, A. and Bhattacharya, S. 2008. Cryopreservation of in vitro grown nodal segments of *Rauvolfia serpentina* by PVS2 vitrification. *Cryo-Letters* 29(4): 321–328.

Rekha, K. and Thiruvengadam, M. 2017. Secondary Metabolite Production in Transgenic Hairy Root Cultures of Cucurbits. In: Transgenesis and Secondary Metabolism. Reference Series in Phytochemistry (Ed. S. Jha S.) (pp. 267–293). Springer, Cham.

Roychowdhury, D., Halder, M. and Jha, S. 2016. *Agrobacterium rhizogenes*-mediated transformation in medicinal plants: genetic stability in long-term culture. In: *Transgenesis and Secondary Metabolism. Reference Series in Phytochemistry* (Ed. S. Jha) (pp. 1–23). Springer, Cham.

Sabzehzari, M. and Naghavi, M.R. 2019. Phyto-miRNAs-based regulation of metabolites biosynthesis in medicinal plants. *Gene* 682: 13–24.

Sadat Noori, S.A., Norouzi, M., Karimzadeh, G., Shirkool, K. and Niazian, M. 2017. Effect of colchicine-induced polyploidy on morphological characteristics and essential oil composition of ajowan (*Trachyspermum ammi* L). *Plant Cell Tiss. Org. Cult.* 130(3): 543–551.

Saito, K. 2013. Phytochemical genomics-a new trend. *Curr. Opi. Plant Biol.* 16(3): 373–380.

Sales, E., Segura, J. and Arrillaga, I. 2003. *Agrobacterium tumefaciens*-mediated genetic transformation of the cardenolide-producing plant *Digitalis minor* L. *Planta Med.* 69(2): 143–147.

Salma, U., Kundu, S. and Mandal, N. 2017. Artificial polyploidy in medicinal plants: Advancement in the last two decades and impending prospects. *J. Crop Sci. Biotech.* 20(1): 9–19.

Scragg, A.H. 1992. Large-scale plant cell culture: Methods, applications and products. *Curr. Opin. Biotech.* 3(2): 105–109.

Shahzad, A., Shaheen, A., Kozgar, M.I., Sahai, A. and Sharma, S. 2013. Phytoactive Compounds from In Vitro Derived Tissues. In *Recent Trends Biotech. and Therapeutic Applications of Medicinal Plants* (Eds. M. Shahid, A. Shahzad, A. Malik and A. Sahai) (pp. 1–23). Springer Netherlands.

Sharma, A., Verma, P., Mathur, A. and Mathur, A.K. 2018. Overexpression of tryptophan decarboxylase and strictosidine synthase enhanced terpenoid indole alkaloid pathway activity and antineoplastic vinblastine biosynthesis in *Catharanthus roseus*. *Protoplasma* 255(5): 1281–1294.

Sharma, K.K., Bhatnagar-Mathur, P. and Thorpe, T.A. 2005. Genetic transformation technology: Status and problems. *In Vitro Cell. Dev. Biol. – Plant* 41(2): 102–112.

Sharma, R. and Arora, R. 2006. *Herbal Drugs (A Twenty First Century Perspective).* Jaypee Brothers Medical Publishers (P) Ltd, New Delhi.

Sharmin, S.A., Alam, M.J., Sheikh, M.M.I., Sarker, K.K., Khalekuzzaman, M., Haque, M.A., Alam, M.F. and Alam, I. 2014. Somatic embryogenesis and plant regeneration in Wedelia calendulacea less. an endangered medicinal plant. *Brazilian Arch. Biol. Techn.* 57(3): 394–401.

Shetty, K. 1998. Comparison of the growth pattern and rosmarinic acid production in rosemary (*Rosmarinus officinalis*) shoots and genetically transformed callus cultures. *Food Biotech.* 12(1–2): 27–41.

Shi, J., Ma, Y., Zhu, J., Chen, Y., Sun, Y., Yao, Y., Yang, Z. and Xie, J. 2018. A review on electroporation-based intracellular delivery. *Molecules (Basel, Switzerland)*, 23(11): 3044.

Siwach, M., Siwach, P., Solanki, P. and Gill, A.R. 2013. Biodiversity conservation of Himalayan medicinal plants in India: A retrospective analysis for a better vision'. *Int J. Biodiver. Conser.* 5(9): 529–540.

Song, G.Q., Prieto, H. and Orbovic, V. 2019. *Agrobacterium*-mediated transformation of tree fruit crops: Methods, progress and challenges. *Front. Plant Sci.* 10: 226.

Sorokina, M. and Steinbeck, C. 2020. Review on natural products databases: Where to find data in 2020. *J. Cheminform.* 12(1): 1–51.

Srinivasan, V., Ciddi, V., Bringi, V. and Shuler, M.L. 1996. Metabolic inhibitors, elicitors and precursors as tools for probing yield limitation in taxane production by *Taxus chinensis* cell cultures. *Biotech. Prog.* 12(4): 457–465.

Srivastava, V., Mehrotra, S. and Verma, P.K. 2017. Biotechnological interventions for production of therapeutic secondary metabolites using hairy root cultures of medicinal plants. In *Current Developments in Biotechnology and Bioengineering: Crop Modification, Nutrition and Food Production* (Eds. S.K.Dubey, A. Pandey and R.S. Sangwan) (pp. 259–282). Cambridge: Elsevier.

Sultan, P., Shawl, A.S., Ramteke, P.W., Jan, A., Chisti, N., Jabeen, N. and Shabir, S. 2006. In vitro propagation for mass multiplication of *Podophyllum hexandrum*: A high value medicinal herb. *Asian J. Plant Sci.* 5(2): 179–184.

Tang, K.X., Xing, S.H., Guo, X.B., Wang, Q., Pan, Q.F., Tian, Y.S., Liu, P., Zhao, J.Y., Wang, G.F. and Sun, X.F. 2011. Induction and flow cytometry identification of tetraploids from seed-derived explants through colchicine treatments in *Catharanthus roseus* (L.) G. Don. *J. Biomed. Biotech.* 2011: 793198.

Tasheva, K. and Kosturkova, G. 2013. Role of Biotechnology for Protection of Endangered Medicinal Plants. In: *Environmental Biotechnology – New Approaches and Prospective Applications* (Ed. M. Petre). IntechOpen, London.

Tatsis, E.C. and O'Connor, S.E. 2016. New developments in engineering plant metabolic pathways. *Curr. Opin. Biotech.* 42: 126–132.

Tavan, M., Mirjalili, M.H. and Karimzadeh, G. 2015. In vitro polyploidy induction: Changes in morphological, anatomical and phytochemical characteristics of *Thymus persicus* (Lamiaceae). *Plant Cell Tiss. Org. Cult.* 122(3): 573–583.

Tavassoli, P. and Safipour Afshar, A. 2018. Influence of different *Agrobacterium rhizogenes* strains on hairy root induction and analysis of phenolic and flavonoid compounds in marshmallow (*Althaea officinalis* L.). 3 *Biotech* 8(8): 351.

Teli, N.P. and Timko, M.P. 2004. Recent developments in the use of transgenic plants for the production of human therapeutics and biopharmaceuticals. *Plant Cell Tiss. Org. Cult.* 79(2): 125–145.

Teske, D., Peters, A., Möllers, A. and Fischer, M. 2020. Genomic profiling: the strengths and limitations of chloroplast genome-based plant variety authentication. *J. Agric. Food Chem.* 68(49): 14323–14333.

Thakur, V., Bains, S., Pathania, S., Sharma, S., Kaur, R. and Singh, K. 2020. Comparative transcriptomics reveals candidate transcription factors involved in costunolide biosynthesis in medicinal plant-*Saussurea lappa*. *Int. J. Biol. Macromol.* 150: 52–67.

Tiwari, R.K., Trivedi, M., Guang, Z.C., Guo, G.Q. and Zheng, G.C. 2007. Genetic transformation of *Gentiana macrophylla* with *Agrobacterium rhizogenes*: Growth and production of secoiridoid glucoside gentiopicroside in transformed hairy root cultures. *Plant Cell Rep.* 26(2): 199–210.

Touno, K., Yoshimatsu, K. and Shimomura, K. 2006. Characteristics of *Atropa belladonna* hairy roots cryopreserved by vitrification method. *Cryo-Letters* 27(2): 65–72.

Tripathi, L. and Tripathi, J.N. 2003. Role of biotechnology in medicinal plants. *Trop. J. Pharmaceu. Res.* 2(2): 243–253.

Utami, E.S.W., Hariyanto, S. and Manuhara, Y.S.W. 2018. *Agrobacterium tumefaciens*-mediated transformation of *Dendrobium lasianthera* J.J.Sm: An important medicinal orchid. *J. Gene. Eng. Biotech.* 16(2): 703–709.

van Uden, W., Pras, N. and Maingré, T. M. 1990. The accumulation of podophyllotoxin-β-d-glucoside by cell suspension cultures derived from the conifer *Callitris drummondii*. *Plant Cell Rep.* 9(5): 257–260.

Vanegas, P.E., Valdez-Morales, M., Valverde, M.E., Cruz-Hernández, A. and Paredes-López, O. 2006. Particle bombardment, a method for gene transfer in marigold. *Plant Cell Tiss. Org. Cult.* 84(3): 359–363.

Verma, P.C., Singh, D., Rahman, L.U., Gupta, M.M. and Banerjee, S. 2002. In vitro-studies in *Plumbago zeylanica*: Rapid micropropagation and establishment of higher plumbagin yielding hairy root cultures. *J. Plant Physiol.* 159(5): 547–552.

Vines, G. 2004. Herbal harvests with a future: towards sustainable sources for medicinal plants. *Plantlife Int.* 1–12.

Walsh, C.T. and Tang, Y. 2017. *Natural Product Biosynthesis: Chemical Logic and Enzymatic Machinery*. Croydon, UK: The Royal Society of Chemistry.

Wang, C.T., Liu, H., Gao, X.S. and Zhang, H.X. 2010. Overexpression of G10H and ORCA3 in the hairy roots of *Catharanthus roseus* improves catharanthine production. *Plant Cell Rep.* 29(8): 887–894.

Wang, H.M. and Zu, Y.G. 2007. *Agrobacterium*-mediated genetic transformation of *Camptotheca acuminata*. *J. Forestry Res.* 18(4): 316–318.

Wang, J., Nie, J., Pattanaik, S. and Yuan, L. 2016. Efficient *Agrobacterium*-mediated transformation of *Artemisia annua* L. using young inflorescence. *In Vitro Cell. Dev. Biol. – Plant* 52(2): 204–211.

Wei, T., Gao, Y., Deng, K., Zhang, L., Yang, M., Liu, X., Qi, C., Wang, C., Song, W., Zhang, Y. and Chen, C. 2019. Enhancement of tanshinone production in *Salvia miltiorrhiza* hairy root cultures by metabolic engineering. *Plant Meth.* 15(1): 53.

Whitmer, S., Canel, C., Van Der Heijden, R. and Verpoorte, R. 2003. Long-term instability of alkaloid production by stably transformed cell lines of *Catharanthus roseus*. *Plant Cell Tiss. Org. Cult.* 74(1): 73–80.

World Health Organization. 2019. WHO | This year's World malaria report at a glance (who.int).

Wijesekera, R.O.B. 2017. The medicinal plant industry. In *The Medicinal Plant Industry* (Ed. R. O. B. Wijesekera). UK: Routledge.

Wilson, S.A. and Roberts, S.C. 2012. Recent advances towards development and commercialization of plant cell culture processes for the synthesis of biomolecules. *Plant Biotech. J.* 10(3): 249–268.

Woerdenbag, H.J., van Uden, W., Frijlink, H.W., Lerk, C.F., Pras, N. and Malingré, T.M. 1990. Increased podophyllotoxin production in *Podophyllum hexandrum* cell suspension cultures after feeding coniferyl alcohol as a β-cyclodextrin complex. *Plant Cell Rep.* 9(2): 97–100.

Wójcik, A. and Rybczyński, J.J. 2015. Electroporation and morphogenic potential of *Gentiana kurroo* (Royle) embryogenic cell suspension protoplasts. *Biotechnologia* 96(1): 19–29.

World Health Organization. 2019. WHO | This year's World malaria report at a glance.

Wu, C.H., Murthy, H.N., Hahn, E.J. and Paek, K.Y. 2008. Establishment of adventitious root co-culture of Ginseng and Echinacea for the production of secondary metabolites. *Acta Physiol. Plant.* 30(6): 891–896.

Wu, M., Ge, Y., Xu, C. and Wang, J. 2020. Metabolome and transcriptome analysis of hexaploid solidago canadensis roots reveals its invasive capacity related to polyploidy. *Genes* 11(2).

Xing, B., Yang, D., Liu, L., Han, R., Sun, Y. and Liang, Z. 2018. Phenolic acid production is more effectively enhanced than tanshinone production by methyl jasmonate in *Salvia miltiorrhiza* hairy roots. *Plant Cell Tiss. Org. Cult.* 134(1): 119–129.

Yan, Y.P. and Wang, Z.Z. 2007. Genetic transformation of the medicinal plant *Salvia miltiorrhiza* by *Agrobacterium tumefaciens*-mediated method. *Plant Cell Tiss. Org. Cult.* 88(2): 175–184.

Yuan, Y., Liu, W., Zhang, Q., Xiang, L., Liu, X., Chen, M., Lin, Z., Wang, Q. and Liao, Z. 2015. Overexpression of artemisinic aldehyde Δ11 (13) reductase gene-enhanced artemisinin and its relative metabolite biosynthesis in transgenic *Artemisia annua* L. *Biotech. App. Biochem.* 62(1): 17–23.

Yun, D.J., Hashimoto, T. and Yamada, Y. 1992. Metabolic engineering of medicinal plants: transgenic *Atropa belladonna* with an improved alkaloid composition. *Proc. Nat. Acad. Sci. USA* 89(24): 11799–11803.

Zahra, W., Rai, S.N., Birla, H., Singh, S. Sen, Rathore, A.S., Dilnashin, H., Keswani, C. and Singh, S. P. 2019. Economic importance of medicinal plants in Asian countries. In *Bioeconomy for Sustainable Development* (Eds. C. Keswani), pp. 359–377. Springer, Singapore.

Zhang, K., Wang, X. and Cheng, F. 2019. Plant polyploidy: Origin, evolution and its influence on crop domestication. *Hort. Plant J.* 5(6): 231–239.

Zhang, L., Ding, R., Chai, Y., Bonfill, M., Moyano, E., Oksman-Caldentey, K.M., Xu, T., Pi, Y., Wang, Z., Zhang, H., Kai, G., Liao, Z., Sun, X. and Tang, K. 2004. Engineering tropane biosynthetic pathway in *Hyoscyamus niger* hairy root cultures. *Proc. Nat. Acad. Sci. USA* 101(17): 6786–6791.

Zhi, B.H. and Alfermann, A.W. 1993. Diterpenoid production in hairy root cultures of *Salvia miltiorrhiza. Phytochemistry* 32(3): 699–703.

Zhou, Z., Tan, H., Li, Q., Chen, J., Gao, S., Wang, Y., Chen, W. and Zhang, L. 2018. CRISPR/Cas9-mediated efficient targeted mutagenesis of RAS in *Salvia miltiorrhiza. Phytochemistry* 148: 63–70.

Zhu, T. and Salmeron, J. 2007. High-definition genome profiling for genetic marker discovery. *Trends Plant Sci.* 12(5): 196–202.

4 Big Omics Data and Medicinal Plants

Ankita Chatterjee, Geetanjali Baruah and Jagajjit Sahu

CONTENTS

4.1 INTRODUCTION

Understanding the biology of a particular system under specific times or conditions and so on needs collective information by observing the whole system. This is the basic principle of any omics technology, which requires understanding all the components such as genes, proteins, and metabolites in a system. Omics technologies have clearly changed the traditional view of the central dogma (Figure 4.1). Omics-based studies provide important biological insights into levels of biological components (such as gene expression, protein levels, and modification, metabolite levels), with each data type being biologically connected with others. Integrating these omics datasets through data-mining and statistical/computational tools provides a comprehensive understanding of relationships and interactions between different biological components. The shift of interest toward omics technologies in most research areas is evident in recent times, as seen in the increase in the publication volume of omics data and articles.

Along with the massive data, omics technologies have also created the data deluge, commonly known as the Big Data issue (Morrison et al. 2006). Reinforcement of the knowledge obtained from omics analysis provides an efficient means of studying the harmonious regulation of each biological component in a particular biological system (Rai, Saito and Yamazaki 2017). Nearly 50% of drugs are derived from chemical compounds isolated from various plants (Fabricant and Farnsworth 2001, Yarnell and Abascal 2002). The development of sequencing technology and synthetic biology has enabled obtaining molecular information from the transcriptomic and genomic data of plants and then utilizing bacteria to synthesize desired chemical compounds (Atanasov et al. 2015, Smanski et al. 2016).

4.2 OMICS DATA FOR MEDICINAL PLANTS

Like most research areas, the research field of medicinal plants has been greatly influenced by the advancement of omics technologies. Certain powerful data analytics approaches have provided a successful way of inferring the patterns of publications by accessing the keywords or

DOI: 10.1201/9781003179139-4

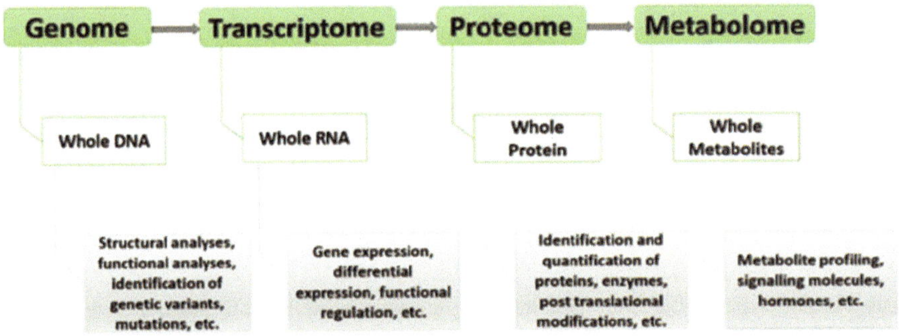

FIGURE 4.1 Omics driven central dogma.

frequently used words. To gain insight into the same, we performed a basic analysis by taking data from PubMed for medicinal plants. We used "Plants, Medicinal"[Mesh] as the search term and filtered the retrieval with only English articles dated up to 31 December 2020. This fetched a total of 41,523 articles related to medicinal plant research with no particular focus on omics. Since there can be numerous keywords to search for omics-related literature, we restricted the keywords to widely used Mesh terms only along with the "Plants, Medicinal"[Mesh]. The search term was "Plants, Medicinal"[Mesh] AND ("Genomics"[Mesh] OR "Proteomics"[Mesh] OR "Metabolomics"[Mesh] OR "Genome"[Mesh] OR "Proteome"[Mesh] OR "Metabolome"[Mesh] OR "RNA-Seq"[Mesh] OR "Transcriptome"[Mesh] OR "Chromatin Immunoprecipitation Sequencing"[Mesh] OR "Microarray Analysis"[Mesh] OR "Whole Exome Sequencing"[Mesh] OR "Gene Expression Profiling"[Mesh]) and we were able to retrieve 1,790 records with the same filters. The bibliometrics data files were then simply loaded onto VOSviewer to create a Density Visualization of the frequent keywords for both the files. The density graph for medicinal plant research as a whole is a clear representation of scant research in medicinal plants using omics technologies (Figure 4.2).

Further, Figure 4.3 shows the density visualization for the keywords of the possible omics studies in medicinal plants. There is presence of certain omics-related keywords such as "metabolomics," "gene expression regulation," "computational biology," "genome, chloroplast," and so on. The difference between the frequent keywords in the case of medicinal plant research as a whole and omics in medicinal plants (Figure 4.2 vs. Figure 4.3) represents the fact that there is wide scope in exploring omics technology for medicinal plants.

4.2.1 Genomics Data

In simple terms, genomics can be defined as analyzing complete DNA sequences in an organism. The science of the genomes, or "genomics," initially dedicated to the determination of DNA sequences (the nucleotide order on a given fragment of DNA), has promptly expanded toward a more functional level – studying the expression profiles and the roles of both genes and proteins. Genomics is an emerging area in biotechnology, having DNA sequencing as its core technology. Since scientists had access to genomics technology, the genomics data have been increasing on various public domains for various plants (Bevan and Uauy 2013). Advancement in technologies has led to a drastic reduction in the time required for whole-genome sequencing, which has revolutionized genomics on a wide scale. In reality, genome sequencing projects have been carried out frequently on model organisms rather than medicinal plants specifically. This may be to standardize the model organism's genetic information first and then go for other plants that can then be fed the available

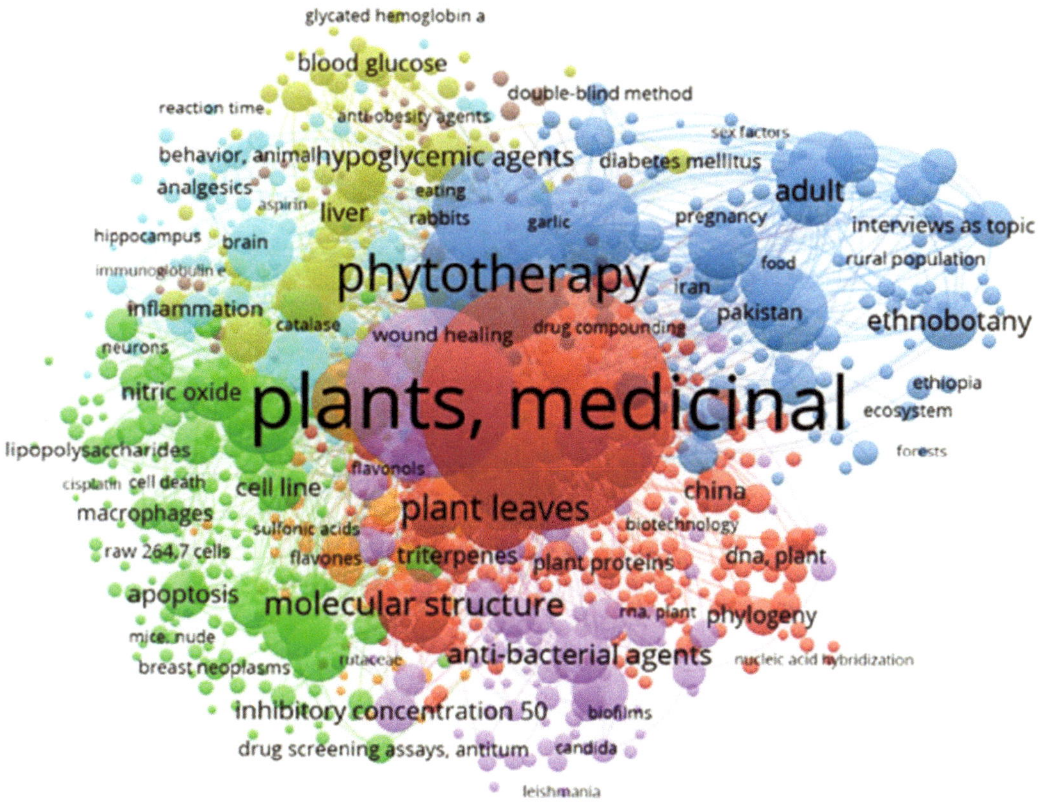

FIGURE 4.2 Density visualization of reports in medicinal plants.

information. Most medicinal plants have not been subjected to this high-throughput sequencing for this generic cause. The most important examples of the medicinal plant genome projects are *Panax ginseng*, *Artemisia annua*, Salvia, *Glycyrrhiza uralensis*, *Lycium chinense*, and so on (Chen et al. 2011, Yao et al. 2011, Hao et al. 2012, Hao and Xiao 2015a, Hao et al. 2015b, Hao, Gu, and Xiao 2015c, Moses et al. 2015).

However, Hao and Xiao have nicely put some important principles before selecting a medicinal plant for whole-genome sequencing projects (Hao and Xiao 2015a). These are established drugs from the plant which is in high demand, known as crucial secondary metabolism pathways, the plant belonging to an important medicinal genus/family, the plants for which there is a considerable amount of biological data indicating their medicinal value, and medicinal plants with little known genetic information (preferably with a small diploid genome).

The application of genomics to plants has been shown to promote the development of natural medicines and the selection of cultivars with good agricultural traits and high levels of desired metabolites of medicinal and pharmaceutical importance (Hao and Xiao 2015, Unamba Nag and Sharma 2015). A study on terpene diversification across sequenced plant genomes reported different mechanisms of pathway assembly in eudicots and monocots (Boutanaev et al. 2015). Bunsupa et al. (2016) reported positive selection sites in the ornithine decarboxylase coding genes that grant the promiscuous function of lysine decarboxylase. In another study analyzing genome sequences for 25 plant species, Tohge et al. (2014) reported orthologous genes associated with terpenoid metabolism, alkaloid metabolism, phenolic secondary metabolism, and glucosinolate metabolism. Together with several others, these studies have used genomic resources to gain insight into the biosynthesis and

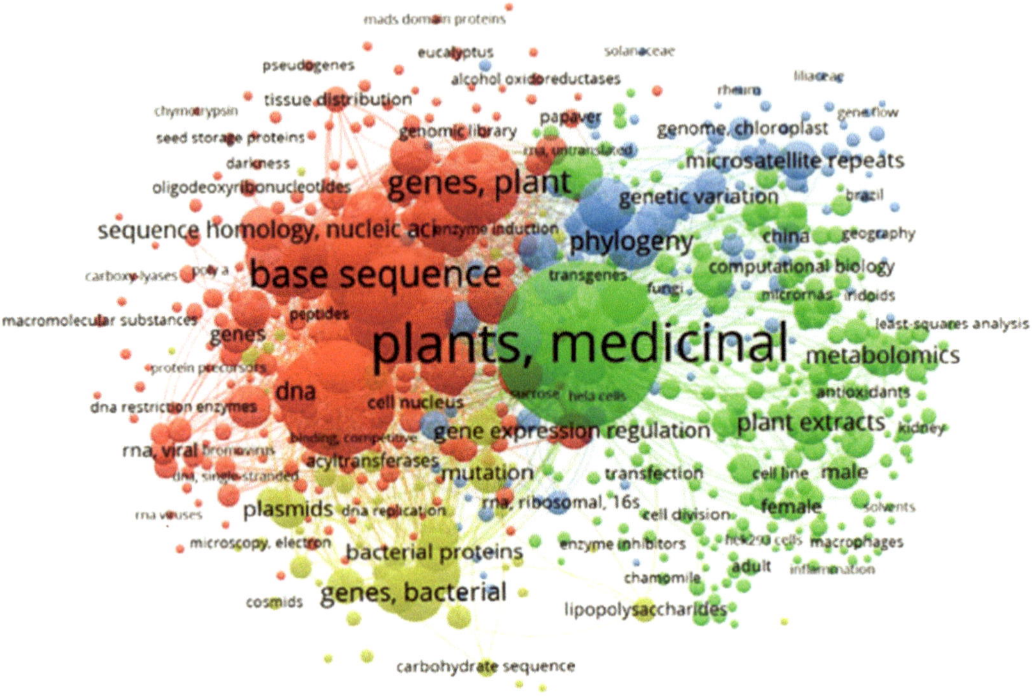

FIGURE 4.3 Density visualization of reports in medicinal plants related to omics studies.

evolution of specialized secondary metabolites and are vital for pathway discovery of important phytochemicals.

Pathway discovery in medicinal plants is now increasingly shifting from the characterization of individual pathway steps to genome-based prediction and validation of the entire pathway. The metabolic gene cluster for several plants has been reported, including rice (Swaminathan et al. 2009), *Solanum lycopersicum* (Matsuba et al. 2013), *Sorghum bicolor* (Takos et al. 2011), and *Arabidopsis thaliana* (Field and Osbourn 2008). These findings, therefore, make the mining of genomic sequences for the identification of metabolic gene clusters a promising approach for the prediction and subsequent characterization of components of a biosynthetic pathway.

4.2.2 Transcriptomics Data

The transcriptome is the complete set of RNA transcripts that are produced by the genome in a specific cell. Transcriptomics is the study of the transcriptome using high-throughput methods, such as microarray analysis. Comparison of transcriptomes allows the identification of genes that are differentially expressed in distinct cell populations or in response to different treatments. There are two key techniques in the field: microarrays, which quantify a set of pre-determined sequences, and RNA sequencing (RNA-Seq), which uses high-throughput sequencing to capture all sequences. In fact, the major breakthrough in transcriptome analysis came with next-generation sequencing (NGS)-based RNA sequencing (RNA-seq). Unlike microarray analysis, RNA-seq-based transcriptome profiling does not require prior genomic information thus providing a vital method for expediting transcriptome profiling and functional characterization of transcripts for non-model plants, including medicinal plants (Han et al. 2016, Rai, Saito and Yamazaki 2017). Over the last few years, RNA-seq-based transcriptome profiling and *de novo* transcriptome assembly, co-expression, and comparative transcriptome analysis of various medicinal plants have become an attractive alternative for generating genetic resources for plants of biological importance (Unamba, Nag and Sharma 2015, Wang et al. 2016a).

The main objectives of transcriptome research include classifying all transcriptional products, determining the transcription structure of genes (such as starting sites, 5′ and 3′ terminals, splicing patterns, and post-transcriptional modifications), and the changes in the expression levels of individual transcripts during the development. This technology has been widely used in biological research, medical and clinical research, and drug development. Specifically, transcriptomics in medicinal plants has been reported in various studies. For example, TF prediction and their regulatory role in costunolide synthesis in *Saussurea lappa* (Thakur et al. 2020), a large number of putative genes involved in the biosynthesis of polysaccharides (Shen et al. 2017), alkaloids (Li et al. 2017), and flavonoids (Lei et al. 2018) have been identified in Dendrobium through transcriptome sequencing. With an increasing number of transcriptome datasets being available from numerous plants, comparative transcriptome analysis across multiple plant species could accelerate the effort to understand the biosynthesis and evolution of specialized metabolites.

4.2.3 Proteomics Data

Proteins are responsible for the functioning and regulation of all biological processes. Protein synthesis and modification are highly regulated, multi-step processes that serve as the basis of the tight homeostasis that characterizes any biological system. Understanding protein-protein and protein-nucleotide interactions, the dynamic state of protein abundance and localization, and post-translational modification provides vital clues in explaining cellular functionality and regulation. In recent years, advances in mass spectrometry (MS)-based protein identification with increased resolution and accuracy, sample preparation, and computational analysis have led to the identification and quantification of nearly all of the expressed proteins (Di Girolamo et al. 2013, Hu, Rampitsch and Bykova 2015). Further, an exponential increase in the number of RNA-seq-based transcriptome profiles and *de novo* transcriptome assemblies for diverse plant species has provided a rich database for proteome studies and thus extended protein identification to non-model plants without known genomic sequences (Lopez-Casado et al. 2012).

Several studies report the usability of proteomics in medicinal plants. For example, Desgagne-Penix et al. (2010) used transcriptome and proteome analysis to propose components of alkaloid metabolism in opium poppy cell cultures. Oldham et al. (2010) used a shotgun proteomics approach on yeast-elicited *Eschscholzia californica* suspension cultures to identify candidate proteins involved in the benzophenanthridine alkaloids. Miettinen et al. (2014) used proteomics approaches on epidermal, and mesophyll protoplasts isolated from *C. roseus* leaves to explore the localization of biosynthetic enzymes.

Further, recent technological advances and an explosion of proteomics workflows have significantly expanded proteomics. The approach promises advanced study of different aspects of proteins or protein-protein interactions and in the identification of enzymes involved in secondary metabolic pathways (Martinez-Esteso et al. 2015). In the near future more studies using proteomics-based approaches together with transcriptome and metabolome analysis in decoding the secondary metabolism in medicinal plants are expected.

4.2.4 Metabolomics Data

Metabolomics studies metabolites such as carbohydrates, lipids, amino acids, and organic acids present in a biological sample (Dixon and Strack 2003). The collection of all metabolites within a cell is called the metabolome. Metabolomics tools include chromatography for separating metabolites and spectroscopy techniques for their identification. Within the last decade, metabolomics has been extensively used as the principal tool for functional genomics, systems biology, and molecular breeding (Quanbeck et al. 2012, Rhee 2013), the reasons being advancement in improved mass resolution and sensitivity, high-throughput analysis, chromatography, and analytical methodologies, and increasingly available bioinformatics tools, databases, and other resources.

Metabolomics uses bioinformatics and statistical techniques (Madsen, Lundstedt and Trygg 2010). Metabolomics is a multidisciplinary topic involving biologists, computer scientists, and organic/analytical chemists. Metabolomics tools separate, identify, and quantify metabolites. These include chromatography, such as High-Performance Liquid Chromatography (HPLC), for separating metabolites and spectroscopy techniques, mainly nuclear magnetic resonance (NMR), and MS, for their identification. The metabolomics approach has the advantage of scoring the phenotype of two interacting organisms – host and pathogen. Phenotype is the expression of the genome modulated by the environment. When an organism gets sick or stressed, its phenotype is altered, thereby triggering specific molecular changes. This change may then, in principle, be measured using metabolomics (Lindon, Nicholson and Holmes 2011). Metabolomics has contributed to our present knowledge about metabolite diversity and bioactive compounds enriched in different medicinal plant species. Identification of bioactive metabolites such as paclitaxel (isolated from the bark of the Pacific yew tree and having anti-cancer properties), artemisinin (isolated from *Artemisia annua* and used as an anti-malarial drug), vincristine and vinblastine (isolated from *C. roseus* and used to treat leukemia and lymphoma), camptothecin (isolated from the bark of *Camptotheca accuminata* and having anti-cancer properties), and morphine (isolated from *P. somniferum*, indispensable in the treatment of severe pain) was made possible by metabolomics approaches.

A combined transcriptomic and metabolomic analysis of the medicinal and edible plant, *Chrysanthemum morifolium*, revealed the phenylpropanoid biosynthetic pathway and the phenylalanine metabolic pathway had the highest degree of differentially expressed genes and significantly changed metabolites enrichment, respectively (Zou et al. 2021). The development of metabolomics resources, databases, and bioinformatics toolsets has been one of the important catalysts for the increasing application of metabolomics in functional genomics and metabolic modeling. This includes the development of open-source bioinformatics tools, including XCMS (Mahieu, Genenbacher and Patti 2016), MS-Dial (Tsugawa et al. 2016), and MS-FINDER (Tsugawa et al. 2016), Mass++ (Tanaka et al. 2014).

4.3 META-ANALYSIS

To attain high-dimension biological insights using omics technologies, meta-analyses have played vital roles (Field and Hughes 2005, Bhandary et al. 2018). These analyses are nothing but collecting the available omics data and extracting a combined meaning out of them. In the last few decades, there have been several reports of meta-analysis in the form of scientific articles (Mentzen and Wurtele 2008, Sun et al. 2017, Sahu et al. 2019). With the invention of microarray technologies, the need for efficient meta-analysis techniques has increased. Initially, the main focus was to combine similar studies carried out at various times at different laboratories. However, gradually it became a powerful method for downstream analysis. Meta-analysis certainly created a different way of looking at the omics data. This part of the study was mainly categorized under the systems biology domain, and the analysis is mostly done with the help of computational approaches.

There are several databases that store a vast amount of high-throughput data, and these are freely accessible. Such an important database is the Sequence Read Archive (SRA). It stores primary next-generation sequence data as projects with all the experimental details (Leinonen, Sugawara and Shumway 2011). It is being functioned by the partnership between the three premier institutes, the National Center for Biotechnology Information (NCBI), European Bioinformatics Institute (EBI), and DNA Data Bank of Japan (DDBJ) under the International Nucleotide Sequence Database Collaboration (INSDC). The Gene Expression Omnibus (GEO) also contains transcriptome data, however, it started with the need to store microarray datasets (Barrett et al. 2013). The NCBI maintains this and the basic content of this database is storing gene expression datasets and the experimental details. Though initially the deposition of the actual data was missing for a number of datasets, gradually this database retained its importance, and researchers submitted their data. To date, GEO contains 148,194 numbers of series that have run on 22,054 platforms. GEO also started providing tools such as GEO2R and other interactive facilities to perform basic gene expression data analysis such as identifying differentially expressed genes (DEG). A parallel database is also maintained at EBI named ArrayExpress and contains functional genomics data (Parkinson et al. 2007). It mainly contains microarray datasets, and the current statistics of the data is with 74,159 experiments and 2,509,649 assays.

Metabolite profiling with the help of powerful LC-MS and GC-MS technologies has also created a good number of experimental data for the metabolites. Unlike other omics data, the metabolomics data are scattered across several databases. There are important databases which have either carried out excellent meta-analysis or sometimes provide scope for the same such as RiceFREND and AraNet (Sato et al. 2012, Lee and Lee 2017). RiceFREND provides gene co-expression network in the form of HyperTree for both single gene and multiple gene search options. These have been basically derived from the available rice microarray data meta-analysis. Similarly, the platform ArenaNet describes itself as a genome-scale functional system-level molecular network. In the case of protein data, databases such as STRING have been developed with the help of combining lots of resources with several statistical methods into one to provide protein-protein interaction data (Szklarczyk et al. 2011). Meta-analysis for metabolomics data has also been proven to be very informative.

Furthermore, the advancement of the computational approaches has enabled researchers to perform various multivariate analyses in order to gather maximum possible information from several omics data. There has also been an increased use of free source tools such as R in handling these data. A package such as Weighted Gene Coexpression Network Analysis (WGCNA) has taken the center stage for performing co-expression analysis for each type of omics data (Langfelder and Horvath 2008, Pei, Chen, and Zhang 2017). XCMS is yet another important package which has been of a great use for the metabolomics data analysis (Smith et al. 2006, Domingo-Almenara and Siuzdak 2020).

Most pre-processing steps consider the normalization of the data across samples and expressions of entities. The most common practice for omics studies is to find the differential expression patterns which involve two conditions. Most of the cases compare a control group and a treated/diseased/conditioned group. Interestingly, the differential analysis finds the value such as fold change considering the solo entity across different sample groups. In a comparative study between the two tulsi variants of *Ocimum tenuiflurum* (subtypes Krishna Tulsi and Rama Tulsi), the relative gene expressions were studied via transcriptome analysis (Upadhyay et al. 2015). This study suggested that minute changes in the amino acids of the key sites of genes of the metabolite synthesis pathways are the reasons for the special medicinal properties of tulsi. Certain genes, for example ursolic acid-producing genes, a likely drug against cancer, in young leaves of the Rama tulsi showed higher expression than the other variant. In a study, the transcriptomes of *A. indica* and *M. azedarach* were analyzed for the genes involved in the terpenoid synthesis pathway with special emphasis on azadirachtin, a well-known insecticide. It showed that all six DEGs of the terpenoid backbone biosynthesis pathway were up-regulated in the case of *A. indica* (Wang et al. 2016b).

Performing meta-analysis sometimes considers the published information for the differential expression if there is a specific aim or group of interest entities. However, most of the time, the meta-analysis tries to bring all the different datasets to the same experimental conditions and recompute the differential expression parts again. Several statistical methods have been used in the case of meta-analysis, and an integral part of the analysis is a correlation. It should conceptually provide conserved patterns for the biological entities such as genes, protein, and metabolites. There are also many cases where the correlation between single entities has been extended up to correlation between entity clusters. Primarily the correlation is being used as an initial step to understanding the nature of the data or compatibility between datasets. However, in many studies, it has been used for co-expression studies where researchers are interested in creating co-expression networks, mostly in the case of genes (Sanchita and Sharma 2016, Sun et al. 2017, Sahu et al. 2019, Tai et al. 2020, Sheng et al. 2020). These networks show the connection between genes having similar expression patterns in a given biological condition. The network represents genes (also other entities) as nodes and edges define if they are co-expressed. Often, the nodes in the networks carry weights based on the number of connections they have and/or other characteristics such as interconnectedness, eigen centrality, and so on.

A good number of studies documented similar kinds of statistical approaches that of meta-analysis with in-house data (Kim et al. 2011, Bhandari et al. 2018, Gonulalan et al. 2020). A study used quantitative trait locus mapping and genome-wide association mapping to find the count of polymorphic genetic loci within *Arabidopsis* (Joseph et al. 2014). The results of the study could not find the count, however, it concluded that there might be a need for at least a 1000 line bi-parental population to find the number of polymorphic loci that control metabolic variation. Interestingly, there are quite a good number of works reported where using transcriptome analysis, researchers have tried to infer metabolite information or carried out both the omics data analysis for better insight (Chang et al. 2019, Xu et al. 2019, Xu et al. 2020, Wei et al. 2020, Sun et al. 2020). An excellent report describes the distribution of saponin in the root tissues of two well-known medicinal plants, *Panax quinquefolius* and *Panax notoginseng* (Wei et al. 2020). A study by Sun et al. (2020) reported a beautiful co-expression network analysis to identify the transcriptional activator CaMYB48, which belongs to the MYB transcription factor family that regulates capsaicinoid in hot pepper. Another weighted gene co-expression network analysis by Chang and colleagues investigated the differential expressions for the biosynthesis of tanshinone in *Salvia miltiorrhiza* (Chang et al. 2019). They used WGCNA to construct the network and clustered modules based on the differential expression intensities and found the possible role of four newly identified cytochromes P450 (CYPs) in the tanshinone biosynthesis.

Functional annotation plays a significant role in several meta-analyses where the main aim is to classify or group the entities with similar functions. Gene Ontology is the most widely used functional annotation method for expression datasets. It offers functional information for the gene products into three categories: biological process, cellular component, and molecular function based on structured vocabularies in ontologies (GO Consortium 2001). In addition to these annotations, it has been common to identify the pathways for a cluster of entities and sometimes a single entity, depending on the need. Several extensive databases provide the pathway information and tools or access to link the entities to their pathways, such as the Kyoto encyclopedia of genes and genomes (KEGG) or MetaCyc (Caspi et al. 2010, Kanehisa et al. 2012). In addition, servers such as DAVID bioinformatics resources are uniquely implemented functional annotation platforms, which provide enriched biological knowledgebase and efficient data-mining environments (Huang da, Sherman and Lempicki 2009).

Since meta-analysis aims at drawing a holistic view from the available data, many researchers started combining differential expression and co-expression to interpret system-wide differential co-expression networks (Sahu et al. 2019). There are various efforts to connect various omics studies for similar biological samples to infer more into the genetic makeup and system-wide understanding of various regulatory circuits and network motifs. These are much more sophisticated ways of analyzing various data and often make it difficult to handle due to various dependencies and issues.

4.4 ISSUES IN OMICS DATA ANALYSIS

Though these technologies create huge data of potent, a lack of proper planning, skilled analysis, and scattered data often add to the ambiguity, issues, and even false information. Several issues include data unavailability, different experimental setups, platform-dependent datasets, required data analytic tools, and many more (Kauffmann et al. 2017, Čuklina, Pedrioli and Aebersold 2020). For example, several platforms from various companies provide their own probes, chips, kits, and so on, which are hugely difficult to analyze in a combined way. In most cases, these companies provide their own tools and pipelines to analyze data that preferably work well in their data format. However, performing a meta-analysis or even reproducing the results with other standard tools may provide different results. Different experimental setups are the worst kind of issue, making it hard to collate all data of specific interest. Microarray experiments have shown an unnecessary bias with respect to the quantity of samples loaded. Similarly, in the case of transcriptome analyses also, many researchers addressed the issue of variation in the data due to different experimental conditions such as the depth of RNA-Seq. Another issue is simply caused by the data in terms of the sample structure, such as the number of replicates. In some cases, there are important studies for which scientists have not considered enough biological replicates due to some reasons. This brings a notable variation in the basic analysis, such as normalization and differential expression. Analysis such as co-expression, in general, requires a certain minimum number of samples as it is not useful to calculate the correlation between only two samples. Sometimes, while combining datasets, the lack of enough common entities may not allow proceeding further. Additionally, the inability to bear the high expenses of the experiments also forces the researchers to yield low-quality data.

Proteomics data are comparatively complex and therefore are flimsier with the variations in sample preparation, data processing methods, and other experimental criteria (Shen et al. 2015). Though the MS technology has provided a tremendous amount of advancement in proteomics, it is often limited to quantitative analysis. The compromised quality is mostly a result of the process being a label-free method that does not have the facility of internal standards (Shen et al. 2014, Margolin et al. 2009).

As far as medicinal plants are concerned, these high-throughput technologies have been applied to only some of them. There are quite a few numbers of plants with important medicinal properties which have not been subjected to these kinds of technologies. This indicates the lack of proper data to analyze and the need for new data required at this moment to get into the genetic makeup of most medicinal plants. One of the reasons for this may be the huge number of species.

The most widely used analysis in the case of omics data is finding differential expression. Though the samples are prepared considering specific conditions, the differential expression is performed on the expression change of single entities between groups of samples. Most of the omics studies in the case of medicinal plants have tried to identify the differentially expressed genes and then moved on to a different analysis to confirm the DEGs, and so on. There are several omics data still lying on various public domains and waiting to be analyzed in a deeper means so that system-wide understanding can be increased.

Microarray-generated data have always been affected hugely by missing value. It is the most challenging step of the pre-processing omics data, especially while performing a meta-analysis. Though statisticians have developed a few suitable methods apart from the available methods at some point, they all become obsolete. It can be said that there is no correct way of performing missing value imputation, but choosing one from the available ones may help to evaluate the data in a better way. Another observation in medicinal plants is that the availability of omics data in the case of genomics and transcriptomics are more abundant when compared to proteomics and metabolomics. This may be why there have not been many meta-analyses carried out for the later ones in the case of medicinal plants.

4.5 FUTURE PROSPECTS

Omics technologies appeared as a boon to a scientific society, especially computational biologists. The individual experimental outputs will have the future scope of being considered in meta-analyses as long as they are available on public domains or even for collaborations. With advanced data analytics techniques such as artificial intelligence, the available omics data can be mined to offer enormous insights into uncovered information. Looking at the improvement in data analytics methods, we can aim to handle the omics data in a sophisticated manner and even join the dots between different types of omics data. Although new technologies may take over current omics techniques, such as microarray, the omics data will still have potency for getting new information. There must be extensive omics data creation as well as analysis in the medicinal plants. We hope to see technologies such as single-cell genomics employed to extract very important information which is still hidden.

The area of medicinal plant functional genomics can only be improved by developing new efficient methods to connect the different types of omics data which are genomics, transcriptomics, proteomics, and metabolomics, to infer and interpret the extracted information. It will be great if the omics studies by various research groups will maintain the scope of combining different studies. However, it is not that simple; for example, a research group or groups with similar interests (in case of a collaboration) can at least keep the biological sample the same for various omics studies, keeping its usefulness in the near future. It will eventually help them and other researchers to connect these studies to extract a piece of extensive, meaningful information.

REFERENCES

Atanasov, A.G., Waltenberger, B., Pferschy-Wenzig, E. M., Linder, T., Wawrosch, C., Uhrin, P., Temml, V., Wang, L., Schwaiger, S., Heiss, E.H., Rollinger, J. M., Schuster, D., Breuss, J. M., Bochkov, V., Mihovilovic, M. D., Kopp, B., Bauer, R., Dirsch, V. M. and Stuppner, H. 2015. Discovery and resupply of pharmacologically active plant-derived natural products: a review. *Biotechnol. Adv.* 33: 1582–1614.

Barrett, T., Wilhite, S. E., Ledoux, P., Evangelista, C., Kim, I. F., Tomashevsky, M., Marshall, K. A., Phillippy, K. H., Sherman, P. M., Holko, M., Yefanov, A., Lee, H., Zhang, N., Robertson, C. L., Serova, N., Davis, S., and Soboleva, A. 2012. NCBI GEO: archive for functional genomics data sets-update. *Nucleic Acids Res.* 41: D991–D995.

Bevan, M.W. and Uauy, C. 2013. Genomics reveals new landscapes for crop improvement. *Genome biology* 14: 206.

Bhandary, P., Seetharam, A. S., Arendsee, Z. W., Hur, M. and Wurtele, E. S. 2018. Raising orphans from a metadata morass: A researcher's guide to re-use of public 'omics data. *Plant Sci.* 267: 32–47.

Boutanaev, A.M., Moses, T., Zi, J., Nelson, D.R., Mugford, S.T., Peters, R.J., Osbourn, A. 2015. Investigation of terpene diversification across multiple sequenced plant genomes. *Proc Natl Acad Sci U S A.* 112: E81–8.

Bunsupa, S., Hanada, K., Maruyama, A., Aoyagi, K., Komatsu, K., Ueno, H., Yamashita, M., Sasaki, R., Oikawa, A., Saito, K. and Yamazaki, M. 2016. Molecular evolution and functional characterization of a bifunctional decarboxylase involved in lycopodium alkaloid biosynthesis. *Plant Physiol.* 171: 2432–2444.

Caspi, R., Altman, T., Dale, J.M., Dreher, K., Fulcher, C.A., Gilham, F., Kaipa, P., Karthikeyan, A.S., Kothari, A., Krummenacker, M., Latendresse, M., Mueller, L.A., Paley, S., Popescu, L., Pujar, A., Shearer, A.G., Zhang, P. and Karp, P.D. 2010. The MetaCyc database of metabolic pathways and enzymes and the BioCyc collection of pathway/genome databases. *Nucleic Acids Res.* 38: D473–D479.

Chang, Y., Wang, M., Li, J. and Lu, S. 2019. Transcriptomic analysis reveals potential genes involved in tanshinone biosynthesis in *Salvia miltiorrhiza*. *Sci. Rep.* 9: 1–12.

Chen, S., Luo, H., Li, Y., Sun, Y., Wu, Q., Niu, Y., Song, J., Lv, A., Zhu, Y., Sun, C., Steinmetz, A. and Qian, Z. 2011. 454 EST analysis detects genes putatively involved in ginsenoside biosynthesis in *Panax ginseng*. *Plant Cell Rep.* 30: 1593–1601.

Čuklina, J., Pedrioli, P.G. and Aebersold, R. 2020. Review of batch effects prevention, diagnostics, and correction approaches. *Methods Mol. Biol.* 2051: 373–387.

Desgagne-Penix, I., Khan, M.F., Schriemer, D.C., Cram, D., Nowak, J. and Facchini, P.J. 2010. Integration of deep transcriptome and proteome analyses reveals the components of alkaloid metabolism in opium poppy cell cultures. *BMC Plant Biol.* 10: 252.

Di Girolamo, F., Lante, I., Muraca, M. and Putignani, L. 2013. The role of mass spectrometry in the "Omics" era. *Curr. Org. Chem.* 17: 2891–2905.
Dixon, R.A. and Strack, D. 2003. Phytochemistry meets genome analysis, and beyond. *Phytochemistry* 62: 815–816.
Domingo-Almenara, X. and Siuzdak, G. 2020. Metabolomics data processing using XCMS. *Methods Mol. Biol.* 2104:11–24.
Fabricant, D.S. and Farnsworth, N.R. 2001. The value of plants used in traditional medicine for drug discovery. *Environ. Health Perspect.* 109: 69–75.
Field, D. and Hughes, J. 2005. Cataloguing our current genome collection. *Microbiology* 151: 1016–1019.
Field, B. and Osbourn, A.E. 2008. Metabolic diversification – independent assembly of operon-like gene clusters in different plants. *Science* 320: 543–547.
Gene Ontology Consortium. 2001. Creating the gene ontology resource: design and implementation. *Genome Res.* 11: 1425–1433.
Gonulalan, E.M., Nemutlu, E., Bayazeid, O., Koçak, E., Yalçın, F.N. and Demirezer, L.O. 2020. Metabolomics and proteomics profiles of some medicinal plants and correlation with BDNF activity. *Phytomedicine* 74: 152920.
Han, R., Rai, A., Nakamura, M., Suzuki, H., Takahashi, H., Yamazaki, M. and Saito, K. 2016. De novo deep transcriptome analysis of medicinal plants for gene discovery in biosynthesis of plant natural products. *Methods Enzymol.* 576: 19–45.
Hao, D.C., Chen, S.L., Xiao, P.G. and Liu, M. 2012. Application of high-throughput sequencing in medicinal plant transcriptome studies. *Drug Development Research* 73: 487–498.
Hao, D.C. and Xiao, P.G. 2015a. Genomics and evolution in traditional medicinal plants: road to a healthier life. *Evol. Bioinform. Online* 11: 197–212.
Hao, D.C., Chen, S.L., Osbourn, A., Kontogianni, V.G., Liu, L.W. and Jordán, M.J. 2015b. Temporal transcriptome changes induced by methyl jasmonate in *Salvia sclarea*. *Gene* 558: 41–53.
Hao, D.C., Vautrin, S., Song, C., Zhu, Y.J., Berges, H., Sun, C., and Chen, S. 2015c. The first insight into the *Salvia* Lamiaceae genome via BAC library construction and high-throughput sequencing of target BAC clones. *Pakistan J. Bot.* 47: 1347–1357.
Hao, D.C., Gu, X.J. and Xiao, P.G. 2015. Medicinal Plants: Chemistry, Biology and Omics, 1st ed. Oxford: Elsevier-Woodhead.
Hu, J., Rampitsch, C., and Bykova, N.V. 2015. Advances in plant proteomics toward improvement of crop productivity and stress resistance. *Front. Plant Sci.* 6: 209.
Huang da, W., Sherman, B.T., and Lempicki, R.A. 2009. Systematic and integrative analysis of large gene lists using DAVID bioinformatics resources. *Nature Protocols* 4: 44–57.
Joseph, B., Atwell, S., Corwin, J.A., Li, B. and Kliebenstein, D.J. 2014. Meta-analysis of metabolome QTLs in Arabidopsis: trying to estimate the network size controlling genetic variation of the metabolome. *Front. Plant Sci.* 5: 461.
Kanehisa, M., Goto, S., Sato, Y., Furumichi, M. and Tanabe, M. 2012. KEGG for integration and interpretation of large-scale molecular data sets. *Nucleic Acids Res.* 40: D109–D114.
Kauffmann, H.M., Kamp, H., Fuchs, R., Chorley, B.N., Deferme, L., Ebbels, T., Hackermüller, J., Perdichizzi, S., Poole, A., Sauer, U.G., Tollefsen, K.E., Tralau, T., Yauk, C. and van Ravenzwaay, B. 2017. Framework for the quality assurance of 'omics technologies considering GLP requirements. *Regulatory Toxicol. Pharmacol.* 91: S27–S35.
Kim, J., Choi, J.N., Ku, K.M., Kang, D., Kim, J.S., Park, J.H.Y. and Lee, C.H. 2011. A correlation between antioxidant activity and metabolite release during the blanching of *Chrysanthemum coronarium* L. *Biosci. Biotech. Biochem.*75: 674–680.
Langfelder, P. and Horvath, S. 2008. WGCNA: an R package for weighted correlation network analysis. *BMC Bioinfor.* 9: 1–13.
Lee, T. and Lee, I. 2017. AraNet: a network biology server for *Arabidopsis thaliana* and other non-model plant species. *Methods Mol. Biol.* 1629: 225–238.
Lei Z., Zhou C., Ji X., Wei G., Huang Y., Yu W., Luo, Y. and Qiu, Y. 2018. Transcriptome analysis reveals genes involved in flavonoid biosynthesis and accumulation in *Dendrobium catenatum* from different locations. *Sci. Rep.* 8: 6373.
Leinonen, R., Sugawara, H. and Shumway, M. International Nucleotide Sequence Database Collaboration. 2011. The sequence read archive. *Nucleic Acids Res.* 39: D19–D21.

Li Q., Ding G., Li B. and Guo S.X. 2017. Transcriptome analysis of genes involved in dendrobine biosynthesis in *Dendrobium nobile* Lindl. infected with mycorrhizal fungus MF23 (*Mycena* sp.). *Sci. Rep.* 7: 316.

Lindon, J.C., Nicholson, J.K. and Holmes, E. 2011. *The Handbook of Metabonomics and Metabolomics.* Amsterdam: Elsevier Science.

Lopez-Casado, G., Covey, P.A., Bedinger, P.A., Mueller, L.A., Thannhauser, T.W., Zhang, S., Fei, Z., Giovannoni, J.J. and Rose, J.K. 2012. Enabling proteomic studies with RNA-Seq: the proteome of tomato pollen as a test case. *Proteomics* 12: 761–774.

Madsen, R., Lundstedt, T. and Trygg, J. 2010. Chemometrics in metabolomics – A review in human disease diagnosis. *Anal. Chimica Acta* 659: 23–33.

Mahieu, N.G., Genenbacher, J.L. and Patti, G.J. 2016. A roadmap for the XCMS family of software solutions in metabolomics. *Curr. Opin. Chem. Biol.* 30: 87–93.

Margolin, A.A., Ong, S.E., Schenone, M., Gould, R., Schreiber, S.L., Carr, S.A. and Golub, T.R. 2009. Empirical Bayes analysis of quantitative proteomics experiments. *PloS One* 4: e7454.

Martinez-Esteso, M.J., Martinez-Marquez, A., Selles-Marchart, S., Morante-Carriel, J.A. and Bru-Martinez, R. 2015. The role of proteomics in progressing insights into plant secondary metabolism. *Front. Plant Sci.* 6: 504.

Matsuba, Y., Nguyen, T.T., Wiegert, K. Falara, V., Gonzales-Vigil, E., Leong, B., Schäfer, P., Kudrna, D., Wing, R.A., Bolger, A.M., Usadel, B., Tissier, A., Fernie, A.R., Barry, C. S. and Pichersky, E. 2013. Evolution of a complex locus for terpene biosynthesis in *Solanum*. *Plant Cell* 25: 2022–2036.

Mentzen, W.I.and and Wurtele, E.S. 2008. Regulon organization of *Arabidopsis*. *BMC Plant Biol.* 8: 99.

Miettinen, K., Dong, L., Navrot, N., Schneider, T., Burlat, V., Pollier, J., Woittiez, L., van der Krol, S., Lugan, R., Ilc, T., Verpoorte, R., Oksman-Caldentey, K. M., Martinoia, E., Bouwmeester, H., Goossens, A., Memelink, J. and Werck-Reichhart, D. 2014. The seco-iridoid pathway from *Catharanthus roseus*. *Nat. Commun.* 5: 3606.

Morrison, N., Cochrane, G., Faruque, N., Tatusova, T., Tateno, Y., Hancock, D. and Field, D. 2006. Concept of sample in OMICS technology. *Omics: A Journal Integ. Biol.* 10: 127–137.

Moses, T., Pollier, J., Shen, Q., Soetaert, S., Reed, J., Erffelinck, M.L., Van Nieuwerburgh, F. C., Vanden Bossche, R., Osbourn, A., Thevelein, J. M., Deforce, D., Tang, K. and Goossens, A. 2015. OSC2 and CYP716A14v2 catalyze the biosynthesis of triterpenoids for the cuticle of aerial organs of *Artemisia annua*. *Plant Cell* 27: 286–301.

Oldham, J.T., Hincapie, M., Rejtar, T., Wall, P.K., Carlson, J.E. and Lee-Parsons, C.W. 2010. Shotgun proteomic analysis of yeast-elicited California poppy *Eschscholzia californica* suspension cultures producing enhanced levels of benzophenanthridine alkaloids. *J. Proteome Res.* 9: 4337–4345.

Parkinson, H., Kapushesky, M., Shojatalab, M., Abeygunawardena, N., Coulson, R., Farne, A., Holloway, E., Kolesnikov, N., Lilja, P., Lukk, M., Mani, R., Rayner, T., Sharma, A., William, E., Sarkans, U. and Brazma, A. 2007. ArrayExpress – a public database of microarray experiments and gene expression profiles. *Nucleic Acids Res.* 35: D747–D750.

Pei, G., Chen, L. and Zhang, W. 2017. WGCNA application to proteomic and metabolomic data analysis. *Meth. Enzymol.* 585: 135–158.

Quanbeck, S.M., Brachova, L., Campbell, A.A. Guan, X., Perera, A., He, K., Rhee, S.Y., Bais, P., Dickerson, J.A., Dixon, P., Wohlgemuth, G., Fiehn, O., Barkan, L., Lange, I., Lange, B.M., Lee, I., Cortes, D., Salazar, C., Shuman, J., Shulaev, V., Huhman, D.V., Sumner, L.W., Roth, M.R., Welti, R., Ilarslan, H., Wurtele, E.S. and Nikolau, B.J. 2012. Metabolomics as a hypothesis-generating functional genomics tool for the annotation of *Arabidopsis thaliana* genes of "Unknown Function." *Front. Plant Sci.* 3: 15.

Rai, A., Nakamura, M., Takahashi, H., Suzuki, H., Saito, K. and Yamazaki, M. 2016. High-throughput sequencing and de novo transcriptome assembly of Swertia japonica to identify genes involved in the biosynthesis of therapeutic metabolites. *Plant Cell Rep.* 35: 2091–2111.

Rai, A., Saito, K. and Yamazaki, M. 2017. Integrated omics analysis of specialized metabolism in medicinal plants. *Plant J.* 90: 764–787.

Rhee, K. 2013. Minding the gaps: metabolomics mends functional genomics. *EMBO Rep.* 14: 949–950.

Sahu, J., Panda, D., Baruah, G., Patar, L., Sen, P., Borah, B.K. and Modi, M.K. 2019. Revealing shared differential co-expression profiles in rice infected by virus from Reoviridae and sequiviridae group. *Gene* 698: 82–91.

Sanchita, Sharma, A. 2016. Computational gene expression profiling under salt stress reveals patterns of co-expression. *Genomics Data* 7: 214–221.

Sato, Y., Namiki, N., Takehisa, H., Kamatsuki, K., Minami, H., Ikawa, H., Ohyanagi, H., Sugimoto, K., Itoh, J., Antonio, B.A. and Nagamura, Y. 2013. RiceFREND: a platform for retrieving co-expressed gene networks in rice. *Nucleic Acids Res.* 41: D1214–D1221.

Shen, X., Young, R., Canty, J.M. and Qu, J. 2014. Quantitative proteomics in cardiovascular research: global and targeted strategies. *Proteomics Clin. Applications* 8: 488–505.

Shen, X., Hu, Q., Li, J., Wang, J. and Qu, J. 2015. Experimental null method to guide the development of technical procedures and to control false-positive discovery in quantitative proteomics. *J. Proteome Res.* 14: 4147–4157.

Shen, C., Guo, H., Chen, H., Shi, Y., Meng, Y., Lu, J., Feng, S. and Wang, H. 2017. Identification and analysis of genes associated with the synthesis of bioactive constituents in *Dendrobium officinale* using RNA-Seq. *Sci. Rep.* 7: 187.

Sheng, M., She, J., Xu, W., Hong, Y., Su, Z. and Zhang, X. 2020. HpeNet: co-expression network database for de novo transcriptome assembly of *Paeonia lactiflora* Pall. *Front. Gen.* 11: 570138.

Smanski, M.J., Zhou, H., Claesen, J., Shen, B., Fischbach, M.A. and Voigt, C.A. 2016. Synthetic biology to access and expand nature's chemical diversity. *Nat. Rev. Microbiol.* 14: 135–149.

Smith, C.A., Want, E. ., O'Maille, G., Abagyan, R. and Siuzdak, G. 2006. XCMS: Processing mass spectrometry data for metabolite profiling using nonlinear peak alignment, matching, and identification. *Anal. Chem.* 78: 779–787.

Sun, W., Wang, B., Yang, J., Wang, W., Liu, A., Leng, L., Xiang, L., Song, C. and Chen, S. 2017. Weighted gene co-expression network analysis of the dioscin rich medicinal plant *Dioscorea nipponica. Front. Plant Sci.* 8: 789.

Sun, B., Zhou, X., Chen, C., Chen, C., Chen, K., Chen, M., Liu, S., Chen, G., Cao, B., Cao, F., Lei, J. and Zhu, Z. 2020. Co-expression network analysis reveals an MYB transcriptional activator involved in capsaicinoid biosynthesis in hot peppers. *Hort. Res.* 7: 162.

Swaminathan, S., Morrone, D., Wang, Q., Fulton, D.B. and Peters, R.J. 2009. CYP76M7 is an ent-cassadiene C11alpha-hydroxylase defining a second multifunctional diterpenoid biosynthetic gene cluster in rice. *Plant Cell* 21: 3315–3325.

Szklarczyk, D., Franceschini, A., Kuhn, M., Simonovic, M., Roth, A., Minguez, P., Doerks, T., Stark, M., Muller, J., Bork, P., Jensen, L.J. and von Mering, C.V. 2010. The STRING database in 2011: functional interaction networks of proteins, globally integrated and scored. *Nucleic Acids Res.* 39: D561–D568.

Tai, Y., Ling, C., Wang, C., Wang, H., Su, L., Yang, L., Jiang, W., Yu, X., Zheng, L., Feng, Z., Liu, C. and Yuan, Y. 2020. Analysis of terpenoid biosynthesis pathways in German chamomile Matricaria recutita and Roman chamomile *Chamaemelum nobile* based on co-expression networks. *Genomics* 112: 1055–1064.

Takos, A.M., Knudsen, C., Lai, D. Kannangara, R., Mikkelsen, L., Motawia, M.S., Olsen, C.E., Sato, S., Tabata, S., Jørgensen, K., Møller, B.L. and Rook, F. 2011. Genomic clustering of cyanogenic glucoside biosynthetic genes aids their identification in *Lotus japonicus* and suggests the repeated evolution of this chemical defence pathway. *Plant J.* 68: 273–286.

Tanaka, S., Fujita, Y., Parry, H.E. Yoshizawa, A.C., Morimoto, K., Murase, M., Yamada, Y., Yao, J., Utsunomiya, S.I., Kajihara, S., Fukuda, M., Ikawa, M., Tabata, T., Takahashi, K., Aoshima, K., Nihei, Y., Nishioka, T., Oda, Y. and Tanaka, K. 2014. Mass++: a visualization and analysis tool for mass spectrometry. *J. Proteome Res.* 13: 3846–3853.

Thakur, V., Bains, S., Pathania, S., Sharma, S., Kaur, R. and Singh, K. 2020. Comparative transcriptomics reveals candidate transcription factors involved in costunolide biosynthesis in medicinal plant-*Saussurea lappa. Int. J. Biol. Macromolecules* 150: 52–67.

Tohge, T., Wendenburg, R., Ishihara, H. Nakabayashi, R., Watanabe, M., Sulpice, R., Hoefgen, R., Takayama, H., Saito, K., Stitt, M. and Fernie, A.R. 2016. Characterization of a recently evolved flavonol-phenylacyltransferase gene provides signatures of natural light selection in *Brassicaceae. Nat. Commun.* 7: 12399.

Tsugawa, H., Kind, T., Nakabayashi, R., Yukihira, D., Tanaka, W., Cajka, T., Saito, K., Fiehn, O. and Arita, M. 2016. Hydrogen rearrangement rules: computational MS/MS fragmentation and structure elucidation using MS-FINDER software. *Anal. Chem.* 88: 7946–7958.

Unamba, C.I., Nag, A. and Sharma, R.K. 2015. Next-generation sequencing technologies: the doorway to the unexplored genomics of non-model plants. *Front. Plant Sci.* 6: 1074.

Upadhyay, A.K., Chacko, A.R., Gandhimathi, A., Ghosh, P., Harini, K., Joseph, A.P., Joshi, A.G., Karpe, S.D., Kaushik, S., Kuravadi, N., Lingu, C.S., Mahita, J., Malarini, R., Malhotra, S., Malini, M., Mathew, O.K., Mutt, E., Naika, M., Nitish, S., Pasha, S.N., Raghavender, U.S., Rajamani, A., Shilpa, S., Shingate, P.N., Singh, H.R., Sukhwal, A., Sunitha, M.S., Sumathi, M., Ramaswamy, S., Gowda, M. and Sowdhamini, R. 2015. Genome sequencing of herb Tulsi (*Ocimum tenuiflorum*) unravels key genes behind its strong medicinal properties. *BMC Plant Biol.* 15: 212.

Wang, Y., Chen, X., Wang, J., Xun, H., Sun, J. and Tang, F. 2016a. Comparative analysis of the terpenoid biosynthesis pathway in *Azadirachta indica* and *Melia azedarach* by RNA-seq. *Springerplus* 5: 819.

Wang, Y.N., Tang, L., Hou, Y., Wang, P., Yang, H. and Wei, C.L. 2016b. Differential transcriptome analysis of leaves of tea plant (*Camellia sinensis*) provides comprehensive insights into the defense responses to *Ectropis oblique* attack using RNA-Seq. *Func. Integr. Genomics* 16: 383–398.

Wei, G., Yang, F., Wei, F., Zhang, L., Gao, Y., Qian, J., Chen, Z., Jia, Z., Wang, Y., Su, H., Dong, L., Xu, J. and Chen, S. 2020. Metabolomes and transcriptomes revealed the saponin distribution in root tissues of *Panax quinquefolius* and *Panax notoginseng*. *J. Ginseng Res.* 44: 757–769.

Xu, Y., Zhu, C., Xu, C., Sun, J., Grierson, D., Zhang, B. and Chen, K. 2019. Integration of metabolite profiling and transcriptome analysis reveals genes related to volatile terpenoid metabolism in finger citron (*C. medica* var. sarcodactylis). *Molecules* 24: 2564.

Xu, P., Su, H., Zhao, S., Jin, R., Cheng, H., Xu, A., Lai, W., Yin, X. and Wang, Y. 2020. Transcriptome and phytochemical analysis reveals the alteration of plant hormones, characteristic metabolites, and related gene expression in tea (*Camellia sinensis* L.) leaves during withering. *Plants (Basel).* 9: 204.

Yao, X., Peng, Y., Xu, L.J., Li, L., Wu, Q.L. and Xiao, P.G. 2011. Phytochemical and biological studies of Lycium medicinal plants. *Chem. Biodiversity* 8: 976–1010.

Yarnell, E. and Abascal, K. 2002. Dilemmas of traditional botanical research. *Herbal Gram.* 55: 46–54.

Zhao, Y., Yin, J., Guo, H., Zhang, Y., Xiao, W., Sun, C., Wu, J., Qu, X., Yu, J., Wang, X. and Xiao, J. 2015. The complete chloroplast genome provides insight into the evolution and polymorphism of *Panax ginseng*. *Front. Plant Sci.* 5: 696.

Zou, Q., Wang, T., Guo, Q., Yang, F., Chen, J., and Zhang, W. 2021. Combined metabolomic and transcriptomic analysis reveals redirection of the phenylpropanoid metabolic flux in different colored medicinal *Chrysanthemum morifolium*. *Ind. Crops Prod.* 164: 113343.

5 An Overview of Genomics, Transcriptomics and Proteomics Approaches for Medicinal Plants Research

Anindita Sengupta, Faizia Bano, Neha Arora, Saima Wajid and Mohd Mughees

CONTENTS

5.1 INTRODUCTION

With the invention of genotyping arrays, high-throughput technologies revolutionized medical research and enabled large-scale genome-wide association studies and methods for examining global transcript levels, giving rise to the field of "integrative genetics." Other omics technologies, such as transcriptomics and proteomics, have emerged and revolutionized biological researchers' everyday methodology.

Genomics is defined as the comprehensive study of genetic information of a cell or organism, including gene sequences in living organisms, the function of specific genes, the interactions among them and the control and regulation of gene expression through activation and suppression of target genes. Biotechnological techniques are regarded as one of the most trustworthy resources for elucidating the processes that underpin stress tolerance and instilling desired features in essential crops that confer stress resistance. Recent scientific advancements have greatly altered gene identification and functional genomics capabilities, allowing for comprehensive transcriptomic, proteomic and metabolomic profiling of a cell (Cramer et al. 2011). This type of biotechnological approach has pushed forward the identification of stress-responsive and tolerant genes.

In most circumstances, controlling the external environment in favor of better plant production is extremely difficult, if not impossible. As a result, we try to improve conventional crop plants by

DOI: 10.1201/9781003179139-5

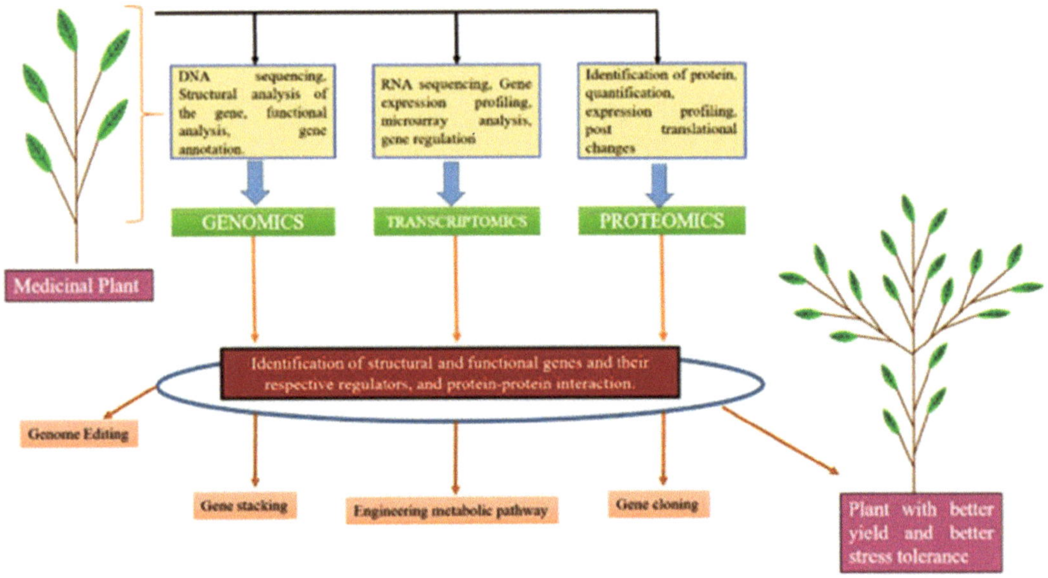

FIGURE 5.1 Diagram indicating the role of genomics, transcriptomics and proteomics and their role in enhancing medicinal plant yield.

breeding procedures or biotechnology approaches, or we use our limited competence in cultural practices. Plant breeding's main goal is to create plants that produce bigger yields and have superior quality, stability and stress resistance (Bray 2004). Stress tolerance, according to the current consensus, is multigenic and quantitative in nature, making it difficult to instill in plants using currently accessible approaches (Collins, Tardieu and Tuberosa 2008). The molecular mechanisms behind plant abiotic stress tolerance are being uncovered, but a complete understanding of how plants interpret environmental signals and how these signals are relayed to cellular machinery to activate adaptive responses is still a work in progress.

Sequencing and functional genomics are currently thought to be critical for understanding stress signal reception and transduction in relation to molecular regulatory networks involved in stress responses (Heidarvand and Amir 2010). Polyomics methods such as transcriptomics, proteomics, metabolomics, ionomics and micromics are vital for uncovering new genes, proteins and secondary plant metabolites that are involved in plant stress adaption (Mochida and Shinozaki 2011) (Figure 5.1).

5.2 GENOMICS

With the advent of technologies that permitted whole-genome sequencing, the complete genomes of several important plant species have been published, leading to the improvement in crop yield and the discovery of new targets for further improvement. In the past years, many plants having medicinal importance have been sequenced, leading to new information regarding the metabolites that are produced. The Arabidopsis Genome Initiative established the framework for a more extensive analysis of conserved processes across all eukaryotes by sequencing the model plant *Arabidopsis thaliana* and identifying many genes with plant-specific activities (Kaul 2000).

The characterization of many genes and their functions in metabolic and cellular processes is being revealed (Bevan and Uauy 2013) using whole-genome sequencing and RNA-sequencing to characterize the genes and elucidate their functions in the regulation of various aspects of cellular and metabolic processes. Whole-genome sequencing of a number of medicinal plants has been accomplished so far, with examples such as *Calotropis gigantea*, an important anticancer and anti-malarial drugs. About 18,917 genes were annotated using an ab initio, evidence-driven pipeline (Hoopes et al. 2018). Another plant that has been put under the trial of the omics approach: *Capsicum anuum*, an important source of capsaicinoid complex, which includes capsaicin and dihydrocapsaicin by two groups working independently of each, which led to the identification of candidate genes involved in their biosynthesis (Kim et al. 2014, Qin et al. 2014).

The next critical step, given the quantity of information collected by these genome projects, is to figure out what all of the newly identified genes perform. The field of functional genomics is a young one, with the goal of determining the function of all the transcribed sequences in the genome (the transcriptome) and all the proteins that are produced (the proteome).

5.3 TRANSCRIPTOMICS

The transcriptome is the complete set of RNA transcripts found in cells or tissues, and it reflects the genes that are expressed at different stages of life, tissue types, physiological states and environmental factors (Wang et al. 2015). The transcriptomics study of medicinal plants is gradually becoming the most active space in the genome research of medicinal plants. The transcriptome not only provides a perspective on the encoded genes, it also answers questions pertaining to subjects such as the evolution of the genome, ecology, cross-breeding, and so on, as the profiling of the Spatio-temporal expression of genes in plants provides subtle hints pertaining to the functional aspect of genomics. The first method employed for the transcriptomic study was the use of expressed sequence tag (EST) (Zhao et al. 2013), which had limited coverage and was unable to identify the weakly expressed genes, and was soon used in tandem with probe hybridization-based microarray technologies which provide increased coverage for gene discovery and analysis of expression (Kiyama 2017). With the advent of NGS (Next Generation Sequencing), many projects were commissioned where the examination of an organ or tissue-specific expression pattern was undertaken (Vashisht et al. 2016, Amini et al. 2019). The transcriptome can also be employed to study changes in gene expression induced by an external stimulus or an environmental factor, as it can accurately quantify the RNA expression level, thereby carrying out a comparative analysis between species or cells for the expression of a metabolite. Moreover, the process of gene annotation becomes easier for identification in plants that may have similar activities. By comparing transcripts to reference genomes, possible SNPs or SSRs can be identified, reducing the expense of research (González-Díaz et al. 2008).

An application of transcriptomics for medicinal plants can be the mining of functional genes with the transcriptome, representing a gene expression profile. Further, the isolation and cloning of a novel functional gene can be carried out, without the need for a complete analysis of the genome. For most medicinal plants a genome-wide study is not possible therefore transcriptomics-based approach acts as a faster alternative via the construction of cDNA libraries for EST analysis. These libraries are then used to identify secondary metabolite genes of medicinal plants for their function and identity. One such example is the study conducted by Guo et al. in 2014 on the roots of *Salvia miltiorrhiza*, analyzing the tanshione biosynthesis pathway by employing a combinatorial approach using transcriptomics and metabolomics to identify 39 genes for cytochrome P450 that were found upregulated in tanshione biosynthesis in roots. The study, coupled with a previous study, showed that 2,863 unigenes were found to be overexpressed in roots which included the copalyl diphosphate synthase (SmCPS), kaurene synthase-like (SmKSL) and CYP76AH1, all of which are key enzymes in the tanshione biosynthesis pathway (Yang et al. 2013).

5.3.1 Targets of Research

Once the genome sequencing is completed, the next target in identifying potential functionality is the correct annotation of the genes, accomplished via the bioinformatics-based approach. Important targets for such a genome analysis are homologous genes and duplication events. Homologous genes act as sights that have been subjected to mutations and thus do not have similar nucleotide sequences but are located on similar sites denoting some level of functional similarity. The first stage in functional genomic research is genome sequencing, which is followed by methods that rely on bioinformatics analysis. In the field of gene annotation, homology is a crucial notion. As homologous genes are subjected to independent random mutations after their occurrence, they do not have the same nucleotide sequence in general. Homologous genes, on the other hand, have comparable sequences and most nucleotides with no mutations are in the same place, so they can be used as references when searching and comparing different plant genomes for homologous genes. We can identify gene families and the number of copies they have by grouping these homologous genes, which is a key aspect in understanding the various levels of secondary metabolites produced in medicinal plant sections and their expression in tissues. Moreover, the genes and gene families can be employed as tools to analyze the phylogenetic analysis between species, and the time of species divergence can be further analyzed to reveal the evolution of species. Although in order to understand the gene products, we need to develop the transcriptome, which provides splicing forms in transcription, which will generate sequences of mRNA as per the different exon combinations that may exist in the cell to form different proteins.

5.3.2 Medicinal Plants Genomes

5.3.2.1 *Panax ginseng*

Belonging to the family *Araliaceae* the plant *Panax ginseng* has been associated with medicinal properties in old texts and has been attributed to act as an antioxidant, anti-inflammatory, vasorelaxor, antiallergic, antidiabetic and an anticancer agent (Kim 2018). Of all the probable bioactive compounds, ginsenosides are thought to be the most important class of molecules related to the bioactivity of the plant metabolites. The complete sequenced data of the plant was published by Gigascience in 2017 (Xu, Choi and Huang 2017), with the genome size found to be 3.2–3.6 GB (Choi et al. 2014), with approximately 42,000 possible proteins, having a high number of repeats, with around 62% of the genes repeated and a high degree of complexity. On combining the data with ESTs and combinatorial analysis with transcriptome data from tissues, it was reported that for the production of the ginsenosides, the enzymes of the MVA (mevalonate) pathway exist in multiple copies. The MVA's rate-limiting enzyme HMGR was reported to have eight copies in the ginseng genome (Xu, Choi and Huang 2017). Moreover, the glycosyltransferases involved in the synthesis of ginsenosides were modified, with a total of 225 forms identified in ginseng, making it the largest enzyme family in the plant. The glycosyltransferases genes have tissue-specific expression; therefore, they can be used to screen the glycosyltransferases genes implicated in ginsenoside synthesis. The targeting of the pathways involved in the production of ginsenosides can thus be targeted to increase the yield of the metabolites (Chu, Montecillo and Bae 2020).

5.3.2.2 *Papaver somniferum*

Papaver somniferum (also known as poppy) is a medicinal plant whose alkaloids, such as morphine, thebaine and codeine, have a relaxing and pain-relieving effect. The poppy genome was published in 2018 and was estimated to be 2.72 GB in size, with a 7.8-million-year-old doubling event (Hu et al. 2018).

At the same time, the researchers discovered a benzylisoquinoline alkaloid (BIA) gene cluster in a 584-kb region on chromosome 11, which includes a noscapine synthetic gene cluster, a (S)-reticuline to (R)-reticuline catalytic enzyme gene and four genes involved in the morphine biosynthesis pathway, all of which are co-published in the stem to produce morphine (Guo et al. 2018). The emergence of benzylisoquinoline alkaloid anabolism appears to be aided by chromosomal rearrangement, according to the findings.

5.3.3.3 *Artemisia annua*

Artemisia annua has been one of the most studied plants worldwide due to the wide array of bioactive properties it possesses, confirmed to have immune effects, antimalarial, antiparasitic, antibacterial and antipyretic. The active component artemisinin has been used to develop many therapeutic drugs and has a series of derivatives, with its role as an antimalarial being indispensable.

With the publishing of a very high-quality genome of *A. annua* by Shen et al. in 2019, it was shown that the genome of *A. annua* had a high number of repetitions, estimated to be around 61.75%, with 63,226 projected protein-coding genes (Shen et al. 2018). The plant's transcriptome research revealed that there were 871 distinct gene families in the Compositae family, with the genetic code serving as a crucial tool for obtaining information on the artemisinin. The study also obtained transgenic lines of Artemisia with elevated levels of artemisinin by overexpressing multiple gene families HMGR, FPS and DBR2 located at different positions in the artemisinin biosynthetic pathway (Shen et al. 2019). Another gene that was reported by the study by Shen et al. (2019) was the AaMYB2 transcription factor, which, when overexpressed in a close cluster with the artemisinin biosynthesis specific genes, yields artemisinin and dihydroartemisinic acid content.

5.3.3.4 *Aloe vera*

A perennial tropical plant with succulent elongated leaves from the family Asphodelaceae, *Aloe vera*, was completely sequenced by Jaiswal et al. (2021). The plant that has been termed as the "King of medicinal plants" in Ayurveda was found to have a 13.83 GB genome, the largest for an Angiosperm and fifth largest overall, and a total of 86,177 genes having unique characteristics, such as drought resistance and possessing numerous medicinal properties. The revelation of signs of adaptive evolution, specifically in the genes involved in drought stress tolerance, CAM pathway and circadian rhythm in plants, all contribute to the unique phenotype of the species. The study also reported the genes responsible for DNA damage repair to be positively selected in the plant, which aids the better survival of the plant in conditions of stress (Jaiswal et al. 2021).

5.4 PROTEOMICS

The term proteomics can be described as the study of all proteins expressed in a genome, cell, or tissue (Graves and Haystead 2002). Proteomics is a field that helps us understand the expression and functional patterns of all proteins in organisms, including protein expression, post-translational modification, protein structure, protein-protein interactions, and so on (Amiri-Dashatan et al. 2018). Plant proteomics came into the fray with the advancement in genomics: as the number of sequenced plants increased, plant proteomics slowly became an important hotspot as the proteome can aid us in understanding the metabolism and the development of plants. In reference to medicinal plants, proteomics lends importance to identifying how these plants respond to biotic and abiotic stresses, their repair mechanism and ecological adaptations. The proteome also aids in singling out the enzymes that are responsible for the production of a particular secondary metabolite, specifically one which may have some bioactive ability (Wang et al. 2012).

The identification and analysis of proteins on a wide scale utilizing bidirectional electrophoresis and other technologies is referred to as expression proteomics. It is utilized to investigate protein abundance and modification under various settings, as well as to create expression maps in tissues and cells. It involves a differentiation study of protein expression profiles in various developmental stages and settings, as well as the total protein expression profile of cells or tissues. The occurrence of large-scale protein three-dimensional structures with high flux is referred to as structural proteomics (Manzoni et al. 2018). It can more precisely anticipate binding sites and structural domains by analyzing the spatial structure of protein molecules. This allows researchers to predict protein catalytic processes and pathways, as well as protein interactions and, as a result, protein-supermolecule interactions. Because the study object analyzes and expounds the pathways or modes of action of particular proteins, functional proteomics focuses on specific protein groups. The proteomic techniques which confer the analysis of bioactive compounds target the measurement of systemic changes in cellular metabolism and thereby decode the synthesis of proteins and the pathways that produce the bioactive compounds and their health-promoting properties (Hashiguchi, Tian and Komatsu 2017).

5.5 NEW RESEARCHES

Zhu et al. (2015) analyzed the cape periwinkle or *Catharanthus roseus*, a plant in the family *Apocynaceae*, producing many indole-ringed-alkaloids with significant pharmaceutical relevance, using proteomic techniques under conditions of stress. According to Zhu, proteins related to the tricarboxylic acid cycle (TCA) and the cell wall were elevated, particularly the enzyme 10-hydroxygeraniol oxidoreductase, which is implicated in the manufacture of indole alkaloids.

Another plant of medicinal importance is the *Panax ginseng*, a perennial herb used to treat physiological disorders like diabetes, cardiovascular diseases, cancer and neurodegeneration (Xu et al. 2017). As discussed earlier, the active components in ginseng are ginsenosides found almost exclusively in *Panax* spp. (Leung and Wong 2010). Due to the long life cycle of ginseng increasing the ginsenoside yield becomes important cultivation (Mathur et al. 1999), and understanding the biosynthesis routes of ginsenosides is essential. Using proteome analysis, Ma et al. (2013) discovered that ginsenosides are generated when the plant enters a sluggish growth phase. A more recent proteomic analysis also showed increased ginsenoside content in older plants (Ma et al. 2016), with a correlation established via metabolic profiling, which showed a correlation with energy metabolism (Liu et al. 2017). The quantity of the 6-phosphogluconate dehydrogenase family protein was also linked to the radical scavenging action of ginseng fruits, demonstrating the utility of proteomic approaches in identifying essential enzymes important for medicinal plants' biological functions (Kim et al. 2016). Other medicinal plants that have been studied in recent years using proteomics include *Rabdosia rubrescens* which produces the bioactive compound oridonin, known to act against a protein stathmin involved in hepatocellular carcinoma cells via 2D- electrophoresis (Wang et al. 2011) and *Cucurma aromatica* which gives curdione which has antithrombotic activities (Zhang et al. 2017).

5.6 CONCLUSION

Medicinal plants have been a potential target for new drugs and therapeutics since ancient times when the nature of their therapeutic quality was not known. However, with the advances in science, we now know the action of some plants against various diseases and disorders was due to the fact that they could synthesize special chemical compounds termed bioactive molecules which have the ability to counter these diseases. As knowledge expanded and the use of modern science progressed, we found a way to harvest these compounds, but due to the nature of the biosynthetic pathway they were involved in they were either limited in yield or remained unidentified. Thus, the use of the plant genetic resources had to be undertaken as a study in order to identify these hidden pathways and increase the yield. The omics concept thus helps in identifying their hidden traits and the functionality that they possess with genomics using its power of sequencing to unravel the complete coded set of nucleotides, transcriptomics aiding in identifying the functional portion from the non-functional, and finally proteomics identifying the end products that play a role directly in the pathways. Plant genome sequencing and identification of genes, transcripts and enzymes have aided in elucidating the multistep, complicated biosynthetic pathways of plant bioactive compounds. External factors that affect the pathways, as well as the physiological changes caused by the factors, are described using omics analyses of plants. By expanding the number of instances of multidisciplinary collaborative study, these research disciplines can be combined to help advance medicinal plant research.

REFERENCES

Amini, H., Naghavi, M. R., Shen, T., Wang, Y., Nasiri, J., Khan, I. A., Fiehn, O., Zerbe, P. and Maloof, J. N. 2019. Tissue-specific transcriptome analysis reveals candidate genes for terpenoid and phenylpropanoid metabolism in the medicinal plant *Ferula assafoetida. G3 (Bethesda)* 9(3): 807–816.

Amiri-Dashatan, N., Koushki, M., Abbaszadeh, H. A., Rostami-Nejad, M. and Rezaei-Tavirani, M. 2018. Proteomics applications in health: Biomarker and drug discovery and food industry. *Iran J. Pharm. Res.* 17(4): 1523–1536.

Bevan, M.W. and Uauy, C. 2013. Genomics reveals new landscapes for crop improvement. Genome Biol. 14: 206.

Bray, E. A. 2004. Genes commonly regulated by water-deficit stress in *Arabidopsis thaliana. J. Exp. Bot.* 55(407): 2331–2341.

Choi, H. I., Waminal, N. E., Park, H. M., Kim, N. H., Choi, B. S., Park, M., Choi, D., Lim, Y. P., Kwon, S. J., Park, B. S., Kim, H. H. and Yang, T. J. 2014. Major repeat components covering one-third of the ginseng (*Panax ginseng* C.A. Meyer) genome and evidence for allotetraploid. *Plant J.* 77(6): 906–916.

Chu, L. L., Montecillo, J. A. V. and Bae, H. 2020. Recent advances in the metabolic engineering of yeasts for ginsenoside biosynthesis. *Front. Bioeng. Biotechnol.* 8: 139.

Collins, N. C., Tardieu, F. and Tuberosa, R. 2008. Quantitative trait loci and crop performance under abiotic stress: where do we stand? *Plant Physiol.* 147(2): 469–486.

Cramer, G. R., Urano, K., Delrot, S., Pezzotti, M. and Shinozaki, K. 2011. Effects of abiotic stress on plants: a systems biology perspective. BMC Plant Biology. 11: 163.

Graves, P. R. and Haystead, T. A. 2002. Molecular biologist's guide to proteomics. *Microbiol Mol. Biol. Rev.* 66(1): 39–63.

González-Díaz, H., Gonzâlez-Díaz, Y., Santana, L., Ubeira, F. M. and Uriarte, E. 2008. Proteomics, networks and connectivity indices. *Proteomics*. 8 (4):750–778.

Guo, Y., Li, Y., Xue, L., Severino, R. P., Gao, S., Niu, J., Qin, L. P., Zhang, D. and Brömme, D. 2014. Salvia miltiorrhiza: An ancient Chinese herbal medicine as a source for anti-osteoporotic drugs. Journal of ethnopharmacology 155(3): 1401–1416.

Guo, L., Winzer, T., Yang, X., Li, Y., Ning, Z., He, Z., Teodor, R., Lu, Y., Bowser, T. A., Graham, I. A. and Ye, K. 2018. The opium poppy genome and morphinan production. *Science* 362(6412): 343–347.

Hashiguchi, A., Tian, J. and Komatsu, S. 2017. Proteomic contributions to medicinal plant research: from plant metabolism to pharmacological action. *Proteomes* 5(4): 35.

Heidarvand, L. and Amiri, R. M. 2010. What happens in plant molecular responses to cold stress? *Acta Physiol. Plant.* 32(3): 419–431.

Hoopes, G. M., Hamilton, J. P., Kim, J., Zhao, D., Wiegert-Rininger, K., Crisovan, E. and Buell, C. R. 2018. Genome assembly and annotation of the medicinal plant *Calotropis gigantea*, a producer of anticancer and antimalarial cardenolides. *G3 (Bethesda)* 8(2): 385–391.

Hu, Y., Zhao, R., Xu, P. and Jiao, Y. 2018. The genome of opium poppy reveals evolutionary history of morphinan pathway. *Genomics Proteomics Bioinformatics* 16(6): 460–462.

Jaiswal, S. K., Mahajan, S., Chakraborty, A., Kumar, S. and Sharma, V. K. 2021. The genome sequence of Aloe vera reveals adaptive evolution of drought tolerance mechanisms. *iScience* 24(2): 102079.

Kaul, S. et al. 2000. Analysis of the genome sequence of the flowering plant Arabidopsis thaliana. *Nature* 408(6814): 796–815.

Kim, S., Park, M., Yeom, S. I., Kim, Y. M., Lee, J. M., Lee, H. A. ... and Jung, K. 2014. Genome sequence of the hot pepper provides insights into the evolution of pungency in *Capsicum* species. *Nature Gen.* 46(3): 270–278.

Kim, J. H. 2018. Pharmacological and medical applications of *Panax ginseng* and ginsenosides: a review for use in cardiovascular diseases. *J. Ginseng Res.* 42(3): 264–269.

Kim, S. W., Gupta, R., Lee, S. H., Min, C. W., Agrawal, G. K., Rakwal, R., Kim, J. B., Jo, I. H., Park, S. Y., Kim, J. K., Kim, Y. C., Bang, K. H. and Kim, S. T. 2016. An integrated biochemical, proteomics, and metabolomics approach for supporting medicinal value of *Panax ginseng* fruits. *Front. Plant Sci.* 7: 994.

Kiyama, R. 2017. DNA Microarray-Based Screening and characterization of traditional. Chinese Medicine. *Microarrays (Basel)* 6(1): 4.

Leung, K. W. and Wong, A. S. 2010. Pharmacology of ginsenosides: A literature review. *Chin. Med.* 5: 20.

Liu, J., Liu, Y., Wang, Y., Abozeid, A., Zu, Y. G. and Tang, Z. H. 2017. The integration of GC-MS and LC-MS to assay the metabolomics profiling in *Panax ginseng* and *Panax quinquefolius* reveals a tissue- and species-specific connectivity of primary metabolites and ginsenosides accumulation. *J. Pharm. Biomed. Anal.* 135: 176–185.

Ma, R., Sun, L., Chen, X., Jiang, R., Sun, H. and Zhao, D. 2013. Proteomic changes in different growth periods of ginseng roots. *Plant Physiol. Biochem.* 67: 20–32.

Ma, R., Sun, L., Chen, X., Mei, B., Chang, G., Wang, M. and Zhao, D. 2016. Proteomic analyses provide novel insights into plant growth and ginsenoside biosynthesis in forest cultivated *Panax ginseng* (F. ginseng). *Front. Plant Sci.* 7: 1.

Manzoni, C., Kia, D. A., Vandrovcova, J., Hardy, J., Wood, N. W., Lewis, P. A. and Ferrari, R. 2018. Genome, transcriptome and proteome: the rise of omics data and their integration in biomedical sciences. *Brief Bioinform.* 19(2): 286–302.

Mathur, A., Mathur, A. K., Pal, M. and Uniyal, G. C. 1999. Comparison of qualitative and quantitative in vitro ginsenoside production in callus cultures of three *Panax* species. *Planta Med.* 65(5): 484–486.

Mochida, K. and Shinozaki, K. 2011. Advances in omics and bioinformatics tools for systems analyses of plant functions. *Plant Cell Physiol.* 52(12): 2017–2038.

Qin, C., Yu, C., Shen, Y., Fang, X., Chen, L., Min, J. ... and Yang, Y. 2014. Whole-genome sequencing of cultivated and wild peppers provides insights into *Capsicum* domestication and specialization. *Proc. Nat. Acad. Sci. USA* 111(14): 5135–5140.

Shen, Q., Huang, H., Zhao, Y., Xie, L., He, Q., Zhong, Y., Wang, Y. and Tang, K. 2019. The transcription factor aabzip9 positively regulates the biosynthesis of artemisinin in *Artemisia annua*. *Front. Plant Sci.* 10: 1294.

Shen, Q. et al. 2018. The genome of *Artemisia annua* provides insight into the evolution of asteraceae family and artemisinin biosynthesis. *Mol. Plant* 11(6): 776–788.

Vashisht, I., Pal, T., Sood, H. and Chauhan, R. S. 2016. Comparative transcriptome analysis in different tissues of a medicinal herb, *Picrorhiza kurroa* pinpoints transcription factors regulating picrosides biosynthesis. *Mol. Biol. Rep.* 43(12): 1395–1409.

Wang, H., Ye, Y., Pan, S. Y., Zhu, G. Y., Li, Y. W., Fong, D. W. and Yu, Z. L. 2011. Proteomic identification of proteins involved in the anticancer activities of oridonin in HepG2 cells. *Phytomedicine* 18(2–3): 163–169.

Wang, Y. D., Wang, X. and Wong, Y. S. 2012. Proteomics analysis reveals multiple regulatory mechanisms in response to selenium in rice. *J. Proteomics* 75(6): 1849–1866.

Wang, Y. L., Huang, L. Q., Yuan, Y. and Zha, L. P. 2015. Research advances on analysis of medicinal plants transcriptome. *Zhongguo Zhong Yao Za Zhi* 40(11): 2055–2061.

Xu, J. et al. 2017. *Panax ginseng* genome examination for ginsenoside biosynthesis. *Gigascience* 6(11): 1–15.

Xu, W., Choi, H. K. and Huang, L. 2017. State of *Panax ginseng* research: A global analysis. *Molecules:* 22(9.

Yang, L. et al. 2013. Transcriptome analysis of medicinal plant *Salvia miltiorrhiza* and identification of genes related to tanshinone biosynthesis. *PLoS One* 8(11): e80464.

Zhang, D., Qiao, W., Zhao, Y., Fang, H., Xu, D. and Xia, Q. 2017. Curdione attenuates thrombin-induced human platelet activation: β1-tubulin as a potential therapeutic target. *Fitoterapia* 116: 106–115.

Zhao, M. M., Zhang, G., Zhang, D. W., Hsiao, Y. Y. and Guo, S. X. 2013. ESTs analysis reveals putative genes involved in symbiotic seed germination in *Dendrobium officinale*. *PLoS One* 8(8): e72705.

Zhu, W., Yang, B., Komatsu, S., Lu, X., Li, X. and Tian, J. 2015. Binary stress induces an increase in indole alkaloid biosynthesis in *Catharanthus roseus*. *Front. Plant Sci.* 6: 582.

6 Genomic Profiling of Medicinal Plants

Kritika Shukla, Mohd Ashraf Ashfaq and Md. Salik Noorani

CONTENTS

6.1 INTRODUCTION

Plants act as natural factories, synthesizing a large number of compounds. Medical plants or medicinal herbs are plants that have been shown to have pharmacological benefits for humans and are thus utilized medicinally. The majority of natural compounds derived from plants with pharmacological activity are secondary metabolites. For over 4000 years, Unani Hakims and Ayurveda Vaiyds have used herbs as medicines. India is the world's largest producer of medicinal herbs, and so is rightly dubbed the World's Botanical Garden (Ahmedullah and Nayar 1999). Chemically produced pharmaceuticals have transformed the healthcare business over the last few decades, but due to their serious side effects, the herbal sector has re-emerged. Medicinal plants are critical to emerging countries' modern economies. Ayurveda is predicted to contribute roughly US $813 million to India's economy each year (Joshi et al. 2004). They are typically more inexpensive and substantially safer to use in prescribed doses due to their non-toxic nature. Over 60% of on-the-market cancer treatments are derived from natural sources. Palcitaxel was isolated from *Taxus brevifolia*, Camptothecin was isolated from *Camptotheca accuminata*, Combretastatin was isolated from *Combretum caffrum* and Vinblastine and Vincristine were isolated from *Catharanthus roseus* (Brower 2008). Numerous plant-derived chemicals, including alkaloids, flavonoids, coumarins, phenolics, terpenes and tannins have been shown to be anti-HIV-1. *Vernonia amygdalina*, *Calendula officinalis*, *Artemisia annua* and *Galanthus nivalis* are only a few of the therapeutic herbs that exhibit anti-HIV efficacy (Laila et al. 2019). Using the current COVID-19 pandemic as an example, chemicals derived from herbs such as *Tinospora cordifolia*, *Cassia alata*, *Lycoris radiate* and *Polygonum multiflorum* have been demonstrated to inhibit coronavirus replication (Adhikari et al. 2020).

In ancient times, herbal remedies were prepared, stored and distributed by indigenous physicians. Their understanding of medicinal herbs was extensive, and they maintained an intimate relationship with nature. However, the current condition has shifted dramatically as a result of urbanization.

DOI: 10.1201/9781003179139-6

Their connection to nature has been severed, and as a result, knowledge about medicinal plant identification has diminished significantly. As a result, there is currently much misunderstanding regarding their naming and identification. There are homonyms (the same name applied to two distinct plants) and synonyms (different names for the same plants). These issues also develop as a result of vernacular nomenclature. Along with the nomenclature conflict, there is another issue about adulteration of herbal plant-based pharmaceuticals on the market. They are either accidentally or purposefully swapped with morphologically identical herbs. Substitution can take numerous forms and be motivated by a variety of factors, including resemblance in morphology, color, chemical composition and action. The original plant is replaced by a different section of the same plant, a different species within the same genus, or a completely different plant. For example, *Vitex negundo* is sold as Tulsi (*Ocimum sanctum*). *Anemone* flowers are sold as "Banafsha" (*Viola* spp.) and "Saffron" (*Crocus sativa*). *Malva rotundifolia* is sold as "Brahmi" (*Centella asiatica*). *Leea alata* is sold as "Manjith" (*Rubia cordifolia*) (Ganie et al. 2015). To avoid such misunderstandings and abuses, it is critical to standardize medicinal herbs. Due to the fact that DNA is stable and does not change with time, physiological conditions, or environment, genomic profiling of medicinal plants is more reliable than macroscopic, microscopic, or chemical examination. They are more efficient, precise and time-saving than traditional methods. Genomic profiling is also used for the conservation of medicinal plants (Kumar et al. 2014).

6.2 MOLECULAR MARKERS FOR GENOMIC PROFILING

Genomic profiling or DNA fingerprinting or DNA profiling is a technique invented in 1984 by Alec Jeffrey, a British biologist. It is a process of identifying cultivars through the use of molecular marker techniques (Archak 2000). There is a distinction between the fingerprinting of plants and humans. Individuals can be identified in the case of humans, whereas populations (a species, a variety, or a breeding line) are identified in the case of plants. Markers are described as allelic variation at a particular locus in the genome that can be seen at the morphological, biochemical (isozyme), or DNA levels. DNA polymorphism can be visualized using a variety of DNA-based marker systems. While morphological and biochemical indicators are typically scarce and highly impacted by their surroundings, DNA markers are unaffected by age, environmental circumstances, or physiological conditions. They are more efficient, precise and time-saving than traditional methods. Ideal DNA markers should be highly polymorphic, co-dominant in expression (allow to discriminate hetero- and homozygotes), high reproducible, economic to use, easy and fast assay and a frequent occurrence in the genome. Application of molecular marker based genomic profiling of some medicinal plants is given in Table 6.1. These molecular markers can be hybridization-based markers such as Restriction Fragment Length Polymorphism (RFLP), PCR-based markers like arbitrary or random markers, Random Amplification of Polymorphic DNA (RAPD), Amplified Fragment Length Polymorphism (AFLP), Inter-Simple Sequence Repeat (ISSR) and DNA Amplification Fingerprinting (DAF) and Sequence-based markers such as Simple Sequence Repeats (SSR), Sequence Characterized Amplified Region (SCAR), Cleaved Amplified Polymorphic Sequences (CAPS) and Single Nucleotide Polymorphism (SNP).

6.2.1 Restriction Fragment Length Polymorphism (RFLP)

It is the first DNA marker developed by Botstein et al. (1980) to genetically differentiate between different organisms based on polymorphism created due to changes in restriction sites in DNA between sites. RFLPs are differences in restriction fragment lengths caused by single nucleotide

TABLE 6.1
Versatile Application of Molecular Marker in Medicinal Plants

Markers	Name of Plant	Purpose	References
RFLP	*Musa* spp.	Phylogenetic analysis	Gawel et al. 1992
RFLP	*Hedysarum* spp.	Phylogenetic analysis	Neila and Mohamed 2001
PCR-RFLP	*Citrus* spp.	Taxonomic and Phylogenetic analysis	Narayan et al. 2009
PCR-RFLP, RAPD, ISSR	*Matricaria recutita*	Phylogenetic analysis	El-Twab and Zahran 2010
PCR-RFLP, RAPD, ISSR	*Achillea fragrantissima*	Phylogenetic analysis	El-Twab and Zahran 2010
PCR-RFLP, RAPD, ISSR	*Artemisia arborescens*	Phylogenetic analysis	El-Twab and Zahran 2010
RFLP	*Glehnia littoralis*	To study genetic diversity	Mizukami et al. 1993
RFLP, RAPD, AFLP, SSR	*Olea europea*	Construction of genetic maps	Rosa et al. 2003
RFLP, AFLP	*Brassica juncea*	Construction of the genetic map	Pradhan et al. 2003
PCR-RFLP	*Dendrobium chrysanthum, D. fimbriatum*	Molecular identification	Zhang et al. 2005
AFLP	*Dendrobium thyrsiflorum*	To study genetic and	Bhattacharyya et al. 2017
ISSR & DAMD	*Dendrobium nobile*	phytochemical diversity	Bhattacharyya et al. 2015
PCR-RFLP	*Sinopodophyllum hexadendrum*	Comparative study and identification	Gong et al. 2006
PCR-RFLP	*Dysosma* spp.	Comparative study and identification	Gong et al. 2006
RFLP & RAPD	*Piper nigrum*	Genetic analysis for comparison	Fatma et al. 2016
PCR-RFLP	*Peganum harmala*	To study genetic diversity	Bakatoushi and Amhed 2018
PCR-RFLP	*Griffonia simplicifolia*	For identification	Vigliante et al. 2019
PCR-RFLP & ARMS	*Alisma orientale*	For authentication	Le et al. 2007
PCR-RFLP, ARMS, RAPD, AP-PCR, SCAR,	*Angelica sinensis*	For molecular authentication	Feng et al. 2010
PCR-RFLP	*Pueraria candollei*	For authentication	Wiriyakarun et al. 2012
PCR-RFLP	*Butea superba*	For authentication	Wiriyakarunet al. 2012
PCR-RFLP	*Mucuna collettii*	For authentication	Wiriyakarun et al. 2012
PCR-RFLP	*Desmodium giganicum*	For authentication	Biswas and Biswas 2013
PCR-RFLP	*Aegle marmelos*	For authentication	Biswas and Biswas 2013
PCR-RFLP	*Solanum xanthocarpum*	For authentication	Biswas and Biswas 2013
PCR-RFLP	*Solanum indicum*	For authentication	Biswas and Biswas 2013
PCR-RFLP	*Tribulus terresteris*	For authentication	Biswas and Biswas 2013
PCR-RFLP	*Oroxylum indicum*	For authentication	Biswas and Biswas 2013
PCR-RFLP	*Boerhavia diffusa*	For authentication	Biswas et al. 2013
RFLP & ARMS	*Dracocephalum moldavica*	For authentication	Horn et al. 2014
RFLP	*Pulsatilla chinensis*	For authentication	Shi et al. 2017
PCR-RFLP	*Cinnamomum* spp.	For authentication	Yang et al. 2020
RAPD	*Silybum marianum*	For studying genetic diversity	Hamouda 2019 AbouZid 2014
RAPD, ISSR	*Nilgirianthus ciliates*	For studying genetic diversity & phylogenetic relationship	Rameshkumar 2019

(*continued*)

TABLE 6.1 (Continued)
Versatile Application of Molecular Marker in Medicinal Plants

Markers	Name of Plant	Purpose	References
RAPD, ISSR	*Justicia adhatoda*	For studying genetic variability	Kumar et al. 2014
RAPD	*Cassia tora*	For studying genetic variability	Tilwari et al. 2016
SCAR	*Cassia angustifolia* *Cassia acutifolia*	For authentication	Abdin 2013
RAPD	*Pterolobium hexapetalum*	For studying genetic and ecological diversity	Sharmila et al. 2019
RAPD, ISSR	*Clitoria ternatea*	To study genetic variability	Ali et al. 2013
RAPD	*Convolvulus pluricaulis*	For authentication	Ganie et al. 2015
RAPD	*Lonicera* spp.	To study genetic variability	Cehula et al. 2020
RAPD, AFLP, SSR, ISSR, SNP, ITS, SCAR	*Codonopsis* spp.	To study molecular diversity	Yang et al. 2020
RAPD	*Pluricaria* spp.	To study genetic relationship and variability	El-Kamali et al. 2010
RAPD	*Scutellaria* spp.	For identification or differentiation	Hosokawa et al. 1999
RAPD	*Croton tiglium*	For authentication	Vekariya 2017
RAPD	*Cuscuta* spp.	For authentication	Khan et al. 2010
RAPD	*Glycyrrhiza glabra*	For authentication	Khan et al. 2009
SSR ISSR	*Glycyrrhiza uralensis*	To study genetic diversity	Liu et al. 2019, Yao at el. 2008
RAPD	*Zilla spinosa*	For identification and conservation	Arif et al. 2010
RAPD	*Achillea fragrantissima*	For identification and conservation	Arif et al. 2010
RAPD	*Lycium shawii*	For identification and conservation	Arif et al. 2010
RAPD	*Moricandia sinaica*	For identification and conservation	Arif et al. 2010
RAPD	*Withania somnifera*	For identification and conservation	Arif et al. 2010
RAPD	*Sonchus oleraceus*	For identification and conservation	Arif et al. 2010
RAPD	*Zygophyllum propinquum*	For identification and conservation	Arif et al. 2010
RAPD	*Rumex vesicarius*	For identification and conservation	Arif et al. 2010
RAPD	*Bassia eriophora*	For identification and conservation	Arif et al. 2010
RAPD-PCR	*Corylus avellana*	For identification and authentication	Felbinger et al. 2020
RAPD & ISSR	*Cissampelos pereira* (Patha)	For authentication	Viyayan et al. 2013
RAPD	*Rauwolfia tetraphylla*	For conservation	Rohela et al. 2019
ISSR	*Eruca sativa*	For detection of genotoxicity	Al-Qurainy et al. 2010
AFLP	*Aloe vera*	To study genetic diversity	Tripathi et al. 2011
AFLP	*Echinacea* spp.	For authentication	Russi et al. 2009
AFLP	*Plectranthus* spp.	For authentication	Passinho-Soares et al. 2006
AFLP	*Curcuma comosa*	For authentication and phylogenetic analysis	Keeratinijakal et al. 2010

TABLE 6.1 (Continued)
Versatile Application of Molecular Marker in Medicinal Plants

Markers	Name of Plant	Purpose	References
AFLP	*Aconitum heterophyllum*	For authentication	Misra et al. 2010b
SCAR			Seethapathy et al. 2014
AFLP	*Zanthoxylum* spp.	For authentication	Gupta and Mandi 2014
AFLP	*Zingiber officinale* (Ginger)	For authentication	Ghosh et al. 2011
AFLP	*Swertia* spp.	For authentication	Misra et al. 2010
ISSR	*Swertia chirayita*		Tamhankar et al. 2009
AFLP	*Origanum* spp.	To study genetic diversity	El-Demerdash et al. 2019
AFLP	*Thymus* spp.	To study genetic diversity	El-Demerdash et al. 2019
AFLP	*Paris polyphylla*	To study genetic diversity	Huang et al. 2019
AFLP, SAMPL, ISSR, RAPD	*Tribulus terrestris*	To study genetic diversity	Sarwat et al. 2007
cDNA-AFLP	*Lavandula angustifolia*	To identify salinity responsive genes	Banikamali et al. 2020
AFLP & ISSR	*Papaver bracteatum* (Persian Poppy)	To study genetic diversity	Hadipour et al. 2020
AFLP	*Prunus padus*	To study genetic diversity and structure	Young Ahn et al. 2020
AFLP	*Adhatoda vasica*	To study genetic diversity	Varma and Shrivastava 2018
AFLP	*Andrographis paniculata*	To study genetic diversity	Varma and Shrivastava 2018
AFLP, SSR, InDel	*Allium sativum*	To study genetic diversity	Wang et al. 2016
AFLP, ISSR, RAPD	*Dactylis glomerata*	For genotyping and studying genetic variation	Costa et al. 2016
SSR	*Artemesia dracunculus*	To study genetic diversity	Karimi et al. 2015
SSR	*Camellia huana*	To study genetic diversity	Li et al. 2020
SSR	*Camellia sinensis*	Construction of genetic maps and identification of QTLs	Qiang-Ma et al. 2014
SSR	*Ziziphus jujube*	To Study genetic diversity	Wang et al. 2014
AFLP		& cultivar identification	Bai et al. 2009
SSR	*Panax ginseng, P. quinquefolius*	For authentication	Hon et al. 2003
SSR	*Capsicum* spp.	To study genetic diversity	Shirasawa et al. 2013
RAPD		and comparative study	Oyama et al. 2006
SNP			Jung et al. 2010
SSR & ISSR	*Etlingera elatior*	To study genetic diversity	Ismail et al. 2019
SSR and RAPD	*Trigonella* spp.	To study genetic diversity	Sindhu et el. 2017
SSR	*Origanum compactum*	To study genetic diversity for its conservation	Aboukhalid et al. 2017
SSR	*Rheum tanguticum*	To study genetic diversity	Chen et al. 2009
ISSR		To study genetic diversity and authentication	Hu et al. 2011
ISSR	*R. palmatum* *R. tanguticum*	To study genetic variation	Wang et al. 2012
RAPD, STS, ISSR, AFLP	*Humulus lupulus*	To study genetic variation	Josef Patzak 2001
ISSR	*Psychotria ipecacuanha*	To study genetic diversity and geographic differentiation	Rossi et al. 2009
RAPD & ISSR	*Benincasa hispida*	To study genetic diversity	Verma et al. 2007

(*continued*)

TABLE 6.1 (Continued)
Versatile Application of Molecular Marker in Medicinal Plants

Markers	Name of Plant	Purpose	References
ISSR	*Nelumbo nucifera*	To study clonal diversity	Chuang-Han et al. 2007
ISSR	*Vanilla planifolia*	To study genetic diversity	Verma et al. 2009
ISSR	*Malva* spp.	Species identification	Jedrzejczyk and Rewers 2020
ISSR	*Ferula pseudalliacea* *Ferula assa foetida*	For species differentiation	Jafari et al. 2019
ISSR SCoT	*Salvia* spp.	To study genetic diversity and compare ISSR and SCoT	Etminan et al. 2018
ISSR	*Satureja khuzistanica*	To study genetic diversity	Hadian et al. 2017
SCAR	*Curcuma alismatifolia*	For authentication	Anuntalabhochai et al. 2007
SCAR	*Jatropha curcus*	For authentication	Basha et al. 2009
SCAR	*Panax japonicas*	For authentication	Choi et al. 2008
SCAR	*Atractylodes* spp.	For authentication	Huh and Bang 2006
SCAR	*Phyllanthus* spp.	For authentication	Dnyaneshwaret al. 2006
SCAR	*Echinacea* spp.	For authentication	Adinolfi et al. 2007
SCAR	*Artemisia* spp.	For authentication	Lee et al. 2006
RT-SCAR	*Catharanthus roseus*	For authentication	Chaudhary et al. 2013
SCAR	*Anthricus sylvestris*	For authentication	Choo et al. 2009
SCAR	*Lycium chinese*	For identification and authentication	Liu et al. 2020
SCAR	*Viola serpenes*	For identification and authentication	Jha et al. 2020
SCAR	*Ocimum tenuiflorum*	For identification and authentication	Kumar et al. 2018
SCAR	*Dipsacus asper* *Phlomoides umbrosa*	For authentication	Park et al. 2018
SCAR	*Pueraria tuberosa*	For authentication	Devaiah and Venkatasubramanian 2008
SCAR	*Angelica dahurica*	For authentication	Noh et al. 2018
SCAR	*Lonicera japonica*	For identification and authentication	Yang et al. 2014
SCAR	*Litchi chinensis*	For identification and authentication	Cheng et al. 2015
SCAR	*Podophyllum hexandrum*	For identification and authentication	Al-shaqha et al. 2014
SCAR	*Fritillariae* spp.	For identification and authentication	Xin et al. 2014
SCAR	*Knema andamanica*	For identification and classification	Sheeja et al. 2013
LAMP	*Zingiber officinale*	For identification and authentication	Chaudhary et al. 2014
LAMP	*Curcuma longa*	For identification and authentication	Sasaki and Nugumo 2007
LAMP	*Panax ginseng*	For identification and authentication	Sasaki et al. 2008
LAMP	*Catharanthus roseus*	For identification and authentication	Chaudhary et al. 2012

TABLE 6.1 (Continued)
Versatile Application of Molecular Marker in Medicinal Plants

Markers	Name of Plant	Purpose	References
LAMP	*Nigella sativa*	For identification and authentication	Ganie et al. 2013
LAMP	*Morinda officinalis*	For identification and authentication	Wang et al. 2020
LAMP	*Polygonum multiflorum*	For identification and authentication	Lee and Hxiao 2019
LAMP	*Dendrobium officinale*	For identification and authentication	Lu et al. 2019
LAMP	*Crocus sativus*	For identification and authentication	Zhao et al. 2016

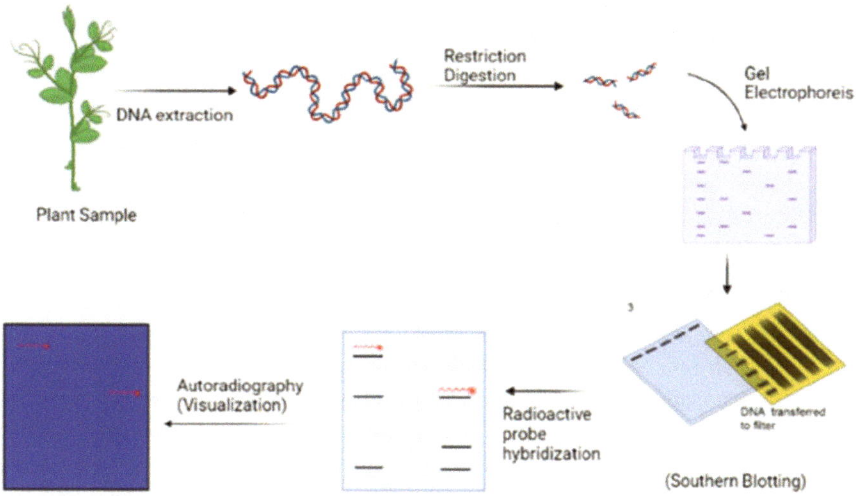

FIGURE 6.1 An overview of the steps involved in RFLP.

polymorphism (SNPs) and insertions and deletions (INDELs) that create or abolish restriction endonuclease recognition sites. A specific RFLP pattern emerges on electrophoretic separation of digested DNA, producing variable length of cleaved fragments which are then blotted onto a membrane, hybridized and exposed to a labeled probe. An overview of the steps involved in RFLP is given in Figure 6.1. The main advantages of RFLPs are that they are co-dominant, highly reproducible and do not require prior sequence information. Disadvantages incudes its time-consuming procedure, labor intensive, expensive and requires a large quantity of good quality DNA.

Numerous medicinal plants have been analyzed using RFLPs for their evolutionary relationships. For instance, a phylogenetic analysis of 19 *Musa* species and subspecies placed *M. boman* in the *Australimusa* genus and indicated that *M. beccarii* is closely related to *M. acuminate* (Jarret et al. 1992). Additionally, the evolutionary link between *Hedysarum* species was investigated. The similarity in banding patterns between species was utilized to determine their genetic distance and to construct phylogenetic trees (Neila and Mohamed 2001). *Matricaria recutita*, *Achillea fragrantissima* and *Artemisia arborescens* were also analyzed phylogenetically (El-Twab and Zahran 2010).

The analysis of *Glehnia littoralis* strains from various geographical origins revealed little difference in their RFLP profiles, implying that their genetic diversity is quite limited (Mizukami et al.

1993). RFLPs were also employed to create genetic maps for a variety of medicinal plants, including *Olea europea* (De la Rosa et al. 2003) and *Brassica juncea* (Pradhan et al. 2003). *Dendrobium chrysanthum*, *Dendrobium fimbriatum* and their morphologically related species were identified molecularly utilizing PCR-RFLP analysis of the rDNA-ITS region (Zhang et al. 2005). *Sinopodophyllum hexadendrum* and *Dysosma* species were identified using cpDNA sequences, and PCR-amplified fragments of trnT-trnL revealed a length difference between the two (Gong et al. 2006). Two samples of *Piper nigrum* were genetically analyzed using the RFLP technique for the purpose of distinguishing the two samples (Fatma, Amarnath and Suneetha 2016). The genetic diversity of 12 *Peganam harmala* genotypes was determined using PCR-RFLP of rDNA-ITS. ITS-RFLP reveals a polymorphism rate of 66.67%, which is fairly low in comparison to other markers (El-Bakatoushi and Amhed 2018). *Griffonia simplicifolia* was identified using PCR-RFLP using three restriction enzymes (*Msp*I, *Hha*I and *Hae*III) to establish a unique fingerprinting pattern (Vigliante, Mannino and Maffei 2019).

The RFLPs also play a significant part in the molecular authentication of numerous medicinal plants, such as the authentication of *Alisma orientale*, by sequencing the internal transcribed spacer (ITS) regions of the six species of the genus *Alisma*'s nuclear ribosomal DNA (nrDNA) (Li et al. 2007). To distinguish *Angelica sinensis* from its hybrid *Angelica* species by the use of internal transcribed spacer (ITS) sequences (Feng, Liu and He 2010). The Thai rejuvenating herbs *Pueraria candollei* (White Kwao Khruea), *Butea superba* (Red Kwao Khruea) and *Mucuna collettii* (Black Kwao Khruea) all have unique medicinal properties, but due to their similar characteristics and confusing names, proper discrimination is necessary. This was accomplished using PCR-RFLP (Wiriyakarun et al. 2012). The roots of six plant species, namely *Desmodium giganicum*, *Aegle marmelos*, *Solanum xanthocarpum*, *Solanum indicum*, *Tribulus terresteris* and *Oroxylum indicum*, were investigated and identified using PCR-RFLP (Biswas and Biswas 2013). To authenticate *Trianthema portulacastrum* and *Trianthema monogyna* by PCR-RFLP employing the *Msp*I, *Hinf*I, and *Mbo*I enzymes (Biswas, Kapoor and Biswas 2013). RFLP can be used to detect single nucleotide exchanges between *Dracocephalum moldavica* L. (Moldavian dragonhead) and putative surrogate species *Melissa officinalis* L. and *Nepeta cataria* L. (Horn et al. 2014). Using the ITS2 barcode library, this technique was also used to differentiate *Pulsatilla chinensis* from its common adulterants (Shi et al. 2017). *Cinnamomum osmophloeum* (Co), a medicinal plant used in Taiwan, contains a higher concentration of cinnamaldehyde than other cinnamon species. *Cinnamomum burmannii* (Cb) is a frequent adulterant. Using ITS sections of nuclear ribosomal DNA from diverse *Cinnamomum* species, PCR-RFLP was used to rapidly authenticate Co from Cb, and the sequences were evaluated and characterized (Yang et al. 2020).

6.2.2 Random Amplified Polymorphic DNA (RAPD)

Random Amplified Polymorphic DNA (RAPD) is a PCR-based approach that employs arbitrary or random primers (10 nucleotides with a G-C content of 50%) that anneal and amplify genomic DNA at numerous random sites (Williams et al. 1990). These PCR amplicons are subsequently separated on an agarose or polyacrylamide gel, allowing for visualization of the variation in banding patterns. An overview of the steps involved in RAPD is presented in Figure 6.2. RAPDs are DNA polymorphisms that result from rearrangements or deletions at or near oligonucleotide primer binding sites in the genome. Only one random oligonucleotide primer is used in this case, and no prior information of the genome being analyzed is necessary (Hadrys, Balick and Schierwater 1992). A very small amount of DNA (nanograms) may be used. There are two variants of RAPD. DNA Amplification Fingerprinting (DAF) was proposed by Caetano-Anolles Bassam and Gresshoff (1991), in which very brief primers (8 nucleotides) are utilized at quite high concentrations. Welsh and McClelland (1990) introduced the concept of arbitrarily primed PCR (AP-PCR). It employs somewhat longer primers, radioactively labels the amplification products and also resolves them

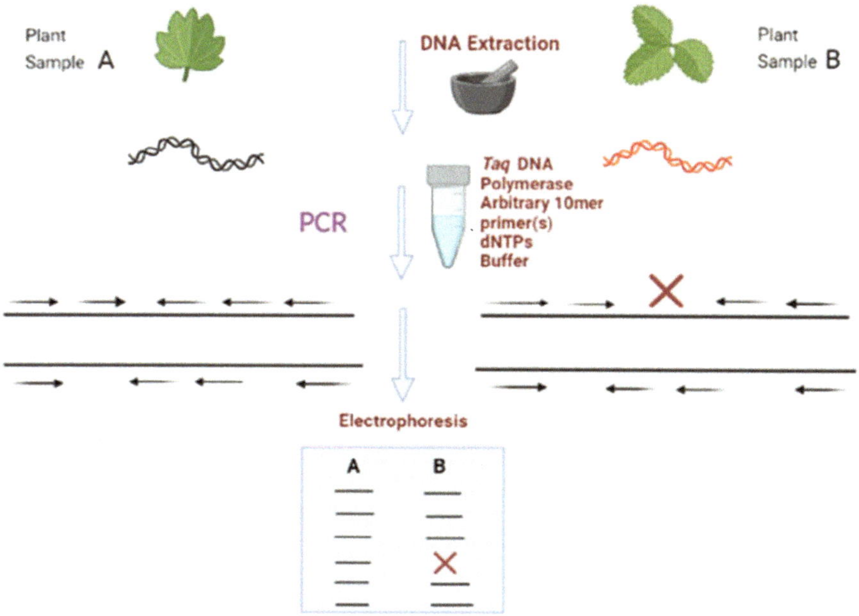

FIGURE 6.2 An overview of the steps involved in RAPD.

using polyacrylamide gel electrophoresis. The technique's primary disadvantages include its dominating nature, low reproducibility and sensitivity to laboratory changes.

The RAPD approach was used to determine the genetic diversity of 14 *Silybum marianum* specimens. Using 12 RAPD primers, the results demonstrated a significant level of polymorphism (73.2%) (Hamouda 2019). Additionally, it was used to characterize the genetic diversity and phylogenetic linkages within the *Nilgirianthus ciliatus* population in order to aid in its conservation. The results indicated that out of a total of 99, 89 (89.9%) were polymorphic (Rameshkumar 2019). The wild population of *Justicia adhatoda*, a medicinal herb, is rapidly depleting. RAPD was utilized to decipher the genetic diversity and connections among the wild 32 *J. adhatoda* accessions (Kumar et al. 2014). Tilwari et al. (2016) investigated the genetic variability of 13 distinct *Cassia tora* cultivars grown in various locations of central India. The findings indicated a high level of genetic variation among *Cassia tora* accessions. *Pterolobium hexapetalum* is a medicinal plant whose ecological and genetic diversity has been studied in two hills, Maruthamalai (arid) and Chennimalai (very arid) (Sharmila et al. 2019). Ali et al. (2013) used 25 RAPD primers to examine *Clitoria ternatea* for genetic variability and variance in Kaempferol concentration in 11 accessions. To separate *Convolvulus pluricaulis* from its adulterants, RAPD analysis was undertaken. As a result of the results, it was determined that primer OPN-09 is the species-specific primer (Ganie et al. 2015). Cehula et al. (2020) examined the genetic diversity of several cultivars and clones of *Lonicera* spp. from two countries with contrasting climatic and soil conditions. Numerous molecular markers, including RAPD, were used to determine the molecular diversity of various species in the genus *Codonopsis* (Yang et al. 2020). Using RAPD, the genetic connection and dissimilarity between two *Pluricaria* species, *P. undulate* and *P. crispa* and their three putative hybrids were evaluated (El-Kamali et al. 2010). Two samples of *Piper nigrum* were analyzed genetically utilizing the RAPD approach for the purpose of differentiating the two samples (Fatma, Amarnath and Suneetha 2016). RAPD analysis was used to differentiate three species of the medicinal plant *Scutellaria*, namely *Scutellaria galericulata*, *Scutellaria lateriflora* and *Scutellaria baicalensis*. The results indicated that primers A02 and A06 yielded RAPD markers that may be used to rapidly identify these three *Scutellaria* species (Hosokawa et al. 2000).

Croton tiglium was authenticated using RAPD analysis to distinguish it from adulterants such as seeds of *Baliospermum montanum*, *Ricinus communis* and *Croton roxburghii* (Vekariya et al. 2017). Two *Silybum marianum* variants, *S. marianum* var. *album* and *S. marianum* var. *purple*, were also differentiated using polymorphic bands generated by RAPD (Abou Zid 2014). Similarly, the same technique was used to distinguish *Cuscuta reflexa* from *Cuscuta chinensis*. OPC-1, OPC-02, OPC-03, OPC-04, OPC-05, OPC-06, OPC-07 and OPC-08 were used to differentiate both species (Khan et al. 2010). The RAPD method was used to distinguish *Glycyrrhiza glabra* from its adulterant *Abrus precatorius* (Khan et al. 2009). Rapid characterization of nine medicinal plants was carried out using RAPD in the Saudi Arabian deserts. These plants included *Zilla spinosa*, *Achillea fragrantissima*, *Lycium shawii*, *Moricandia sinaica*, *Withania somnifera*, *Sonchus oleraceus*, *Zygophyllum propinquum*, *Rumex vesicarius* and *Bassia eriophora* (Arif et al. 2010). Using 16 RAPD primers for identification and authenticity, RAPD-PCR analysis was undertaken on 13 distinct *Corylus avellana* cultivars (Felbinger et al. 2020). The genuine plant *Cissampelos pereira* was differentiated from its adulterants *Cyclea peltata* and *Stephania japonica* using RAPD profiling. Between them, RAPD profiles revealed 95.68% polymorphism (Vijayan et al. 2013). Recently, in 2019, an experiment was conducted to determine the viability of *Rauwolfia tetraphylla* using *in vitro* clonal multiplication via indirect regeneration using stem and leaf-based callus. The regenerated plantlets are subsequently genotyped using RAPD, and the results indicate that they lack polymorphism with the mother plant, indicating that they are true-to-type (Rohela et al. 2019).

6.2.3 Amplified Fragment Length Polymorphism (AFLP)

AFLP is a PCR-based approach that combines RFLP and RAPD (Vos et al. 1995). AFLP begins with the digestion of DNA using restriction endonuclease and the ligation of oligonucleotide adaptors, followed by the selective amplification of restriction fragments using polymerase chain reaction (PCR), and finally the gel analysis of the amplified fragments. An overview of the steps involved in AFLP is given in Figure 6.3. It is an efficient method of genotyping, particularly when little is known about an organism's genome or genetics (Foster, Bull and Keim 2020). The fingerprints

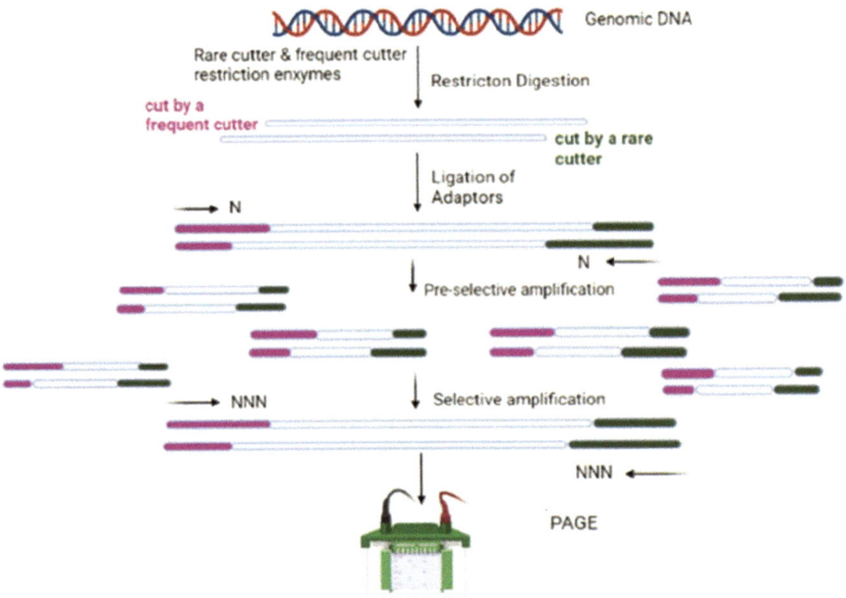

FIGURE 6.3 An overview of the steps involved in AFLP.

are generated using a limited selection of primers without any prior knowledge of the sequence. It necessitates a tiny amount of high-quality DNA. As with RAPD, they inherit in a dominating manner. Earlier PCR primers had radioactive labels to facilitate viewing of amplified fragments, while fluorescent AFLP primers were created recently. Selectively Amplified Microsatellite Polymorphic Locus (SAMPL) is a variant of the AFLP technique. It is a technique for amplifying microsatellite loci that combines an AFLP primer with a primer complementary to the microsatellite sequences (Witsenboer, Michelmore and Vogel 1997).

Eight AFLP primer combinations were used to analyze a collection of 12 *Aloe vera* germplasm accessions. The results indicated that two primer combinations, E-ATC + M-AG and E-GCA + M-GT, were particularly successful in detecting a large number of distinct fragments (Tripathi, Saini and Tiwari 2011). Three *Echinacea* species were identified using the primers E+CAC/ M+ AAT or E+CAC/ M+AGC (Russi et al. 2009). Three distinct species of the genus *Plectranthus* were identified using AFLP. These are *P. barbatus*, *P. grandis* and *P. ornatus* (Passinho-Soares et al. 2006). *Curcuma comosa* is distinguished from related species using AFLP. The E-ACC/M-CTG primer combination resulted in the highest polymorphism % age (100%). Additionally, the AFLP data were utilized to construct a phylogenetic tree, which classified the 97 accessions into five groups (Keeratinijakal, Kladmook and Laosatit 2010). With over 300 species worldwide, the genus *Aconitum* is quite complex. The AFLP technique was used to create fingerprints for four *Aconitum* species and to differentiate *A. heterophylla* from its adulterants. The results indicated that 99.9% of the species evaluated were polymorphic (Misra et al. 2010a). To eliminate the possibility of adulteration, an AFLP analysis was performed on two *Zanthophylum* species, *Z. acanthopodium* and *Z. oxyphyllum* (Gupta and Mandi 2014). Due to the frequent adulteration of *Zingiber officinale* with less effective *Zingiber* species, AFLP was employed to generate DNA fingerprints for three *Zingiber* species (Ghosh Majumder and Mandi 2011). Numerous *Swertia* species are frequently employed as adulterants in place of *Swertia chirayita*. To circumvent this issue, AFLP analysis was performed, which revealed 99% polymorphism between the species (Misra et al. 2010b).

Origanum and *Thymus*, two genera of the Labiatae family, were genetically analyzed using AFLP. The primer pair E-AGG/M-CTC generated an abundance of amplicons (El-Demerdash et al. 2019). *Paris polyphylla* var. yunnanensis is under danger of extinction. The genetic analysis of 15 wild and 17 cultivated populations using AFLP markers found that cultivated populations contained greater genetic diversity than wild populations (Huang et al. 2019). *Dendrobium thyrsiflorum* was analyzed using AFLP to determine its genetic and phytochemical diversity and to determine the connection of AFLP markers with antioxidant activity in *D. thyrsiflorum* in order to conserve it (Bhattacharyya et al. 2017). For the first time, AFLP and SAMPL markers were utilized to examine the genetic diversity of *Tribulus terrestris* (Sarwat, Das and Srivastava 2008). The demand for lavender essential oils is expanding globally. Salinity stress can alter the content of natural products by inducing oxidative damage. The cDNA-AFLP technique was used to investigate the alterations in its transcriptome in response to NaCl treatment (Banikamali et al. 2020). *Papaver bracteatum* is a rare natural source of the opiate thebaine. The genetic diversity of Poppy populations was analyzed using AFLP and ISSR markers, and the results indicated that AFLP markers are more effective than ISSR markers at detecting population variance (Hadipour et al. 2020). AFLP analysis was used to investigate the genetic diversity, differentiation and structure of *Prunus padus*. The results indicated that the species had a low genetic variability when compared to other *Prunus* species (Ahn et al. 2020). AFLP analysis was used to determine the intra-specific genetic diversity of two plants, *Adhatoda vasica* and *Andrographis paniculata*. In *A. vasica*, there is a strong link between populations, as the five populations can be classified into two large clusters, whereas in *A. paniculata*, genetic variability is more heterogeneous (Varma and Shrivastava 2018). To assess the genetic diversity of *Allium sativum* at the molecular level, 121 garlic accessions were analyzed using AFLP, SSR and InDel markers (Wang et al. 2016). To explore genetic variation and to assess the level of information offered by these three marker systems, DNA profiling of *Dactylis glomerata* was performed using

AFLP, ISSR and RAPD. Due to its high degree of polymorphism (91 %), AFLP demonstrated to be the most effective method for genotype discrimination (Costa et al. 2016).

6.2.4 Simple Sequence Repeats (SSR)

The SSRs, also known as Microsatellites and Short Tandem Repeats (STRs), are short DNA motifs (1–6 bp in length) that are tandemly repeated at a specific locus. Litt and Luty (1989) discovered and developed these, which were first used to analyse polymorphism in soyabean by Akkaya, Bhagwat and Cregan (1992). They have the benefit of being co-dominant, locus-specific and widely distributed and abundant in all genomes with high levels of polymorphism. It requires a negligible amount of DNA and is easily exchanged between laboratories. To amplify SSRs using PCR, primer design requires knowledge of unique flanking DNA sequences. After amplification, the products are sorted by size using electrophoresis and visualized using silver staining or fluorescent dyes (Gebhardt 2007). An overview of the steps involved in SSR is given in Figure 6.4. The SSR polymorphism results from variations in the number of repeats of the motif produced by polymerase strand slippage during DNA replication or from recombination mistakes (Vieira et al. 2016). The technique's shortcomings include its high development cost and labor-intensive nature. SSRs can be utilized for a variety of purposes, including the study of genetic diversity in plants, evolutionary studies, the building of linkage maps and determining the degree of kinship between genotypes.

Microsatellite markers have been employed to investigate genetic variability in a variety of medicinal plant genera, including *Artemesia dracunculus* (Karimi et al. 2015), *Camellia sativa* (Li et al. 2020). 1118 distinct SSR markers derived from SSR-enriched genomic libraries have been found and described in *Ziziphus jujuba.* The genetic diversity of six cultivars was determined using 31 polymorphic primer pairs (Wang et al. 2014). For the purpose of determining the genetic diversity of *Etlingera elatior*, 57 accessions were collected and examined using SSR and ISSR. The results indicated that they had a significant degree of genetic variability and that SSR markers were more informative than ISSR markers in determining gene diversity (Ismail et al. 2019). *Glycyrrhiza*'s genetic diversity was examined in order to conserve it. Seven hundred and thirty-three samples of each of its four species were collected, and nine SSR markers were produced. The results indicated that *G. uralensis* possesses the greatest genetic diversity (Liu et al. 2019). Ninety *Trigonella* genotypes

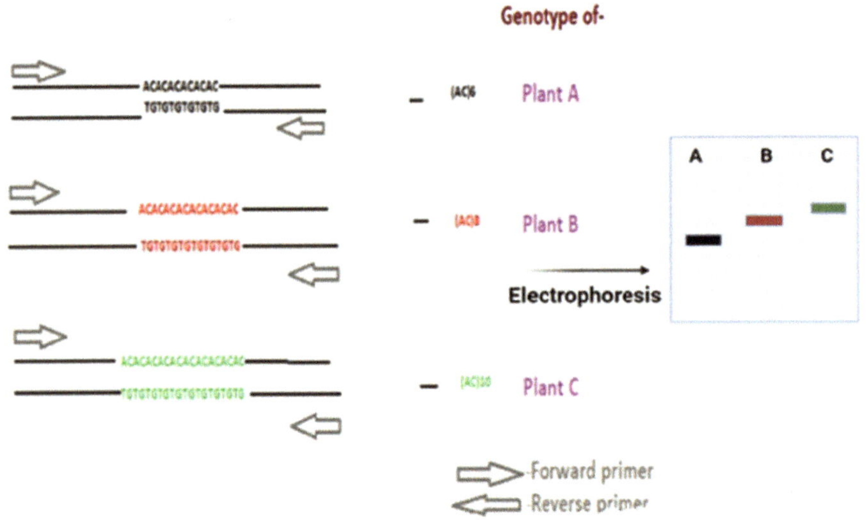

FIGURE 6.4 An overview of the steps involved in SSR.

were genotyped using RAPD (49) and SSR (13) primers and an average of 55.6 and 50.16% polymorphism was observed, respectively (Sindhu, Tehlan and Chaudhury 2017). Six hundred and seventy individuals of *Origanum compactum* were genotyped using 15 microsatellite loci. Moroccan oregano has a greater genetic diversity in isolated and declining groups (Aboukhalid et al. 2017).

One hundred and sixty *Panax quinquefolius* (American ginseng) and 40 *Panax ginseng* (Oriental ginseng) roots were analyzed using 16 microsatellites. The results indicated that nine out of 16 microsatellites could tell the difference between these two (Hon et al. 2003). Microsatellites were used to evaluate 800 *Capsicum* species lines collected from various places. SSR primers (5751 pairs) were developed, and the genome of *C. annuum* was compared to that of tomato (also a member of the Solanaceae) by mapping 2245 *C. annuum* markers onto the tomato genome (Shirasawa et al. 2012).

6.2.5 Inter-Simple Sequence Repeats (ISSR)

Inter-simple sequence repeats (ISSRs) are microsatellite-encoded sections of the genome. It is a version of polymerase chain reaction that was separately published by Meyer et al. (1993), Gupta et al. (1994), Wu et al. (1994) and Zietkiewicz, Rafalski and Labuda (1994). Random DNA segments (RAMs) in the genome can be PCR-amplified and used as a marker (Ng and Tan 2015) for studying genetic variation, phylogeny, gene mapping and DNA fingerprinting using arbitrarily designed primers that contain repetitive sequences complementary to microsatellite regions in the genome (ISSR primers). An overview of the steps involved in ISSR is given in Figure 6.5. In comparison to RAPDs, ISSR primers are significantly greater in length (18–25 mers). It is a straightforward, low-cost technique that combines the majority of the benefits of SSRs and AFLP with the universality of RAPD (Reddy, Sarla and Siddiq 2002). Primer creation does not require prior sequence information. ISSR offers certain advantages than RAPD in terms of reliability, as the primers used are longer and more specific. The disadvantages of this strategy are its dominating inheritance style and repeatability concerns. Single primer amplification reaction (SPAR) and directed amplification of minisatellite-region DNA (DAMD) are two techniques linked to ISSR.

ISSR was used to determine the genetic diversity of four *Rheum tanguticum* populations: 340 of the 342 bands generated were polymorphic (PPB= 99.42%) (Hu et al. 2011). ISSR-PCR was used

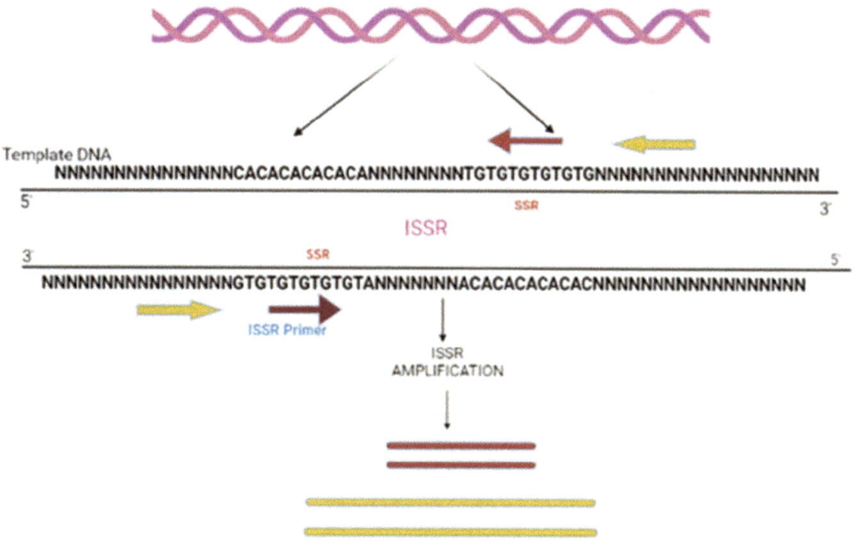

FIGURE 6.5 An overview of the steps involved in ISSR.

to rapidly and accurately identify species belonging to the genus. Twelve *Malva* genotypes were analyzed using 13 ISSR primers. The ISSR-4 and ISSR-2 primers were found to be suitable for identifying all species examined (Jedrzejczyk and Rewers 2020). ISSR markers were utilized to distinguish bitter asafoetida (*Ferula pseudalliacea*) from sweet asafetida (*Ferula assafoetida*). ISSR-16 and ISSR-55 were critical in establishing (Jafari et al. 2019). ISSR markers were used to examine the genetic diversity of *Satureja khuzistanica* populations. One hundred and twenty bands were polymorphic out of 157 (76.4%). In the instance of primer IS15, a maximum of 16 amplified bands were seen (Hadian et al. 2017). *Eruca sativa* is a valuable medicinal herb. The RAPD approach was employed to determine the genotoxicity of three heavy metals in *E. sativa*, namely Zn, Cd and Pb. Cd2+>Pb2+>Zn2+ were the metals with the highest genotoxic potentials (Al-Qurainy 2010).

ISSR markers were used to authenticate samples of the Chirayat complex, which contains several adulterants in addition to *Swertia chirayita*. The complex contains numerous species of the genus *Swertia*, including *Swertia angustifolia*, *Swertia chirayita*, *Swertia cordata*, *Swertia densifolia*, *Swertia lurida*, *Swertia ciliate*, *Swertia paniculata*, *Swertia alata* and *Swertia bimaculata*, as well as other non-*Swertia* species (*Andrographis paniculata*, *Enicostemma axillare*, *Exacum tetragonum*). The market sample's DNA and that of three legitimate species were amplified using ISSR primers and the resulting profile was examined for authentication (Tamhankar et al. 2009).

6.2.6 Sequence Characterized Amplified Regions (SCAR)

The SCARs are a single-locus, co-dominant RAPD-derived molecular marker created by Paran and Michelmore (1993). These are constructed by cloning and sequencing desired RAPD segments and constructing long (24-mer) oligonucleotide primers corresponding to the ends of the original RAPD fragment (Weising 2017). An overview of the steps involved in SCAR is given in Figure 6.6. SCAR markers have the potential to function as both dominant and co-dominant markers. It is the primary indicator for medicinal plant authentication (Ganie et al. 2015). Apart from RAPD, SCAR markers can be converted from AFLP, SSR and ISSR sequences. While RAPD markers are simple to use, quick and inexpensive, there are issues with sensitivity and reproducibility. To circumvent these difficulties and enhance the reliability of RAPD markers, they can be transformed to SCAR

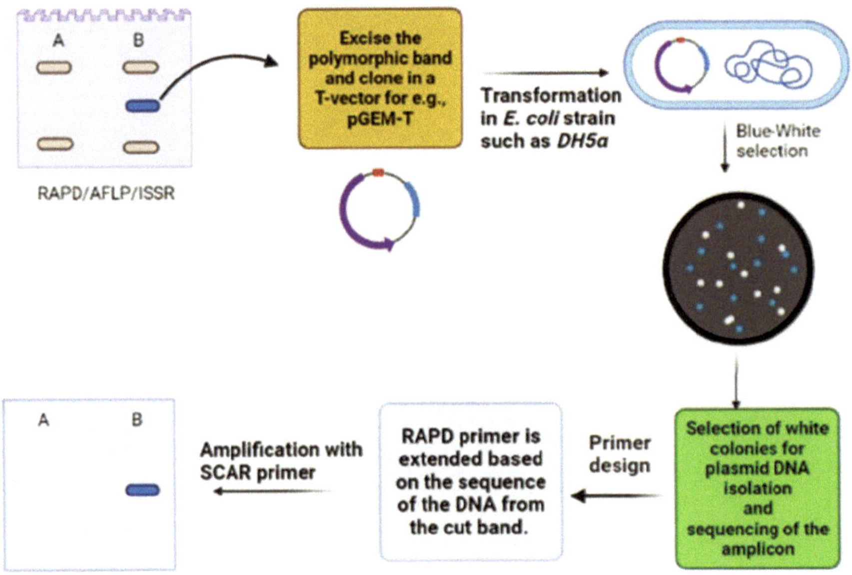

FIGURE 6.6 An overview of the steps involved in SCAR.

markers. These markers have drawbacks in that they require prior sequencing information and have a relatively modest amount of polymorphism.

The SCAR markers have been established to verify the authenticity of a variety of medicinal plants, including *Curcuma alismatifolia* (Anuntalabhochai et al. 2007), *Jatropha curcus* (Basha et al. 2009), *Panax japonicas* (Choi et al. 2008) and *Atractylodes* (Huh and Bang 2006) *Phyllanthus sylvestris* (Dnyaneshwar et al. 2006), *Echinacea* (Adinolfi et al. 2007), *Artemisia* (Lee et al. 2006), *Peucedanum decursivum* (Choo et al. 2009). Goji 10–5, a novel SCAR molecular marker for *Lycium chinense*, was created. It can be utilized to genetically identify and authenticate *L. chinense* in comparison to other species (Liu et al. 2020). A SCAR marker was constructed utilizing a 1135 bp amplicon acquired from RAPD for the purpose of identifying genuine *Viola serpenes* samples (Jha et al. 2020). For the authentication of *Ocimum tenuiflorum*, an ISSR-based species-specific SCAR marker was created (Kumar et al. 2018). A SCAR marker based on RAPD was created to differentiate genuine *Cassia angustifolia* and *C. acutifolia* from their adulterants *Cassia sophera* and *Cassia tora*. The SCAR markers developed were then utilized to examine market samples, with the results indicating that more than 50% of market samples were tainted (Abdin 2013). *Dipsacus japonicas* and *Phlomoides umbrosa* are frequently seen infected with *Dipsacus asper*. SCAR markers were designed to differentiate DNA barcodes (Park et al. 2018). Devaiah and Venkatasubramanian (2008) identified and differentiated *Pueraria tuberosa* from suspected adulterants *Ipomoea mauritiana*, *Adenia hondala* and *Cycas circinalis* using a RAPD-based SCAR marker. *Angelica dahurica* ITS regions of nrRNA genes were sequenced. Using these sequences, a SCAR marker exclusive to *A. dahurica* was generated to differentiate it from its adulterants (Noh et al. 2018). To differentiate *Bacopa monnieri* from its adulterants *Centella asiatica*, *Eclipta alba* and *Malva rotundifolia*, a RAPD-based SCAR marker was developed. A 589 bp polymorphic band identified with the OPAA-3 primer was discovered to be unique to *B. monnieri* and not present in any of its adulterants. SCAR markers based on nrDNA ITS sequences were created to authenticate *Aconitum heterophyllum* (Ativisha) and *Cyperus rotundus* (Musta). When market samples were tested, *Ativisha* could not be recognized using SCAR markers, implying that the herbal medication formulation contains no legitimate species (Seethapathy, Balasubramani and Venkatasubramanian 2014). Different samples of *Lonicera japonica* were collected and studied using the RAPD-SCAR marker, and it was concluded that the marker JYH3-3 was *L. japonica*-specific (Yang et al. 2014). For the identification of *Litchi chinensis*, a SCAR marker based on RAPD was created. SCAR marker L9-6 was discovered to be universally specific for *L. chinensis* samples (Cheng et al. 2015). Three therapeutic plants were identified using RAPD-SCAR: *Akebia quinata*, *Aristolochia manshuriensis* and *Clemetis armandii* (Moon et al. 2015). A SCAR marker based on RAPD was created for the purpose of identifying and authenticating *Podophyllum hexandrum* (Al-Shaqha, Mohsin and Chaudhary 2014). Using a RAPD-based SCAR marker, *Fritillariae cirrhosae* may be differentiated from other Fritillariae sp (Xin et al. 2014). To enable accurate identification and classification of *Knema andamanica*, RAPD, SCAR and conserved 18S rDNA markers were generated (Sheeja et al. 2013).

6.2.7 Loop-mediated Isothermal Amplification (LAMP)

Notomi et al. (2000) devised a simple and rapid technology for DNA amplification called loop-mediated isothermal amplification. Unlike PCR, which uses a sequence of alternate temperatures, isothermal amplification uses a constant temperature (60–65°C) and does not require a thermal cycler. It makes use of four primers that recognize six separate sections of the target base sequence and requires just *Bacillus stearothermophilus* DNA (Bst DNA) polymerase. The four primers are as follows: 1) Forward inner primer (FIP), 2) Forward outer primer (FOP), 3) Backward inner primer (BIP) and 4) Backward outer primer (BOP) (BOP). Gene amplification occurs in the LAMP reaction by repeatedly performing two types of elongation reactions via loop regions (i.e., template self-elongation from the stem-loop structure generated at the 3'-terminals and subsequent binding and elongation of fresh primers to the loop region) (Girish and Karabasanavar 2020). By observing the turbidity created by a by-product of amplification, magnesium pyrophosphate, LAMP products can be viewed with the naked eye. Additionally, other fluorescent dyes such as SYBR green, calcein and others can be used for detection. RT-LAMP is a method for detecting RNA that combines reverse transcription with LAMP. According to some reports, this marker has the ability to amplify a specific gene from the entire genome by detecting a single nucleotide change (Parida et al. 2008). The technique's advantages are its quickness, simplicity, specificity, low cost and efficiency, but its disadvantage is the lengthy process of primer design. LAMP assay provides good results in a variety of disciplines, including the development of the LAMP approach (mRT-LAMP-LFB) for the diagnosis of COVID-19 (Zhu et al. 2020).

A LAMP marker based on RAPD was created to differentiate *Nigella sativa*, a medical plant, from potential adulterants (Ganie et al. 2013). *Morinda officinalis* was differentiated from its adulterants *Morinda umbellata*, *Damnacanthus indicus* and *Schisandra propinqua* utilizing LAMP in conjunction with bioinformatics approaches and DNA barcoding based on the ITS2 sequence of the nuclear ribosomal DNA (Wang et al. 2020). A set of LAMP primers was constructed based on ribosomal DNA ITS sequences for the purpose of authenticating the *Polygonum multiflorum* (Lee and Hxiao 2019). The LAMP approach was used to build a quick identification protocol for *Dendrobium officinale* in order to accurately separate it from other *Dendrobium* species (Lu et al. 2019). Additionally, the LAMP approach was employed to distinguish Saffron (*Crocus sativus*) from its adulterants. *Carthamus tinctorius* petals (safflower), *Curcuma longa* powdered rhizomes (turmeric), *Calendula officinalis*, *Daucus carota*, *Zea mays* and *Nelumbo nucifera* are the most common adulterants. Six distinct LAMP primers were developed based on the ITS2 nucleotide sequence of *C. sativus* nrDNA (Zhao et al. 2016). Additionally, LAMP markers have been created for a number of additional medicinal plants, including *Catharanthus roseus* (Chaudhary et al. 2012), *Panax ginseng* (Sasaki, Komatsu and Nagumo 2008), *Curcuma longa* (Sasaki and Nugumo 2007) and *Zingiber officinale* (Chaudhary et al. 2014).

6.3 CONCLUSION

Looking at the current scenario of increasing demand for herbal products, there is a need for high-quality and authentic products for ensuring therapeutic effectiveness. DNA-based markers have many advantages and are superior which makes them suitable for the identification of medicinal plants and their products but there are certain limitations as well. Some of the DNA markers require good quality and quantity of DNA for analysis but fluctuations in temperature and pH while processing herbal products might lead to degradation or fragmentation of the DNA. Also, the presence of secondary metabolites may affect PCR analysis and DNA isolation by inhibiting certain enzymes. Usually more than one herb is used in the preparation of a drug. This also makes the process cumbersome and complicated. Genetic profiling alone is not sufficient for the complete analysis of a herbal drug. Sometimes the adulterants are from the same plant like, for example, if the roots of a given plant is responsible for its therapeutic effects but instead its flowers are being sold. In such cases, analysis by molecular markers may not be sufficient and require macroscopic and microscopic identification. Along with this, for judging the efficiency of a drug, standardized phytochemical analysis by chromatographic and spectroscopic methods is also required. Dominant markers like RAPD, ISSR and AFLP are a good choice for studying the genetic diversity of plants but should be converted to SCAR or LAMP markers for authentication. The best choice out of the markers discussed above for authentication is SCAR, LAMP and DNA barcoding along with morphological, anatomical and chemical analysis. Unfortunately, there is no ideal marker which can assure 100% authentication of the drug, rather it involves set of methods, in a correct order to achieve this goal. For future purposes, a comprehensive database consisting of plant species, its genetic profile, macro- and microscopic properties along with its chemical profile may prove to be useful for authentication.

REFERENCES

Abd El-Twab, M.H. and Zahran, F.A. 2010. RAPD, ISSR and RFLP analysis of phylogenetic relationships among congeneric species (Anthemideae, Asteraceae) in Egypt. *Int. J. Bot.* 6 (1): 1–10.

Abdin, M.Z. 2013. Authentication of *Cassia angustifolia* and *Cassia acutifolia* in herbal medicines employing SCAR markers. *Planta Med.* 79 (5): P120.

Aboukhalid, K., Machon, N., Lambourdière, J., Abdelkrim, J., Bakha, M., Douaik, A., Korbecka-Glinka, G., Gaboun, F., Tomi, F., Lamiri, A. and Al-Faiz, C. 2017. Analysis of genetic diversity and population structure of the endangered *Origanum compactum* from Morocco, using SSR markers: Implication for conservation. *Biol. Conser.* 212: 172–182.

Abou Zid, S. 2014. Authentication of *Silybum marianum* varieties using RAPD analysis. *Plant Tissue Cult. Biotechnol.* 24: 57–63.

Adhikari, S.P., Meng, S., Wu, Y.J., Mao, Y.P., Ye, R.X., Wang, Q.Z., Sun, C., Sylvia, S., Rozelle, S., Raat, H. and Zhou, H. 2020. Epidemiology, causes, clinical manifestation and diagnosis, prevention and control of coronavirus disease (COVID-19) during the early outbreak period: a scoping review. *Infect. Dis. Poverty* 9 (1): 1–12.

Adinolfi, B., Chicca, A., Martinotti, E., Breschi, M.C. and Nieri, P. 2007. Sequence characterized amplified region (SCAR) analysis on DNA from the three medicinal *Echinacea* species. *Fitoterapia* 78 (1): 43–45.

Ahmedullah, M. and Nayar M.P. 1999. *Red Data Book of Indian Plants*, Vol. 4. Botanical Survey of India, Calcutta, India.

Ahn, J.Y., Lee, J.W., Lim, H.I. and Hong, K.N. 2020. Genetic diversity and structure of *Prunus padus* populations in South Korea based on AFLP markers. *For. Sci. Technol.* 16 (4): 171–179.

Akkaya, M.S., Bhagwat, A.A. and Cregan, P.B. 1992. Length polymorphisms of simple sequence repeat DNA in soybean. *Genetics* 132 (4): 1131–1139.

Ali, Z., Ganie, S.H., Narula, A., Sharma, M.P. and Srivastava, P.S. 2013. Intra-specific genetic diversity and chemical profiling of different accessions of *Clitoria ternatea* L. *Ind. Crops Prod.* 43: 768–773.

Al-Qurainy, F. 2010. Application of inter simple sequence repeat (ISSR marker) to detect genotoxic effect of heavy metals on *Eruca sativa (L). Afr. J. Biotechnol.* 9 (4): 467–474.

Al-Shaqha, W.M., Mohsin, K. and Chaudhary, A.A. 2014. SCAR molecular markers for identification and authentication of medicinal plants *Podophyllum hexandrum* Royle. *Asian J. Biochem. Pharm. Res.* 4: 66–75.

Anuntalabhochai, S., Sitthiphrom, S., Thongtaksin, W., Sanguansermsri, M. and Cutler, R.W. 2007. Hybrid detection and characterization of *Curcuma* spp. using sequence characterized DNA markers. *Sci. Hortic.* 111 (4): 389–393.

Archak, S. 2000. Plant DNA fingerprinting: An overview. *AgBiotechNet* 2: 1–5.

Arif, I.A., Bakir, M.A., Khan, H.A., Al Farhan, A.H., Al Homaidan, A.A., Bahkali, A.H., Sadoon M.A. and Shobrak, M. 2010. Application of RAPD for molecular characterization of plant species of medicinal value from an arid environment. *Genet. Mol. Res.* 9 (4): 2191–2198.

Bai, R.X., Peng, J.Y., Li, L., Zhang, Y. and Li, X.Y. 2009. Infraspecific classification of *Ziziphus jujuba Mill.* Using AFLP marker technique. *Acta Hortic.* 840: 89–96.

Banikamali, M., Soltanloo, H., Ramezanpour, S.S., Yamchi, A. and Sorahinobar, M. 2020. Identification of salinity responsive genes in lavender through cDNA-AFLP. *Biotechnol. Rep.* 28: DOI: https://doi.org/10.1016/j.btre.2020.00520

Basha, S.D., Francis, G., Makkar, H.P.S., Becker, K. and Sujatha, M. 2009. A comparative study of biochemical traits and molecular markers for assessment of genetic relationships between *Jatropha curcas* L. germplasm from different countries. *Plant Sci.* 176 (6): 812–823.

Bhattacharyya, P., Ghosh, S., Mandi, S.S., Kumaria, S. and Tandon, P. 2017. Genetic variability and association of AFLP markers with some important biochemical traits in *Dendrobium thyrsiflorum*, a threatened medicinal orchid. *S. Afr. J. Bot.* 109: 214–222.

Biswas, K. and Biswas, R. 2013. Identification of medicinal plants using PCR-RFLP in Dasamula – an Ayurvedic drug. *J. Pharm. BioSci.* 3: 94–99.

Biswas, K., Kapoor, A. and Biswas, R. 2013. Authentication of Herbal Medicinal Plant- *Boerhavia diffusa L.* Using PCR-RFLP. *Curr. Trends Biotechnol. Pharm.* 7 (3): 716–724.

Botstein, D., White, R.L., Skolnick, M. and Davis, R.W. 1980. Construction of a genetic linkage map in man using restriction fragment length polymorphisms. *Am. J. Hum. Genet.* 32 (3): 314–331.

Brower, V. 2008. Back to nature: extinction of medicinal plants threatens drug discovery. *JNCI* 100 (12): 838–839.
Caetano-Anollés, G., Bassam, B.J. and Gresshoff, P.M. 1991. DNA amplification fingerprinting: A strategy for genome analysis. *Plant Mol. Biol. Rep.* 9: 294–307.
Cehula, M., Mlček, J., Juríková, T., Žiarovská, J., Paulen, O., Dokoupil, L., Adamkova, A. and Babosova, R. 2020. Assessment of genetic diversity of edible honeysuckle monitored through RAPD in relative to bioactive substances. *Agronomy* 10 (6): 868.
Chaudhary, A.A., Mohsin, M. and Ahmad, A. 2012. Application of loop-mediated isothermal amplification (LAMP)-based technology for authentication of *Catharanthus roseus* (L.) G. Don. *Protoplasma* 249 (02): 417–422.
Chaudhary, A.A., Khan, M., Al-Shaqha, W.M., Alharbi, M. and Al-Khamees, O.A. 2014. Rapid and easy molecular authentication of medicinal plant *Zingiber officinale* Roscoe by loop-mediated isothermal amplification (LAMP)-based marker. *J. Med. Plants. Res.* 8 (20): 756–762.
Chen, F., Wang, A., Chen, K., Wan, D. and Liu, J. 2009. Genetic diversity and population structure of the endangered and medically important *Rheum tanguticum* (Polygonaceae) revealed by SSR Markers. *Biochem. Syst. Ecol.* 37 (5): 613–621.
Cheng, J., Long, Y., Khan, A., Wei, C., Fu, S. and Fu, J. 2015. Development and significance of RAPD-SCAR markers for the identification of *Litchi chinensis Sonn.* By improved RAPD amplification and molecular cloning. *Electron. J. Biotechnol.* 18: 35–39.
Choi, Y.E., Ahn, C.H., Kim, B.B. and Yoon, E. S. 2008. Development of species specific AFLP-Derived SCAR marker for authentication of *Panax japonicus* C. A. MEYER. *Bio. Pharm. Bull.* 31(1): 135–138.
Choo, B.K., Moon, B.C., Ji, Y., Kim, B.B., Choi, G., Yoon, T. and Kim, H.K. 2009. Development of SCAR markers for the discrimination of three species of medicinal plants, *Angelica decursiva (Peucedanum decursivum), Peucedanum praeruptorum and Anthricus sylvestris,* based on the internal transcribed spacer (its) sequence and random amplified polymorphic dna (RAPD). *Bio. Pharm. Bull.* 32 (1): 24–30.
Costa, R., Pereira, G., Garrido, I., Tavares-de-Sousa, M.M. and Espinosa, F. 2016. Comparison of RAPD, ISSR, and AFLP molecular markers to reveal and classify orchard grass (*Dactylis glomerata L.*) germplasm variations. *PLoS ONE* 11 (4): DOI: https://doi.org/10.1371/journal.pone.0152972
De la Rosa, R., Angiolillo, A., Guerrero, C., Pellegrini, M., Rallo, L., Besnard, G., Bervillé, A., Martin, A. and Baldoni, L. 2003. A first linkage map of olive (*Olea europaea* L.) cultivars using RAPD, AFLP, RFLP and SSR markers. *Theor. Appl. Genet.* 106: 1273–1282.
Devaiah, K.M. and Venkatasubramanian, P. 2008. Development of SCAR Marker for Authentication of *Pueraria Tuberosa* (Roxb. Ex. Willd.) DC. *Curr. Sci.* 94 (10): 1306–309.
Dnyaneshwar, W., Preeti, C., Kalpana, J. and Bhushan, P. 2006. Development and application of RAPD-SCAR marker for identification of *Phyllanthus emblica* L. *Biol. Pharm. Bull.* 29 (11): 2313–2316.
El-Bakatoushi, R. and Ahmed, D.G.A. 2018. Evaluation of genetic diversity in wild populations of *Peganum harmala* L., a medicinal plant. *JGEB* 16 (1): 143–151.
El-Demerdash, E.S.S., Elsherbeny, E.A., Salama, Y.A.M. and Ahmed, M.Z. 2019. Genetic diversity analysis of some Egyptian *Origanum* and *Thymus* species using AFLP markers. *J. Genet. Eng. Biotechnol.* 17 (1): 13.
El-Kamali, H.H., Habeballa, R., Abdalla, I., Mohammed, A.Y., Abdelkarim, N.D., Abbas, I.M. and Ali, S. M. 2010. Genetic relationships of two *Pulicaria* species and identification of their putative hybrids using RAPD markers. *J. Appl. Sci* 8 (6): 687–693.
Etminan, A., Pour-Aboughadareh, A., Noori, A., Ahmadi-Rad, A., Shooshtari, L., Mahdavian, Z. and Yousefiazar-Khanian, M. 2018. Genetic relationships and diversity among wild *Salvia* accessions revealed by ISSR and SCoT markers. *Biotechnol. Biotechnol. Equip.* 32(3): 610–617.
Fatma, F., Amarnath, K.P. and Suneetha, T.B. 2016. DNA Fingerprinting of Medicinal plant by RAPD and RFLP of *Piper nigrum. Res. J. Chem. Environ.* 4 (4): 47–51.
Felbinger, C., Kutzsche, F., Mönkediek, S. and Fischer, M. 2020. Genetic profiling: Differentiation and identification of hazelnut cultivars (*Corylus avellana* L.) using RAPD-PCR. *Food Control* 107: DOI: https://doi.org/10.1016/j.foodcont.2019.106791
Feng, T., Liu, S. and He, X. J. 2010. Molecular authentication of the traditional Chinese medicinal plant *Angelica sinensis* based on internal transcribed spacer of nrDNA. *Electron. J. Biotechnol.* 13 (1): DOI: 10.2225/vol13-issue1-fulltext-13

Foster, J.T., Bull, R.L. and Keim, P. 2020. Ricin forensics: comparisons to microbial forensics. *Microbial Forensics*: 241–250. Academic Press.

Ganie, S.H., Yadav, D., Ahmad, A., Chadhry, A. and Asif, M. 2013. Authentication of traditional crop Kalongi (*Nigella sativa L.*) by LAMP marker. *Ind. J. Res. Pharm. Biotechnol.* 1: 765–771.

Ganie, S.H., Upadhyay, P., Das, S. and Sharma, M.P. 2015. Authentication of medicinal plants by DNA markers. *Plant Gene* 4: 83–99.

Gawel, N. J., Jarret, R.L. and Whittemore, A. P. 1992. Restriction fragment length polymorphism (RFLP)-based phylogenetic analysis of *Musa. Theor. Appl. Genet.* 84 (3): 286–290.

Gebhardt, C. 2007. Molecular Markers, Maps and Population Genetics. In *Potato Biology and Biotechnology*: 77–89. Amsterdam: Elsevier.

Ghosh, S., Majumder, P.B. and Mandi, S.S. 2011. Species-specific AFLP markers for identification of *Zingiber officinale, Z. montanum and Z. zerumbet* (Zingiberaceae). *Genet. Mol. Res.* 10: 218–229.

Girish, P.S. and Karabasanavar, N.S. 2020. Molecular techniques for speciation of meat. Meat Quality Analysis. *Advanced Evaluation Methods, Techniques, and Technologies*: 133–151. Oxford, UK: Academic Press.

Gong, W., Fu, C. X., Luo, Y.P. and Qiu, Y.X. 2006. Molecular Identification of *Sinopodophyllum hexandrum* and *Dysosma* Species using cpDNA Sequences and PCR-RFLP Markers. *Planta Med.* 72 (7): 650–652.

Gupta, M., Chyi, Y. S., Romero-Severson, J. and Owen, J. L. 1994. Amplification of DNA markers from evolutionarily diverse genomes using single primers of simple-sequence repeats. *Theor. Appl. Genet.* 89 (7): 998–1006.

Gupta, D.D. and Mandi, S.S. 2014. Species specific AFLP markers for authentication of *Zanthoxylum acanthopodium* and Zanthoxylum *oxyphyllum. J. Med. Plants. Stud.* 1: 1–9.

Hadian, J., Raeisi, S., Azizi, A., Pezhmanmehr, M. and Sarkhosh, A. 2017. Genetic diversity of natural populations of medicinally valuable plant *Satureja khuzistanica* Jamzad based on ISSR markers. *Braz. J. Bot* 40: 771–781.

Hadipour, M., Kazemitabar, S.K., Yaghini, H. and Dayani, S. 2020. Genetic diversity and species differentiation of medicinal plant Persian Poppy (*Papaver bracteatum L.*) using AFLP and ISSR markers. *Ecol. Genet. Genom.* 16: DOI: https://doi.org/10.1016/j.egg.2020.100058

Hadrys, H., Balick, M. and Schierwater, B. 1992. Applications of random amplified polymorphic DNA (RAPD) in molecular ecology. *Mol. Ecol.* 1 (1): 55–63.

Hamouda, M. 2019. Molecular analysis of genetic diversity in population of *Silybum marianum* (L.) Gaertn in Egypt. *J. Genet. Eng. Biotechnol.* 17 (1): 12.

Han, Y.C., Teng, C.Z., Zhong, S., Zhou, M.Q., Hu, Z.L. and Song, Y. C. 2007. Genetic variation and clonal diversity in populations of *Nelumbo nucifera* (Nelumbonaceae) in central China detected by ISSR markers. *Aquat. Bot.* 86 (1): 69–75.

Hon, C.C., Chow, Y.C., Zeng, F.Y. and Leung, F.C.C. 2003. Genetic authentication of *Ginseng* and other traditional Chinese medicine. *Acta Pharmacol. Sin.* 24: 841–846.

Horn, T., Völker, J., Rühle, M., Häser, A., Jürges, G. and Nick, P. 2014. Genetic authentication by RFLP versus ARMS? The case of Moldavian dragonhead (*Dracocephalum moldavica* L.). *Eur. Food Res. Technol.* 238: 93–104.

Hosokawa, K., Minami, M., Kawahara, K., Nakamura, I. and Shibata, T. 2000. Discrimination among three species of medicinal *Scutellaria* plants using RAPD markers. *Planta Med.* 66: 270–272.

Hossein Jafari, S., Sepehry, A., Soltanloo, H. and Karimian, A.A. 2019. Genetic differentiation between bitter and sweet asafetida plants using ISSR markers. *Mol Biol Rep* 46: 1069–1078.

Hu, Y., Wang, L. and Li, Y. 2011. Inter-simple sequence repeats (ISSR) primer screening and preliminary evaluation of genetic diversity in *Rheum tanguticum. J. Med. Plant Res.* 5 (24): 5907–5911.

Huang, Y., Zhou, N., Yang, M., Shen, Y. and Zhang, D. 2019. A comparative study of the population genetics of wild and cultivated populations of *Paris polyphylla var. yunnanensis* based on amplified fragment length polymorphism markers. *Ecol. Evol.* 9 (18): 10707–10722.

Huh, M.K., Bang, K.H. and Choi, J.S. 2007. Identification of *Atractylodes japonica* and *A. macrocephala* by AFLP and SCAR markers. *Hort. Environ. Biotechnol.* 48 (3): 201–205.

Ismail, N.A., Rafii, M.Y., Mahmud, T.M.M., Hanafi, M.M. and Miah, G. 2019. Genetic diversity of torch ginger (*Etlingera elatior*) germplasm revealed by ISSR and SSR markers. *BioMed Res. Int.* 4: 1–14.

Jarret, R.L., Gawel, N., Whittemore, A. and Sharrock, S. 1992. RFLP-based phylogeny of *Musa* species in Papua New Guinea. *Theo. Appl. Gen.* 84: 579–584.

Jedrzejczyk, I. and Rewers, M. 2020. Identification and genetic diversity analysis of edible and medicinal *Malva* species using flow cytometry and ISSR molecular markers. *Agronomy* 10 (5): 650.

Jena, S.N., Kumar, S. and Nair, N.K. 2009. Molecular phylogeny in *Indian Citrus L.* (Rutaceae) inferred through PCR-RFLP and trnL-trnF sequence data of chloroplast DNA. *Sci. Hort.* 119 (4): 403–416.

Jha, S.R., Naz, R., Asif, A., Okla, M.K., Soufan, W., Al-Ghamdi A.A. and Ahmad, A. 2020. Development of an in vitro propagation protocol and a Sequence Characterized Amplified Region (SCAR) Marker of *Viola serpens* Wall. ex Ging. *Plants* 9 (2): 246.

Joshi, K., Chavan, P., Warude, D. and Patwardhan, B. 2004. Molecular markers in herbal drug technology. *Curr. Sci.* 87: 159–165.

Jung, J.K., Park, S.W., Liu, W.Y. and Kang, B. C. 2010. Discovery of single nucleotide polymorphism in *Capsicum* and SNP markers for cultivar identification. *Euphytica* 175: 91–107.

Karimi, A., Hadian, J., Farzaneh, M. and Khadivi-Khub, A. 2015. Evaluation of genetic variability, rust resistance and marker-detection in cultivated *Artemisia dracunculus* from Iran. *Gene* 554 (2): 224–232.

Keeratinijakal, V., Kladmook, M. and Laosatit, K. 2010. Identification and characterization of *Curcuma comosa roxb*. Phytoestrogens-producing plant, using AFLP markers and morphological characteristics. *J. Med. Plants. Res.* 4 (24): 2651–2657.

Khan, S., Mirza, K.J., Tayaab, M. and Abdin, M.Z. 2009. RAPD Profile for authentication of medicinal plant *Glycyrrhiza glabra* L. *J. Food Safety.* 11: 24–28.

Khan, S., Mirza, K.J. and Abdin, M.Z. 2010. Development of RAPD markers for authentication of medicinal plant *Cuscuta Reflexa*. *EurAsia J. BioSci.* 4: 1–7.

Kumar, A., Mishra, P., Singh, S.C. and Sundaresan, V. 2014. Efficiency of ISSR and RAPD markers in genetic divergence analysis and conservation management of *Justicia adhatoda* L., a medicinal plant. *Plant Syst Evol* 300: 1409–1420.

Kumar, A., Rodrigues, V., Mishra, P., Baskaran, K., Shukla, A.K., Shasany, A.K. and Sundaresan, V. 2018. ISSR-derived species-specific SCAR marker for rapid and accurate authentication of *Ocimum tenuiflorum* L. *Planta Med.* 84 (02): 117–122.

Laila, U., Akram, M., Shariati, M.A., Hashmi, A.M., Akhtar, N., Tahir, I.M., Ghauri, A.O., Munir, N., Riaz, M., Akhter, N., Shaheen, G., Ullah, Q., Zahid, R. and Ahmad, S. 2019. Role of medicinal plants in HIV/AIDS therapy. *Clin. Exp. Pharmaco.l Physiol.* 46: 1063–1073.

Lee, M.Y., Doh, E.J., Park, C.H., Kim, Y.H., Kim, E.S., Ko, B.S. and Oh, S.E. 2006. Development of SCAR marker for discrimination of *Artemisia princeps* and *A. argyi* from other *Artemisia* Herbs. *Biol. Pharm. Bull.* 29 (4): 629–633.

Lee, M.S. and Hxiao, H.J. 2019. Rapid and sensitive authentication of *Polygonum multiflorum* (He-Shou-Wu) of Chinese medicinal crop using specific isothermal nucleic acid amplification. *Ind. Crop. Prod.* 129: 281–289.

Li, X., Ding, X., Chu, B., Ding, G., Gu, S., Qian, L., Wang, Y. and Zhou, Q. 2007. Molecular authentication of *Alisma orientale* by PCR-RFLP and ARMS. *Planta Med.* 73 (1): 67–70.

Li, S., Liu, S.L., Pei, S.Y., Ning, M.M. and Tang, S.Q. 2020. Genetic diversity and population structure of *Camellia huana* (Theaceae), a limestone species with narrow geographic range, based on chloroplast DNA sequence and microsatellite marker. *Plant Divers.* 2 (5): 343–350.

Litt, M. and Luty, J.A. 1989. A hypervariable microsatellite revealed by in vitro amplification of a dinucleotide repeat within the cardiac muscle actin gene. *Am. J. Human Genet.* 44: 397–401.

Liu, X., Cheng, J., Mei, Z., Wei, C., Khan, M., Peng, J. and Fu, J. 2020. SCAR marker for identification and discrimination of specific medicinal *Lycium chinense* Miller from *Lycium* species from ramp-PCR RAPD fragments. *3 Biotech.* 10 (8).

Liu, Y., Geng, Y., Song, M., Zhang, P., Hou J. and Wang, W. 2019. Genetic structure and diversity of *Glycyrrhiza* populations based on transcriptome SSR Markers. *Plant Mol Biol Rep.* 37: 401–412.

Lu, Y., Wen-Ru, W., Hua, Z., Hui-Li, L. and Fei, F. 2019. Rapid identification of *Dendrobium officinale* using Loop-Mediated Isothermal Amplification (LAMP) method. *Chin. J. Nat. Med.* 17 (5): 337–345.

Ma, J.Q., Yao, M.Z., Ma, C.L., Wang X.C., Jin, J.Q., Wang, X.M. and Chen, L. 2014. Construction of an SSR-based genetic map and identification of QTLs for catechins content in tea plant (*Camellia sinensis*). *PLoS ONE* 9 (3): DOI: https://doi.org/10.1371/journal.pone.0093131

Meyer, W., Mitchell, T.G., Freedman, E.Z. and Vilgalys, R. 1993. Hybridization probes for conventional DNA fingerprinting used as single primers in the polymerase chain reaction to distinguish strains of *Cryptococcus neoformans*. *J. Clin. Microbiol.* 31 (9): 2274–2280.

Misra, A., Shasany, A.K., Shukla, A.K., Darokar, M.P., Singh, S.C., Sundaresan, V., Singh, J., Bagchi, G.D., Jain, S.P., Saikia, D. and Khanuja, S.P. 2010a. AFLP markers for identification of *Swertia* species (Gentianaceae). *Genet. Mol. Res.* 9 (3): 1535–1544.

Misra, A., Shukla, A.K., Shasany, A.K., Sundaresan, V., Jain, S.P., Singh, S.C., Bagchi, G.D. and Khanuja, S.P. 2010b. AFLP markers for identification of *Aconitum* species. *Med. Aromat. Plant Sci. Biotechnol.* 4: 15–19.

Mizukami, H., Ohbayashi, K., Umetsu, K. and Hiraoka, N. 1993. Restriction fragment length polymorphisms of medicinal plants and crude drugs. II. Analysis of *Glehnia littoralis* of different geographical origin. *Biol. Pharm. Bull.* 16: 611–612.

Moon, B.C., Ji, Y., Lee, Y.M., Kang, Y.M. and Kim, H.K. 2015. Authentication of *Akebia quinata* DECNE. from its common adulterant medicinal plant species based on the RAPD-derived SCAR markers and multiplex-PCR. *Genes Genom.* 37: 23–32.

Neila, T.F. and Mohamed, M. 2001. *Hedysarum* phylogeny mediated by RFLP analysis of nuclear ribosomal DNA – RFLP and *Hedysarum* phylogeny. *Genet. Resour. Crop Evol.* 48: 339–345.

Ng, W.L. and Tan, S.G. 2015. Inter-Simple Sequence Repeat (ISSR) markers. *ASM Sci. J.* 9 (1): 30–39.

Noh, P., Kim, W.J., Yang, S., Park, I. and Moon, B. C. 2018. Authentication of the herbal medicine *Angelicae Dahuricae* Radix using an ITS sequence-based multiplex SCAR Assay. *Molecules* 23 (9): 2134.

Notomi, T., Okayama, H., Masubuchi, H., Yonekawa, T., Watanabe, K., Amino, N. and Hase, T. 2000. Loop-mediated isothermal amplification of DNA. *Nuc. Acids Res.* 28 (12): e63.

Oyama, K., Hernandez-Verdugo, S., Sanchez, C., González-Rodríguez, A., Sanchez-Pena, P., Garzon-Tiznado, J.A., and Casas, A. 2006. Genetic structure of wild and domesticated populations of *Capsicum annuum* (Solanaceae) from northwestern Mexico analyzed by RAPDs. *Gen. Resour. Crop Evol.* 53: 553–562.

Paran, I. and Michelmore, I.W. 1993. Development of reliable PCR-based markers linked to downy mildew resistance genes in lettuce. *Theo. Appl. Gen.* 85: 985–993.

Parida, M., Sannarangaiah, S., Dash, P.K., Rao, P.V.L. and Morita, K. 2008. Loop mediated isothermal amplification (LAMP): a new generation of innovative gene amplification technique; perspectives in clinical diagnosis of infectious diseases. *Rev. Med. Virol.* 18: 407–421.

Park, I., Yang, S., Kim, W. J., Noh, P., Lee, H.O. and Moon, B.C. 2018. Authentication of Herbal Medicines *Dipsacus asper* and *Phlomoides umbrosa* Using DNA barcodes, chloroplast genome, and sequence characterized amplified region (SCAR) marker. *Molecules* 23 (7): 1748.

Passinho-Soares, H., Felix, D., Kaplan, M.A., Margis-Pinheiro, M. and Margis, R. 2006. Authentication of medicinal plant botanical identity by amplified fragmented length polymorphism dominant DNA marker: inferences from the *Plectranthus* genus. *Planta Med.* 72: 929–93.

Patzak, J. 2001.Comparison of RAPD, STS, ISSR and AFLP molecular methods used for assessment of genetic diversity in hop (*Humulus lupulus* L.). *Euphytica* 121: 9–18.

Pradeep Reddy, M., Sarla, N. and Siddiq, E.A. 2002. Inter simple sequence repeat (ISSR) polymorphism and its application in plant breeding. *Euphytica* 128: 9–17.

Pradhan, A., Gupta, V., Mukhopadhyay, A., Arumugam N., Sodhi, Y. and Pental, D. 2003. A high-density linkage map in *Brassica juncea* (Indian mustard) using AFLP and RFLP markers. *Theor Appl Genet* 106: 607–614.

Rameshkumar, R., Pandian, S., Rathinapriya, P., Selvi, C. T., Satish, L., Gowrishankar, S., David W.M.L. and Ramesh, M. 2019. Genetic diversity and phylogenetic relationship of *Nilgirianthus ciliatus* populations using ISSR and RAPD markers: Implications for conservation of an endemic and vulnerable medicinal plant. *Biocatal. Agric. Biotechnol.* 18: DOI: 10.1016/j.bcab.2019.101072

Rohela, G.K., Jogam, P., Bylla, P. and Reuben, C. 2019. Indirect regeneration and assessment of genetic fidelity of acclimated plantlets by SCoT, ISSR, and RAPD markers in *Rauwolfia tetraphylla* L.: An endangered medicinal plant. *BioMed Res. Int.* (1): 1–14.

Rossi, A.A.B., de Oliveira, L.O., Venturini, B.A. and dos Santos Silva, R. 2009. Genetic diversity and geographic differentiation of disjunct Atlantic and Amazonian populations of *Psychotria ipecacuanha* (Rubiaceae). *Genetica* 136: 57–67.

Russi, L., Moretti, C., Raggi, L., Albertini, E. and Falistocco, E. 2009. Identifying commercially relevant *Echinacea* species by AFLP molecular markers. *Genome* 52: 912–918.

Sarwat, M., Das, S. and Srivastava, P.S. 2008. Analysis of genetic diversity through AFLP, SAMPL, ISSR and RAPD in *Tribulus terrestris* a medicinal herb. *Plant Cell Rep.* 27: 519–528.

Sasaki, Y. and Nagumo, S. 2007. Rapid identification of *Curcuma longa* and *C. aromatica* by LAMP. *Biol. Pharm Bulletin* 30: 2229–2230.

Sasaki, Y., Komatsu, K. and Nagumo, S. 2008. Rapid detection of *Panax ginseng* by loop-mediated isothermal amplification and its application to authentication of *ginseng*. *Biol. Pharm. Bull.* 31 (09): 1806–1808.

Seethapathy, G.S., Balasubramani, S.P. and Venkatasubramanian, P. 2014. NrDNA ITS sequence-based SCAR marker to authenticate *Aconitum heterophyllum* and *Cyperus rotundus* in ayurvedic raw drug source and prepared herbal products. *Food Chem.* 145: 1015–1020.

Sharmila, S., Akilandeswari, D., Ramya, E.K. and Mowni-ka, S. 2019. Ecology and the RAPD techniques used to assess the genetic diversity in *Pterolobium hexapetalum*, a scrambling medicinal shrub in Maruthamalai and Chennimalai hills, the Western Ghats. *Int. J. Pharm. Sci. Rev. Res.* 11 (3): 72–77.

Sheeja, T.E., Rosana, O.B., Swetha, V.P., Shalini, R.S., Siju, S., Dhanya, R., Rahul, P.R. and Krishnamoorthy, B. 2014. The 18S rDNA gene discriminates between red-listed and unexplored ethnomedicinal species of Myristicaceae restricted to humid tropics of India. *Genet. Resour. Crop. Evol.* 61 (2): 523–535.

Shi, Y., Zhao, M., Yao, H., Yang, P., Xin, T., Li, B., Sun, W. and Chen, S. 2017. Rapidly discriminate commercial medicinal *Pulsatilla chinensis* (Bge.) Regel from its adulterants using ITS2 barcoding and specific PCR-RFLP assay. *Sci Rep* 7 (1): 1–12.

Shirasawa, K., Ishii, K., Kim, C., Ban, T., Suzuki, M., Ito, T., Muranaka, T., Kobayashi, M., Nagata, N., Isobe, S. and Tabata, S. 2013. Development of *Capsicum* EST-SSR markers for species identification and in silico mapping onto the tomato genome sequence. *Mol. Breed.* 31 (1): 101–110.

Sindhu, A., Tehlan, S.K. and Chaudhury, A. 2017. Analysis of genetic diversity among medicinal therapist *Trigonella foenum-graecum* L. genotypes through RAPD and SSR markers. *Acta Physiol. Plant.* 39 (4): 1–15.

Tamhankar, S., Ghate, V., Raut, A. and Rajput, B. 2009. Molecular profiling of Chirayat complex using inter simple sequence repeat (ISSR) markers. *Planta Med.* 75: 1266–1270.

Tilwari, A., Chauhan, D., Sharma R. and Singh, R.K. 2016. Assessment of genetic variations among medicinal plant *Cassia tora* from different geographic regions of Central India using RAPD markers. *Med Aromat Plants* 5 (6). DOI: 10.4172/2167–0412.1000276

Tripathi, N., Saini, N. and Tiwari, S. 2011. Assessment of genetic diversity among *Aloe vera* accessions using amplified fragment length polymorphism. *Int. J. Med. Arom. Plants.* 1: 115–121.

Varma, A. and Shrivastava, N. 2018. Genetic structuring in wild populations of two important medicinal plant species as inferred from AFLP markers. *Plant Biosyst.* 152 (5): 1088–1100.

Vekariya, S., Taviad, K. and Acharya, R.N. 2017. Development of random amplified polymorphic DNA markers for authentication of *Croton tiglium Linn. J. Phytopharmacol.* 6 (3): 164–166.

Verma, V.K., Behera, T.K., Munshi, A.D., Parida, S.K. and Mohapatra, T. 2007. Genetic diversity of ash gourd [*Benincasa hispida (Thunb.) Cogn.*] inbred lines based on RAPD and ISSR markers and their hybrid performance. *Sci. Hort.* 113 (3): 231–237.

Verma, P.C., Chakrabarty, D., Jena, S.N., Mishra, D.K., Singh, P.K., Sawant, S.V. and Tuli, R. 2009. The extent of genetic diversity among *Vanilla* species: Comparative results for RAPD and ISSR. *Ind. Crop Prod.* 29 (2–3): 581–589.

Vieira, M.L.C., Santini, L., Diniz, A.L. and Munhoz, C.D.F. 2016. Microsatellite markers: what they mean and why they are so useful. *Genet. Mol. Biol.* 39 (3): 312–328.

Vigliante, I., Mannino, G. and Maffei, M.E. 2019. Chemical characterization and DNA fingerprinting of *Griffonia simplicifolia* Baill. *Molecules* 24 (6): 1032

Vijayan, D., CP, A., SP, G. and Balachandran, I. 2013. Developing RAPD markers for identification of three source plants of ayurvedic raw drug "Patha". *Med. Plants – Int. J. Phytomed. Relat. Ind.* (1): 55–58.

Vos, P., Hogers, R., Bleeker, M., Reijans, M., van de Lee, T., Hornes, M., Friters, A., Pot, J., Paleman, J. and Kuiper, M. 1995. AFLP: a new technique for DNA fingerprinting. *Nucleic Acids Res.* 23: 4407–4414.

Wang, X., Yang, R., Feng, S., Hou, X., Zhang, Y., Li, Y. and Ren, Y. 2012. Genetic variation in *Rheum palmatum* and *Rheum tanguticum* (Polygonaceae), two medicinally and endemic species in China using ISSR markers. *PLoS ONE* 7(12): DOI: 10.1371/journal.pone.0051667

Wang, S., Liu, Y., Ma, L., Liu, H., Tang, Y., Wu, L., Wang, Z., Li, Y., Wu, R. and Pang, X. 2014. Isolation and Characterization of Microsatellite Markers and Analysis of Genetic Diversity in Chinese Jujube (*Ziziphus jujuba Mill.*). *PLoS ONE* 9 (6): e99842.

Wang, H., Li, X., Liu, X., Oiu, Y., Song, J. and Zhang, X. 2016. Genetic diversity of garlic (*Allium sativum* L.) germplasm from China by fluorescent-based AFLP, SSR and InDel markers. *Plant Breed.* 135 (6): 743–750.

Wang, S., Sui, A., Han, Y., Zhou, Q., Xu, H. and Yao, R. 2020. Rapid and reliable distinguish of *Morinda officinalis* How by loop-mediated isothermal amplification (LAMP) assay. *Res. Sq.*: DOI: 10.21203/rs.3.rs-120446/v1

Weising, K. 2017. *DNA Fingerprinting in Plants*. Boca Raton: CRC Press.

Welsh, J. and McClelland, M. 1990. Fingerprinting genomes using PCR with arbitrary primers. *Nucleic acids res.* 18 (24): 7213–7218.

Williams, J.G., Kubelik, A.R., Livak, K.J., Rafalski, J.A. and Tingey, S.V. 1990. DNA polymorphisms amplified by arbitrary primers are useful as genetic markers. *Nucleic Acids Res.* 18 (22): 6531–6535.

Wiriyakarun, S., Yodpetch, W., Komatsu, K., Zhu, S., Ruangrungsi, N. and Sukrong, S. 2013. Discrimination of the Thai rejuvenating herbs *Pueraria candollei* (White Kwao Khruea), *Butea superba* (Red Kwao Khruea), and *Mucuna collettii* (Black Kwao Khruea) using PCR-RFLP. *J Nat Med* 67: 562–570.

Witsenboer, H., Michelmore, R.W. and Vogel, J. 1997. Identification, genetic localization, and allelic diversity of selectively amplified microsatellite polymorphic loci in lettuce and wild relatives (*Lactuca* spp.). *Genome* 40 (6): 923–936.

Wu, K.S., Jones, R., Danneberger, L. and Scolnik, P.A. 1994. Detection of microsatellite polymorphisms without cloning. *Nucleic Acids Res* 22: 3257–3258.

Xin, G.Z., Lam, Y.C., Maiwulanjiang, M., Chan, G.K., Zhu, K.Y., Tang, W.L., Dong, T.T., Shi, Z.Q., Li, P. and Tsim, K.W. 2014. Authentication of *Bulbus fritillariae cirrhosae* by RAPD-derived DNA markers. *Molecules* 19: 3450–3459.

Yang, L., Khan, M., Mei, Z., Yang, M., Zhang, T., Wei, C., Yang, W., Zhu, L., Long, Y. and Fu, J. 2014. Development of RAPD-SCAR markers for *Lonicera japonica* (Caprifolicaceae) variety authentication by improved RAPD and DNA cloning. *Revista de Biología Tropical* 62 (4): 1649–1657.

Yang, B.C., Lee, M.S., Sun, F.C., Chao, H.H., Chang, W.T., Lin, M.K., Chen, H.J. and Lee, M.S. 2020. Rapid identification of the indigenous medicinal crop *Cinnamomum osmophloeum* from various adulterant *Cinnamomum* species by DNA polymorphism analysis. *Phcog Mag* 16 (68): 64–68.

Yang, M., Abdalrahman, H., Sonia, U., Mohammed, A.I., Vestine, U., Wang, M., Ebadi, A.G. and Toughani, M. 2020. The application of DNA molecular markers in the study of *Codonopsis* species genetic variation, a review. *Cell mol. Biol.* 66 (2): 23–30.

Yao, H., Zhao, Y., Chen, D.F., Chen, J.K. and Zhou, T.S. 2008. ISSR primer screening and preliminary evaluation of genetic diversity in wild populations of *Gycyrrhiza uralensis*. *Biol. Plant* 52 (1): 117–120.

Zhang, Y.B., But, P.P.H., Wang, Z.T. and Shaw, P.C. 2005. Current approaches for the authentication of medicinal *Dendrobium* species and its products. *Plant Genet. Res.* 3(2): 144–148.

Zhao, M., Shi, Y., Wu, L., Guo, L., Liu, W., Xiong, C., Yan, S., Sun, W. and Chen, S. 2016. Rapid authentication of the precious herb saffron by loop-mediated isothermal amplification (LAMP) based on internal transcribed spacer 2 (ITS2) sequence. *Sci Rep* 6 (1): 1–9.

Zhu, X., Wang, X., Han, L., Chen, T., Wang, L., Li, H., Li, S., He, L., Fu, X., Chen, S., Xing, M. and Wang, Y. 2020. Multiplex reverse transcription loop-mediated isothermal amplification combined with nanoparticle-based lateral flow biosensor for the diagnosis of COVID-19. *Biosens. Bioelectron.* 166 (04): DOI: https://doi.org/10.1016/j.bios.2020.112437

Zid, S. 2014. Authentication of *Silybum marianum* varieties using RAPD analysis. *Plant Tissue Cult. Biotechnol.* 24: 57–63.

Zietkiewicz, E., Rafalski, A. and Labuda, D. 1994. Genome fingerprinting by simple sequence repeat (SSR)-anchored polymerase chain reaction amplification. *Genomics* 20: 176–183.

7 Genome-based Barcoding for the Authentication of Medicinal Plants and Their Products

Samia Khan

CONTENTS

7.1 INTRODUCTION

There are approximately 300,000 plant species on our earth (Bertzky et al. 2013), but only a fraction of these can be taxonomically identified using morphological-based methods (Hebert et al. 2003; Bickford et al. 2007; Chase and Fay 2009), which are time-consuming (i.e., Linnean). Adding to the challenge, most of the medicinal plants and their products sold in the markets are either in the dried or powdered form, and then it becomes an arduous task to identify them in the absence of diagnostic morphological traits. Moreover, several plant species, in spite of possessing high genetic divergence, do not display any morphological disparity; hence it becomes difficult to identify such cryptic species under modern taxonomy (Heinrichs et al. 2011).

As a result of a drive to overcome these challenges, the last two decades have witnessed an emergence of a molecular technique termed as "DNA barcoding". Paul Hebert of the University of

DOI: 10.1201/9781003179139-7

Guelph first coined this term in 2003 (Hebert et al. 2003), introducing it as a method for taxon identification. This technique was able to overcome all the challenges faced by traditional taxonomy, also providing more rapid and accurate results than the latter (Gregory 2005). DNA barcoding has been regarded as a "renaissance of taxonomy" (Miller 2007) with its ability to discern specimens even in the absence of diagnostic morphological characteristics.

The targets of DNA barcoding are highly conserved short DNA sequences in which minor nucleotide polymorphisms have occurred (Ali et al. 2014). Hence, through this technique, such species can be identified that are morphologically similar but have genetic dissimilarity (Fišer Pečnikar and Buzan 2014). The variations in their nucleotides help create a DNA barcode that is unique to the species and capable of identifying unknown samples (Khan et al. 2012).

7.2 BASIC CRITERIA OF A DNA BARCODE

DNA barcoding emerged as a powerful tool for species level identification in lepidopteran insects, fishes and birds (Hebert et al. 2003, 2004). Kress and Erickson (2008) have recommended the following as basic criteria for choosing a plant barcode:

1. The proposed barcode must be capable of distinguishing the species.
2. It should possess conserved flanking regions that can aid in developing universal primer sets, which can be PCR amplified.
3. A barcode should have short sequences for easy DNA extractions and PCR amplification.
4. Last, it should be capable of being handled efficiently with bioinformatics tools and analysis.

7.3 USES OF DNA BARCODING

DNA barcoding is aiding in the rapid detection of thousands of new species (Ali et al. 2014). It has also made prominent contributions like in forensics studies, investigation of food adulteration (Valentini, Pompanon and Taberlet 2009), facilitating biodiversity inventories (Lahaye et al. 2008; Valentini, Pompanon and Taberlet 2009) and identifying medicinal plants (Chen et al. 2010, Zheng et al. 2015) and detecting cryptic species

7.4 DNA BARCODING IN PLANTS

In the early works of DNA barcoding, plants were not given much importance as cytochrome oxidase (*COX1*), the universal barcode for animals (Hebert et al. 2003), was not a suitable marker for plants due to its slow evolution (Palmer and Herbon 1988). Henceforth, the search for plant barcodes shifted to chloroplast and nuclear genomes, as they have high substitution rates. Tremendous efforts were carried out to find a potential DNA barcode that was universally functional while also having a high discriminatory power (Kress et al. 2005, Lahaye et al. 2008, Hollingsworth et al. 2011, Li et al. 2011). But, owing to various complex evolutionary processes in plants, like hybridization and polyploidy, it became challenging to identify definite species boundaries (Fazekas et al. 2009). Even after a lot of research, none of the available barcode loci could work across all the species (Chase and Fay 2009) (Chen et al. 2010); thus there is no single universal DNA barcode marker for plants. It was found that single-locus DNA barcodes lack adequate variations in closely related taxa (Li et al. 2014), and so a combination of two or more regions could be considered (Fišer Pečnikar and Buzan 2014). Thus, to enhance the discriminatory power of a DNA barcode, taxonomists suggested a multi-locus method for identifying plant species (Kress and Erickson, 2007, Lahaye et al. 2008, Kane et al. 2012).

Various regions from chloroplast and nuclear genome were suggested for barcoding plant taxa. The Consortium for the Barcode of Life (CBOL) Plant Working Group (Hollingsworth et al. 2009a)

recommended chloroplast regions *rbcL* and *matK* as a two-locus combination due to the ease in the recovery of the *rbcL* sequences and the authentication capability of the *matK* sequence. This two-locus region was supplemented with another plastidial region *psbA-trnH*, and a nuclear region *ITS* for barcoding plants (CBOL 2009). Thus, the regions *rbcL* and *matK* are identified as core barcoding regions, while *psbA-trnH* was considered an important supplementary marker (Fazekas et al. 2012). Some studies established that various multiple independent loci paired with the often-faster mutation rates of nuclear over plastid regions make nuclear markers a better choice for DNA barcoding. Endorsing this fact, Chen et al. (2010) suggested *ITS2* as a promising marker for authenticating plants. A combination of 23 genomic regions (Escapa and Catalano 2013) and 11 plastid and rDNA *ITS2* sequence data were proposed for the study of *Araucaria* (Kranitz et al. 2014). Four markers (*rbcL*, *matK*, *psbA-trnH* and *ITS2*) were tested for species identification for 13 species of *Sida*; *psbA-trnH* and *ITS2* were found to be the best marker or combination of markers for species level identification (Vassou, Kusuma and Parani 2015).

7.4.1 DNA Barcoding and Medicinal Plants

The past three decades have witnessed a resurgence in the use of medicinal plants and their products, with nearly 80% of the world population depending on these products for some part of their primary healthcare (Ekor 2014). Although some of these plants can be accessed from their natural habitat, a vast majority of them and their products are procured from the herbal markets, where they are found in dried or powdered form. Hence, their authentication becomes challenging in the absence of distinct phenotypic traits. The technique of DNA barcoding is able to overcome these challenges of validating medicinal plants and their products unambiguously (Li et al. 2014, Zheng et al. 2015). An example highlighting this is the case of the traditional Chinese medicinal herb "Mu-Tong", an effective drug in the treatment of cardiovascular diseases and commonly used in the treatment of hypertension. Through DNA barcoding, it was found that the herb's by-products and parts available in the market were found to be adulterated. As the plants' wild sources were being depleted, the addition of adulterants ensured their regular supply (Zhang et al. 2015). Mahadani, Sharma and Ghosh (2013) proposed *matK* as the best molecular marker for authentication of *Rauvolfioideae* species. Then, Chen et al. (2010) recommended nr*ITS2* as a standard barcode to identify medicinal plants and their products. This locus could also identify medicinal species of Apocynaceae (Selvaraj et al. 2015), as seen in our studies on arid medicinal species *Rhazya stricta*, a two-locus marker combination *psbK-psbI* + *atpF-atpH* from the chloroplast genome (Khan et al. 2017) and *ITS2* from nuclear genome was proposed (Khan et al. 2019).

7.5 CHARACTERISTICS OF POTENTIAL BARCODE REGIONS

Higher plants contain three genomes: nuclear, mitochondrial and chloroplast genomes. Most of the proposed DNA barcode loci belong to the nucleus or the chloroplast genome. However, at present, DNA barcoding technology relies more on the latter.

7.5.1 Chloroplast Genome

Chloroplast DNA is a circular molecule greatly exceeding the length of commonly used DNA barcodes; however, it is small compared to the nuclear genome and is uniparental in inheritance. With a few notable exceptions, chloroplast genomes are highly conserved (Sugiura 1992). They lack large deletions, insertions and SNPs (single nucleotide polymorphism), thus making them favorable for phylogenetic studies. The low nucleotide substitution (Dong et al. 2012) in their protein-coding genes (Clegg et al. 1994), their presence in each plant cell in a high number of copies, uniparental inheritance, non-recombination and structural stability in both the genic and intergenic regions of the chloroplast makes them favorable targets to be explored as DNA barcoding loci. It provides enough variation to discriminate closely related plants.

There is a lookout for rapidly evolving regions in the chloroplast genome, which are present in all land plants and can be good candidates for robust, universal primers. For various phylogenetic studies of medicinal species of Apocynaceae, the chloroplast genome has been assessed for genus *Asclepias* (Fishbein et al. 2011), *Hoya* (Wanntorp, Gotthardt and Muellner 2011), Rauvolfioideae (Mahadani, Sharma and Ghosh 2013) and *Rhazya* (Khan et al. 2017). Some commonly investigated chloroplast genes for DNA barcoding with their characteristics are mentioned below.

7.5.1.1 Maturase K (*matK*)

The *matK* gene of chloroplast codes for Maturase-like protein, which is involved in Group II intron splicing. In plant systematics, *the matK* gene stands out among other plastid genes as one of the most rapidly evolving coding sections of the plastid genome, showing high levels of discrimination in angiosperm species (Hollingsworth et al. 2011, Vassou et al. 2015). In *matK*, the rate of substitution is much higher than the other coding gene *rbcL* (Barthet and Hilu 2007), thus being a rapidly evolving gene (Soltis and Soltis 2004) it can be a potential candidate for molecular systematics and evolution studies (Notredame, Higgins and Heringa 2000). It is considered to be the closest plant barcode equivalent to the CO1 animal barcode. *MatK* is very useful for the identification of plant families (Xing et al. 2008), and could accurately identify medicinal species of Rauvolfioideae (Mahadani, Sharma and Ghosh 2013).

7.5.1.2 Ribulose Bisphosphate Carboxylase (*rbcL*)

The *rbcL* gene codes for "RUBISCO" ribulose-1, 5-bisphosphate-carboxylase/oxygenase, a free enzyme present in the stroma of the chloroplast genome (Savolainen et al. 2000). In higher plants, this gene is present as a single copy per chloroplast genome, but numerous copies of the genome exist in each plastid; hence the actual *rbcL* copy number per chloroplast can be high. It is conserved in nature (Albert et al. 1994), but has substitutions in sites of known functional importance (Kellogg and Juliano 1997). Due to both of these characteristics, *rbcL* has great phylogenetic importance. It stands next to the *matK* in the identification capability and can be easily amplified and sequenced in most land plants (Kress et al. 2007). It is able to resolve many taxonomic issues at the higher (family) and as well at the lower (inter/intrageneric) level in Cupressaceae, Cornaceae, Ericaceae and Geraniaceae (Khan et al. 2015).

7.5.1.3 Chloroplast RNA Polymerase Genes *-rpoB* and *rpoC*

Plastid genomes contain *rpoA*, *rpoB*, *rpoC1* and *rpoC2* genes coding for α, β, β', and β" of RNA polymerase subunits. During plant evolution, genes for core subunits are typically retained in the chloroplast genome as *rpoA*, *rpoB*, *rpoC1* and *rpoC2* (Yagi and Shiina 2014). The genes *rpoB* and *rpoC1* have a high level of resolving power, due to which they have been relied upon for plant DNA barcoding. Hollingsworth et al. (2011) investigated the key seven candidate plastid regions, *rpoB*, *rpoC1*, *matK*, *rbcL*, *atpF-atpH*, *psbK-psbI* and *trnH-psbA*, among three different groups of land plants, and *rpo1* was found to be the best amplified universal locus.

7.5.1.4 Intergenic Spacer *psbA-trnH*

The non-coding intergenic spacer *psbA-trnH* has evolved relatively rapidly, displaying most sequence divergence due to high rates of insertion/deletion (Kress and Erikson 2007) in larger families of angiosperms (Chase et al. 2007). Its short length (500 bp) and presence of highly conserved coding sequences on both the sides of this intergenic spacer make it feasible to get amplified by universal primers (Shaw et al. 2005). Also, owing to these characteristics, it can get amplified from some degraded DNA, which makes it a suitable barcode candidate (Chase et al. 2005, Kress et al. 2005, Erickson et al. 2008). The alignment of the *psbA-trnH* spacer can be tremendously indistinct because of its complicated molecular evolution and considerable varying length (Chang et al. 2006). Limitations related to the use of *psbA-trnH* as a standard barcode are the recurrent inversions in some plant lineages, leading to overestimating genetic divergence (Whitlock, Hale and Groff 2010). Also, its short length will not have enough sequence variation for species discrimination, as observed in the genera Solidago (Kress et al. 2005). Consequently, it has been proposed that it can be used in two-locus or three-locus barcode systems to provide adequate resolution (Kress et al. 2005, Chase et al. 2007).

The plastid marker *psbA-trnH* intergenic spacer was considered efficient in barcode recovery and species discrimination in many plant groups including *Roscoea* (Zingiberaceae) (Chen et al. 2014). It could resolve the authenticity of market samples of medicinal plant *S. cordifolia* (Vassou et al. 2015) and in newly recorded medicinal plant, *Osmunda japonica* (Zheng et al. 2015).

7.5.1.5 Intergenic Spacer *psbK-psbI*

This intergenic spacer is located in the large size copy (LSC) of the plastid genome. Both genes *psbK* and *psbI* encode two low molecular mass polypeptides, K and I, respectively, of the photosystem II (Meng, Wakasugi and Sugiura 1991). It is conserved from algae to land plants (McNeal et al. 2007). This locus has displayed extremely promising results in the barcoding of the Kruger National Park's flora in South Africa (Lahaye et al. 2008) *and Orchidaceae* in Korea (Kim et al. 2014). CBOL-Plant Working (Hollingsworth et al. 2009b) endorsed this locus over *matK* and other loci owing to its species discriminatory power.

7.5.1.6 Intergenic Spacer *atpF-atpH*

The genes *atpF* and *atpH* encode ATP synthase subunits CFO III, and I, respectively (Drager and Hallick 1993). This intergenic spacer *atpF-atpH* is localized in the large single copy (LSC) of the plastid genome and is also conserved from algae to terrestrial plants (Knauf and Hachtel 2002). Hollingsworth et al. (2009b) indicated it as a rapidly evolving but length variable intergenic spacers. Because of its low capability of species identification, the CBOL-Plant Working (Hollingsworth et al. 2009a) suggested that it can be used as a supplementary loci along with other standard barcodes for better resolution (Vijayan and Tsou 2010). In *Landoltia* and *Spirodela*, it has shown promising results by identifying three species of taxa (Bog et al. 2015).

7.5.2 Nuclear Genomes

The nuclear genome is located in the nucleus of a cell. The nuclear DNA is biparentally inherited, due to which it can provide more information as compared to the chloroplast genome (Hagemann 2004). From the nuclear DNA regions, the *ITS* regions of the ribosomal DNA (rDNA) have been extensively tested for suitability as plant barcodes (Kress et al. 2005).

7.5.2.1 Characteristics of Nuclear Genes

It was found that the nuclear genome has a higher rate of mutation (Wolfe, Li and Sharp 1987), due to which this region has a higher number of species-diagnostic haplotypes, some of which were distinct by more than one substitution. This allowed a clear delineation of a large number of species (Pillon et al. 2013). This region was recommended as a DNA barcode for various flowering plants (Hollingsworth et al. 2011), particularly medicinal plants (Chen et al. 2010).

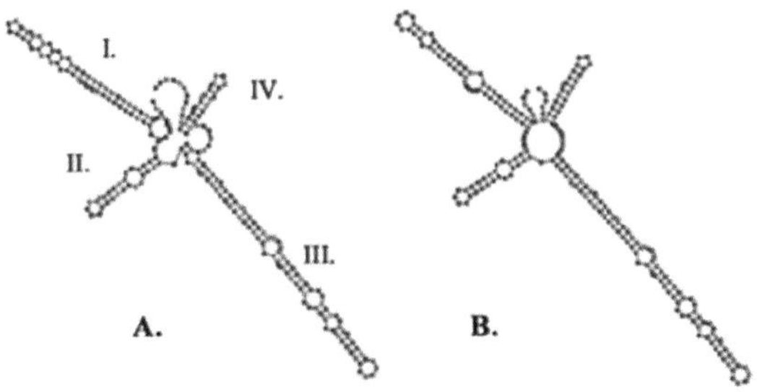

FIGURE 7.1 Secondary structures of *ITS2* sequences from the genus *Rhazya* (A) *Rhazya stricta* and (B) *Rhazya orientalis* (*Amsonia orientalis*) displaying four helices.

7.5.2.1.1 Internal Transcribed Spacer (ITS)

The ribosomal DNA cistron is a multi-gene family which encodes the nucleic acid core of the ribosome. Inside the cell, the rDNA is arranged as tandem repeated units comprising *18S*, *5.8S* and *26S* coding regions and two internal transcribed spacers (*ITS1* and *ITS2*). The *ITS*1 is present between 18S and 5.8S rRNA, and *ITS*2 is present between 5.8S and 26S rRNA (Gerbi 1985). Nuclear ribosomal DNA possesses several advantages over other loci. It is a combination of highly conserved and variable regions, which allow phylogenetic inference across a broad range of evolutionary time scales. Its conserved regions allow the use of "universal" PCR primers across a large range of taxa, and a reliable amplification is possible from low-quality DNA extractions and from plant species that have lost their plastid genomes (Smith and Lee 2014). There is a similarity in its copies within individuals due to the process of concerted evolution (Baldwin et al. 1995).

The length of *ITS1* and *ITS2* of all flowering plants varies, and they range from 200–300 bp for *ITS1* and nearly 180–250 bp for *ITS2* (Gonzalez et al. 1990). Discrimination at species level and technical ease has made the *ITS* region a powerful phylogenetic marker. Another feature favoring it to be used in phylogenetic studies is that both *ITS1* and *ITS2* regions can be amplified separately by anchoring primers in the conserved coding genes. Hence, the *ITS* region can be amplified even from poor quality of DNA (Kress et al. 2005); preferring it to be suitable barcode loci for authenticating market available medicinal plant products. The rate of successful identification with the *ITS2* was 92.7% at the species level in medicinal plants of China (Chen et al. 2010), where more than 6600 plant samples belonging to 4800 species from 753 distinct genera were studied. Chinese medicinal herb *Bupleurum* species could be differentiated on the basis of polymorphism in its *Internal Transcribed Spacer* (Kim et al. 2012). Yao et al. (2010) also evaluated 50,790 plants and 12,221 animals and projected *ITS2* as the probable universal barcode. Thus, *ITS2* has been recommended as a suitable barcode (Kress and Erikson 2007, Chen et al. 2010). The last two decades have witnessed extensive attention to the *ITS* region regarding the molecular phylogeny and taxonomy (Ali et al. 2014). The *ITS2* region could discriminate between *Citrus* species (Sun et al. 2015), and some herbal medicinal products (Michel et al. 2016). It could also authenticate xerophytic medicinal plant *Rhazya stricta* in its processed powered state, resolving it from its sister species *R. orientalis* (*Amsonia orientalis*) (Khan et al. 2019). The secondary structure of *ITS2* sequences shed extra morphological characteristics to a species. These structures are unique to a species with respect to its central ring and helices. Also, they differ on the basis of stem loop number, position, length and screw angle from the center of the spiral arm (Figure 7.1).

7.6 GENERATION OF DNA BARCODES

The process of developing DNA barcodes for medicinal plants is discussed here, starting from the extraction of DNA from fresh samples to uploading our barcodes to the Gene Data Banks.

Occasionally, the example of *Rhazya stricta* is taken as an example to demonstrate how the methods could be applied practically. To put into context the relevance of these techniques, the scenario of authenticating medicinal plant products available in local markets is assumed. Many of the plants being utilized for such medicinal use are usually available in dried or powdered form. Unfortunately, such derived products have a higher chance of adulteration (Newmaster et al. 2013). Hence, to safeguard consumers' health, it is imperative to develop DNA barcodes for our plant of interest, which could detect any substitution or contamination in its market products. To generate the DNA barcode, the following steps are carried out:

1) Collection of fresh plant material samples (preferably young fresh leaves) from a variety of populations.
2) Collection of dried powdered plant material samples from the local herbal market.
3) Extraction and quantification of genomic DNA from both samples.
4) Amplification of DNA barcode target regions by PCR.
5) Checking of PCR product size on 1.5% Agarose Gel.
6) Analysis and documentation of Gel images.
7) Purification and sequencing of verified PCR products.
8) Alignment and analysis of nucleotide sequences by bioinformatics software programs.
9) Appropriate regions selected for use as barcodes (i.e., those showing maximum diversion for sample sequences).
10) Authentication of dried powered plant samples by the proposed barcodes.

For macerating plant samples, a sterile mortar and pestle under liquid nitrogen is used. Genomic DNA for barcoding purposes is generally extracted using the DNeasy Plant Mini TM kit (Kress and Erikson 2007, Newmaster et al. 2008, Hollingsworth et al. 2009b) and DNeasy 96 Plant Kit (Fazekas et al. 2008). Slight modifications may be carried out for DNA extraction while following the manufacturer's instructions to maximize genomic DNA yield. The concentration and quality of the extracted DNA is then determined by gel electrophoresis and a NanoDrop 2000 spectrophotometer.

DNA barcode sequences are amplified by the polymerase chain reaction (PCR), following previously published specifications for each primer. The Plant Working Group at Kew has suggested one thermal cycle for amplification of all the loci (www.kew.org/protocols); this will help in reducing amplification time and money. While some have suggested modification in thermal cycles for amplification (Kress and Erikson 2007, Hollingsworth et al. 2009b). Details of primers and thermal cycle conditions used for the plant *Rhazya stricta* have been included as an example in Table 7.1.

To get the maximum sequence length, assembled bands are to be sequenced in both forward and reverse directions. There is a larger probability of getting more barcodes with higher sequence lengths, so it is advisable to choose the locus with the highest size (Ruhsam et al. 2015). After sequencing a particular locus, the sequences are aligned to identify the base sequence. The sequences are then studied in relation to the base sequence to identify the nucleotide variations common for all the samples.

The PCR products are checked for their presence and sizes by gel electrophoresis, the products are purified using a QIAquick PCR Purification Kit (Qiagen). Bidirectional sequencing was done by Sanger's di-deoxy chain termination (Sanger, Nicklen and Coulson 1992). An ABI3730XL automated sequencer (Applied Biosystems) working on the principle of Sanger's technique (using the same PCR primers) was used for sequencing. A sequencer is used to determine the nucleotide sequence through a process known as Base Calling. This is done by identifying the nucleotide base present at a given location in the sequence. The output of a sequencer is in the form of chromatograms and sequence files. The chromatograms give ideas about the base call quality and

TABLE 7.1
Primer List and Thermal Cycles

DNA Type	Region Name	Primer Sequence	Thermal Cycle	Amplified Product Size
rDNA	Inter-transcribed Spacer *(HSI)*	F 5'-ACGAATTCATGGTCCGGTGAAGTGTTCG-3' R 5 '-TAGAATTCCCCGGTTCGCTCGCCGTTAC-3'	95°C-5 min 95°C-1 min 52°C-30 sec 38 x 72°C-1 min 72°C-10 min 4°-Hold	~200bp
rDNA	Inter-transcribed Spacer *(ITS2)*	F 5'- GCGATACTTGGTGTGAAT-3' R 5 '- GACGCTTCTCCAGACTACAAT-3	94°C-5 min 94°C-30 sec 56°C-30 sec 40 x 72°C-1 min 72°C-10 min 4°-Hold	~500 bp
cpDNA	Maturase K *(MatK)*	F 5'- TAATTTACGATCAATTCATTC-3 ' R 5'- ACAAGAAAGTCGAAGTAT-3 '	94°C- 4 min 94°C-50 sec 52°C-1 min 40 x 72°C-1 min 72°C-10 min 4°-Hold	~900 bp
cpDNA	*rbcL*	F 5 '-ATGTCACCACAAACAGAGACTAAAGC -3' R 5'- CTTCTGCTACAAATAAGAATCGATCTC-3'	94°C-1 min 94°C-45 sec 51°C-45 sec 30 x 72°C-5 sec 4°-Hold	~650bp

consist of peaks corresponding to each of the four nucleotides. The nucleotide whose peak is of the highest amplitude is the nucleotide present at that location in the sequence (*DNA and Sequencing* 2021). Finch TV (Finch TV 1.4.0 Geospiza, Inc., Seattle, WA, USA) is an example of a chromatogram viewer software that can be used to study base calls and quality values. The sequence files are then used as the input into the Geneious software (Kearse et al. 2012) for further analysis.

Generally, three methods are used for discriminating the species as well as identifying a suitable DNA barcode region; they are genetic distance, phylogenetic tree method and BLAST analysis. Of these three, the most commonly used is BLAST analysis (Ross 2008), and it is considered to be one of the best methods for Genus to species level identification (Casiraghi et al. 2010). BLAST is used to identify the species accurately by comparing and providing the percentage of correlation to the sample sequence against the sequences in the database. This database has reference barcodes of correctly identified species, and the first hit with a maximum score (higher score indicating higher similarity) is considered as the species to which the specimen belongs. BLAST also produces the gaps present in compared sequences. In an alignment, a gap is introduced to take the place of an insertion or deletion from the two or more sequences being aligned ("NCBI," n.d.). BLAST analysis also provides the expected value or E-value. "The Expect value (E) is a parameter that describes the number of hits one can 'expect' to see by chance when searching a database of a particular size. The closer the E-value is to zero, the higher the significance of the match" ("BLAST Frequently Asked Questions" 2021). Genetic distance is a quantity that denotes the extent of genetic divergence between species. The genetic distance can be calculated by evaluating intra- and inter-specific distances. For accurate identification of species and production of DNA barcodes, intra- and inter-specific variations assessments are significant (Lahaye et al. 2008). An efficient barcode can differentiate between closely related species. This is possible when there is ample genetic difference between the species rather than within species, that is, when inter-specific distance is considerably higher than intraspecific distance (Hebert et al. 2004, Lahaye et al. 2008), and the difference between the two is expressed as barcode gap (Meyer and Paulay 2005). A perfect barcode should exhibit a

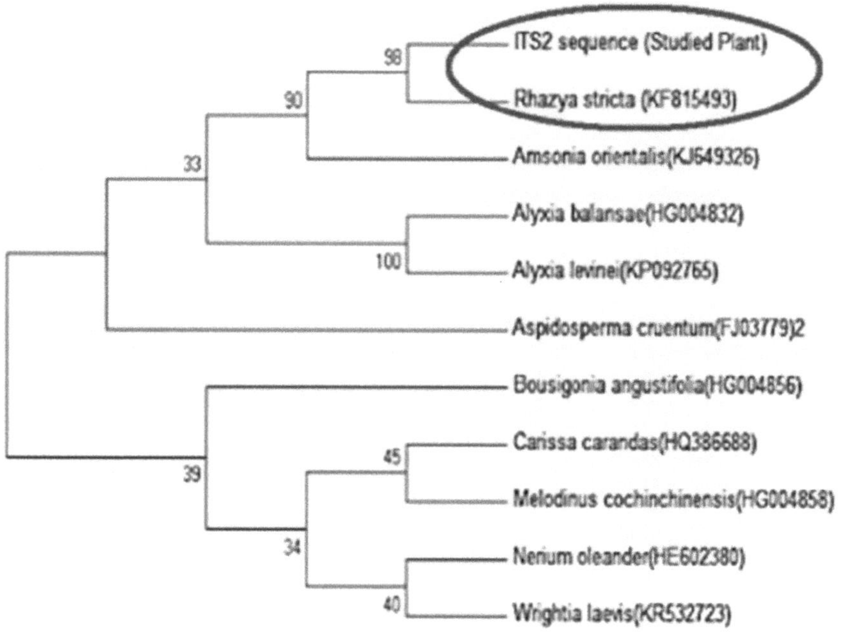

FIGURE 7.2 ITS2 sequences the maximum likelihood (ML) tree.

barcode gap, due to which the distribution of intra- and inter-specific divergences do not overlap (Lahaye et al. 2008). An ideal barcoding-gap aids in allocating an unknown sample to its respective species accurately (Meyer and Paulay 2005) and in identifying potential new species. To construct a phylogenetic tree, sequences of the candidate locus are used, and the percent species are determined by cluster analysis (China Plant BOL Group et al. 2011). A species is said to be "resolved" when all the entities cluster together in a single clade, while the ones which cluster with the other species are termed as unresolved (Figure 7.2). In the image below, the maximum likelihood (ML) method based on K2P distance has been used to construct phylogenetic trees for the sample plant, *Rhazya stricta*. In phylogenetic analyses, for successful identification of genus, all the sample sequences together should form a monophyletic group.

Genetic divergence between the species can be analyzed using software-MEGA, MUSCLE, CLUSTALW and BLAST, and so on. These can be used as standalone software and as well as utilized as plugins on the platform GENEIOUS. GENEIOUS is also capable of carrying out numerous tasks like genome browsing, reference mapping, sequence assembly and tree building. Evolutionary relationships between the species can be deduced by the MEGA software also. It can estimate the evolutionary distance between the sequence pair, intra- and interspecies average distance. The proportion (p) of nucleotide sites where two sequences being compared are dissimilar is termed as p-distance. The number of nucleotide differences divided by the total number of nucleotides compared gives its p-distance (Nei and Kumar 2000). A species is considered distinct if all inter-specific p-distances are greater than zero. Besides this, gaps formed in sequence alignment are also an important diagnostic tool in the context of DNA barcoding; they are the result of polymorphism. Figure 7.3 shows the polymorphism of ITS1 sequences of *Rhazya stricta* and *R. orientalis* (syn. *Amsonia orientalis*).

The DNA barcoding sequences and other related details have to be submitted at NCBI (www.ncbi.nlm.nih.gov/BLAST/) or BOLD systems for the generation of barcodes for every specimen. Upon entering the relevant details for the project, BOLD generates the DNA barcode for a specimen.

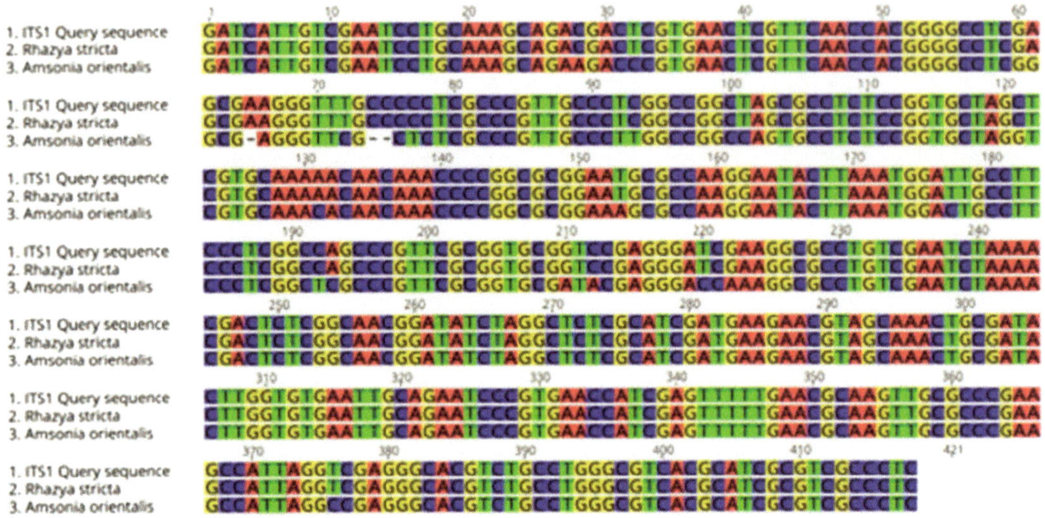

FIGURE 7.3 ITS1 sequences of *Rhazya stricta* and *R. orientalis* (syn. *Amsonia orientalis*) showing polymorphism.

(The details on submission are available in the Bold handbook at www.boldsystems.org/libhtml_v3/static/BOLD_ Handbook_Oct2013.pdf.) The sequence details can also be submitted to GenBank. After submission, accession numbers are provided for all the target regions. GenBank is part of the International Nucleotide Sequence Database Collaboration, which consists of GenBank at NCBI, the European Nucleotide Archive (ENA), and the DNA Data Bank of Japan (DDBJ). Data is exchanged on a daily basis by these three organizations.

7.7 APPLICATIONS OF DNA BARCODING AND RECOMMENDATIONS

The DNA barcoding technique is being employed for the routine market analysis of herbal products (Sucher and Carles 2008), as it can identify medicinal plants, their substitutes and adulterants (Newmaster et al. 2013). The proposed DNA barcode sequences can be converted to QR codes. This will aid consumers in easily and efficiently identifying medicinal plants and their products irrespective of their physical state. Customs officers and other related agencies could use these generated barcodes for checking their illegal trade and thus carry out their conservation. In this way, DNA barcoding can also help in the conservation of endangered plant species (Muellner, Schaefer and Lahaye 2011) and protect biodiversity. With the rising demand for herbal medicines and the significance of their authentication for consumers' health, it is vital to provide a sole, extensive database with DNA data for swift and easy identification. With species-specific barcode databases, chances of adulterations, accidents related to herbal medication and over-exploitation of flora will reduce. En masse, this will be a big step toward the conservation of resources.

ACKNOWLEDGMENTS

I am thankful to Ms Iman Aqeel for her contributions toward editing and helping draft this manuscript. I am also grateful to Prof Nabih Baeshen for his support during my research at the King AbdulAziz University. I thank Dr. Ahmed Al Higin for providing me with the research facilities, equipment and lab for carrying out my work.

REFERENCES

Ali, A., Gyulai, M.G., Hidvégi, N., Kerti, B., Al Hemaid, F.M.A., Pandey, A.K. and Lee, J. 2014. The changing epitome of species identification – DNA barcoding. *Saudi J. Biol. Sci.* 21(3): 204–31.

Albert, V.A., Backlund, A., Bremer, K., Chase, M.W., Manhart, J.R., Mishler, B.D. and Nixon, K.C. 1994. Functional constraints and rbcL evidence for land plant phylogeny. *Ann. Missouri Bot. Garden* 81: 534–67.

Baldwin, B.G., Sanderson, M.J., Porter, J.M., Wojciechowski, M.F., Campbell, C.S. and Donoghue, M.J. 1995. The its region of nuclear ribosomal DNA: A valuable source of evidence on angiosperm phylogeny. *Ann. Missouri Bot. Garden* 82: 247–77.

Barthet, M.M. and Hilu, K.W. 2007. Expression of matK: Functional and evolutionary implications. *American J. Bot.* 94: 1404–12.

Bertzky, B., Shi, Y., Hughes, A., Engels, B., Ali, M.K. and Badman, T. 2013. *Terrestrial Biodiversity and the World Heritage List: Identifying broad gaps and potential candidate sites for inclusion in the natural World Heritage network.* IUCN, Gland, Switzerland and UNEP-WCMC, Cambridge, UK.

Bickford, D., Lohman, D.J. Sodhi, N.S., Ng, P.K.L., Meier, R., Winker, K., Ingram, K.K. and Das, I. 2007. Cryptic species as a window on diversity and conservation. *Trends in Ecol. Evol.* 22(3): 148–55.

"BLAST Frequently Asked Questions." 2021. Accessed October 7. https://blast.ncbi.nlm.nih.gov/Blast.cgi?CMD=Web&PAGE_TYPE=BlastDocs&DOC_TYPE=FAQ#expect.

Bog, M., Lautenschlager, U., Landrock, M.F., Landolt, E., Fuchs, J., Sree, K.S., Oberprieler, C. and Appenroth, K. 2015. Genetic characterization and barcoding of taxa in the genera *Landoltia* and *Spirodela* (Lemnaceae) by three plastidic markers and Amplified Fragment Length Polymorphism (AFLP). *Hydrobiologia.* 749: 169–82.

Casiraghi, M., Labra, M., Ferri, E., Galimberti, A. and De Mattia, F. 2010. DNA barcoding: A six-question tour to improve users' awareness about the method. *Briefings in Bioinformatics* 11: 440–53.

CBOL Plant Working Group. 2009. A DNA Barcode for Land Plants. *Proceedings of the National Academy of Sciences of the United States of America*, 106, 12794–7.

Chang, C.C., Lin, H.C., Lin, I., Chow, T.Y., Chen, H.H., Chen, W.H., Cheng, C.H. et al. 2006. The chloroplast genome of *Phalaenopsis aphrodite* (Orchidaceae): Comparative analysis of evolutionary rate with that of grasses and its phylogenetic implications. *Mol. Biol. Evol.* 23(2): 279–91.

Chase, Mark W., Salamin, N., Wilkinson, M., Dunwell, J.M., Prasad Kesanakurthi, R., Haider, N. and Savolainen, V. 2005. Land plants and DNA barcodes: Short-term and long-term goals. *Phil. Trans. Royal Soc. London. Series B, Biol. Sci.* 360(1462): 1889–95.

Chase, M.W., Cowan, R.S., Hollingsworth, C., Madriñán, S., Petersen, G., Seberg, O. et al. 2007. A proposal for a standardised protocol to barcode all land plants. *TAXON* 56: 295–9.

Chase, Mark W. and Fay, M.F. 2009. Ecology. barcoding of plants and fungi. *Science* 325(5941): 682–3.

Chen, Shilin, Yao, H., Han, J., Liu, C., Song, J., Shi, L. Zhu, Y. et al. 2010. Validation of the its2 region as a novel DNA barcode for identifying medicinal plant species. *PloS One* 5(1): e8613.

Chen, Shilin, Pang, X., Song, J., Shi, L., Yao, H., Han, J. and Leon, C. 2014. A renaissance in herbal medicine identification: From morphology to DNA. *Biotech. Adv.* 32(7): 1237–44.

China Plant BOL Group, Li, D.Z., Gao, L.M., Li, H.T., Wang, H., Ge, X.J., Liu, J.Q., Chen, Z.D., Zhou, S.L., Chen, S.L., Yang, J.B., Fu, C.X., Zeng, C.X., Yan, H.F., Zhu, Y.J., Sun, Y.S., Chen, S.Y., Zhao, L., Wang, K., Yang, T. and Duan, G.W.. 2011. Comparative analysis of a large dataset indicates that internal transcribed spacer (ITS) should be incorporated into the core barcode for seed plants. *Proc. Natl. Acad. Sci. USA.* 108(49):19641–6.

Clegg, M.T., Gaut, B.S., Learn, G.H. Jr. and Morton, B.R. 1994. Rates and patterns of chloroplast DNA evolution. *Proc. Nat. Acad. Sci. USA* 91 (15): 6795–801.

DNA and Sequencing. 2021. Accessed October 11. https://media.ed.ac.uk/media/DNA+and+Sequencing/1_dvu56a7f

Dong, W., Liu, J. Yu, J., Wang, L. and Zhou, S. 2012. Highly variable chloroplast markers for evaluating plant phylogeny at low taxonomic levels and for DNA barcoding. *PloS One* 7(4): e35071.

Drager, R.G. and Hallick, R.B. 1993. A novel *Euglena gracilis* chloroplast operon encoding four atp synthase subunits and two ribosomal proteins contains 17 introns. *Curr. Gene.* 23(3): 271–80.

Ekor, M. 2014. The growing use of herbal medicines: Issues relating to adverse reactions and challenges in monitoring safety. *Front. Pharmacol.* 4(January): 177.

Erickson, D.L., Spouge, J., Resch, A., Weigt, L.A. and Kress, W.J. 2008. DNA barcoding in land plants: Developing standards to quantify and maximize success. *Taxon* 57(4): 1304–16.

Escapa, I.H. and Catalano, S.A. 2013. Phylogenetic analysis of *Araucariaceae*: Integrating molecules, morphology, and fossils. *Int. J. Plant Sci.* 174: 1153–70.

Fazekas, A.J., Burgess, K.S., Kesanakurti, P.R., Graham, S.W., Newmaster, S.G., Husband, B.C., Percy, D.M., Hajibabaei, M. and Barrett, S.C.H. 2008. Multiple multi-locus DNA barcodes from the plastid genome discriminate plant species equally well. *PloS One* 3(7): e2802.

Fazekas, A.J., Kesanakurti, P.R., Burgess, K.S., Percy, D.M., Graham, S.W., Barrett, S.C.H., Newmaster, S.G., Hajibabaei, M. and Husband, B.C. 2009. Are plant species inherently harder to discriminate than animal species using DNA barcoding markers? *Mol. Ecol. Res.* 9 Suppl s1 (May): 130–9.

Fazekas, A.J., Kuzmina, M.L., Newmaster, S.G. and Hollingsworth, P.M. 2012. DNA barcoding methods for land plants. *Meth. Mol. Biol.* 858: 223–52.

Fišer Pečnikar, Ž. and Buzan, E.V. 2014. 20 years since the introduction of DNA barcoding: From theory to application. *J. Appl. Gen.* 55(1): 43–52.

Fishbein, M., Chuba, D., Ellison, C., Mason-Gamer, R.J. and Lynch, S.P. 2011. Phylogenetic relationships of *Asclepias* (Apocynaceae) inferred from non-coding chloroplast DNA sequences. *Syst. Bot.* 36: 1008–23.

Gerbi, S.A. 1985. Evolution of ribosomal DNA. *Mol. Evol. Gene.* doi:10.1007/978–1-4684–4988–4_7

Gonzalez, I.L., Chambers, C., Gorski, J.L., Stambolian, D., Schmickel, R.D. and Sylvester, J.E. 1990. Sequence and structure correlation of human ribosomal transcribed spacers. *J. Mol. Biol.* 212: 27–35.

Gregory, T.R. 2005. DNA barcoding does not compete with taxonomy. *Nature* 434: 1067.

Hagemann, R. 2004. The sexual inheritance of plant organelles. In: Daniell, H. and Chase, C. (eds) *Molecular Biology and Biotechnology of Plant Organelles*, pp. 93–113, Springer, Dordrecht.

Hebert, P.D.N., Cywinska, A., Ball, S.L. and deWaard, J.R. 2003. Biological identifications through DNA barcodes. *Proc. Biol. Sci. Royal Soc.* 270(1512): 313–21.

Hebert, P.D.N., Penton, E.H., Burns, J.M., Janzen, D.H. and Hallwachs, W. 2004. Ten species in one: DNA barcoding reveals cryptic species in the neotropical skipper butterfly *Astraptes fulgerator*. *Proc. Nat. Acad. Sci. USA* 101(41): 14812–17.

Heinrichs, J., Kreier, H., Feldberg, K., Schmidt, A.R., Zhu, R., Shaw, B., Shaw, A.J. and Wissemann, V. 2011. Formalizing morphologically cryptic biological entities: New insights from DNA taxonomy, hybridization, and biogeography in the leafy liverwort *Porella platyphylla* (Jungermanniopsida, Porellales). *Am. J. Bot.* 98: 1252–62.

Hollingsworth, P.M., Forrest, L.L., Spouge, J.L., Hajibabaei, M., Ratnasingham, S. et al. 2009a. A DNA barcode for land plants. *Proc. Nat. Acad. Sci. USA* 106(31):12794–7.

Hollingsworth, M.L., Clark, A.A., Forrest, L.L., Richardson, J., Pennington, R.T., Long D.G., Cowan, R., Chase, M.W., Gaudeul, M. and Hollingsworth, P.M. 2009b. Selecting barcoding loci for plants: Evaluation of seven candidate loci with species-level sampling in three divergent groups of land plants. *Mol. Ecol. Res.* 9 (2): 439–57.

Hollingsworth, P.M., Graham, S.W. and Little, D.P. 2011. Choosing and using a plant DNA barcode. *PloS One* 6 (5): e19254.

Kane, N., Sveinsson, S., Dempewolf, H., Yang, J.Y., Zhang, D., Engels, J.M.M. and Cronk, Q. 2012. Ultra-Barcoding in Cacao (*Theobroma* spp.; Malvaceae) using whole chloroplast genomes and nuclear ribosomal DNA. *Am. J. Bot.* 99(2): 320–9.

Kearse, M., Moir, R., Wilson, A., Stones-Havas, S., Cheung, M., Sturrock, S., Buxton, S. et al. 2012. Geneious basic: An integrated and extendable desktop software platform for the organization and analysis of sequence data. *Bioinformatics* 28(12): 1647–9.

Kellogg, E.A. and Juliano, N.D. 1997. The structure and function of RuBisCO and their implications for systematic studies. *Am. J. Bot.* 84: 413–28.

Khan, S., Al-Qurainy, F., Nadeem, M. and Tarroum, M. 2012. Development of genetic markers for *Ochradenus arabicus* (Resedaceae), an endemic medicinal plant of Saudi Arabia. *Gen. Mol. Res. GMR* 11(2): 1300–8.

Khan, M.Q., Khalil, A.T., Rehman, K. and Shinwari, Z.K. 2015. Searching for DNA barcodes in plants. *Global J. Biotechnol. Biochem.* 10(1): 1–10.

Khan, S.A., Baeshen, M.N., Ramadan, H.A. and Baeshen, N.A. 2017. Emergence of plastidial intergenic spacers as suitable DNA barcodes for arid medicinal plant *Rhazya stricta*. *Am. J. Plant Sci.* 8: 1774–89.

Khan, S.A., Baeshen, M.N., Ramadan, H.A. and Baeshen, N.A. 2019. ITS2: An ideal DNA barcode for the arid medicinal plant *Rhazya stricta*. *Pharmaceu. Med.* 33: 53–61.

Kim, Y.H., Kim, E.S., Ko, B.S., Uddin, M.R., Oh, S.E., Choi, G.Y., Chae, S.W., Lee, H.W., Lee, J.H., Park, J.Y. and Lee, M.Y. 2012. Internal transcribed spacer-based identification of Bupleurum species used as sources of medicinal herbs. *J. Med. Plants Res.* 6(5): 841–8.

Kim, H.M., Oh, S.-H., Bhandari, G.S., Kim, C.-S. and Park, C.-W. 2014. DNA barcoding of *Orchidaceae* in Korea. *Mol. Ecol. Res.* 14(3): 499–507.

Knauf, U. and Hachtel, W. 2002. The genes encoding subunits of ATP synthase are conserved in the reduced plastid genome of the heterotrophic alga *Prototheca wickerhamii*. *Mol. Gene. Genomics* 267(4): 492–7.

Kranitz, M.L., Biffin, E., Clark, A., Hollingsworth, M.L., Ruhsam, M., Gardner, M.F., Thomas, P., Mill, R.R., Ennos, R.A., Gaudeul, M. and Lowe, A.J. 2014. Evolutionary diversification of new Caledonian Araucaria. *PLoS One* 9(10): e110308.

Kress, W.J., Wurdack, K.J., Zimmer, E.A., Weigt, L.A. and Janzen, D.H. 2005. Use of DNA barcodes to identify flowering plants. *Proc. Nat. Acad. Sci. USA* 102(23): 8369–74.

Kress, W.J. and Erickson, D.L. 2007. A two-locus global DNA barcode for land plants: The coding rbcL gene complements the non-coding trnH-psbA spacer region. *PLoS ONE.* doi:10.1371/journal.pone.0000508

Kress, W.J. and D.L. Erickson. 2008. DNA barcodes: Genes, genomics, and bioinformatics. *Proc. Nat. Acad. Sci. USA* 105: 2761–2.

Lahaye, R., Van der Bank, M., Bogarin, D., Warner, J., Pupulin, F., Gigot, G., Maurin, O., Duthoit, S., Barraclough, T.G. and Savolainen, V. 2008. DNA barcoding the floras of biodiversity hotspots. *Proc. Nat. Acad. Sci. USA* 105(8): 2923–8.

Li, De-Zhu, Gao, L.-M., Li, H.-T., Wang, H., Ge, X.-J., Liu, J.-Q. et al. 2011. Comparative analysis of a large dataset indicates that Internal Transcribed Spacer (ITS) should be incorporated into the core barcode for seed plants. *Proc. Nat. Acad. Sci. USA* 108(49): 19641–6.

Li, M., Au, K.Y., Lam, H., Cheng, L., But, P.P.H. and Shaw, P.C. 2014. Molecular identification and cytotoxicity study of herbal medicinal materials that are confused by *Aristolochia* herbs. *Food Chem.* 147: 332–9.

Mahadani, P., Sharma, G.D. and Ghosh, S.K. 2013. Identification of ethnomedicinal plants (Rauvolfioideae: Apocynaceae) through DNA barcoding from northeast India. *Pharmacognosy Magazine* 9(35): 255.

McNeal, J.R., Kuehl, J.V., Boore, J.L. and De Pamphilis, C.W. 2007. Complete plastid genome sequences suggest strong selection for retention of photosynthetic genes in the parasitic plant genus Cuscuta. *BMC Plant Biol.* 7(1): 1–22.

Meng, B.Y., Wakasugi, T. and Sugiura, M. 1991. Two promoters within the psbK-psbI-trnG gene cluster in tobacco chloroplast DNA. *Curr. Gen.* 20(3): 259–64.

Meyer, C.P. and Paulay, G. 2005. DNA barcoding: Error rates based on comprehensive sampling. *PLoS Biol.* 3(12): e422.

Michel, C.I., Meyer, R.S., Taveras, Y. and Molina, J. 2016. The nuclear internal transcribed spacer (ITS2) as a practical plant DNA barcode for herbal medicines. *J. Appl. Res. Med. Aromatic Plants* 3(3): 94–100.

Miller, S.E. 2007. DNA Barcoding and the renaissance of taxonomy. *Proc. Nat. Acad. Sci. USA* 104(12): 4775–6.

NCBI (n.d.). *Blast: Basic Local Alignment Search Tool* (no date) National Center for Biotechnology Information. U.S. National Library of Medicine. Available at: https://blast.ncbi.nlm.nih.gov/Blast.cgi (Accessed: November 21, 2022).

Muellner, A.N., Schaefer, H. and Lahaye, R. 2011. Evaluation of candidate DNA barcoding loci for economically important timber species of the Mahogany family (*Meliaceae*). *Mol. Ecol. Resour.* 11(3): 450–60.

Nei, M. and Kumar, S. 2000. *Molecular Evolution and Phylogenetics*. Oxford University Press, New York.

Newmaster, S.G., Fazekas, A.J., Steeves, R.A.D. and Janovec, J. 2008. Testing candidate plant barcode regions in the Myristicaceae. *Mol. Ecol. Resour.* 8(3): 480–90.

Newmaster, S.G., Grguric, M., Shanmughanandhan, D., Ramalingam, S. and Ragupathy, S. 2013. DNA barcoding detects contamination and substitution in North American herbal products. *BMC Med.* 11(1): 1–13.

Notredame, C., Higgins, D.G. and Heringa, J. 2000. T-Coffee: A novel method for fast and accurate multiple sequence alignment. *J. Mol. Biol.* 302: 205–17.

Palmer, J.D. and Herbon, L.A. 1988. Plant mitochondrial DNA evolved rapidly in structure, but slowly in sequence. *J. Mol. Evol.* 28(1): 87–97.

Pillon, Y., Johansen, J., Sakishima, T., Chamala, S., Barbazuk, W.B., Roalson, E.H., Price, D.K. and Stacy, E.A. 2013. Potential use of low-copy nuclear genes in DNA barcoding: A comparison with plastid genes in two Hawaiian plant radiations. *BMC Evol. Biol.* 13(1): 1–11.

Ross, H.A., Murugan, S. and Sibon Li, W.L. 2008. Testing the reliability of genetic methods of species identification via simulation. *Syst. Biol.* 57(2): 216–30.

Ruhsam, M., Rai, H.S., Mathews, S., Ross, T.G., Graham, S.W., Raubeson, L.A., Mei, W., Thomas, P.I., Gardner, M.F., Ennos, R.A. and Hollingsworth, P.M. 2015. Does complete plastid genome sequencing improve species discrimination and phylogenetic resolution in *Araucaria*?. *Mol. Ecol. Resour.* 15(5): 1067–78.

Sanger, F., Nicklen, S. and Coulson, A.R. 1992. DNA sequencing with chain-terminating inhibitors. *Biotechnology* 24: 104–8.

Savolainen, V., Chase, M. W., Hoot, S. B., Morton, C. M., Soltis, D. E., Bayer, C., Fay, M. F., de Bruijn, A. Y., Sullivan, S. and Qiu, Y. L. 2000. Phylogenetics of flowering plants based on combined analysis of plastid atpB and rbcL gene Sequences. *Syst. Biol.* 49(2): 306–62.

Selvaraj, D., Sarma, R.K., Shanmughanandhan, D., Srinivasan, R. and Ramalingam, S. 2015. Evaluation of DNA barcode candidates for the discrimination of the large plant family *Apocynaceae*. *Plant Syst. Evol.* 301: 1263–73.

Shaw, J., Lickey, E.B., Beck, J.T., Farmer, S.B., Liu, W., Miller, J., Siripun, K.C., Winder, C.T., Schilling, E.E. and Small, R.L. 2005. The tortoise and the hare II: relative utility of 21 noncoding chloroplast DNA sequences for phylogenetic analysis. *American J. Bot.* 92(1): 142–66.

Smith, D.R. and Lee, R.W. 2014. A plastid without a genome: Evidence from the nonphotosynthetic green algal genus *Polytomella*. *Plant Physiol.* 164(4): 1812–9.

Soltis, P.S. and Soltis, D.E. 2004. The origin and diversification of angiosperms. *Am. J. Bot.* 91(10): 1614–26.

Sucher, N.J. and Carles, M.C. 2008. Genome-based approaches to the authentication of medicinal plants. *Planta Med.* 74(6): 603–23.

Sugiura, M. 1992. The chloroplast genome. *Plant Mol. Biol.* 19(1): 149–68.

Sun, Y.L., Kang, H.M., Han, S.H., Park, Y.C. and Hong, S.K. 2015. Taxonomy and phylogeny of the genus citrus based on the nuclear ribosomal DNA its region sequence. *Pak. J. Bot.* 47(1): 95–101.

Valentini, A., Pompanon, F. and Taberlet, P. 2009. DNA barcoding for ecologists. *Trends Ecol. Evol.* 24(2): 110–17.

Vassou, S.L., Kusuma, G. and Parani, M. 2015. DNA barcoding for species identification from dried and powdered plant parts: A case study with authentication of the raw drug market samples of Sida cordifolia. *Gene* 559(1): 86–93.

Vijayan, K. and Tsou, C.H. 2010. DNA barcoding in plants: taxonomy in a new perspective. *Curr. Sci.* 99(11): 1530–41.

Wanntorp, L., Gotthardt, K. and Muellner, A.N. 2011. Revisiting the wax plants (*Hoya,* Marsdenieae, Apocynaceae): Phylogenetic tree using the matK gene and psbA-trnH intergenic spacer. *Taxon* 60(1): 4–14.

Whitlock, B.A., Hale, A.M. and Groff, P.A. 2010. Intraspecific inversions pose a challenge for the trnH-psbA plant DNA barcode. *PloS One* 5(7): e11533.

Wolfe, K.H., Li, W.H. and Sharp, P.M. 1987. Rates of nucleotide substitution vary greatly among plant mitochondrial, chloroplast, and nuclear DNAs. *Proc. Nat. Acad. Sci. USA* 84 (24): 9054–8.

Xing, G.A.O., Yu-Ping, Z.H.U., Bao-Cheng, W.U., Ya-Mei, Z.H.A.O., Jian-Qun, C. H.E.N. and Yue-Yu, H.A.N.G. 2008. Phylogeny of *Dioscorea* sect. *Stenophora* based on chloroplast matK, rbcL and trnL-F sequences. *J. Syst. Evol.* 46(3): 315–21.

Yagi, Y. and Shiina, T. 2014. Recent advances in the study of chloroplast gene expression and its evolution. *Front. Plant Sci.* 5: 61.

Yao, H., Song, J., Liu, C., Luo, K., Han, J., Li, Y., Pang, X., Xu, H., Zhu, Y., Xiao, P. and Chen, S. 2010. Use of ITS2 region as the universal DNA barcode for plants and animals. *PloS one*, 5(10): p.e13102.

Zhang, W., Yuan, Y., Yang, S., Huang, J. and Huang, L. 2015. ITS2 secondary structure improves discrimination between medicinal "Mu Tong" species when using DNA barcoding. *PLoS One* 10(7): e0131185.

Zheng, S.H., Ren, W.G., Wang, Z.H. and Huang, L.F. 2015. Use of chloroplast DNA barcodes to identify *Osmunda japonica* and its adulterants. *Plant Syst. Evol.* 301(7): 1843–50.

8 Genome-wide Transcriptome Profiling of the Medicinal Plants

Sana Basri

CONTENTS

8.1 INTRODUCTION

Medicinal plants have been utilized in traditional medicine for millennia all across the world, with India and China being the leading manufacturers. The genesis and development of herbal compositions are based on these plants (Sangwan, Yadav and Sangwan 2003). For thousands of years, plants and raw compounds obtained from them have been utilized therapeutically. Even now, natural compounds or similar derivatives account for more than half of all new medications developed, and many of these are derived from plants. Plants offer a rich supply of specialized metabolites that have been employed as fragrances, tastes, pigments, pesticides and medicinal substances for millions of years (Facchini et al. 2012). Plants can synthesize a wide range of organic compounds. These molecules are divided into two categories according to their usefulness for plant life: primary and secondary metabolites (Sangwan et al. 1994). Many investigations have been carried out to discover the usable importance of secondary metabolites in herbal plants; nevertheless, the knowledge accessible in this regard is dispersed (Chowhan et al. 2013, Lee et al. 2015, Sunohara et al. 2015). Plant-based herbal therapy is gaining popularity due to the fact that it has fewer or no adverse effects than allopathic and artificial drugs. These molecules are made through a variety of complicated and interrelated metabolism reactions, and little is known about their biosynthesis processes. As a result, research into biosynthetic pathways involving genetic mechanisms for the synthesis of these chemicals is greatly needed (Kushwaha et al. 2013, Srivastava et al. 2015, Mishra et al. 2016). However, due to insufficient

DOI: 10.1201/9781003179139-8

genetic knowledge and little investigation into their viable genes and genetic processes, multiple therapeutic plants are not model plants (Liu 2019). As a result, transcriptome profiling of medicinal plants is critical. Until recently, this technique has mostly been applied in model plants like *Arabidopsis* (Guo et al. 2009), maize (Xu B. et al. 2014) and rice (Li et al. 2012). Proteomics, genomics and transcriptomics are now regarded as popular methods in the post-genomic era. The most widely investigated and used of these is transcriptomics, as it allows for the analysis of both viable and differentially expressed genes (DEGs) (Jia et al. 2015). Transcriptomics is defined as the study of the transcriptome (the whole library of RNA) as well as expression profiling (the analysis of transcriptional levels in a given cell). With the exception of mutations, the genome is basically constant for a specific cell line. However, the transcriptome is alive because it is a reflection of the genes that are expressed at any given time undergoing various circumstances (Jeanette and Emon 2016). There are some major contrasts between transcriptomics and genomics in medicinal plant investigation. First, assembling a genome is more crucial and high-priced in plant research than transcriptome analysis; the transcriptome may be utilized to track an organism's overall transcriptional activity in the truancy of a corresponding genome. The transcriptome, on the other hand, as it varies depending on observing time and space, not only represents changes in gene expression at different times of the day but also reflects seasonal changes in genetic expression. It not only stores information on transitory and geographical locations, but it also retains information on metabolomic processes with a secondary purpose. Gene expression patterns show temporal and spatial variances, showing unique accumulation patterns of therapeutic components, according to studies. Although species were the same, gene expression patterns show temporal and spatial differences, showing the distinct deposition paths of therapeutic substances. As a result, the transcriptome is better for determining genes involved in therapeutic plant components (Wang et al. 2009). A flow chart of transcriptome analysis is given in Figure 8.1.

Transcriptomics study methods have progressed from the classic chip hybridization platform to RNA-sequencing technology recently, thanks to continual developments in techniques for sorting (RNA-seq) (Mironova, Weinholdt and Grosse 2015). Transcriptome sequencing (RNA-seq) is a technology for studying the model and non-model organisms in the absence of corresponding genetic sequences, and it has several benefits, including increased efficiency, electronic signals and reactivity. In the realm of molecular biology, transcriptome profiling technology is a frequently utilized sequencing approach (Sun and Wei 2018). The capacity to comprehend the operative parts of the genome is provided by high-efficiency methods such as next-generation sequencing (Valdés et al. 2013). Transcriptome sequencing is a valuable tool for examining the genomes of medicinal plants, a subject about which there is little knowledge. Transcriptome sequencing is favored because it delivers crucial information by being influenced by various developmental stages as well as a variety of physiological factors such as environmental alterations in contrast to the genome. As a result, the method enables the creation of a thorough perspective of the numerous pathways engaged in the secondary metabolism of therapeutic and scented plants (Gupta et al. 2013, Sangwan et al. 2013). Researchers can use transcriptome analysis to investigate medicinal plants in order to better understand their functional genes and regulatory pathways, as well as improve breeding selection and cultivation methods. Genome-wide transcription studies help us learn more about genetic functionality and the complicated mechanisms that affect genetic expressions. Genetic expressions data provide for the vast bulk of genome-wide information (viz. alterations in mRNA quantity). This chapter deals with the transcriptome profiling technique and also sequencing analysis techniques and summarizes the growth in the implementation of transcriptomics in herbal plant synthesis to investigate secondary metabolite biosynthesis pathways, functional gene mining and recognition of expression patterns in therapeutic plants in order to generate plans for traditional medicinal viable gene extraction and breeding new varieties.

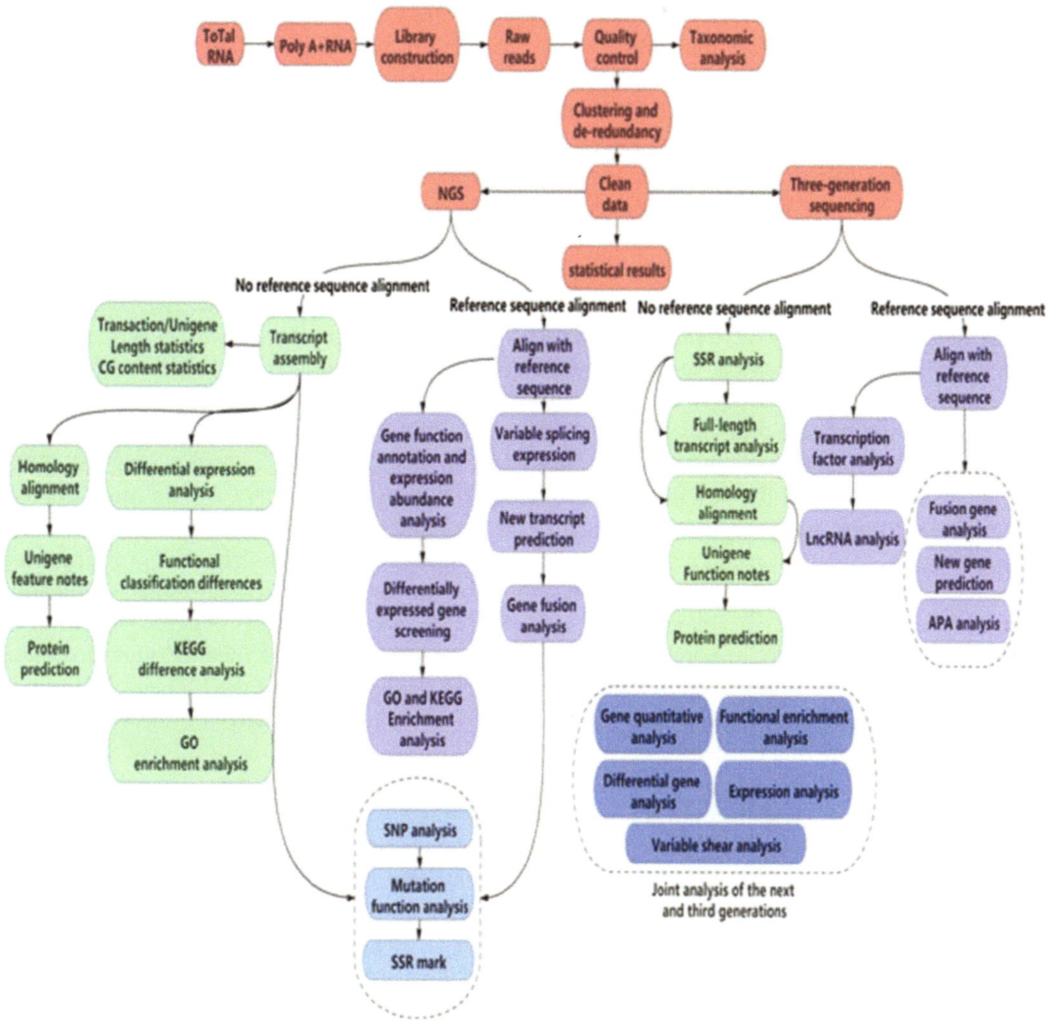

FIGURE 8.1 Flow chart of transcriptome analysis.

Source: Guo et al. 2021.

8.2 MEDICINAL PLANTS TRANSCRIPTOMICS: AN EVOLVING CONCEPT

Transcriptomics, or universal transcriptome profile analysis, is the initial level of a biological system's dynamic state, and it gives a glance into the vital processes that may be present in a given organic state.Transcriptomic analysis under various concerns has allowed for a better knowledge of plant feedback mechanisms and the control of various metabolic reactions, as well as the functional identification of enzymes responsible for the biosynthesis of active compounds (Aharoni and Vorst 2002, Goossens et al. 2003, Parkinson et al. 2005). Plants create hundreds of organic chemicals, and they are commonly categorized as either central or specialized metabolites. Central metabolites are required for growth and development of plants, and the bulk of them seems to be shared by all plants. Secondary metabolites, which were previously assumed to be unnecessary for the growth and development of plants, frequently operate as defence molecules and help protect plants under a variety of unfavorable environments. Formerly, these substances were referred to as secondary metabolites since their exclusion or variation in plant levels did not result in deadly consequences, as opposed to primary metabolites, that are required for progress and expansion. This ancient idea of secondary metabolites has subsequently been refined, and it has been widely thought, and increasingly confirmed, that specialized metabolites take part in plant growth, adaptability and protection (Bose et al. 2013, Yadav et al. 2014, Nick 2015). Much research has now been undertaken to uncover the functional relevance of secondary metabolites in herbal medicines, however, the information available in this regard is dispersed (Chowhan et al. 2013, Lee et al. 2015, Sunohara et al. 2015). In certain plants, such as *Withania somnifera*, *Azadirachta indica* and Mentha species, all sections exhibit therapeutic characteristics, while in others, these characteristics are tissue- and even cell-specific, such as the formation of Vinblastine and Vincristine, which is only found in foliage (Kushwaha et al. 2013, Narnoliya et al. 2014, Pan et al. 2016, Sangwan et al. 2014). Scented plant oils, such as *Geranium*, *Cymbopogon*, *Ocimum* species and *Eucalyptus*, have therapeutic qualities. For that reason, reflexology makes use of them. Initially started in several places throughout the world, the therapy has lately emerged as a contemporary therapy to reduce stress and despair, resulting in increased comfort (Cooke and Ernst 2000, Fakari et al. 2015). These oils are also utilized as ingredients in numerous beauty products (Sangwan, Yadav and Sangwan 2003, Wang et al. 2016). One of the first initiatives for Apiaceae medicinal plants, *Centella asiatica*'s leaf transcriptome was recently declared to have undergone fresh sequencing by a group of scientists (Sangwan et al. 2013). A further significant advancement in de novo transcriptome profiling from medicinal plants is the ability to process information for *Azadirachta indica*, an Indian therapeutic plant (Krishnan et al. 2012). A genome-wide and four transcriptomes analysis of neem (*Azadirachta indica*) indicates numerous important stages included in the specialized metabolomic processes (Krishnan et al. 2012). *Azadirachta indica* is the very first representative of the Meliaceae family to have its DNA sequenced, examined for phylogenetic analysis and evaluated in terms of sequence similarity with several other higher-level plants.

Transcriptome profiling has been characterized by a series of technical breakthroughs that have transformed the field. Transcriptomics has also been defined by the emergence of new approaches that, every decade or more, redefine what is feasible and leave old technologies outdated. There seem to be two main modern approaches in the subject: microarrays, which measure a series of preset patterns, and RNA-sequencing (RNA-seq), wherein all transcripts are retrieved using high-throughput sequencing.

Microarray-based transcriptome characterization resulted in the detection and characterization of many genes as well as their affiliation with biosynthetic processes, and influenced the growth of hundreds of different computational methods, database systems and assets for co-expression and relative expression profiling (Aoki et al. 2007, Rung and Brazma 2013). DNA Microarray techniques have been widely employed for comprehensive expression profiling of *Arabidopsis*, the model plant and numerous other species with relevant genetic data. Since the invention of microarray, increased measurement of expression level was carried out utilizing a microarray chip to identify 45 *Arabidopsis thaliana* transcripts (Aharoni and Vorst 2002, Goossens et al. 2003., Parkinson et al. 2005). Libraries of microarray resources, which include ATTED-II (Obayashi et al. 2007, Aoki et al. 2016), OryzaExpress (Hamada et al. 2011), Genevestigator (Zimmermann et al. 2004, Zimmermann et al. 2008), PRIMe (Akiyama et al. 2008), PathoPlant (Bülow et al. 2007), BAR (Patel et al. 2012), At-TAX (Laubinger et al. 2008), PLEXdb (Dash et al. 2012), CORNET (De Bodt et al. 2010, De Bodt et al. 2012), ePlant (Fucile et al. 2011) are among the various databases created to do co-expression assessment on millions of DNA microarrays data from *Arabidopsis* as well as other model plants. Such DNA chip tools have aided in the detection and depiction of several *Arabidopsis* as well as other plant genes (Mao et al. 2009, Yonekura-Sakakibara and Saito 2009, Amrine, Blanco-Ulate and Cantu 2015, Costa et al. 2015, Ransbotyn et al. 2015).

Next-generation sequencing (NGS)-based RNA-seq provided a significant leap in transcriptome profiling. With the exception of microarray, RNA-seq-based transcriptional analysis does not necessitate genetic data, making it an essential tactic that accelerates transcriptional study and biological characterization of genes in non-model plants, notably therapeutic plants (Wang et al. 2009, Jain 2012, Xiao et al. 2013, Han et al. 2016, Rai et al. 2016). RNA-sequencing has shown to be a strong technology with a variety of uses, involving reliable measurement of variable gene expression, and even limited transcripts. However, it can also detect exon/exon borders, alternative splicing, translated areas and simple sequence repetitions in a single session (Martin et al. 2013). RNA-seq readings may be constructed de novo without being mapped to standard genome sequences, making them an essential tool for gene characterization and expression study in non-model medicinal plants (De Luca et al. 2012, Hao and Xiao 2015, Rai and Saito 2016). RNA-seq-based transcriptome profiling and de novo transcriptional arrangement, co-expression as well as correlative transcriptome profiling of different plants, such as herbal plants, have received much attention in recent years that they have become an excellent replacement to produce generic assets for plants of medical significance (Martin et al. 2013, Xiao et al. 2013, Han et al. 2015, Unamba, Nag and Sharma 2015, Wang et al. 2016). The counting of RNA-seq-based transcriptomic profiling conducted by plant scientists has increased exponentially. Alleles-related variability and RNA processing activities are also important to consider when developing superior re-construction methods for using the data kept by the translated segments retrieved by RNA-sequencing. Furthermore, as a result of advances in new transcriptomics, metatranscriptomics is a growing field. The field investigates millions of transcriptomes from the entire microbial population at the same time. Ongoing advancements in RNA-seq studies and associated methodologies will give consistent answers to informatics difficulties.

8.3 APPLICATION OF SEQUENCING TECHNOLOGIES IN MEDICINAL PLANT RESEARCH

Plants produce a plethora of secondary metabolites, the molecular analysis of which should be regarded as critical for the identification of new molecules that might be produced as medications. Through breeding programs, metabolomics and transgenic plants, research on their manufacturing and the associated genetic process could assist large pharmaceutical manufacturers. RNA-sequencing (RNA-seq), which is less easy and less costly to do than entire genetic analysis, has become a critical tool in medicinal and aromatic plants and has advanced the examination of plant gene regulation.

8.3.1 Next-Generation Sequencing Technology

Classical Sanger sequencing is tedious, labor-intensive and costly. Moreover, the invention of next-generation sequencing (NGS) in 2006 met numerous major criteria of advanced molecular research. NGS is a cheaper, quick, elevated and profound approach that overcomes the disadvantages of first-generation sequencing methods. In comparison to conventional sequencing techniques, NGS can concurrently sequence lakhs of oligonucleotides, allowing for the investigation of every species' transcriptome and genome. NGS is now being used to construct several references and non-model plant and animal genomes in order to gather all specific genes of a specific species, like *Chosenia arbutifolia* (Chen, Liu and Zeng 2014, Mei et al. 2016, Feng et al. 2019) and SARS-CoV-2 (Li et al. 2020). NGS is largely employed in transcriptomics for the sequencing and characterization of mRNA and small RNA. NGS systems that are often used include the Roche 454, Illumina Solexa and ABI SOLiD (Zhang, Zhang and Wang 2016).

The Roche 454 sequencing technology, announced in the year 2005, is a pyrosequencing-based elevated sequencing framework (Margulies et al. 2005). When correlated with different methods, Roche sequencing has benefits like extended reading periods and high run speed. This approach does not need the use of fluorescent probes or oligonucleotide probes, but it may result in the removal or addition of sequencing mistakes during the sequencing method (Liang et al. 2017).

Illumina Solexa sequencing technology is based on the notion of synthesis sequencing, which involves randomly attaching Fragments of DNA to a flow cell. Upon elongation and multiplication, many million clusters emerge on the glass substrate, each with many thousand similar DNA molecular fragments. Whenever the Target DNA is stretched, fluorescently tagged dNTPs are transcribed. This technology is suitable for genome-wide expression experiments that do not require standard sequences or artificial probes (Liu et al. 2012). The short read duration and the complexity of building the readings from scratch are the techniques' key drawbacks.

SOLiD sequencing technique, which was introduced in the year 2007, is a highly parallel sequencing system based on magnetic beads. This approach, which is based on the constant joining and formation of fluorescently tagged nucleotide sequences, may be utilized for massive multiplication and high-throughput parallel analysis of ssDNA fragments (Li and Xu 2019). The main advantage of this technique is its high precision; nevertheless, the biggest downside is that when a mistake happens, it is susceptible to chain interpretation mistakes.

8.3.2 Third-Generation Sequencing Technology

Third-generation sequencing has been made possible by advancements in transcriptomics. Because NGS systems need PCR amplification, among other procedures, template movement, nucleotide mismatches and GC preference are typical issues with these approaches, reducing the consistency and validity of sequencing data. Because of the long read length, third-generation sequencing has now been frequently utilized in genomic studies, structural variation identification, transcriptome profiling and methylation identification. *PacBio single-molecule real-time* (SMRT) sequencing technique and *Oxford Nanopore* sequencing technique are two third-generation sequencing approaches (Ma et al. 2019).

The SMRT sequencing technique relies on the synthesis-by-sequencing concept. In comparison to conventional approaches, SMRT technology includes two innovative features. First, to fix the problems of a noisy environment, the fluorescent group is first linked to the phosphate group. Second, amplification is not required, and SMRT allows for more precise measurement by self-correction (Li et al. 2018). SMRT sequencing technique, on the other hand, has significant disadvantages, along with the insertion of randomized mistakes. By using CCS sequencing approach and improving the sequence coverage, the precision of a single read may now be enhanced to 99.8 percent (Wenger et al. 2019).

Nanopore single-molecule sequencing technology is a next-generation single-molecule real-time sequencing device that estimates base composition mostly through variable electrical impulses. The benefits of nanopore sequencing are high throughput, extended read duration and cheap cost. It does, unfortunately, have downsides like as a high single-base error margin and random error. A comparison of the benefits and drawbacks of next-generation (NGS) and third-generation sequencing technologies is given in Table 8.1.

Conventional means for acquiring and analyzing gene expression profiling data involve expressed sequence tag (EST), complementary DNA-amplified fragment length polymorphism (cDNA-AFLP), hybridization technology-based chip technology (such as cDNA chip), serial gene analysis (SAGE) and massively parallel signature sequencing (Cheng 2004, Xu and Wu 2005, Hang and Peng 2006, Zhang and Sheng 2008, Wu et al. 2009, Yang et al. 2009, Simkin et al. 2011, Qian et al. 2012, Zhang et al. 2012, Zhuo et al. 2015, Suet al. 2016). The procedures described here can be costly and tedious in the lack of a standard genome sequence. As a result of the advancement of rising sequencing technique, RNA-seq technology based on NGS and third-generation sequencing technique has become the standard approach of transcriptomic study. The principles, benefits and drawbacks of the conventional transcriptome profiling method are presented in Table 8.2.

TABLE 8.1
Advantages and Disadvantages of Next-generation and Third-generation Sequencing Technologies

Sequencing technique	Principle	Advantages	Disadvantages	References
Roche 454	Pyrosequencing technology	Extended read length, a huge number of produced pieces and high running speed.	Might introduce deletions or inserting sequencing mistakes, as well as limited throughput.	Liang et al. 2017
Illumina Solexa	Technique based on single-molecule array	High throughput, cheap cost, and appropriate for identifying repeated sequences.	Short read duration and complexity in building the readings.	Liu et al. 2012
ABI SOLiD	ligation-based sequencing	High precision, simple to differentiate between SNPs and sequencing mistakes.	Short read duration, susceptible to chain decoding mistakes.	Li and Xu 2019
PacBio single-molecule real-time (SMRT)	Synthesis-based sequencing	It takes little time and there is no requirement for PCR amplification.	Expensive and low throughput.	Wenger et al. 2019
Nanopore single-molecule sequencing technology	Technology for electrical signal sequencing	High throughput, long read duration, cheap and identify epigenetic regulation point.	A high number of single-base errors and a lengthy sequencing duration.	Ma et al. 2019

8.4 TRANSCRIPTOME PROFILING AND SECONDARY METABOLISM PATHWAYS

Herbal medicines are high in phytochemicals, and a few are complex to tackle for molecular testing. This sort of investigation might be carried out using whole-genome sequencing or a gene expression technique (Sangwan et al. 2013). Presently, the transcriptomic procedure utilizing the RNA-sequencing technology has drawn increasing curiosity than whole-genome analysis because of its speedy information outputs, reduced sophistication, relatively short assembly duration and low cost. The additional reason for transitioning to the RNA-seq approach might be the large genome size and different chromosome sets. Plants contain around 20,000–60,000 genes, with only 15–25% of such genes contributing to secondary metabolite synthesis pathways (Góngora-Castillo et al. 2012). Formerly, the DNA chip technique was utilized to identify the transcriptional regulation of genes in different categories of organisms, tissues or similar species over a period; however, RNA-seq has recently surpassed the microarray approach. RNA-sequencing allows for constant monitoring of thousands of genes, and unlike microarray, there is no requirement for prior knowledge of the genetic makeup to create probes. Advancement in generation sequencing is occurring on a daily basis, as evidenced by the latest transition from Sanger's sequencing technique to elevated screening systems such as Roche pyrosequencing (Margulies et al. 2005), Illumina (Bentley et al. 2008) and SoLiD™ (McKernan et al. 2009) and data analytics instruments allow us all to decipher the metabolomic intricacies of herbal medicines to a certain point at the genome level. In the context of traditional medicines, the most often employed platform for generating transcriptome data of numerous potential therapeutic plants are Roche pyrosequencing and Illumina high-throughput sequencing (Hao et al. 2012, Sangwan et al. 2013, Yang et al. 2013, Bose and Chattopadhyay 2016, Deng et al. 2016, Jain et al. 2016). Traditional medicines provide a vast supply of active substances

TABLE 8.2
Principles, Benefits, and Drawbacks of Conventional Transcriptome Profiling Method

Sequencing technology	Principle	Advantages	Disadvantages	References
Expressed Sequence Tag (EST)	Based on Next-generation Sequencing technology	Maximum gene separation efficiency, adaptability, and conservatism, as well as cheap cost and ease of operation.	Certain data errors were discovered, as well as insufficient genetic information and library quality standards.	Zhang et al. 2012
cDNA-AFLP	RT-PCR and AFLP technology	Good repeatability, minimal false positives, detects limited mRNA, and properly reflects gene expression differences.	Many stages, tedious and expensive, difficulties in operation and poor recovery rate, tight standards, and tough to obtain desired outcomes.	Qian et al. 2012, Hang et al. 2006
cDNA Chip technology	Hybridization technique	Rapid, high throughput, diversification, automated, and miniaturization are all advantages.	Large false positives and low stability, the chip is costly, the experimental costs are high, it is only suited for detecting known sequences, and there is no research of novel genes.	Mironova et al. 2015, Wu et al. 2009, Su et al. 2016, Zhai and Lu 2017
SAGE	Based on Sanger sequencing technology	Data is similar, high in resolution, with little background noise, and is simple to handle and scale.	Tags are prone to interference, and processing is both tedious and costly.	Zhang and Sheng 2008, Zhuo et al. 2015
MPSS	Based on Sanger sequencing technology	Good reproducibility and responsiveness, applicable to any organisms and any characteristic, great research productivity, and large genome coverage.	Working cooperatively necessitates more costly gear and software.	Yang et al. 2009, Simkin et al. 2011
RNA-seq technology	Next-generation and third-generation technology	Appropriate for entire expression profiling of any organism, digital output, fast response, precision, and detection limit.	Sequencing sequence read duration is often brief, with a higher false-positive rate and also an amount of information that is too huge to store.	Mironova et al. 2015, Cheng 2004

that can serve as a target molecule for pharmaceutical research and forecasting models against a wide range of life-threatening illnesses. There is a significant possibility of developing novel medications derived from plants (Mondal et al. 2012). As a result, it is critical for scholars to create a system that organizes knowledge on all traditional medicines so that it may be utilized for the betterment of humankind. The majority of biologically active compounds generated from flora come under the purview of secondary metabolites. Secondary metabolites are typically the main compounds of herbal remedies, which not only possess therapeutic importance but they also play an important part in plants' adjustments to changes in their surroundings. Secondary metabolite deposition differs depending on the growing environment, development stage and organs. Gene expression analysis is used to investigate the biosynthesis routes of metabolites and genes involved in chemical synthesis in various environments, development stages and organs. The knowledge gathered could be used as a scholarly base for the time ahead of research into the collection and effective usage of bioactive substances in herbal medicines. Therefore, secondary metabolites are a diverse set of molecules with diverse biosynthesis and structural skeletons, so for the sake of understanding their genesis, production and functional characteristics, these molecules are roughly categorized into three groups: terpenoid, alkaloid and phenylpropanoids.

8.4.1 Terpenoid Pathway Explication Using Transcriptomics

Terpenes seem to be the most abundant class of plant secondary metabolites, and because the basic element of terpene is the 5-carbon chemical isoprene, they are also known as isoprenoids. Their numbers may vary between 5 (hemiterpene) to 1000 (rubber), and the geometry might be straight or aromatic rings. Terpenes could be categorized into several classes determined by the number of carbon atoms available in the molecule, like hemiterpene (C5), monoterpene (C10), sesquiterpene (C15), diterpene (C20), triterpene (C30), tetraterpene (C40) and polyterpene (more than eight isoprene unit). Terpenes help the plant's defense system by emitting volatile substances. They are also a huge category of pharmacologically significant substances that have been utilized by people from prehistoric days, such as artemisinin from *A. annua*, which has been utilized as an antimalarial medication, *Mentha* species are used to produce peppermint, as are withanolides from *W. somnifera* and asiaticosides from *C. asiatica* (Sangwan et al. 2010, Chaurasiya et al. 2009). Several facts point to the existence of 2 terpene biosynthetic pathways: cytosolic mevalonate (MVA) and plastidial non-mevalonate (MEP) (Sangwan et al. 2008). However, there is relatively little research on the biosynthetic pathways as well as genetic control of terpenoid biosynthesis. As a result, there is a tremendous need to learn much more about metabolic paths, and the issue has been handled to a significant extent by using NGS techniques. There have been various therapeutic plants where the transcriptome profiling technique has been used to understand the terpenoids biosynthesis, such as *Ilex asprella* (Zheng et al. 2014) *C. asiatica* (Sangwan et al. 2013), *W. somnifera* (Gupta et al. 2013), *Panax notoginseng* (Luo et al. 2012), *S. miltiorrhiza* (Yang et al. 2013) and *Costus pictus* (Annadurai et al. 2012). Sangwan et al. (2013) mentioned that KEGG analysis of the *C. asiatica* transcriptome profiling reveals the occurrence of enzymes associated with both biosynthesis pathway (MVA and MEP) of terpenes and vital genes of terpenoid metabolism, including limonene synthase for monoterpenoids, farnesyl diphosphate synthase (FPPS) for sesquiterpenoids and beta-amyrin synthases (BAS), cycloartenol synthases (CAS), squalene epoxidase (SQE), squalene synthases (SS), and so on, linked with triterpenoid/steroids were present in large number. Further instances of how transcriptomic profiling has been used to identify gene-related terpene pathways include saikosaponins synthesis in *Bupleurum chinense* (Sui et al. 2011), tanshinone synthesis in *S. miltiorrhiza* (Yang et al. 2013), ginkosides synthesis in *P. ginseng* (Chen et al. 2011) and taxol, a probable anticancerous drug, synthesis in *Taxsus capsidata* (Wu et al. 2011).

8.4.2 Alkaloid Pathway Explication Using Transcriptomics

Alkaloids are a class of bitter-tasting polycyclic aromatic hydrocarbons/nitrogen-containing compounds. Almost 10,000 alkaloids have been documented from various sources, the plant kingdom is not the only source of alkaloids, but their prevalence has also been observed in mammals. Alkaloids defend plants from herbivores and pathogens. Alkaloids' molecular mass ranges from simple amino acids to complicated isoquinoline alkaloids. Alkaloids are categorized into numerous categories based on their structure, precursors molecules and biosynthesis process, including protoalkaloid, purine alkaloid, pyridine alkaloid, pseudo alkaloid, quinoline, indole alkaloid, isoquinoline alkaloids, tropine and psedotropine alkaloids (Guo et al. 2013). The alkaloids, such as caffeinated tea and coffee, have a significant influence on human health concerns owing to their direct involvement in the neurological system. Certain alkaloids, like cannabin from *Cannabis sativa*, have already been observed to produce psychedelic effects. Just 1% of the total were employed as medication, probably due to a dearth of considerable study for their extraction and isolation method because these are manufactured in extremely small quantities. Therefore, there must be an alternative approach to manufacture these compounds in some other system, which may be accomplished with thorough insights into the molecular processes involved in their creation. Because there is a scarcity of information for traditional medicinal genome sequences, next-generation sequencing (NGS) is widely employed to build the metabolomic process of therapeutically significant non-model plants (Wang et al. 2009). *Catharanthus roseus* is regarded as a model plant species for alkaloid synthesis and is the primary basis of the antitumor medicines vincristine and vinblastine (Jacobs et al. 2005).

8.4.3 Phenylpropanoid Pathway Explication Using Transcriptomics

Phenolic compounds are a broad collection of bioactive components that act as a building element and provide antioxidant properties to give protection from biotic and abiotic stress (Korkina et al. 2011). They are produced by a highly intricate branched route that begins with phenylalanine and ends with coumaryl CoA. The very first step of this process is performed by phenylalanine ammonia-lyase (PAL), which is identified as the pathway's rate-limiting enzyme (Zhao et al. 2013). Numerous phenyl propanoid by-products have been identified, including flavonoids, isoflavonoids, coumarins and lignins. KEGG analysis was utilized to recognize genes associated with the phenylpropanoid pathway.

8.4.4 Shikimate Pathway

In microorganisms, fungi, apicomplexans and plants, the shikimic acid biosynthetic process produces a predecessor for aromatic compounds, but not invertebrates (Dewick et al. 2009, Mittelstädt et al. 2013). Plants biosynthesize three types of secondary metabolites: phenols, terpenoids/isoprenoids, alkaloids/ glucosinolates (nitrogen- or sulfur-containing molecules, correspondingly) (Aharoni and Galili 2011). Shikimic acid was extracted from the very poisonous Japanese shikimi (*Illicium anisatum*) flower that yielded it (Ghosh, Chisti and Banerjee 2012). The shikimic acid pathway synthesizes phenolic chemicals, which are readily available to plants (Macheroux et al. 1999). Phenolic substances are secondary metabolites present in high concentrations in plants. These organic hydrocarbons have critical functions as pigments, antioxidants, signaling agents, structural elements lignan and defense mechanisms. Stresses, both biotic and abiotic, stimulate the release of phenolic chemicals (e.g., herbivores, pathogens, unfavorable climate and pH, salt stress, thrash metal distress and UVB and UVA radiations). In higher plants and microbes, these chemicals are produced through the shikimic acid pathway. The structural plasticity of phenols is produced naturally through the shikimate route. The enzymes that regulate phenol metabolic activities are well understood, and shikimic acid is a key intermediate. In plants, the shikimic acid process is found in the chloroplast (Macheroux et al. 1999). Through the pioneering investigations of Srinivasan, Rothschild and Sprinson, the shikimic acid pathway has been found as the biosynthesis pathway to the heterocyclic amino acids phenylalanine, tyrosine and tryptophan. The said pathway has only been discovered in microbes as well as plants. Seven stages of the shikimic acid pathway are depicted in Table 8.3. In Table 8.4, some enzymes and genes involved in the aromatic amino acid biosynthetic pathways in plants are shown.

8.4.4.1 DAHP Synthase

DAHP synthases are divalent metal cation-containing metalloenzymes (e.g., Mn^{2+} or Co^{2+}) (Huisman and Kosuge 1974, Rubin and Jensen 1985). Most DAHP synthases have had a central (β/

TABLE 8.3
Stages of the Shikimic Acid Pathway

Reaction step	Substrate	Enzyme/cofactor	Product
1.	Phosphoenolpyruvate (PEP), erythrose-4-phosphate	3-Deoxy-D-*arabino*-heptulosonate-7-phosphate synthase (DAHPS, EC 4.1.2.15, now EC 2.5.1.54)/ Co_2+, Mg_2+ or Mn_2+ (Schmid et al. 1995)	3-Deoxy-D-*arabino*-heptulosonic acid 7-phosphate (DAHP), Pi
2.	3-Deoxy-D-*arabino*-heptulosonic acid 7-phosphate (DAHP)	3-Dehydroquinate synthase DHQS (EC. 4.2.3.4)/Co_2+, NAD+ (Schmid et al. 1995, Bender et al. 1989)	3-Dehydroquinic acid (DHQ), Pi
3.	3-Dehydroquinic acid (DHQ)	3-Dehydroquinate dehydratase (DHQ dehydratase EC 4.2.1.10) (Schmid et al. 1995)	3-Dehydroshikimic acid (DHS), H_2O
4.	3-Dehydroshikimic acid (DHS), NADPH + H+	Shikimate dehydrogenase (SDH, EC 1.1.1.25) (Harris et al. 1999, Singh et al. 2007)	Shikimic acid, NADP+
5.	Shikimic acid, ATP	Shikimate kinase enzyme (SK, EC 2.7.1.71)	Shikimic acid 3-phosphate (S3P), ADP
6.	Shikimic acid 3-phosphate (S3P), PEP	5-*Enol*pyruvylshikimate 3-phosphate synthase, also called aroA enzyme (EPSPS, EC 2.5.1.19) (Lewis et al. 1999)	5-*Enol*pyruvylshikimate 3-phosphate (EPSP), Pi
7.	5-*Enol*pyruvylshikimate 3-phosphate (EPSP)	Chorismate synthase (CS, EC 4.2.3.5)/ FMNH2 (Dewicket al. 2009, Hermann, 1995, Bornemann et al. 2000, Osborne et al. 2000)	Chorismic acid, Pi

TABLE 8.4
Some Enzymes and Genes Involved in the Aromatic Amino Acid Biosynthetic Pathways in Plants

SN.	Enzyme name	Abbreviation	EC Number	Number of isolated genes	Subcellular localization	Cofactor (s)
1.	3-Deoxy-D-arabinoheptulosonate 7-phosphate synthase	DAHP synthase	2.5.1.54	2 (*Lycopersicon esculentum*)	Chloroplast	Mn^{2+} or Co^{2+}
2.	3-Dehydroquinate synthase	DHQS	4.2.3.4	1 (*Lycopersicon esculentum*)	Chloroplast	NAD+, Co2+
3.	5-Enolpyruvylshikimate-3-phosphate synthase	EPSP synthase	2.5.1.19	2 (*Petunia hybrida*)	Chloroplast	_

$\alpha)_8$ barrel monomer configuration and a wide range of diversity in their outer smaller domain that are required for AAA mediated allosteric control. DAHP synthases are categorized into two categories focusing on pattern similarities, with 10 percent amino acid sequence similarity. DAHP synthases from *E. coli* (AroF, AroG and AroH) and *S. cerevisiae* (Aro3 and Aro4) with an N-terminal Phe/Tyr-binding domain (Bentley and Haslam 1990), bifunctional chorismate mutase (CM)– DAHP synthase enzymes (for example, from *B. subtilis*) (Wu et al. 2006) and uncontrolled DAHP synthase (for example, from *Pyrococcus furiosus*) (Schofield et al. 2005) are instances of type I enzymes. Plants and certain microorganisms include type II DAHP synthases (for example, *Mycobacterium tuberculosis*) (Webby et al. 2005). Regardless of the reality, that plant DAHP synthases are distantly correlated to *E. coli* and yeast enzymes, two copies of DAHP synthase genes have indeed been separated from many plant species by complementing *E. coli* or yeast mutants lacking the associated activity (Dyer et al. 1990, Keith et al. 1991). In *Arabidopsis thaliana* (*Arabidopsis*), *Solanum lycopersicum* (tomato) and *Solanum tuberosum* (potato), the 2 DAHP synthase genes (DAHPS1 and DAHPS2) show differential expression. DAHPS2 (shkB in potato) is expressed continuously, but DAHPS1 (shkA in potato) is substantially upregulated in relation to injury and various pathogenic infections (Görlach et al. 1995, Keith et al. 1991). A ferredoxin-thioredoxin mechanism that integrates carbon input into the shikimic acid pathway alongside photosynthesis electron transport regulates *Arabidopsis* DAHP synthase 1 (Entus, Poling and Herrmann 2002). Despite the fact that Mn^{2+} is used by both plant DAHP synthases 1 and 2 (Huisman et al. 1974, Pinto, Suzich and Herrmann 1986). Inside the cytoplasm of numerous plant tissues, an extra Co^{2+}-dependent DAHP synthase activity has now been discovered (Ganson, d'Amato and Jensen 1986, Rubin and Jensen 1985). Nevertheless, the associated genes have yet to be discovered, and their physiological activities remain unclear. This enzyme participates in the first step of the subpathway that produces chorismate from D-erythrose 4-phosphate and phosphoenolpyruvate. This subpathway is a component of the chorismic acid synthesis pathway, which is a component of the Metabolomic intermediate biosynthesis process. The cotyledons have greater concentrations than that the flowers and leaves. Low concentrations can be found in the roots and shoots.

D-erythrose 4-phosphate + H_2O + phosphoenolpyruvate = 7-phospho-2-dehydro-3-deoxy-D-arabino-heptonate + phosphate

The shikimate pathway starts with a condensation reaction of PEP and carbohydrate erythrose-4-P to produce 3-deoxy-D-arabino-heptulosonic acid 7-phosphate (DAHP) (Figure 8.2). The stoichiometry of DAHP synthase's condensation process of (Z)-[3-3H]PEP and D-erythrose-4-phosphate (Floss, Onderka and Carroll 1972).

8.4.4.2 Dehydroquinate Synthase (DHQS)

Dehydroquinate synthase catalyzes the second step in the shikimic acid pathway. The enzymes contain two cofactors binding sites, a divalent metal cation, and NAD^+7-phospho-2-dehydro-3-deoxy-D-*arabino*-heptonate = 3-dehydroquinate + phosphate.

Roots of *Solanum lycopersicum* showed a lot of expression of its gene. Decreased levels of expression are found in trunks, flowers and cotyledons of *Solanum Lycopersicum* but almost never found in foliage (Bischoff et al. 1996). Elicitor causes an increase in the expression of this gene (Bischoff et al. 1996).

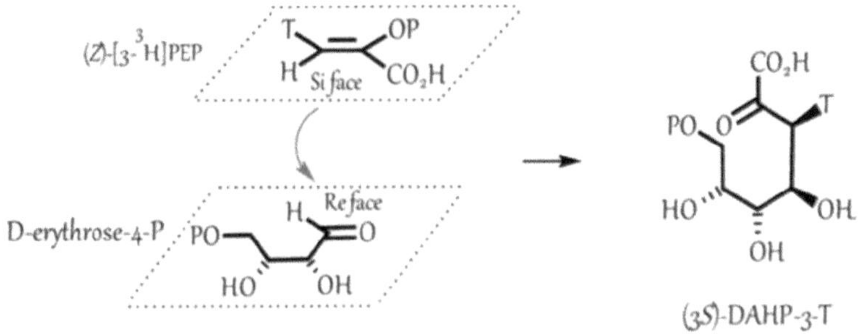

FIGURE 8.2 The stoichiometry of DAHP synthase's condensation process.

3-Dehydroquinate synthase (DHQS) transforms DAHP to 3-dehydroquinate by five sequential chemical processes involving a divalent ion (e.g., Co^{2+}) and NAD^+ cofactors: alcohol oxidation, elimination of inorganic phosphate, carbonyl reduction, ring-opening and intramolecular aldol condensation (Bender, Mehdi and Knowles 1989, Srinivasan, Rothschild and Sprinson 1963). DHQS, according to crystallographic structural elucidation, accomplishes multiple reactions in a single active site without producing by-products (Carpenter et al. 1998). Via complementary base pairing of an *E. coli* mutant, a plant DHQS gene was sequestered from tomato (Bischoff et al. 1996). The single-copy gene is strongly expressed in tomato roots, and elicitor therapy induces its expression in suspension cell culture (Bischoff et al. 1996). 3-Dehydroquinate synthases from *Escherichia coli* (AroB) and plants are monofunctional catalysts, but those from molds are parts of a pentafunctional enzyme (AROM complex) that catalyzes five successive steps in the shikimic acid pathway to convert DAHP to 5- enolpyruvylshikimate 3-phosphate (EPSP) (Duncan, Edwards and Coggins 1987).

8.4.4.3 5-Enolpyruvylshikimate 3-Phosphate (EPSP) Synthase

This enzyme catalyzes the conversion of the phosphoenolpyruvyl moiety of phosphoenolpyruvate (PEP) to the 5-hydroxyl of shikimate-3-phosphate (S3P), resulting in enolpyruvyl shikimate-3-phosphate and inorganic phosphate.

3-phosphoshikimate + phosphoenolpyruvate = 5-O-(1-carboxyvinyl)-3-phosphoshikimate + phosphate.

The protein participates in step 6 of the subpathway that produces chorismate from D-erythrose 4-phosphate and phosphoenolpyruvate. This subpathway is a portion of the chorismic acid synthesis route, and that is part of the Metabolic intermediate biosynthesis process. Expression of the gene of this enzyme is commonly observed in flower petals and to a lesser level in bases, shoots and inflorescences, but not so much in the foliage of *Petunia hybrida* (Petunia). Circadian-regulation, having historic highs toward the finish of the light period in blossoms, this translation is controlled by ODO1-mediated induction as well as LHY binding and inhibition to cis-regulatory evening regions in its promoter.

The EPSP synthase (also known as 3-phosphoshikimate 1-carboxyvinyltransferase) catalyzes the synthesis of EPSP, the second last step in the shikimic acid pathway, by transferring the enolpyruvyl moiety of PEP to the 5-hydroxyl position of shikimate 3-phosphate. Crystallographic investigations have revealed that the attachment of the initial substrate, shikimate 3-phosphate, causes a broad conformation shift that results in the development of the active site in the interdomain cleft of EPSP synthase (Schönbrunn et al. 2001). The non-specific, wide-ranging weed glyphosate [N-(phosphonomethyl)glycine, formerly known as Roundup R] is the major target of EPSP synthase. Glyphosate restricts EPSP synthase efficiently concerning the second substrate, PEP, by filling the PEP binding site of the enzyme-shikimate 3-phosphate complex (Schönbrunn et al. 2001). The EPSP synthases across various species have been classified into two groups depending on glyphosate responsivity: All plants and most bacteria, including *E. coli*, have glyphosate-sensitive class I EPSP synthases; however, other bacteria, such as *Agrobacterium* sp. strain CP4, have class II EPSP synthases that are glyphosate-resistant and have therefore been utilized to make glyphosate-resistant crops (Funke et al. 2006). The EPSP synthase gene is mainly expressed at lower concentrations in most plants, and it is tissue-specific and dynamically controlled in *Petunia hybrida* (petunia) flowers (Benfey and Chua 1989, Gasser et al. 1988), most likely for the generation of Phe-derived volatiles.

REFERENCES

Aharoni, A. and Galili, G. 2011. Metabolic engineering of the plant primary–secondary metabolism interface. *Curr. Opinion Biotec.* 22(2): 239–244.

Aharoni, A. and Vorst, O. 2002. DNA microarrays for functional plant genomics. *Plant Mol. Biol.* 48(1): 99–118.

Akiyama, K., Chikayama, E., Yuasa, H., Shimada, Y., Tohge, T., Shinozaki, K., ... Saito, K. 2008. PRIMe: A Web site that assembles tools for metabolomics and transcriptomics. *In Silico Biol.* 8(3–4): 339–345.

Amrine, K. C., Blanco-Ulate, B. and Cantu, D. 2015. Discovery of core biotic stress-responsive genes in *Arabidopsis* by weighted gene co-expression network analysis. *PLoS One* 10(3): e0118731.

Annadurai, R. S., Jayakumar, V., Mugasimangalam, R. C., Katta, M. A., Anand, S., Gopinathan, S., ... Rao, S. N. 2012. Next-generation sequencing and de novo transcriptome analysis of *Costus pictus* D. Don, a non-model plant with potent anti-diabetic properties. *BMC Genomics* 13(1): 1–15.

Aoki, K., Ogata, Y. and Shibata, D. 2007. Approaches for extracting practical information from gene co-expression networks in plant biology. *Plant & cell physiology,* 48, 381–390.

Aoki, Y., Okamura, Y., Tadaka, S., Kinoshita, K. and Obayashi, T. 2016. ATTED-II in 2016: A plant co-expression database towards lineage-specific co-expression. *Plant Cell Physiol.* 57(1): e5-e5.

Benfey, P. N. and Chua, N. H. 1989. Regulated genes in transgenic plants. *Science* 244(4901): 174–181.

Bentley, D. R., Balasubramanian, S., Swerdlow, H. P., Smith, G. P., Milton, J., Brown, C. G., ... Roe, P. M. 2008. Accurate whole human genome sequencing using reversible terminator chemistry. *Nature* 456(7218): 53–59.

Bentley, R. and Haslam, E. 1990. The shikimate pathway – a metabolic tree with many branches. *Critical Rev. Biochem. Mol. Biol.* 25(5): 307–384.

Bischoff, M., Rösler, J., Raesecke, H. R., Görlach, J., Amrhein, N. and Schmid, J. 1996. Cloning of a cDNA encoding a 3-dehydroquinate synthase from a higher plant, and analysis of the organ-specific and elicitor-induced expression of the corresponding gene. *Plant Mol. Biol.* 31(1): 69–76.

Bornemann, S., Theoclitou, M. E., Brune, M., Webb, M. R., Thorneley, R. N. and Abell, C. 2000. A secondary β deuterium kinetic isotope effect in the chorismate synthase reaction. *Bioorganic Chem.* 28(4): 191–204.

Bose M. A. and Chattopadhyay, S. 2016. Sequencing, de novo assembly, functional annotation and analysis of *Phyllanthus amarus* leaf transcriptome using the Illumina platform. *Front. Plant Sci.* 6: 1199.

Bose, S. K., Yadav, R. K., Mishra, S., Sangwan, R. S., Singh, A. K., Mishra, B., ... Sangwan, N. S. 2013. Effect of gibberellic acid and calliterpenone on plant growth attributes, trichomes, essential oil biosynthesis and pathway gene expression in differential manner in *Mentha arvensis* L. *Plant Physiol. Biochem.* 66: 150–158.

Bülow, L., Schindler, M. and Hehl, R. 2007. PathoPlant®: a platform for microarray expression data to analyze co-regulated genes involved in plant defense responses. *Nucleic Acids Res.* 35(suppl_1): D841–D845.

Carpenter, E. P., Hawkins, A. R., Frost, J. W. and Brown, K. A. 1998. Structure of dehydroquinate synthase reveals an active site capable of multistep catalysis. *Nature* 394(6690): 299–302.

Chaurasiya, N. D., Sangwan, R. S., Misra, L. N., Tuli, R. and Sangwan, N. S. 2009. Metabolic clustering of a core collection of Indian ginseng *Withania somnifera* Dunal through DNA, isoenzyme, polypeptide and withanolide profile diversity. *Fitoterapia* 80(8): 496–505.

Chen, S., Luo, H., Li, Y., Sun, Y., Wu, Q., Niu, Y., ... Qian, Z. 2011. 454 EST analysis detects genes putatively involved in ginsenoside biosynthesis in *Panax ginseng*. *Plant Cell Rep.* 30(9): 1593–1601.

Chen, Y., Liu, Y. S. and Zeng, J. G. (2014). Progresses on plant genome sequencing profile. *Life Sci. Res.* 18: 66–74.

Cheng, J. 2004. A novel gene identification approach: massively parallel signature sequencing. *Prog. Biochem. Biophys.* 8, 761–765. doi: 10.1371/journal.pbio. 0020225

Chowhan, N., Singh, H. P., Batish, D. R., Kaur, S., Ahuja, N. and Kohli, R. K. 2013. β-Pinene inhibited germination and early growth involves membrane peroxidation. *Protoplasma* 250(3): 691–700.

Cooke, B., Ernst, E. 2000. Aromatherapy: A systematic review. *British J. General Practice* 50(455): 493–496.

Costa, M. C. D., Righetti, K., Nijveen, H., Yazdanpanah, F., Ligterink, W., Buitink, J.and Hilhorst, H. W. 2015. A gene co-expression network predicts functional genes controlling the re-establishment of desiccation tolerance in germinated *Arabidopsis thaliana* seeds. *Planta* 242(2): 435–449.

Dash, S., Van Hemert, J., Hong, L., Wise, R. P. and Dickerson, J. A. 2012. PLEXdb: Gene expression resources for plants and plant pathogens. *Nucleic Acids Res.* 40(D1); D1194–D1201.

De Bodt, S., Carvajal, D., Hollunder, J., Van den Cruyce, J., Movahedi, S. and Inzé, D. 2010. CORNET: A user-friendly tool for data mining and integration. *Plant Physiol.* 152(3), 1167–1179.

De Bodt, S., Hollunder, J., Nelissen, H., Meulemeester, N. and Inzé, D. 2012. CORNET 2.0: Integrating plant co-expression, protein-protein interactions, regulatory interactions, gene associations and functional annotations. *New Phytologist* 195(3): 707–720.

De Luca, V., Salim, V., Atsumi, S. M., Yu, F. 2012. Mining the biodiversity of plants: A revolution in the making. *Science* 336(6089): 1658–1661.

Deng, N., Chang, E., Li, M., Ji, J., Yao, X., Bartish, I. V., ... Shi, S. 2016. Transcriptome characterization of *Gnetum parvifolium* reveals candidate genes involved in important secondary metabolic pathways of flavonoids and stilbenoids. *Front. Plant Sci.* 7: 174.

Dewick, P. M. 2009. Secondary metabolism: The building blocks and construction mechanisms. In *Medicinal Natural Products: A Biosynthetic Approach* (pp. 7–38). John Wiley & Sons, Ltd, West Sussex, United Kingdom.

Duncan, K., Edwards, R. M. and Coggins, J. R. 1987. The pentafunctionalarom enzyme of *Saccharomyces cerevisiae* is a mosaic of monofunctional domains. *Biochem. J.* 246(2): 375–386.

Dyer, W. E., Weaver, L. M., Zhao, J. M., Kuhn, D. N., Weller, S. C. and Herrmann, K. M. (1990). A cDNA encoding 3-deoxy-D-arabino-heptulosonate 7-phosphate synthase from *Solanum tuberosum* L. *J. Biol. Chem. 265*(3): 1608–1614.

Entus, R., Poling, M. and Herrmann, K. M. 2002. Redox regulation of *Arabidopsis* 3-deoxy-D-arabino-heptulosonate 7-phosphate synthase. *Plant Physiol.* 129(4): 1866–1871.

Facchini, P. J., Bohlmann, J., Covello, P. S., De Luca, V., Mahadevan, R., Page, J. E., ... Martin, V. J. (2012). Synthetic biosystems for the production of high-value plant metabolites. *Trends Biotech.* 30(3): 127–131.

Fakari, F. R., Tabatabaeichehr, M., Kamali, H., Fakari, F. R. and Naseri, M. 2015. Effect of inhalation of aroma of geranium essence on anxiety and physiological parameters during first stage of labor in nulliparous women: A randomized clinical trial. *J. Caring Sci.* 4(2): 135.

Feng, C. H., Hei, C. Y., Wang, Y., Zeng, Y. F. and Zhang, J. G. 2019. Phylogenetic position of *Chosenia arbutufolia* in the Salicaceae inferred from whole chloroplast genome. *Forest Res.* 32, 73–77. doi: 10.13275/j.cnki.lykxyj. 2019.02.011

Floss, H. G., Onderka, D. K. and Carroll, M. 1972. Stereochemistry of the 3-deoxy-D-arabino-heptulosonate 7-phosphate synthetase reaction and the chorismate synthetase reaction. *J. Biol. Chem.* 247(3): 736–744.

Fucile, G., Di Biase, D., Nahal, H., La, G., Khodabandeh, S., Chen, Y., ... Provart, N. J. 2011. ePlant and the 3D data display initiative: Integrative systems biology on the world wide web. *PLoS One 6*(1): e15237.

Funke, T., Han, H., Healy-Fried, M. L., Fischer, M. and Schönbrunn, E. 2006. Molecular basis for the herbicide resistance of Roundup Ready crops. *Proc. Nat. Aca. Sci. USA* 103(35): 13010–13015.

Ganson, R. J., d'Amato, T. A. and Jensen, R. A. 1986. The two-isozyme system of 3-deoxy-D-arabino-heptulosonate 7-phosphate synthase in *Nicotiana silvestris* and other higher plants. *Plant Physiol.* 82(1): 203–210.

Gasser, C. S., Winter, J. A., Hironaka, C. M. and Shah, D. M. 1988. Structure, expression, and evolution of the 5-enolpyruvylshikimate-3-phosphate synthase genes of petunia and tomato. *J. Biol. Chem.* 263: 4280–87.

Ghosh, S., Chisti, Y. and Banerjee, U. C. 2012. Production of shikimic acid. *Biotech. Adv.* 30(6): 1425–1431.

Góngora-Castillo, E., Fedewa, G., Yeo, Y., Chappell, J., DellaPenna, D., Buell and C. R. 2012. Genomic approaches for interrogating the biochemistry of medicinal plant species. *Methods in Enzymology* 517: 139–159.

Goossens, A., Häkkinen, S. T., Laakso, I., Seppänen-Laakso, T., Biondi, S., De Sutter, V., ... Oksman-Caldentey, K. M. (2003). A functional genomics approach toward the understanding of secondary metabolism in plant cells. *Proc. Nat. Aca. Sci. USA* 100(14): 8595–8600.

Görlach, J., Raesecke H, Rentsch D, Regenass M, Roy P, et al. 1995. Temporally distinct accumulation of transcripts encoding enzymes of the prechorismate pathway in elicitor-treated, cultured tomato cells. *Proc. Natl. Acad. Sci. USA* 92: 3166–70

Guo, H., Jiao, Y., Di, C., Yao, D., Zhang, G., Zheng, X., ... Su, Z. 2009. Discovery of *Arabidopsis* GRAS family genes in response to osmotic and drought stresses. *Chinese Bull. Bot.* 44(03): 290.

Guo, J., Huang, Z., Sun, J., Cui, X. and Liu, Y. 2021. Research progress and future development trends in medicinal plant transcriptomics. *Front. Plant Sci.* 12: 691838.

Guo, X., Li, Y., Li, C., Luo, H., Wang, L., Qian, J., ... Chen, S. 2013. Analysis of the *Dendrobium officinale* transcriptome reveals putative alkaloid biosynthetic genes and genetic markers. *Gene* 527(1): 131–138.

Hamada, K., Hongo, K., Suwabe, K., Shimizu, A., Nagayama, T., Abe, R., ... Yano, K. 2011. OryzaExpress: an integrated database of gene expression networks and omics annotations in rice. *Plant Cell Physiol.* 52(2): 220–229.

Han, R., Rai, A., Nakamura, M., Suzuki, H., Takahashi, H., Yamazaki, M. and Saito, K. 2016. De novo deep transcriptome analysis of medicinal plants for gene discovery in biosynthesis of plant natural products. *Methods Enzymol.* 576: 19–45.

Han, R., Takahashi, H., Nakamura, M., Bunsupa, S., Yoshimoto, N., Yamamoto, H., ... Saito, K. 2015. Transcriptome analysis of nine tissues to discover genes involved in the biosynthesis of active ingredients in *Sophora flavescens*. *Biol. Pharmaceutical Bull.* 38(6): 876–883.

Hang, B. and Peng, J. Y. 2006. cDNA-AFLP and its application in research about gene expressions of plants. *Acta Bot. Sin.* 8, 1753–1758. doi: 10.3321/j.issn: 1000-4025.2006.08.042

Hao, D. C., Xiao, P. G. 2015. Genomics and evolution in traditional medicinal plants: Road to a healthier life. *Evolutionary Bioinformatics* 11: EBO-S31326.

Hao, D., Ma, P., Mu, J., Chen, S., Xiao, P., Peng, Y., ... Sun, C. 2012. De novo characterization of the root transcriptome of a traditional Chinese medicinal plant *Polygonum cuspidatum*. *Sci. China Life Sci.* 55(5): 452–466.

Harris JM, Gonzalez-Bello C, Kleanthous C, Hawkins A, Coggins J, Abell C. 1999. Evidence from kinetic isotope studies for an enolate intermediate in the mechanism of type II dehydroquinases. *Biochemical Journal*. 319(1999): 333–336.

Herrmann, K. M. 1995. The shikimate pathway: Early steps in the biosynthesis of aromatic compounds. *Plant Cell* 7(7): 907.

Huisman, O. C., Kosuge, T. 1974. Regulation of aromatic amino acid biosynthesis in higher plants: II. 3-deoxy-arabino-heptulosonic acid 7-phosphate synthetase from cauliflower. *J. Biol. Chem.* 249(21): 6842–6848.

Jacobs, D. I., Gaspari, M., van der Greef, J., van der Heijden, R. and Verpoorte, R. 2005. Proteome analysis of the medicinal plant *Catharanthus roseus*. *Planta* 221(5): 690–704.

Jain, M. 2012. Next-generation sequencing technologies for gene expression profiling in plants. Briefings in Funct. *Genomics* 11(1): 63–70.

Jain, M., Srivastava, P. L., Verma, M., Ghangal, R. and Garg, R. 2016. De novo transcriptome assembly and comprehensive expression profiling in *Crocus sativus* to gain insights into apocarotenoid biosynthesis. *Sci. Rep.* 6(1): 1–13.

Jeanette M., Emon V. 2016. The omics revolution in agricultural research. *J. Agric. Food Chem.* 64, 2016(1): 36–44.

Jia, C. L., Zhang, Y., Zhu, L. and Zhang, R. 2015. Application progress of transcriptome sequencing technology in biological sequencing. *Mol. Plant Breed.* 10: 2388–2394.

Keith, B., Dong, X. N., Ausubel, F. M. and Fink, G. R. 1991. Differential induction of 3-deoxy-D-arabino-heptulosonate 7-phosphate synthase genes in *Arabidopsis thaliana* by wounding and pathogenic attack. *Proc. Nat. Aca. Sci. USA* 88(19): 8821–8825.

Korkina, L., Kostyuk, V., De Luca, C. and Pastore, S. 2011. Plant phenylpropanoids as emerging anti-inflammatory agents. *Mini-reviews Med. Chem.* 11(10): 823–835.

Krishnan, N. M., Pattnaik, S., Jain, P., Gaur, P., Choudhary, R., Vaidyanathan, S., ... Panda, B. 2012. A draft of the genome and four transcriptomes of a medicinal and pesticidal angiosperm *Azadirachta indica*. *BMC Genomics* 13(1): 1–13.

Kushwaha, A. K., Sangwan, N. S., Tripathi, S. and Sangwan, R. S. 2013. Molecular cloning and catalytic characterization of a recombinant tropine biosynthetic tropinone reductase from *Withania coagulans* leaf. *Gene* 516(2): 238–247.

Laubinger, S., Zeller, G., Henz, S. R., Sachsenberg, T., Widmer, C. K., Naouar, N., ... Weigel, D. 2008. At-TAX: a whole-genome tiling array resource for developmental expression analysis and transcript identification in *Arabidopsis thaliana*. *Genome Biol.* 9(7): 1–16.

Lee, G. W., Lee, S., Chung, M. S., Jeong, Y. S. and Chung, B. Y. 2015. Rice terpene synthase 20 (OsTPS20) plays an important role in producing terpene volatiles in response to abiotic stresses. *Protoplasma* 252(4): 997–1007.

Lewis, J., Johnson, K. A. and Anderson, K. S. 1999. The catalytic mechanism of EPSP synthase revisited. *Biochemistry* 38(22): 7372–7379.

Li, X. L., Bai, B., Wu, J., Deng, Q. Y. and Zhou, B. 2012. Transcriptome analysis of early interaction between rice and *Magnaporthe oryzae* using next-generation sequencing technology. *Hereditas* 34(1): 102–112.

Li, Y. M., Li, S. X., Li, X. S. and Li, C. Y. 2018. Transcriptome studies with the third-generation sequencing technology. *Life Sci. Instrum.* 16: 114–121.

Li, Y. and Xu, X. Y. 2019. Research progress of high-throughput sequencing technology. *China Med. Eng.* 27: 32–37.

Li, W. W., Sun, Y., Yuan, Y., Yu, J. L., Chen, Q. Q., Ge, Y. L. et al. 2020. Isolation and genomic analyses of *SARS-CoV-2* in Anhui Province, China. *Bing Du Xue Bao* 36, 751–757. doi:10.13242/j.cnki.bingduxuebao.003795

Liang, S. B., Liu, J. Y., Liu, J., Yang, J. T., Li, J. L. and Zhang, Y. M. 2017. Next-generation sequencing applications for crop genomes. *China Biotechn.* 37(2): 111–120.

Liu, H. B. 2019. Applications of RNA-seq technology on medicinal plants. *Chinese Trad. Herbal Drugs* 5346–5354.

Liu, L., Li, Y., Li, S., Hu, N., He, Y., Pong, R., ... Law, M. 2012. Comparison of next-generation sequencing systems. *J. Biomed. Biotech.* 2012.

Luo, R., Liu, B., Xie, Y., Li, Z., Huang, W., Yuan, J., ... Wang, J. 2012. SOAPdenovo2: an empirically improved memory-efficient short-read de novo assembler. *Gigascience* 1(1): 2047–217X.

Ma, L. N., Yang, J. B., Ding, Y. F. and Li, Y. K. 2019) Research progress on three generations sequencing technology and its application. *China Anim. Husb. Vet. Med.* 46, 2246–2256. doi: 10.16431/j.cnki.1671-7236.2019. 08.007

Macheroux, P., Schmid, J., Amrhein, N. and Schaller, A. 1999. A unique reaction in a common pathway: Mechanism and function of chorismate synthase in the shikimate pathway. *Planta* 207(3): 325–334.

Mao, L., Van Hemert, J. L., Dash, S. and Dickerson, J. A. 2009. *Arabidopsis* gene co-expression network and its functional modules. *BMC Bioinformatics* 10(1): 1–24.

Margulies, M., Egholm, M., Altman, W. E., Attiya, S., Bader, J. S., Bemben, L. A., ... Rothberg, J. M. (2005). Genome sequencing in microfabricated high-density picolitre reactors. *Nature* 437(7057): 376–380.

Martin, L. B., Fei, Z., Giovannoni, J. J. and Rose, J. K. 2013. Catalyzing plant science research with RNA-seq. *Front. Plant Sci.* 4: 66.

McKernan, K. J., Peckham, H. E., Costa, G. L., McLaughlin, S. F., Fu, Y., Tsung, E. F., ... Blanchard, A. P. 2009. Sequence and structural variation in a human genome uncovered by short-read, massively parallel ligation sequencing using two-base encoding. *Genome Res.* 19(9): 1527–1541.

Mei, C., Wang, H., Zan, L., Cheng, G., Li, A., Zhao, C. and Wang, H. 2016. Research progress on animal genome research based on high-throughput sequencing technology. *J. Northwest A & F University-Natural Science Edition* 44(3): 43–51.

Mironova, V. V., Weinholdt, C. and Grosse, I. 2015. RNA-seq data analysis for studying abiotic stress in horticultural plants. In *Abiotic Stress Biology in Horticultural Plants* (pp. 197–220). Springer, Tokyo.

Mishra, S., Bansal, S., Mishra, B., Sangwan, R. S., Jadaun, J. S. and Sangwan, N. S. 2016. RNAi and homologous over-expression based functional approaches reveal triterpenoid synthase gene-cycloartenol synthase is involved in downstream withanolide biosynthesis in *Withania somnifera*. *PLoS One* 11(2): e0149691.

Mittelstädt, G., Negron, L., Schofield, L. R., Marsh, K. and Parker, E. J. 2013. Biochemical and structural characterisation of dehydroquinate synthase from the New Zealand kiwifruit *Actinidia chinensis*. *Arch. Biochem. Biophy.* 537(2): 185–191.

Mondal, S., Bandyopadhyay, S., K Ghosh, M., Mukhopadhyay, S., Roy, S. and Mandal, C. 2012. Natural products: Promising resources for cancer drug discovery. *Anti-Cancer Agents in Medicinal Chemistry (Formerly Current Medicinal Chemistry-Anti-Cancer Agents)*, 12(1): 49–75.

Narnoliya, L. K., Rajakani, R., Sangwan, N. S., Gupta, V. and Sangwan, R. S. 2014. Comparative transcripts profiling of fruit mesocarp and endocarp relevant to secondary metabolism by suppression subtractive hybridization in *Azadirachta indica* (neem). *Mol. Biol. Rep.* 41(5): 3147–3162.

Nick, P. 2015. Perfumes of survival. *Protoplasma* 252(4): 933–934.

Obayashi, T., Kinoshita, K., Nakai, K., Shibaoka, M., Hayashi, S., Saeki, M., ... Ohta, H. 2007. ATTED-II: a database of co-expressed genes and cis elements for identifying co-regulated gene groups in *Arabidopsis*. *Nucleic Acids Res.* 35(suppl_1): D863-D869.

Osborne, A., Thorneley, R. N., Abell, C., Bornemann, S. 2000. Studies with substrate and cofactor analogues provide evidence for a radical mechanism in the chorismate synthase reaction. *J. Biol. Chem.* 275(46): 35825–35830.

Pan, Q., Saiman, M. Z., Mustafa, N. R., Verpoorte, R. and Tang, K. 2016. A simple and rapid HPLC-DAD method for simultaneously monitoring the accumulation of alkaloids and precursors in different parts and different developmental stages of *Catharanthus roseus* plants. *J. Chromatography B* 1014: 10–16.

Parkinson, H., Sarkans, U., Shojatalab, M., Abeygunawardena, N., Contrino, S., Coulson, R., ... Brazma, A. 2005. ArrayExpress – a public repository for microarray gene expression data at the EBI. *Nucleic Acids Res.* 33(suppl_1), D553-D555.

Patel, R. V., Nahal, H. K., Breit, R. and Provart, N. J. 2012. BAR expressolog identification: Expression profile similarity ranking of homologous genes in plant species. *Plant J.* 71(6): 1038–1050.

Pinto, J. E., Suzich, J. A. and Herrmann, K. M. 1986. 3-Deoxy-D-arabino-heptulosonate 7-phosphate synthase from potato tuber (*Solanum tuberosum L.*). *Plant Physiol.* 82(4): 1040–1044.

Qian, X. L., Ming, H. F. and Li, Y. H. (2012). The research progress, application and prospects of EST. *Life Sci. Res. 16:* 446–450.

Rai, A. and Saito, K. 2016. Omics data input for metabolic modeling. *Curr. Opinion Biotech.* 37: 127–134.

Rai, A., Yamazaki, M., Takahashi, H., Nakamura, M., Kojoma, M., Suzuki, H. and Saito, K. 2016. RNA-seq transcriptome analysis of *Panax japonicus*, and its comparison with other *Panax* species to identify potential genes involved in the saponins biosynthesis. *Front. Plant Sci.* 7: 481.

Ransbotyn, V., Yeger-Lotem, E., Basha, O., Acuna, T., Verduyn, C., Gordon, M., ... Barak, S. (2015). A combination of gene expression ranking and co-expression network analysis increases discovery rate in large-scale mutant screens for novel *Arabidopsis thaliana* abiotic stress genes. *Plant Biotech. J.* 13(4): 501–513.

Rubin, J. L. and Jensen, R. A. 1985. Differentially regulated isozymes of 3-deoxy-D-arabino-heptulosonate-7-phosphate synthase from seedlings of *Vigna radiata* [L.] Wilczek. *Plant Physiol.* 79(3): 711–718.

Rung, J. and Brazma, A. 2013. Reuse of public genome-wide gene expression data. *Nat. Rev. Genetics* 14(2): 89–99.

Sangwan, N. S., Kumar, R., Srivastava, S., Kumar, A., Gupta, A. and Sangwan, R. S. 2010. Recent developments on secondary metabolite biosynthesis in *Artemisia annua* L. *J. Plant Biol.* 37(1): 1–24

Sangwan, N. S., Mishra, L. N., Tripathi, S., Kushwaha, A. K., Sangwan, R. S. 2013. Omics of secondary metabolic pathways in *Withania somnifera Dunal* (ashwagandha). Omics Applications: In: Barh D, Taylor Francis (eds) *Crop Science*, pp. 410–434. CRC Press, Boca Raton.

Sangwan, N. S., Yadav, U., Sangwan, R. S. 2003. Genetic diversity among elite varieties of the aromatic grasses, *Cymbopogon martinii. Euphytica* 130(1): 117–130.

Sangwan, R. S., Chaurasiya, N. D., Lal, P., Misra, L., Tuli, R., Sangwan, N. S. 2008. Withanolide A is inherently de novo biosynthesized in roots of the medicinal plant Ashwagandha (*Withania somnifera*). *Physiol. Plant.* 133(2): 278–287.

Sangwan, R. S., Tripathi, S., Singh, J., Narnoliya, L. K. and Sangwan, N. S. 2013. De novo sequencing and assembly of *Centella asiatica* leaf transcriptome for mapping of structural, functional and regulatory genes with special reference to secondary metabolism. *Gene* 525(1): 58–76.

Schmid, J. and Amrhein, N. 1995. Molecular organization of the shikimate pathway in higher plants. *Phytochem.* 39(4): 737–749.

Schofield, L. R., Anderson, B. F., Patchett, M. L., Norris, G. E., Jameson, G. B. and Parker, E. J. 2005. Substrate ambiguity and crystal structure of *Pyrococcus furiosus* 3-deoxy-D-arabino-heptulosonate-7-phosphate synthase: An ancestral 3-deoxyald-2-ulosonate-phosphate synthase?. *Biochemistry* 44(36): 11950–11962.

Schönbrunn, E., Eschenburg, S., Shuttleworth, W. A., Schloss, J. V., Amrhein, N., Evans, J. N. and Kabsch, W. 2001. Interaction of the herbicide glyphosate with its target enzyme 5-enolpyruvylshikimate 3-phosphate synthase in atomic detail. *Proc. Nat. Aca. Sci. USA* 98(4): 1376–1380.

Simkin, A. J., Guirimand, G., Papon, N., Courdavault, V., Thabet, I., Ginis, O., ... Clastre, M. 2011. Peroxisomal localisation of the final steps of the mevalonic acid pathway in planta. *Planta* 234(5): 903–914.

Singh, S. A., Christendat, D. 2007. The DHQ-dehydroshikimate-SDH-shikimate-NADP (H) complex: Insights into metabolite transfer in the shikimate pathway. *Crystal Growth and Design* 7(11): 2153–2160.

Singh, J., Farzana, S., Sangwan, R. S., Narnoliya, L. K. and Sangwan, N. S. 2014. Enhanced secondary metabolite production and pathway gene expression by leaf explants-induced direct root morphotypes are regulated by combination of growth regulators and culture conditions in *Centel laasiatica* (L.) urban. *Plant Growth Regulation* 75: 55–66.

Sangwan, S., Neelam, S., Abad Farooqi, A. H. and Singh, S.R. 1994. Effect of drought stress on growth and essential oil metabolism in lemon grasses. *New Phytol.* 128(1): 173–179.

Sangwan NS, Mishra LN, Tripathi S, Kushwaha AK and Sangwan RS (2014) Omics of secondary metabolic pathways in *Withania somnifera* Dunal (ashwagandha). *OMICS Applications In Crop Science* 385–408.

Srinivasan, P. R., Rothschild, J. and Sprinson, D. B. 1963. The enzymic conversion of 3-deoxy-D-arabino-heptulosonic acid 7-phosphate to 5-dehydroquinate. *J.Biol.Chem.* 238(10): 3176–3182.

Srivastava, S., Sangwan, R. S., Tripathi, S., Mishra, B., Narnoliya, L. K., Misra, L. N. and Sangwan, N. S. 2015. Light and auxin-responsive cytochrome P450s from *Withania somnifera Dunal*: Cloning, expression and molecular modelling of two pairs of homologue genes with differential regulation. *Protoplasma* 252(6): 1421–1437.

Su, N., Yang, L., Liu, J., Zhou, K. Y., Zhang, S. H., Lei, F. M., ... Zheng, H. Y. 2016. Application research progress of gene chip technology in China. *Biotechnol. Lett.* 27: 289–292.

Sui, C., Zhang, J., Wei, J., Chen, S., Li, Y., Xu, J., ... Xu, Y. 2011. Transcriptome analysis of *Bupleurum chinense* focusing on genes involved in the biosynthesis of saikosaponins. *BMC Genomics* 12(1): 1–16.

Sun, H. J., & Wei, H. J. 2018. The application of RNA-seq technology in the study of the transcriptome. *Chin. Foreign Med. Res.* 16: 184–187.

Sunohara, Y., Baba, Y., Matsuyama, S., Fujimura, K. and Matsumoto, H. 2015. Screening and identification of phytotoxic volatile compounds in medicinal plants and characterizations of a selected compound, eucarvone. *Protoplasma 252*(4): 1047–1059.

Unamba, C. I., Nag, A. and Sharma, R. K. 2015. Next-generation sequencing technologies: The doorway to the unexplored genomics of non-model plants. *Front. Plant Sci.* 6: 1074.

Valdés, A., Ibáñez, C., Simó, C. and García-Cañas, V. 2013. Recent transcriptomics advances and emerging applications in food science. *Trends Anal. Chem.* 52: 142–154.

Wang, L. H., Lin, Y. P. and Lin, Y. C. 2016. Screening nicotinamide in cosmetic and pharmaceutical products and nicotinic acid skin penetration from essential-oil formulations using attenuated total reflectance-infrared spectroscopy. *Biomed. Spectroscopy Imaging* 5(1): 89–97.

Wang, W., Wang, Y., Zhang, Q., Qi, Y. and Guo, D. 2009. Global characterization of *Artemisia annua* glandular trichome transcriptome using 454 pyrosequencing. *BMC Genomics* 10(1): 1–10.

Wang, Y. N., Tang, L., Hou, Y., Wang, P., Yang, H. and Wei, C. L. 2016. Differential transcriptome analysis of leaves of tea plant (*Camellia sinensis*) provides comprehensive insights into the defense responses to Ectropis oblique attack using RNA-Seq. *Funct. Integ. Genomics* 16(4): 383–398.

Wang, Z., Gerstein, M. and Snyder, M. 2009. RNA-Seq: A revolutionary tool for transcriptomics. *Nature Rev. Gen.* 10(1): 57–63.

Webby, C. J., Baker, H. M., Lott, J. S., Baker, E. N. and Parker, E. J. 2005. The structure of 3-deoxy-D-arabino-heptulosonate 7-phosphate synthase from *Mycobacterium tuberculosis* reveals a common catalytic scaffold and ancestry for type I and type II enzymes. *J. Mol. Biol.* 354(4): 927–939.

Wenger, A. M., Peluso, P., Rowell, W. J., Chang, P. C., Hall, R. J., Concepcion, G. T., ... Hunkapiller, M. W. 2019. Accurate circular consensus long-read sequencing improves variant detection and assembly of a human genome. *Nat. Biotech.* 37(10): 1155–1162.

Wu, J. and Woodard, R. W. 2006. New insights into the evolutionary links relating to the 3-deoxy-D-arabino-heptulosonate 7-phosphate synthase subfamilies. *J. Biol. Chem.* 281(7): 4042–4048.

Wu, J. M., Li, Y. R., Yang, L., Wang, A. Q., Yang, L. T. and Gan, Z. K. 2009. Applications of cDNA-AFLP and cDNA-SRAP methods for gene differential expression analysis in plant and the technical comparison. *Biotechnol. Bull.* 11, 52–55. doi:10.13560/j.cnki.biotech.bull.1985.2009.11.009

Wu, Q., Sun, C., Luo, H., Li, Y., Niu, Y., Sun, Y., ... Chen, S. 2011. Transcriptome analysis of *Taxus cuspidata* needles based on 454 pyrosequencing. *Planta Med.* 77(04): 394–400.

Xiao, M., Zhang, Y., Chen, X., Lee, E. J., Barber, C. J., Chakrabarty, R., ... Sensen, C. W. 2013. Transcriptome analysis based on next-generation sequencing of non-model plants producing specialized metabolites of biotechnological interest. *J. Biotech.* 166(3): 122–134.

Xu, F. Z. and Wu, J. Z. (2005). Research progress of Chinese medicine gene chip technology. *Vet.Orientat.*32:300. doi:10.3969/j.issn.1674-0440.2005.05.004

Xu, B., Zhang, W., Feng, X., Wang, C., Zhang, H., Xu, H., ... Zhang, J. 2014. Application progress of transcriptome sequencing technology in maize. *J. Maize Sci.* 22(1): 67–78.

Yadav, R. K., Sangwan, R. S., Sabir, F., Srivastava, A. K. and Sangwan, N. S. 2014. Effect of prolonged water stress on specialized secondary metabolites, peltate glandular trichomes, and pathway gene expression in *Artemisia annua* L. *Plant Physiol. Biochem.* 74: 70–83.

Yang, L., Ding, G., Lin, H., Cheng, H., Kong, Y., Wei, Y., ... Yang, C. 2013. Transcriptome analysis of medicinal plant *Salvia miltiorrhiza* and identification of genes related to tanshinone biosynthesis. *PLoS One* 8(11): e80464.

Yang, X., Liu, L. Y., Guan, F. Z., Li, Z. G., Wu, G. W., Wang, X., ... Chen, H. 2009. Advances of SAGE technology in plant gene expression. *Plant Fiber Sci. China* 31: 233–237.

Yonekura-Sakakibara, K. and Saito, K. 2009. Functional genomics for plant natural product biosynthesis. *Natural Prod. Rep.* 26(11): 1466–1487.

Zhang, C., Qin, Z., Wang, G., Ji, Z. and Wang, J. 2012. Transcriptome and RNA-seq technology. *Biotech. Bull.* (12): 51.

Zhang, D. Y., Zhang, T. X. and Wang, G. X. 2016. Development and application of second-generation sequencing technology. *Environ. Sci. Technol.* 39: 96–102.

Zhang, Q. and Sheng, J. 2008. Development and application of gene chip technology. *Zhongguoyixuekexue Yuan xue bao. Acta Academiae Medicinae Sinicae* 30(3): 344–347.

Zhao, S., Tuan, P. A., Li, X., Kim, Y. B., Kim, H., Park, C. G., ... Park, S. U. 2013. Identification of phenylpropanoid biosynthetic genes and phenylpropanoid accumulation by transcriptome analysis of *Lycium chinense*. *BMC Genomics* 14(1): 1–11.

Zheng, X., Xu, H., Ma, X., Zhan, R., Chen, W. 2014. Triterpenoid saponin biosynthetic pathway profiling and candidate gene mining of the *Ilex asprella* root using RNA-seq. *Int. J. Mol. Sci.* 15(4), 5970–5987.

Zhuo, W. W., Zhang, X. M., Li, F., Li, Y. P. and Wang, Y. B. 2015. Comparison of methods for studying differential expression of genes based on PCR technology. *Biotech World* 5, 8–15.

Zimmermann, P., Hirsch-Hoffmann, M., Hennig, L., Gruissem, W. 2004. GENEVESTIGATOR. *Arabidopsis* microarray database and analysis toolbox. *Plant Physiol.* 136(1): 2621–2632.

Zimmermann, P., Laule, O., Schmitz, J., Hruz, T., Bleuler, S. and Gruissem, W. 2008. Genevestigator transcriptome meta-analysis and biomarker search using rice and barley gene expression databases. *Mol. Plant* 1(5): 851–857.

9 LC-MS-based Metabolomics of Medicinal Plants

Gaurav, Rabea Parveen and Sayeed Ahmad

CONTENTS

9.1 INTRODUCTION

Metabolomics is the comprehensive analytical technique for quality assessment of targeted or untargeted metabolites in a biological system (Zhou et al. 2012). Metabolomic studies have been concomitant with many research fields, including functional genomics, environmental and biological stress studies, integrative systems and biology biomarker discovery (Khoo and Al-Rubeai 2007, Goodacre et al. 2004). Besides, such analytical studies are gaining exponential growth in the field of quality control analysis of phytoconstituents' diversity present in medicinal plants (Salem et al. 2020). Medicinal plants are typically accounted with the complex mixture of different varieties of phytoconstituents such as carotenoids, steroids, terpenoids, alkaloids, phenols, flavonoids and glycosides and so far are still remaining a challenge to validate herbal medicine or products based on their quality, safety and regulatory aspects (Gad et al. 2013, Goodarzi, Russell and Vander Heyden 2013, Senguttuvan, Paulsamy and Karthika 2014). However, the two most common

DOI: 10.1201/9781003179139-9

analytical techniques include nuclear magnetic resonance (NMR) and mass spectrometry (MS) and are used abundantly in quantitative and quantitative estimation of plant metabolites. However, these techniques are the main object of choice with plant metabolomics studies due to their high sensitivity and accuracy when coupled to influential chromatographic techniques (e.g., capillary electrophoresis–mass spectrometry (CE-MS), liquid chromatography-mass spectrometry (LC-MS) and gas chromatography-mass spectrometry (GC-MS)), allows the separation and characterization of the polar and nonpolar compound diversity present in the medicinal plant (Jorge, Mata and António 2016).

Furthermore, LC-MS is one of the advanced and first choice techniques due to its high selectivity and sensitivity and it enables extensive detection of metabolites within a sample. It makes us more familiar in characterizing comprehensive metabolite accretion patterns without having liabilities on authentic reference compounds and or isolation of the individual metabolites. The structural assignment process generally depends on the reference databases search for the interpretation of MS and MS/MS data. Moreover, unknown metabolites characterization is a little challenging since these cannot be consigned to a reference database. Further, the fragmentation pattern/parent ions/daughter ions are the unique parameter in LC-MS/MS-based plant metabolomics, allowing us familiarity for characterization of the targeted and untargeted metabolites based on m/z values and reference database.

For authentication, the m/z values of derived data are matched with the data derived from the acquisition reference data of mass data banks, where the matched data is resumed under the defined tolerance for m/z values (Smith et al. 2006). The MS data acquisition depends on intensities for scoring the probability between query and reference data search. For the untargeted metabolites, the data acquisition tolerance can be both in m/z value and chromatographic retention time, and, further, these implications are accustomed across measurements using alignment software (Pluskal et al. 2010). The advanced web-based platform analyses have been established based on automated workflows from processing to peak alignment of row data, annotations and statistical analysis tools to enable easy access for a broad range of investigators or informatics experts (Sawada and Yokota Hirai 2013).

Through high-speed scanning by LC-Q-TOF-MS analysis, in excess of one million untargeted MS/MS implications are reported, and among them, only few thousand have been confined as reference data. The break between detected and reference evidence of such analytical data will extend further due to their extension of detectable metabolites and sensitivity in progressive MS instruments (Tohge and Fernie 2010). As per the cited protocol, more than 90% of the detectable metabolites are unknown due to the large gap between interrogation and reference MS/MS. However, no specific identification approach is still liable to identify the unknown matrix of the phytoconstituents and establish their metabolic pattern (Sawada and Yokota-Hirai 2013).

Besides, several studies have been conducted on plant metabolite evaluation based on LC-MS and revealed the diversity of different metabolites such as lipid-related compounds, carbohydrates, amino and organic acids, as well as structurally diverse phytohormones. Therefore, the hyphenation of MS-based technologies allows us to validate metabolites qualitatively and quantitatively. Further, the correlation of derived metabolites and target biological assessment has a vast consideration in the screening of the metabolites based on their target biological interaction (Fernie 2003).

In this chapter, an attempt has been made to overview instrumentation, experimental design or a selection of methods and applications of LC-MS as a quantitative tool of plant metabolomics The current scenario of LC-MS techniques in qualitative and quantitative analysis of plant metabolites has also been discussed with special focus on the challenges and prospectives of metabolomic analysis.

9.2 INSTRUMENTATION

The LC is comprised of the physical separation of compounds by distributing between the two immiscible phases: stationary and mobile phases. Chromatographic separation of molecule passes among five phases: chromatographic adsorption of targeted and untargeted compounds, partition chromatography based on compounds polarity, the ion-exchange affinity of compounds, size-exclusion and affinity chromatography (Ho et al. 2003). The reverse-phase (RP) mode of the partition chromatography technique is abundantly used for chromatographic separation of targeted and untargeted metabolites. Chromatographic separation system operates under the high pressure, and the complex matrix of compounds is separated based on their chemical interaction with mobile and stationary phases (Pitt 2009; Smyth and Brooks 2004). During the process, high pressure is required to accomplish a constant flow rate for reproducibility of the experimental study. Depending on the chromatographic separation partitioning, column plays an essential role in the LC system. Typically, the most conventional LC columns are mainly prepared in the form of 100–300 mm long with 6.4 mm (1/4 inch) outer diameter, 3.0–4.6 mm with an internal diameter and 3–5 μm diameter of stationary phase or packing particles for more efficient separation (Cajka and Fiehn 2014).

The MS is generally used for the conversion of molecules into the charged or ionization state, and further, the subsequent identification of charged molecules or their fragments would be done based on their obtained mass or fragmented pattern (*m/z*) (Perreault and Lattová 2019). Many technologies are accessible for ionization and their ion prediction with different combinations for ionization of the analyte system. An LC-MS system is far more adaptable than others which typically emphasizes the ion sources and mass analysis of a wide diversity of phytoconstituents present in the medicinal plants. MS is composed of two different parts as the ion source and mass analyzer (Eliasson et al. 2012; Sawada and Yokota Hirai 2013).

9.2.1 Ion Source

Ion source principally consists of a set of three ion sources such as electrospray ionization source (ESI), atmospheric pressure chemical ionization source and atmospheric pressure photo-ionization source (Niessen 2006, Pitt 2009). The ESI is mainly used for the estimation of moderately polar constituents. It can also be suitable for the analysis of numerous metabolites, peptides and xenobiotics. It works for nebulization and conversion of liquid samples into the fine spray of the charged droplets, and after evaporation, the electrical charge is transferred to the analyte in the presence of applied principle heat and dry nitrogen. The ionized molecule is then transferred to the mass spectrometer through a series of applied voltages. The analysis of the charged analyte operates under two different modes, that is, positive and negative ion mode. Besides, a small analyte with one or two functional groups gives a predominant electrical charge and evolves to coordinate with the proton ($M+H^+$) when operated in positive ion mode, while operating in negative ion mode gives a loss of proton to the analyte ($M+H^-$). The molecules which exhibit multiple charge-carrying functional groups result in the adoption of an ion such as $M+2H2^+$, $M+3H3^+$, and so on.

In the APCI mode of ionization, semi-polar or nonpolar metabolites are analyzed, which are generally not detectable by ESI mode due to their ionization obstacles. For intense, free steroids generally do not ionize easily by the process of ES ionization. Therefore, APCI has been approved to improve the sensitivity and specificity of such metabolites. Furthermore, the technique is used for lipid and fat analysis as well. In the APPI method of ionization, excitation and ionization of photons occur after the nebulization of the analyte. The liberated energy of the photons acted to diminish the concurrent ionization of applied principle solvents along with ion source gases. This approach provides a strong singly-charged ion and is highly advanced for the analysis of neutral steroids etc.

9.2.2 Mass Analyzers

Further, the mass analyzers consist of quadrupole analyzers, time-of-flight analyzers, ion trap analyzers and hybrid analyzers that are highly employed for the determination of ions or fragmented ion of the targeted or untargeted metabolites.

In a quadrupole mass analyzer, a set of four parallel metal rods resides with persistent and fluctuating voltages, which transmit a constricted band of m/z values along the axis of the rods. Under the erratic conditions, the voltages frequency, m/z values of the charged analyte scanned across a defined range of the mass spectrum (Ho et al. 2003). The quadrupole mass analyzers are mostly operated at <4000 m/z with up to 1000 m/z per sec scan speeds. Usually, the operating mass resolution means is to represent the seldom mass accuracy. The tandem MS analysis is an example of a triple quadrupole mass spectrometer in which the analysis of the sample of choice increased the specificity and accuracy over the varieties of constituents with the single mass analysis only. For example, a major M+H+ ion of 401 m/z is produced during the analysis of 25-hydroxy vitamin D3, and during collision-induced dissociation, it loses water and produces a major 383 m/z product ion (Pitt 2009).

In the time-of-flight (TOF), a high voltage is required to operate and accelerate ions analysis (Williamson and Bartlett 2007). The time of flight is taken by the ionized analyte to collapsible down in the flight tube, and the detector depends upon their velocity and m/z values. The induction of hastening voltage and output of the detector can be transformed into a mass spectrum. The TOF spectral analysis of the targeted sample is associated with high sensitivity, mass accuracy and quick spectral outcomes (Bristow 2006).

In ion trap analyzers, hyperbolic electrodes are used to trap ions using static and radiofrequency voltages under their three-dimensional space. The ionized analyte is subsequently translocated to the trap based on its m/z value to generate a mass spectrum. The existing voltage in the trap analyzer can be applied to isolate or even terminate the specific ion, among others, to get the desired spectral information. An inert gas can be acclimatized to the trap for induction of fragmentation or the ability to fragment the desired analyte before the final spectrum (Tozuka et al. 2003).

Besides, different mass analyzers are associated with tandem mass spectrum analysis, particularly two configuration techniques useful in LC-MS. Hybrid analyzed generally refers to the displacement of the third quadrupole of a triple quadrupole MS with TOF analyzer to create hybrid quadrupole time-of-flight (QTOF) (Chernushevich, Loboda and Thomson 2001). Such mass analyzers are strongly hyphenated to the proteomics field but more limited with triple quadrupole instruments due to their scanning functions. In mass spectrum analysis, the design instruments should operate at the different modes of ions trapping and their subsequent translocation based on their m/z values, which are formalized with linear ion trapping and the overall configuration (Hopfgartner et al. 2004). Further, the product ion scanning and additional fragmentation can be considerably enhanced using hybrid mass analyzers so that an additional stage of fragmentation and mass analysis can be achieved for a robust analysis (Rashed et al. 1997).

The principle of LC-MS instrumentation and processing of chromatographic and spectrum data information using current chemometric tools has been described in Figure 9.1.

9.3 EXPERIMENTAL DESIGN AND SELECTION OF METHOD FOR METABOLOMIC ILLUSTRATION

In metabolomic analysis, experimental design and selection of methods play an essential role in validating medicinal plants from their own quality and regulatory aspect. Several steps such as plant harvesting, sample preparation and optimization of an analytical method are associated with versatile chemical or spectrum information in a selective and integrated way. In a diagrammatic flow chart (Figure 9.2) of the study, the road map of steps involved in metabolomics analysis from nature to laboratory assessments has been clearly described.

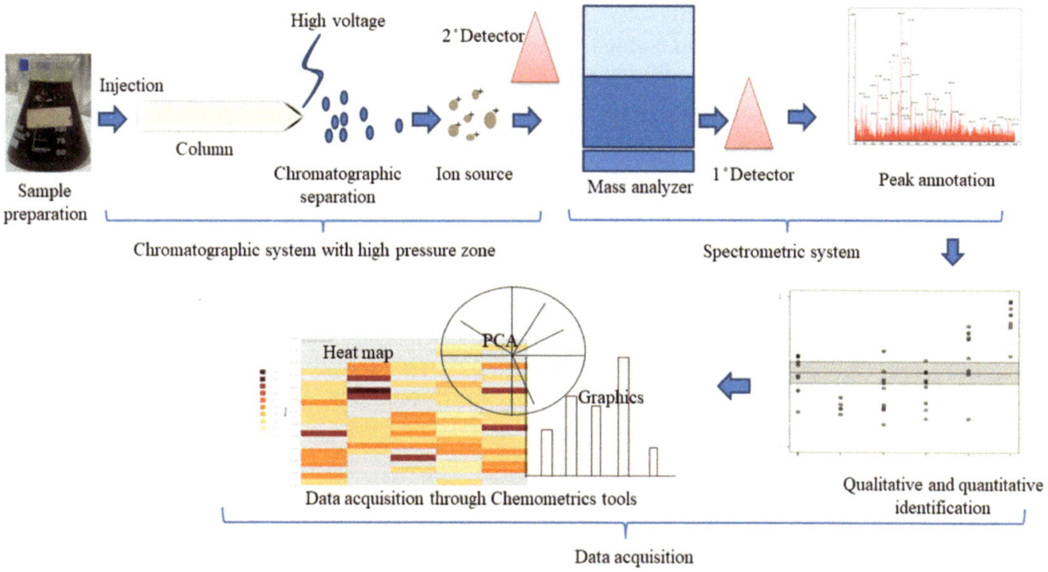

FIGURE 9.1 LC-MS instrumentation, working principle and data acquisition.

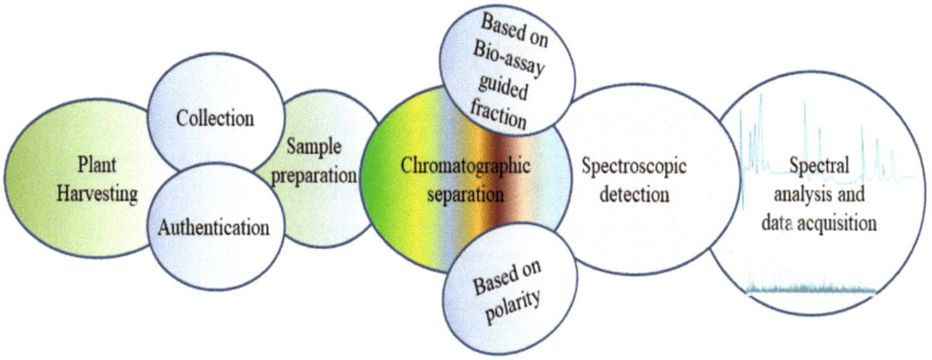

FIGURE 9.2 The road map of steps involved in metabolomic analysis using LC-MS as analytical tool.

Further, the detailed note has been described for analytical methods used in the metabolomic analysis of medicinal plants, which mainly emphasizes the selection of appropriate growth conditions such as relative humidity, photoperiod and temperature during harvesting of the plant of choice or production of secondary. A suitable number of biological replicate, control and randomization trials throughout the entire experimental analysis is necessary to achieve considerable outcomes and minimize experimental variation (Fiehn et al. 2008 2007; Sumner et al. 2007).

Furthermore, it is necessary to select the appropriate analytical method for the analysis of targeted metabolites. In LC-MS analysis, the selection of method generally depends on the diversity of phytoconstituents being analyzed, chromatographic separation, ionization and mass analyzer source employed for the analytical method. During extraction of metabolites, chemical stability and analytical recovery of the metabolites should not be affected; even the variability of metabolites should be minimum during extraction (Kim and Verpoorte 2010; Salem et al. 2020). If polarity base extractions of metabolites or bio-guided fractions are considered for the metabolomic analysis, the nature of compounds and their physiochemical interaction should be known to optimize the analytical method. Polar organic solvents such as methanol and acetonitrile both have a hydrophilic and hydrophobic nature and are able to extract both hydrophilic and hydrophobic compounds. Besides, acetonitrile is a more efficient organic solvent than methanol to improve accuracy and precision in metabolomics experiments. Acetonitrile is often used to precipitate proteins from the biological samples such as serum, plasma and so on, and to improve metabolomic analysis (Rabinowitz and Kimball 2007, Delatte et al. 2009).

During the pre-analytical assessment, selection of ionization and mass analyzing sources is necessary for the desired outcomes of metabolomic studies. LC-ESI/MS or LC-ESI-Q-TOF-MS is highly employed for the analysis of semi-polar or polar constituents (alkaloids, polyphenols, glycosides, proteins) from the complex matrix of the medicinal plant. LC-APCI-Q-TOF-MS method is employed for the analysis of lipids, fat-soluble vitamins and steroids, while the selective technique as LC-APPI–MS is employed for the analysis of steroids and terpenoids (Pitt 2009). A systematic representation of experimental design and selection of methods is represented in Figure 9.3.

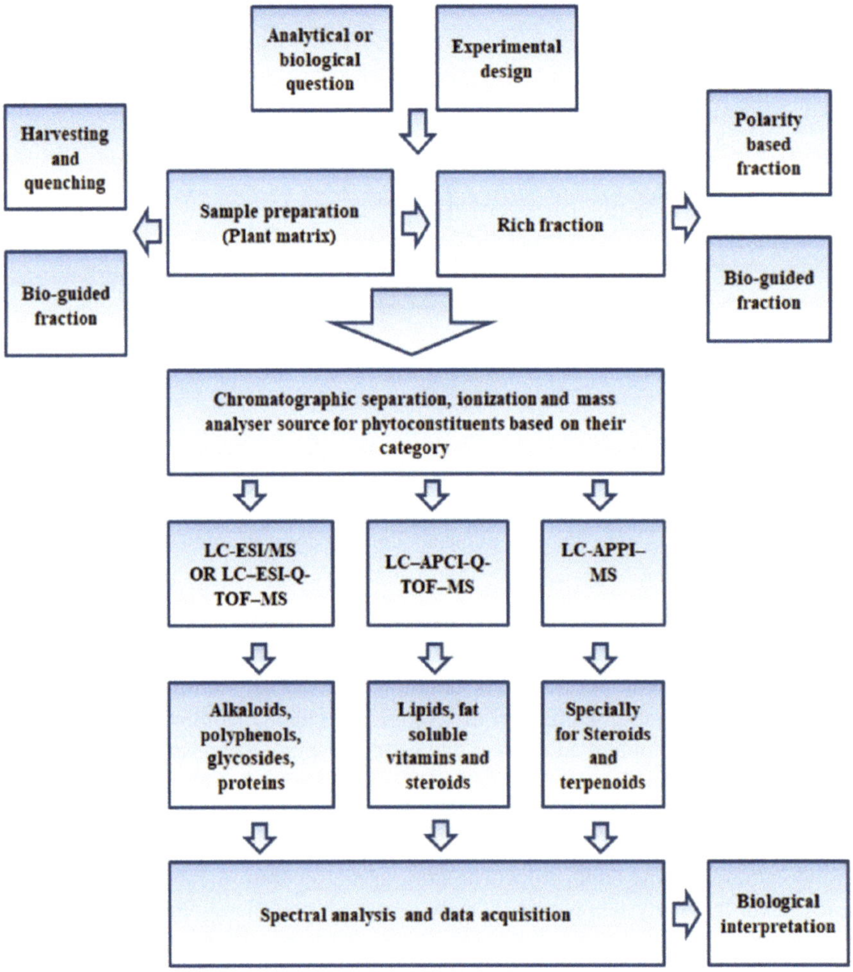

FIGURE 9.3 A systematic representation of experimental design and selection of method for target-based evaluation of metabolites.

9.4 OPTIMIZATION OF CHROMATOGRAPHIC AND IONIZATION PROCESS FOR METABOLOMICS

Flow rates and mobile phase compositions for existing LC separations and their current ion sources can often be directly coupled to the mass spectrometer, which is capable of handling and identification of a wide range of ion sources. However, the quality of MS data can be affected by a number of factors or even LC separation. A successive and precise method can improve assay performance with a choice of modification. Generally, lower flow rates in the range 0.05 to 0.2 mL/min in the columns of 1.0- or 2.1-mm diameters can result in improved sensitivity and resolutions of the mass spectrum. In the chromatographic phase, different ion sources and LC pumps are required for a good separation (Miranda-Osorio et al. 2016, Jorge, Mata and António 2016).

In addition, chloroform, acetonitrile, ethanol, methanol and water are typically used for reverse and normal phase LC associated with ESI. Besides, acetonitrile and water can be used as a conventional mobile phase for LC but are not often suitable for a good separation as they may contain traces of compounds that may affect the quality of LC-MS spectrum analysis. For example, commercial grades of methanol found large differences in their quality (Annesley 2007). The use of inorganic ions in the form of buffers such as phosphate and sodium acetate should be avoided and substituted with ammonium formate, ammonium acetate and ammonium bicarbonate as they significantly suppress the ions, and form MS adduct of phosphate and sodium. To avoid such complications, the minimum required amount of buffers should be used for satisfactory results of chromatographic (Petruczynik et al. 2020, Pitt 2009).

9.5 APPLICATIONS OF LC-MS AS A QUANTITATIVE TOOL OF PLANT METABOLITES

Quantitative analysis of targeted or untargeted metabolites puts major efforts into optimizing extraction, separation and instrumental condition. It is extremely important to understand the physiochemical property of the targeted metabolite and the best suitable analytical condition to validate the method. LC-MS0based quantification of metabolites approached relative and absolute quantification. Relative quantification approaches the instrumental response with respect to the internal standard, while absolute quantification deals with more demanding steps, which includes all the method validation steps for quantification, namely instrument linearity against different concentration for method calibration, the limit of detection (LOD) and limit of quantification (LOQ), method precision, accuracy, specificity and selectivity.

9.5.1 Instrument Linearity

In the assessment of method linearity with respect to calibration line and its corresponding slope against a first-order calibration function ($y = ax+b$), axis intercept (b) and slope (a) are representative of the analytical method sensitivity. To calibrate the sensitivity of the instrument, calibration samples are prepared at varying concentration levels against the corresponding detector response in the form of peak height, area and intensity. The linearity and the correlation coefficient (R2) are then evaluated from the resulting calibration data. However, a good linear relation among the applied concentration is obtained when the value of the correlation coefficient might be very close to one. An appropriate number of standard concentrations is required to define the relationship between the concentration and instrumental response adequately. Although the information on the required specific number of concentrations is scarce for a calibration curve, 5–8 concentration points are most recommended, and each should be analyzed at least in triplicate (Stöckl, D'Hondt and Thienpont 2009).

Further, the choice of the appropriate calibration model is still necessary for reliable quantification studies. In such analysis, the quantification of the targeted metabolite can be achieved via the following three main steps: (i) external standard calibration plots using the standard solution of the analyte; (ii) internal standard calibration plot, in which an internal standard is added to the analyte and analyzed together; (iii) standard addition calibration plot, in which the amount of standard is increased and calculation can be done based on the measured and increased known.

9.5.2 Limit of Detection and Limit of Quantification

The LOD and LOQ represent the lowest amount of the analyte that can be detected and quantified in a sample, but it is not necessary to be quantified the same amount detected by the instrument. The value of detection is often greater than the value of quantification (Krull and Swartz 1999).

9.5.3 Precision and Accuracy

The precision of the analytical/chromatographic method represents the outcome closeness of each experimental measurement of the same analyte under the specified developed condition. The precision of an analytical method is often reported at two different levels, including intra- and inter-day precision, and based on repeatability of retention/migration times or peak areas/height/intensity. The statistical outcomes of precision are expressed as relative standard deviation (RSD, %) or the coefficient of variation (CV, %). Intra-day precision represents the repeatability of the same experiment in a day, while inter-day precision represents the repeatability of the same experiment on different days at the specified operational or experimental condition (Koek et al. 2011; Naz et al. 2014).

The accuracy of the analytical/chromatographic method represents the closeness of the analytical/experimental value to the value obtained theoretically. In metabolomics studies of a medicinal plant, the metabolite concentrations are often represented as the accuracy in recoveries of authentic standards or by determining the isotopically labeled metabolites spiked into the sample (Koek et al. 2011).

9.5.4 Selectivity and Specificity

Selectivity represents the ability to measure and discriminate the interference from other components in the sample during the development of the analytical method for the analyte(s) of interest, while specificity represents the ability for unequivocal measurement of one specific analyte in the presence of other components. Generally, endogenous or exogenous impurities, degradation products, or matrix components and their intensity are being checked to determine the specificity of the developed method (Aboul-Enein 2000, Danzer 2001).

9.5.5 Analytical Recovery

In experimental metabolomics analysis of medicinal plants, the recovery of the analyte influences the specificity of the analytical procedure as well as the developed plant matrix (extract). The analytical recovery or recovery of the analyte is expressed as the percentage of the known amount of analyte spiked to the plant matrix. If the recovery of the analyte is 100% it means the procedure is free from any systematic error. However, in most cases, the percentage recovery of the analyte is observed up to 100% or beyond 100%, which represents the error due to technical or experimental deviation but should be under the limit as specified in standard protocols (González et al. 2014). Quantitative assessment of plant metabolites using LC-MS has been described with the help of diagrammatic representation, as shown in Figure 9.4.

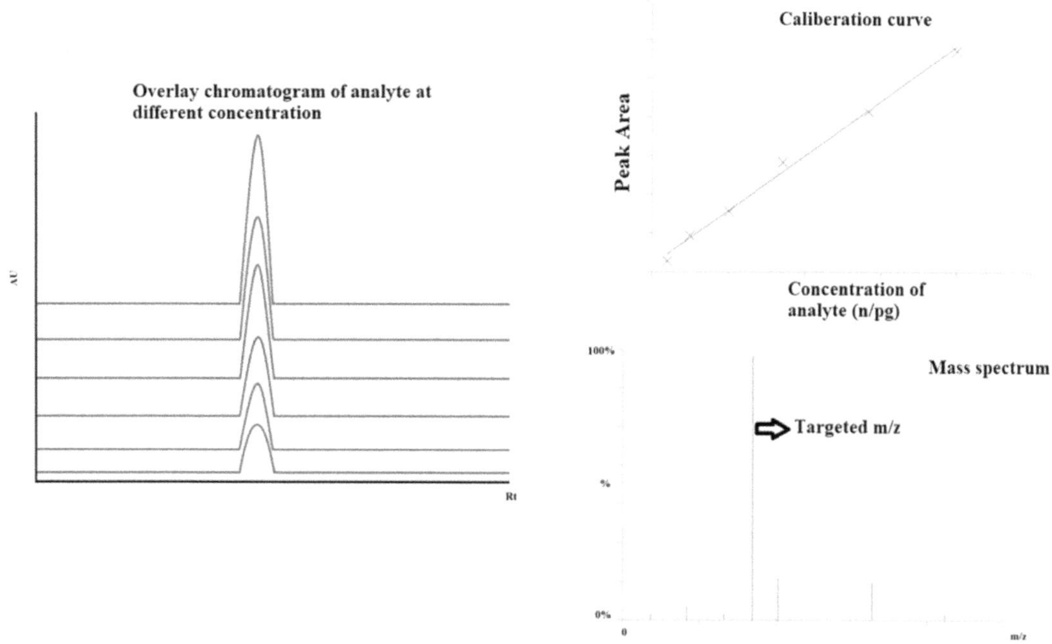

FIGURE 9.4 Quantitative assessment of plant metabolites using LC-MS as analytical tool.

9.6 CURRENT SCENARIO OF LC-MS TECHNIQUES IN QUALITATIVE AND QUANTITATIVE ANALYSIS OF PLANT METABOLITES

Since the advent of metabolomic analysis of medicinal plants, several analytical techniques such as LC-MS, GC-MS, NMR, and so on have been employed for the successive determination of plant metabolites belonging to the different dimensionality. Besides, LC-MS is the growing analytical technique for illustration of metabolomic pattern of medicinal plants, which is normally estimated in positive or negative ion modes. The associated optimized methods, namely LC-ESI/APCI/APPI-Q-TOF-MS, deal with evaluating several polar/semi-polar/nonpolar metabolites present in the complex matrix of medicinal plants. The metabolomic variations due to environmental dissimilarities during growth, harvesting and post-harvesting treatment, the method used for extraction and isolation, are assessed by LC-MS because of its high sensitivity and specificity. Moreover, the quantitative and quantitative estimation of plant metabolites using LC-MS is helpful in regards to the quality and regulatory aspects of metabolites. In the recent past decades, several studies have been done to understand the metabolic pattern and their quantitation, which have been summarized in Table 9.1.

9.7 CHEMOMETRIC TOOLS USED IN METABOLOMICS DATA ACQUISITION

As LC-MS is highly employed in qualitative and quantitative analysis of the complex mixture of compounds with high specificity precision, accuracy and applicability of derived data, it enables us to analyze metabolomic variability of medicinal plants. To extract chemical information and the principal facts of the obtained data, several multivariate analytical tools are accounted as spectral correlative chromatography (SCC), linear discriminate analysis (LDA), information theory (IT), local least square (LLS), heuristic evolving latent projections (HELP) and orthogonal projection analysis (OPA).

TABLE 9.1
Qualitative and Quantitative Identification of Plants Metabolites Using LC-MS/MS (HPLC/MS, UPLC/MS) Analysis

			Type of analysis and metabolites		
S.N.	Medicinal plant/Part used for analysis	Optimized method	Quantitative	Qualitative	References
1.	*Lolium perenne* (seed)	LC-ESI-MS/MS	Ergovaline and lolitrem B	-	(Vassiliadis et al. 2019)
2.	Genus Flaveria	LC-ESI-MS/MS	-	Sulfated hydroxyjasmonates, phytosulfokines, or salicylate	(Kleinenkuhnen et al. 2019)
3.	*Vaccinium angustifolium* (fruit)	LC-IT-TOF-MS	-	Delphinidin-3- galactoside, cyaniding-3- galactoside, Delphinidin-3- glucoside cyaniding-3- glucoside, etc.	(Grace et al. 2019)
4.	*Picrorrhiza kurroa* (rhizomes)	LC-ESI-MS	Picroside I and picroside II	-	(Patil et al. 2013)
5.	Apple peel, Banana peel, Grapefruit peel, Lime peel etc (n=20)	LC-MS/MS	-	Phenols and flavonoids (n=176)	(Suleria et al. 2020)
6.	*Sugarcane molasses*	LC-MS/MS	Chlorogenic acid, caffeic acid, sinapic acid, syringic acid, vanillin, homoorientin, orientin, vitexin, swertisin, diosmin, apigenin, tricin and diosmetin	apigenin-C-glycosides, methoxyluteolin-C-glycosides and tricin-O-glycosides	(Deseo et al. 2020)
7.	*Carica papaya* (leaf)	LC-ESI-MS/MS	Carpaine	-	(Zunjar et al. 2016)
8.	*Theobroma cacao* (seed)	LC-ESI-MS/MS	-	Epicatechin, catechin, quercetin, isoquercitrin (quercetin-3-O-glucoside) and quercetin-3-O-arabinose, hyperoside (quercetin-3-O-galactoside), naringenin, luteolin, apigenin and some O-glucosides and C-glucosides, etc	(Sánchez-Rabaneda et al. 2003)
9.	*Arabidopsis thaliana* (leaf)	LC-ESI-MS	-	Glutathione, deoxyadenosine, sinapate 1-caffeoyl-4-deoxyquinic acid, adenosine 7. 5-methylsufinylpentyl nitrile	(t'Kindt et al. 2009)
10.	*Chrysanthemum morifolium* (flower)	LC–DAD–ESI/MS	-	C-glycosylated flavones and aliphatic acid-containing caffeoylquinic acids	(Lin and Harnly 2010)
11.	*Potentilla anserine* (whole plant)	LC-ESI-MS	-	Chlorogenic acid, kaempferol 3-O-rutinoside, acacetin 7-O-rutinoside, and genistein	(Mari et al. 2013)

12.	*Swertia multicaulis* (leaf)	LC-ESI-MS	-	Xanthones such as 1,3-dihydroxy-5,8-dimethoxyxanthone, 1-hydroxy-3,5,8-trimethoxyxanthone, bellidifolin (1,5,8-tri-hydroxy-3-methoxyxanthone), and decussatin (1-hydroxy-3,7,8-trimethoxyxanthone)	(Timsina et al. 2018)
13.	*Lepidium sativum*	LC-ESI-Q-TOF-MS/MS	-	Kaempferol, coumaroylquinic acid, p-coumaroyl glycolic acid, caffeic acid	(Kadam et al. 2018)
14.	Huangqi-Guizhi-Wuwu (HGW) formulation	LC-ESI-MS/MS	-	Astragaloside I, astragaloside IV, formononetin, paeoniflorin, gingerol, cinnamic acid, carvedilol, tolbutamide	(Lai et al. 2015)
15.	*Morus alba* (leaf)	LC-ESI-MS/MS	1-deoxynojirimycin	-	(Nuengchamnong et al. 2007)
16.	*Swertia chirata* (leaf)	LC-ESI-MS/MS	Mangiferin, amarogentin, amaroswerin, sweroside and swertiamarin	-	(Suryawanshi et al. 2007)
17.	Achyranthes aspera (leaf)	LC/ESI-MS/MS	-	kaempferol, catechin, syringic, quinic, sinapic, caffeic and chlorogenic acids	(Yasir et al. 2016)
18.	*Aerva javanica* (leaf)	LC/ESI-MS/MS	-	p-coumaric, gallic acid, myricetin	
19.	*Mentha pulegium* and *Origanum majorana* (Aerial parts)	LC/ESI-MS/MS	-	Gallocatechin, quercetin dimethyl ether, jaceidin and dihydrokaempferide	(Taamalli et al. 2015)
20.	*Glycyrrhizae radix* (root)	LC/ESI-MS	Glycyrrhizinic acid	-	(Zhang and Ye 2009)
21.	*Tinospora cordifolia* (stem)	HPLC-ESI-QTOF-MS	Berberine, palmatine, jatrorrhizine, magnoflorine, choline	-	(Sharma et al. 2019; Mohan et al. 2017)
22.	*Psidium guajava* (leaf)	UPLC/ESI-MS	-	3-hydroxyanthranilic acid, Cis-Zeatin-9-glucoside, Quercetin, 5-O-Galloylquinic acid, Ellagic acid Kaempferol, Methylsuccinic acid, Oxysterol (R)-3-Amino-4-phenylbutyric acid, etc.	(Ahmad et al. 2019)

(continued)

TABLE 9.1 (Continued)
Qualitative and Quantitative Identification of Plants Metabolites Using LC-MS/MS (HPLC/MS, UPLC/MS) Analysis

S.N.	Medicinal plant/Part used for analysis	Optimized method	Type of analysis and metabolites: Quantitative	Qualitative	References
23.	*Aurantii fructus* (dry whole fruit part)	HPLC-ESI-QTOF-MS	-	Apigenin, saponarin, isovitexin-7-O-xylocoside-2″-O-arabi-noside, quercetin-3-O-rutinoside (rutin), neohesperidin, naringenin-7-O-triglycoside, eriodictol-7-O-glucoside, brutieridin, apigenin-7-O-rutinoside (isorhoifolin), poncirin, naringenin, hesperetin, nobiletin, tangeretin, auraptene etc.	(He et al. 2018)
24.	*Isatis indigotica* and *Strobilanthes cusia* (leaves and roots)	LC-APCI-MS	Tryptanthrin, indigo, and indirubin	-	(Liau et al. 2007)
25.	*Fragaria vesca* (berries)	LC-ESI-MS	-	Phenolic acids, ellagitannins, ellagic acid derivatives, flavonols, monomeric and oligomeric flavanols, dihydrochalcones and anthocyanins (n = 60)	(Del Bubba et al. 2012)
26.	*Gentiana macrophylla* (root)	UPLC-Q-TOF-MS	-	Loganic acid, swertiamarin, gentiopicroside and macrophylloside D	(Li et al. 2020)
27.	*Alectra parasitica* (rhizome)	HR-LC/MS	-	Genipin, Bartsioside, Alectrol, Orobanchol, Naringenin, Ouabagenin, Convallatoxin, Deoxypodophyllotoxin	(Patil 2020)
28.	*Morinda officinalis* (root)	UPLC-QqQ-MS/MS and UPLC-Q/TOF-MSE	Monotropein, deacetyl asperulosidic acid, asperulosidic acid, asperuloside, rubiadin-1-methyl ether, rubiadin, jaceosidin	-	(Zhao et al. 2018)
29.	*Dicentra spectabilis, Fumaria officinalis, Glaucium flavum, Corydalis cava, Berberis thunbergii, Meconopsis cambric, Mahonia aquifolium, Macleaya cordata,* and *Chelidonium majus*	UPLC-MS/MS	Sanguinarine, berberine, protopine, and chelidonine	-	(Och et al. 2017)

30.	*Xinjiang fritillaria* (flower)	UPLC-Q/TOF-MS	Sipeimine-3β-d-glucoside, Sipeimine, Peimisine, and Yibeinoside A	-	(Mohammat et al. 2017)
31.	*Veratrum nigrum* (molasses)	UPLC/ESI-MS/MS	Jervine	-	(Chen et al. 2019)
32.	*Lupinus angustifolius* (Aerial parts)	UPLC/ESI-MS/MS	Lupinine, 13-hydroxylupanine, lupanine, angustifoline, sparteine	-	(Hwang et al. 2020)
33.	*Nelumbo nucifera* (flower buds)	LC/ESI-MS	nuciferine, nornuciferine, N-methylasimilobine, asimilobine, and pronuciferine	armepavine, norarmepavine, N-methylcoclaurine, coclaurine, and norjuziphine	(Morikawa et al. 2016)
34.	*Amaranthus cruentus,* Leaf	LC/ESI-MS	-	Rutin 2-phenylethylamine and amaranthussaponin I, trehalose, trigonelline, lactulose, betaine, valine, alanine, fumarate, formate and kynurenine	(Nkobole and Prinsloo 2021)
35.	*Amaranthus hybridus,* Leaf	LC/ESI-MS	-	Kaempferol, phenylalanine, quercetin, trehalose, trigonelline, lactulose, betaine, valine, alanine, fumarate, formate and kynurenine	(Nkobole and Prinsloo 2021)
36.	*Butea monosperma,* Leaf	UHPLC-QTOF-MS/MS	-	Apigenin, apigenin C-hexoside C-pentoside, apigenin C-hexoside C-hexoside, apigenin-6,8-di-C-pentoside and genistin	(Farooq et al. 2020)
37.	*Gymnema sylvestre,* Leaf	LC/ESI-MS	-	Andrographolide, (–)-epigallocatechin gallate, ajmalicine, ascorbic acid berberine, capillarisin, lobeline, karanjin, linolenic acid, strychnine, hesperetin, 8-geranyloxy psoralen, methionine, dihydrozeatin-9-beta-d-glucoside, linoleic acid, isopentenyl-adenine-7-glucoside-[d6], bufotalin, orcinol, cathine, d-quercitol	(Parveen et al. 2019)
38.	*Azadirachta indica,* Leaf	SPE-LC–Q-TOF-MS	Azadirachtins A, B, D, H and I	-	(Song et al. 2018)
39.	*Swertia chirayit,* Leaf	LC/ESI-MS	Mangiferin, Amarogentin, Swertiamarin	-	(Kumar and Chandra 2015)
40.	*Camellia sinensis, Seeds*	LC-MS/MS or LC-ESI(-ve)-MS/MS	Catechin, epicatechin and2,4-D (2,4-dichlorophenoxyacetic acid)	-	(Sapozhnikova 2014; Mandal et al. 2020)
41.	*Ocimum sanctum,* Leaf	LC-APCI(-ve)-MS/MS	Rosmarinic acid (RA) and ursolic acid (UA)	-	(Shanmuga Sundaram et al. 2012)

(continued)

TABLE 9.1 (Continued)
Qualitative and Quantitative Identification of Plants Metabolites Using LC-MS/MS (HPLC/MS, UPLC/MS) Analysis

S.N.	Medicinal plant/Part used for analysis	Optimized method	Type of analysis and metabolites: Quantitative	Qualitative	References
42.	*Phyllanthus. emblica, fraternus, amarus* and *niruri*, Leaf, bark and fruit	HPLC-ESI-QTOF-MS/MS	-	Quercetin, kaempferol, ellagic acid and their derivatives (n = 51)	(Kumar et al. 2017)
43.	*Cassia occidentalis* Stem	LC-ESI(-ve)-MS/MS	Apigenin, apigenin-6-C-glucopyranoside, luteolin, 3′,4′,7-trihydroxyflavone and emodin	-	(Riyazuddin et al. 2020)
44.	*Cinnamomum zeylanicum*, Bark	LC-DIP-APCI- Q-TOF-MS	Coumarin	-	(Krieger et al. 2013)
45.	*Piper nigrum*, fruit	LC-MS/MS or UFLC-ESI-MS/MS	Piperine, piperlonguminine, Δα,β-dihydropiperlonguminine, pellitorine and piperanine	-	(Chithra et al. 2014; Liu et al. 2015)
46.	*Punica granatum*, Fruit and flower	LC-TOF-MS/MS	-	Valoneic acid bilactone, galloyl-HHDP-hexoside, methyl gallic acid, pomellatannin, valoneic acid bilactone, ellagic acid-hexoside, ellagic acid pentoside, ellagic acid (n= 71 constituents were identified belongs to polyphenols group)	(Liu and Seeram 2018)
47.	*Momordica charantia*, Fruit	HPLC-ESI-MS/MS	Five cucurbitane-type triterpenes and triterpene glycosides	-	(Ma et al. 2012)
48.	*Glycyrrhiza glabra* (Polyherbal formulation)	LC-DAD-MS/MS	Glycyrrhizic acid, glycyrrhizin	-	(De et al. 2012; Gupta et al. 2013)
49.	*Datura taxa (D. innoxia, D. metel, D. stramonium, and D. stramonium var. tatula)*, Leaf	LC-ESI-MS/MS	Hyoscyamine (atropine) and scopolamine	-	(Jakabová et al. 2012)
50.	*Calotropis procera*, Leaf	LC-ESI-MS/MS	-	Calotropagenin and its 10 glycosides were unambiguously identified	(Kanojiya and Madhusudanan 2012)

51.	*Cannabis sativa,* Root	LC-MS/MS	Cannabinoids such as, (–)-trans-Δ9-tetrahydrocannabinol, cannabidiol, cannabigerol, cannabichromene, tetrahydrocannabivarin, cannabinol, (–)-trans-Δ8-tetrahydrocannabinol, cannabidiolic acid, cannabigerolic acid, and Δ9-tetrahydrocannabinolic acid-A		(Gul et al. 2018)
52.	*Terminalia brownie,* Stem wood and bark	RP-HPLC-DAD-MS/MS		Cis-resveratrol-3-o-β-galloyl-glucoside, trans-resveratrol-3-o-β-galloyl-glucoside, methyl-(s)-flavogallonate, gallagic acid dilactone, naringenin-4′-methoxy-7-pyranoside, quercetin-7-β-o-diglucoside, quercetin-7-*o*-galloyl-glucoside, 5,6-dihydroxy-3′,4′,7-trimethoxy-flavon	(Salih et al. 2017)
53.	*Artemisia annua,* Seed	LC/SIM-MS	artemisinin		(Wang et al. 2005)
54.	*Aegle marmelos,* Leaves and fruits	LC/MS/MS		Aeglemarmelosine, marmesiline aegelinoside, shahidine, anhydromarmeline and N-2-methoxy-2-(4-methoxyphenyl) ethylcinnamide, rutin and kaempferol-3-O-rutinoside	(Ibrahim et al. 2018)
55.	*Zingiber officinale* Roscoe, roots/rhizomes	LC-ESI-MS	Gingerol, gingerol, gingerol, shogaol, shogaol, and shogaol		(Tao et al. 2009)
56.	*Foeniculum vulgare Mill*	HPLC/MS	Chlorogenic acid, quercetin-3-O-beta-D-glucuronide, p-anisaldehyde, and trans-anethole		(Bilia et al. 2000)
57.	*Adhatoda vasica,* Leaf	UPLC/Q-TOF-MS	Vasicine		(Madhukar et al. 2014)
58.	*Santalum album L.* Leaves	LC-MS/MS		Vitexin and isovitexin	(Yan et al. 2013)
59.	*Cissampelos pareira, Citrullus colocynthis*	HPLC-QTOF-MS/MS	Cucurbitacin E 2-O-β-D-glucopyranoside, cucurbitacin I 2-O-β-D-glucopyranoside and 22-deoxocucurbitoside B		(Ul Haq et al. 2019)
60.	*Ginkgo biloba*	LC-MS/MS		Bilobalide and the ginkgolides A, B, C and J	(Sun et al. 2005)

9.7.1 Spectral Correlative Chromatography (SCC)

Generally, the chemical components present in different chromatograms are identified as assimilated from the hyphenated instrument, and the analysis is based on key facts for the same chemical components eluted through diverse chromatographic columns. The developed spectral data is to pick up the targeted constituent through comparison by a series of the spectrum at each scan point of other chromatograms. The correlation coefficient versus scan points in the direction of retention time and further validates the information in the local chromatographic cluster where the target exists (Gong et al. 2004).

9.7.2 Linear Discriminate Analysis (LDA)

LDA is one of the common and widely used pattern recognition methods based on the linear function of different variables among multivariate planetary, and it maximizes the ratio between both variances compared to the within-group variance. Despite its advantage, the major limitation of LDA is that it cannot use LDA directly when the obtained multidimensional data of each variable is larger than the number of observations. In such a case, PCA analysis can be employed to compress and transform the original dataset comprising various inter-correlated variables into a compact new set of variables ("Linear algebra done right," 1996, Axler et al. 1997).

9.7.3 Information Theory (IT)

The analysis parameters depend on each constituent's separation degree and concentration distribution in a chromatogram of the applied sample. More precise separation intensity and uniform concentrations provides information content of high value. More chemical information can be gained from such chromatographic data (Gong et al. 2003).

9.7.4 Local Least Square (LLS)

The occurrence of signal shift is one of the major problems in the quality accessibility of herbal medicine, which influence the chromatographic profile of targeted metabolites. To reduce the chromatographic shift, the LLS method is used for precise retention time shift. In pattern recognition analysis, it is necessary to correct retention time in all chromatographic data and make chromatographic profiles with their retention times with respect to all common constituents (Li et al. 2004).

9.7.5 Principal Components Analysis (PCA)

The main objective of PCA is to decrease the dimensionality of a data set comprised of large interrelated variables. The PCA is proficient in transforming to latent variables, and the uncorrelated and ordered principal components (PCs) retain most of the variation in all of the original variables. Further, it is used to accomplish the main source of variability present in the data sets. PCA is further targeted to detect cluster formatting and establish a relationship between object and variable (Jolliffe 2002, Vidal, Ma and Sastry 2016).

9.8 CHALLENGES AND PROSPECTIVE

It can be clearly justified that a single technology-based assessment of metabolomic analysis cannot address the associated issues with dynamic metabolomic illustration of a plant matrix. However, improvement in such techniques is needed to have an actual broadened utility of metabolomics pattern for enhancement of biological understandings (Salem et al. 2020). The main issue in the metabolomic analysis is to quantify all metabolites in a single measurement (Gonda 2020). It is desirable to implement a multifunctional integrated system that may comprehend the quantification of the broad diversity of metabolites. However, the main associated issues are experimental design, selection of method and associated conditions and sources to the analytical method. In most cases in which researchers do not select the appropriate analytical method and their optimized condition, they lacked getting the desired experimental outcomes (Hegeman 2010). The lack of understanding of the chemical nature or physiochemical interaction of the targeted metabolites also impacts the accuracy and precision of data acquired. The factors such as selection of method, sample preparation and condition to the optimized method would need to be controlled to minimize desired data integration (Hall et al. 2002). Moreover, each sample preparation and processing step should be known to maximize the quality and integrity of metabolomic data, while multiple or excessive assessments of analytical steps result in selective loss of the targeted metabolites (Jorge, Mata and António 2016; Salem et al. 2020). Besides, the complexity of multidimensional metabolites can create obstacles to reaching the targeted metabolites or simultaneous estimation of all metabolites using a single analytical technique or LC-MS technique (Ahmed Mueen et al. 2010, Tuli and Ressom 2009). However, a simple, complex mixture of plant matrix such as polarity-based derived fraction or bio-guided fraction considerably reduces the distracting path to reach the exact state of metabolites through a single analytical optimized method (Springfield, Eagles and Scott 2005).

The multiplicity and high dimensional data can be generated by expanding the use of chromatographic ionization and mass analyzer tools to increase the application of the metabolic pattern recognition algorithm (Salem et al. 2020). The superiority of metabolite quantitative and quantitative analysis through advanced analytical technique contributes to its utility in the field of quality control analysis of medicinal plants, which comprehend the metabolomic information of a plant matrix and display data in an authentic, effective and informative way (Fan et al. 2006, Goodarzi, Russell and Vander Heyden 2013). The chemometrics tools such as LLS, LDA and PCA are being exponentially used in the optimization and integration of metabolomic experimental data, extracting useful and unique principal information from an individual or different kinds of analyzed variables. The chemometrics or multivariate analysis was found to be a useful technique to establish the principal quality of phytochemicals among different variables or present in the complex matrix plant (Gad et al. 2013, Tistaert et al. 2011). To assess and explore the quality standard of medicinal plants analysis, new ways should be regularly explored for assessment of simultaneous algorithms on the advancement of the metabolic fingerprint using analytical techniques (Liang et al. 2009, Song et al. 2013). Further, to evaluate and establish the multicomponent data set comprised of large interrelated variables, multivariate statistical analyses such as LLS, LDA, PCA, and so on, should be obliged to reduce such dissimilarities.

9.9 CONCLUSION

Metabolomics analysis of medicinal plants offers an enhanced expedited path for drug discovery from plant sources and is progressively pragmatic in a bioactivity-guided scientific emphasis from plants/natural matrix. With the reference of the present study, it enlightens us about the metabolomic illustration of plant matrix using LC-MS, which is more specific, precise and accurate concerning data selectivity and indignity. Although, in metabolomic analysis, a single optimized method can not be liable for simultaneous estimation of multidimensionality of the plant metabolites, qualitatively and quantitatively. For simultaneous estimation of plant metabolites, the facts such as experimental design, method selection, preparation of the sample and method optimization are required for a proper understanding before proceeding to a forward step for method analysis. Further, a quantitative approach for estimation of metabolites concentration in an unknown matrix of the plant using LC-MS has been clearly described for a better assessment of data integrity. Therefore, quality control analysis associated with the evaluation of multiple constituents using metabolic profiling aids us in avoiding the misconception and adulterants of herbal medicines. Further, to evaluate and establish the multicomponent data set comprised with large interrelated variables, multivariate statistical analyses such as LLS, LDA, PCA and so on should be obliged to reduce such dissimilarities.

REFERENCES

Aboul-Enein, H.Y. 2000. Selectivity versus specificity in chromatographic analytical methods. Accredit. Qual. Assur. https://doi.org/10.1007/s007690050440

Ahmad, A., Ibrahim, M., Chester, K., Khan, W., Ahmad, S. and Ansari, S. 2019. Antithrombocytopenic potential of bioactivity guided fractions of traditionally used Psidium guajava Linn. leaves in busulfan induced-thrombocytopenic rats. Pharmacogn. Mag. https://doi.org/10.4103/pm.pm_608_18

Ahmed Mueen, K., Madkor, H., Attimarad, M. and Khan, T. 2010. Plant metabolomics – A novel method in phytochemical analysis. Pharm. Methods. https://doi.org/10.4103/2229-4708.72235

Annesley, T.M. 2007. Methanol-associated matrix effects in electrospray ionization tandem mass spectrometry. Clin. Chem. https://doi.org/10.1373/clinchem.2007.090811

Axler, S., Gehring, F. and Ribet, K. 1997. Linear Algebra Done Right, Second Edition. Comput. Eng.

Bilia, A.R., Fumarola, M., Gallori, S., Mazzi, G. and Vincieri, F.F. 2000. Identification by HPLC-DAD and HPLC-MS analyses and quantification of constituents of fennel teas and decoctions. J. Agric. Food Chem. https://doi.org/10.1021/jf000171n

Bristow, A.W.T. 2006. Accurate mass measurement for the determination of elemental formula – A tutorial. Mass Spectrom. Rev. https://doi.org/10.1002/mas.20058

Cajka, T. and Fiehn, O. 2014. Comprehensive analysis of lipids in biological systems by liquid chromatography-mass spectrometry. TrAC – Trends Anal. Chem. https://doi.org/10.1016/j.trac.2014.04.017

Chen, L., Weng, Q. and Ma, J. 2019. A new UPLC-MS/MS method validated for quantification of jervine in rat plasma and the study of its pharmacokinetics in rats. J. Anal. Methods Chem. https://doi.org/10.1155/2019/5163625

Chernushevich, I. V., Loboda, A. V. and Thomson, B.A. 2001. An introduction to quadrupole-time-of-flight mass spectrometry. J. Mass Spectrom. https://doi.org/10.1002/jms.207

Chithra, S., Jasim, B., Anisha, C., Mathew, J. and Radhakrishnan, E.K. 2014. LC-MS/MS based identification of piperine production by endophytic *Mycosphaerella* sp. PF13 from Piper nigrum. Appl. Biochem. Biotechnol. https://doi.org/10.1007/s12010-014-0832-3

Danzer, K. 2001. Selectivity and specificity in analytical chemistry. General considerations and attempt of a definition and quantification. Fresenius. J. Anal. Chem. https://doi.org/10.1007/s002160000684

De, A.K., Datta, S. and Mukherjee, A. 2012. Quantitative analysis of glycyrrhizic acid from a polyherbal preparation using liquid chromatographic technique. J. Adv. Pharm. Technol. Res. https://doi.org/10.4103/2231-4040.104711

Del Bubba, M., Checchini, L., Chiuminatto, U., Doumett, S., Fibbi, D. and Giordani, E. 2012. Liquid chromatographic/electrospray ionization tandem mass spectrometric study of polyphenolic composition of four cultivars of Fragaria vesca L. berries and their comparative evaluation. Journal of Mass Spectrometry. https://doi.org/10.1002/jms.3030

Delatte, T.L., Selman, M.H.J., Schluepmann, H., Somsen, G.W., Smeekens, S.C.M. and de Jong, G.J. 2009. Determination of trehalose-6-phosphate in Arabidopsis seedlings by successive extractions followed by anion exchange chromatography-mass spectrometry. Anal. Biochem. https://doi.org/10.1016/j.ab.2009.03.003

Deseo, M.A., Elkins, A., Rochfort, S. and Kitchen, B. 2020. Antioxidant activity and polyphenol composition of sugarcane molasses extract. Food Chem. https://doi.org/10.1016/j.foodchem.2020.126180

Eliasson, M., Rännar, S., Madsen, R., Donten, M.A., Marsden-Edwards, E., Moritz, T., Shockcor, J.P., Johansson, E. and Trygg, J. 2012. Strategy for optimizing LC-MS data processing in metabolomics: A design of experiments approach. Anal. Chem. https://doi.org/10.1021/ac301482k

Fan, X.H., Cheng, Y.Y., Ye, Z.L., Lin, R.C. and Qian, Z.Z. 2006. Multiple chromatographic fingerprinting and its application to the quality control of herbal medicines. Anal. Chim. Acta. https://doi.org/10.1016/j.aca.2005.09.037

Farooq, M.U., Mumtaz, M.W., Mukhtar, H., Rashid, U., Akhtar, M.T., Raza, S.A. and Nadeem, M. 2020. UHPLC-QTOF-MS/MS-based phytochemical characterization and anti-hyperglycemic prospective of hydro-ethanolic leaf extract of Butea monosperma. Sci. Rep. https://doi.org/10.1038/s41598-020-60076-5

Fernie, A.R. 2003. Metabolome characterisation in plant system analysis. Funct. Plant Biol. https://doi.org/10.1071/FP02163

Fiehn, O., Robertson, D., Griffin, J., vab der Werf, M., Nikolau, B., Morrison, N., Sumner, L.W., Goodacre, R., Hardy, N.W., Taylor, C., Fostel, J., Kristal, B., Kaddurah-Daouk, R., Mendes, P., van Ommen, B., Lindon, J.C. and Sansone, S.A. 2007. The metabolomics standards initiative (MSI). Metabolomics. https://doi.org/10.1007/s11306-007-0070-6

Fiehn, O., Wohlgemuth, G., Scholz, M., Kind, T., Lee, D.Y., Lu, Y., Moon, S. and Nikolau, B. 2008. Quality control for plant metabolomics: Reporting MSI-compliant studies. Plant J. https://doi.org/10.1111/j.1365-313X.2007.03387.x

Gad, H.A., El-Ahmady, S.H., Abou-Shoer, M.I. and Al-Azizi, M.M. 2013. Application of chemometrics in authentication of herbal medicines: A review. Phytochem. Anal. https://doi.org/10.1002/pca.2378

Gonda, S. 2020. Special issue: Plant metabolomics. Metabolites. https://doi.org/10.3390/metabo10110467

Gong, F., Liang, Y.Z., Fung, Y.S. and Chau, F.T. 2004. Correction of retention time shifts for chromatographic fingerprints of herbal medicines. J. Chromatogr. A. https://doi.org/10.1016/j.chroma.2003.12.049

Gong, F., Liang, Y.Z., Xie, P.S. and Chau, F.T. 2003. Information theory applied to chromatographic fingerprint of herbal medicine for quality control. J. Chromatogr. A. https://doi.org/10.1016/S0021-9673(03)00648-4

González, O., Blanco, M.E., Iriarte, G., Bartolomé, L., Maguregui, M.I. and Alonso, R.M. 2014. Bioanalytical chromatographic method validation according to current regulations, with a special focus on the non-well defined parameters limit of quantification, robustness and matrix effect. J. Chromatogr. A. https://doi.org/10.1016/j.chroma.2014.03.077

Goodacre, R., Vaidyanathan, S., Dunn, W.B., Harrigan, G.G. and Kell, D.B. 2004. Metabolomics by numbers: Acquiring and understanding global metabolite data. Trends Biotechnol. https://doi.org/10.1016/j.tibtech.2004.03.007

Goodarzi, M., Russell, P.J. and Vander Heyden, Y. 2013. Similarity analyses of chromatographic herbal fingerprints: A review. Anal. Chim. Acta. https://doi.org/10.1016/j.aca.2013.09.017

Grace, M.H., Xiong, J., Esposito, D., Ehlenfeldt, M. and Lila, M.A. 2019. Simultaneous LC-MS quantification of anthocyanins and non-anthocyanin phenolics from blueberries with widely divergent profiles and biological activities. Food Chem. https://doi.org/10.1016/j.foodchem.2018.10.101

Gul, W., Gul, S.W., Chandra, S., Lata, H., Ibrahim, E.A. and Elsohly, M.A. 2018. Detection and Quantification of Cannabinoids in Extracts of Cannabis sativa Roots Using LC-MS/MS. Planta Med. https://doi.org/10.1055/s-0044-100798

Gupta, D.K., Verma, M.K., Anand, R. and Khajuria, R.K. 2013. Development of a validated UPLC-qTOF-MS/MS method for determination of bioactive constituent from Glycyrrhiza glabra. J. Pharm. Anal. https://doi.org/10.1016/j.jpha.2013.01.001

Hall, R., Beale, M., Fiehn, O., Hardy, N., Sumner, L. and Bino, R. 2002. Plant metabolomics: The missing link in functional genomics strategies, Plant Cell. https://doi.org/10.1105/tpc.140720

He, Y., Li, Z., Wang, W., Sooranna, S.R., Shi, Y., Chen, Y., Wu, C., Zeng, J., Tang, Q. and Xie, H. 2018. Chemical profiles and simultaneous quantification of aurantii fructus by use of HPLC-q-of-ms combined with GC-ms and HPLC methods. Molecules. https://doi.org/10.3390/molecules23092189

Hegeman, A.D. 2010. Plant metabolomics-meeting the analytical challenges of comprehensive metabolite analysis. Briefings Funct. Genomics Proteomics. https://doi.org/10.1093/bfgp/elp053

Ho, C.S., Lam, C.W.K., Chan, M.H.M., Cheung, R.C.K., Law, L.K., Lit, L.C.W., Ng, K.F., Suen, M.W.M. and Tai, H.L. 2003. Electrospray ionisation mass spectrometry: principles and clinical applications. Clin. Biochem. Rev.

Hopfgartner, G., Varesio, E., Tschäppät, V., Grivet, C., Bourgogne, E. and Leuthold, L.A. 2004. Triple quadrupole linear ion trap mass spectrometer for the analysis of small molecules and macromolecules. J. Mass Spectrom. https://doi.org/10.1002/jms.659

Hwang, I.M., Lee, H.W., Lee, H.M., Yang, J.S., Seo, H.Y., Chung, Y.J. and Kim, S.H. 2020. Rapid and simultaneous quantification of five quinolizidine alkaloids in lupinus angustifolius l. And its processed foods by UPLC-MS/MS. ACS Omega. https://doi.org/10.1021/acsomega.0c01929

Ibrahim, N.A., Mohammed, M.M.D., Aly, H.F., Ali, S.A. and Al-Hady, D.A. 2018. Efficiency of the leaves and fruits of Aegle marmelos methanol extract (L.) Correa and their relative hepatotoxicity induced by CCL4 and identification of their active constituents by using LC/MS/MS. Toxicol. Reports. https://doi.org/10.1016/j.toxrep.2018.09.005

Jakabová, S., Vincze, L., Farkas, Á., Kilár, F., Boros, B. and Felinger, A. 2012. Determination of tropane alkaloids atropine and scopolamine by liquid chromatography-mass spectrometry in plant organs of Datura species. J. Chromatogr. A. https://doi.org/10.1016/j.chroma.2012.02.036

Jolliffe, I.T. 2002. Principal Component Analysis, Second Edition. Encycl. Stat. Behav. Sci. https://doi.org/10.2307/1270093

Jorge, T.F., Mata, A.T. and António, C. 2016. Mass spectrometry as a quantitative tool in plant metabolomics. Philos. Trans. R. Soc. A Math. Phys. Eng. Sci. https://doi.org/10.1098/rsta.2015.0370

Kadam, D., Palamthodi, S. and Lele, S.S. 2018. LC–ESI-Q-TOF–MS/MS profiling and antioxidant activity of phenolics from L. Sativum seedcake. J. Food Sci. Technol. https://doi.org/10.1007/s13197-017-3031-8

Kanojiya, S. and Madhusudanan, K.P. 2012. Rapid identification of calotropagenin glycosides using high-performance liquid chromatography electrospray ionisation tandem mass spectrometry. Phytochem. Anal. https://doi.org/10.1002/pca.1332

Khoo, S.H.G. and Al-Rubeai, M. 2007. Metabolomics as a complementary tool in cell culture. Biotechnol. Appl. Biochem. https://doi.org/10.1042/ba20060221

Kim, H.K. and Verpoorte, R. 2010. Sample preparation for plant metabolomics. Phytochem. Anal. https://doi.org/10.1002/pca.1188

Kleinenkuhnen, N., Büchel, F., Gerlich, S.C., Kopriva, S. and Metzger, S. 2019. A novel method for identification and quantification of sulfated flavonoids in plants by neutral loss scan mass spectrometry. Front. Plant Sci. https://doi.org/10.3389/fpls.2019.00885

Koek, M.M., Jellema, R.H., van der Greef, J., Tas, A.C. and Hankemeier, T. 2011. Quantitative metabolomics based on gas chromatography mass spectrometry: Status and perspectives. Metabolomics. https://doi.org/10.1007/s11306-010-0254-3

Krieger, S., Hayen, H. and Schmitz, O.J. 2013. Quantification of coumarin in cinnamon and woodruff beverages using DIP-APCI-MS and LC-MS. Anal. Bioanal. Chem. https://doi.org/10.1007/s00216-013-7238-x

Krull, I.S. and Swartz, M. 1999. Analytical method development and validation for the academic researcher. Anal. Lett. https://doi.org/10.1080/00032719908542878

Kumar, S., Singh, A. and Kumar, B. 2017. Identification and characterization of phenolics and terpenoids from ethanolic extracts of Phyllanthus species by HPLC-ESI-QTOF-MS/MS. J. Pharm. Anal. https://doi.org/10.1016/j.jpha.2017.01.005

Kumar, V., Chandra, S. 2015. LC-ESI/MS determination of xanthone and secoiridoid glycosides from in vitro regenerated and in vivo Swertia chirayita. Physiol. Mol. Biol. Plants. https://doi.org/10.1007/s12298-014-0276-9

Lai, K.M., Cheng, Y.Y. and Tsai, T.H. 2015. Integrated LC-MS/MS analytical systems and physical inspection for the analysis of a botanical herbal preparation. Molecules. https://doi.org/10.3390/molecules200610641

Li, B.Y., Hu, Y., Liang, Y.Z., Xie, P.S. and Du, Y.P. 2004. Quality evaluation of fingerprints of herbal medicine with chromatographic data. Anal. Chim. Acta. https://doi.org/10.1016/j.aca.2004.03.041

Li, Z., Du, Y., Yuan, Y., Zhang, X., Wang, Z. and Tian, X. 2020. Integrated quality evaluation strategy for multi-species resourced herb medicine of Qinjiao by metabolomics analysis and genetic comparation. Chinese Med. (United Kingdom). https://doi.org/10.1186/s13020-020-0292-3

Liang, X. miao, Jin, Y., Wang, Y. ping, Jin, G. wa, Fu, Q., Xiao and Y. sheng 2009. Qualitative and quantitative analysis in quality control of traditional Chinese medicines. J. Chromatogr. A. https://doi.org/10.1016/j.chroma.2008.07.026

Liau, B.C., Jong, T.T., Lee, M.R. and Chen, S.S. 2007. LC-APCI-MS method for detection and analysis of tryptanthrin, indigo, and indirubin in Daqingye and Banlangen. J. Pharm. Biomed. Anal. https://doi.org/10.1016/j.jpba.2006.06.029

Lin, L.Z. and Harnly, J.M. 2010. Identification of the phenolic components of chrysanthemum flower (Chrysanthemum morifolium Ramat). Food Chem. https://doi.org/10.1016/j.foodchem.2009.09.083

Linear algebra done right. 1996. Choice Rev. Online. https://doi.org/10.5860/choice.33-6354

Liu, H.L., Luo, R., Chen, X.Q., Ba, Y.Y., Zheng, L., Guo, W.W. and Wu, X. 2015. Identification and simultaneous quantification of five alkaloids in Piper longum L. by HPLC-ESI-MSn and UFLC-ESI-MS/MS and their application to Piper nigrum L. Food Chem. https://doi.org/10.1016/j.foodchem.2015.01.033

Liu, Y. and Seeram, N.P. 2018. Liquid chromatography coupled with time-of-flight tandem mass spectrometry for comprehensive phenolic characterization of pomegranate fruit and flower extracts used as ingredients in botanical dietary supplements. J. Sep. Sci. https://doi.org/10.1002/jssc.201800480

Ma, J., Krynitsky, A.J., Grundel, E. and Rader, J.I. 2012. Quantitative determination of cucurbitane-type triterpenes and triterpene glycosides in dietary supplements containing bitter melon (Momordica charantia) by HPLC-MS/MS. J. AOAC Int. https://doi.org/10.5740/jaoacint.11-511

Madhukar, G., Tamboli, E.T., Rabea, P., Ansari, S.H., Abdin, M.Z. and Sayeed, A. 2014. Rapid, sensitive, and validated UPLC/Q-TOF-MS method for quantitative determination of vasicine in Adhatoda vasica and its in vitro culture. Pharmacogn. Mag. https://doi.org/10.4103/0973-1296.127375

Mandal, S., Poi, R., Bhattacharyya, S., Hazra, D.K. and Karmakar, R. 2020. Method validation, persistence, and safety evaluation of 2,4-D in tea (Camellia sinensis) by LC-MS/MS. Environ. Monit. Assess. https://doi.org/10.1007/s10661-020-08710-7

Mari, A., Lyon, D., Fragner, L., Montoro, P., Piacente, S., Wienkoop, S., Egelhofer, V. and Weckwerth, W. 2013. Phytochemical composition of Potentilla anserina L. analyzed by an integrative GC-MS and LC-MS metabolomics platform. Metabolomics. https://doi.org/10.1007/s11306-01-473-x

Miranda-Osorio, P.H., Castell-Rodríguez, A.E., Vargas-Mancilla, J., Tovilla-Zárate, C.A., Ble-Castillo, J.L., Aguilar-Domínguez, D.E., Juárez-Rojop, I.E. and Díaz-Zagoya, J.C. 2016. Protective action of Carica papaya on β-cells in streptozotocin-induced diabetic rats. Int. J. Environ. Res. Public Health. https://doi.org/10.3390/ijerph13050446

Mohammat, A., Yili, A. and Aisa, H.A. 2017. Rapid quantification and quantitation of alkaloids in Xinjiang fritillaria by ultra-performance liquid chromatography-quadrupole time-of-flight mass spectrometry. Molecules. https://doi.org/10.3390/molecules22050719

Mohan, M.C., Abhimannue, A.P. and Kumar, P.B. 2017. Identification and characterization of berberine in tinospora cordifolia by liquid chromatography quadrupole time of flight mass spectrometry (LC-MS/MS Q-Tof) and evaluation of its anti inflammatory potential. Pharmacogn. J. https://doi.org/10.5530/pj.2017.3.59

Morikawa, T., Kitagawa, N., Tanabe, G., Ninomiya, K., Okugawa, S., Motai, C., Kamei, I., Yoshikawa, M., Lee, I.J. and Muraoka, O. 2016. Quantitative determination of alkaloids in lotus flower (flower buds of Nelumbo nucifera) and their melanogenesis inhibitory activity. Molecules. https://doi.org/10.3390/molecules21070930

Naz, S., Vallejo, M., García, A. and Barbas, C. 2014. Method validation strategies involved in non-targeted metabolomics. J. Chromatogr. A. https://doi.org/10.1016/j.chroma.2014.04.071

Niessen, W.M.A. 2006. Liquid Chromatography-Mass Spectrometry, Liquid Chromatography-Mass Spectrometry. https://doi.org/10.1201/9781420014549

Nkobole, N. and Prinsloo, G. 2021. 1H-NMR and LC-MS Based Metabolomics Analysis of Wild and Cultivated *Amaranthus* spp. Molecules. https://doi.org/10.3390/molecules26040795

Nuengchamnong, N., Ingkaninan, K., Kaewruang, W., Wongareonwanakij, S. and Hongthongdaeng, B. 2007. Quantitative determination of 1-deoxynojirimycin in mulberry leaves using liquid chromatography-tandem mass spectrometry. J. Pharm. Biomed. Anal. https://doi.org/10.1016/j.jpba.2007.03.031

Och, A., Szewczyk, K., Pecio, Ł., Stochmal, A., Załuski, D. and Bogucka-Kocka, A. 2017. UPLC-MS/MS Profile of alkaloids with cytotoxic properties of selected medicinal plants of the berberidaceae and papaveraceae families. Oxid. Med. Cell. Longev. https://doi.org/10.1155/2017/9369872

Parveen, S., Ansari, M.H.R., Parveen, R., Khan, W., Ahmad, S. and Husain, S.A. 2019. Chromatography based metabolomics and in silico screening of gymnema sylvestre leaf extract for its antidiabetic potential. Evidence-based Complement. Altern. Med. https://doi.org/10.1155/2019/7523159

Patil, A.A., Sachin, B.S., Shinde, D.B. and Wakte, P.S. 2013. Supercritical CO_2 assisted extraction and LC-MS identification of picroside i and picroside II from Picrorhiza kurroa. Phytochem. Anal. https://doi.org/10.1002/pca.2383

Patil, S.P. 2020. Proton NMR and HR-LC/MS-based phytochemical analysis of methanolic fraction of Alectra parasitica A. Rich. rhizomes. Heliyon. https://doi.org/10.1016/j.heliyon.2020.e03171

Perreault, H. and Lattová, E. 2019. Mass spectrometry, Comprehensive Biotechnology. https://doi.org/10.1016/B978-0-444-64046-8.00039-2

Petruczynik, A., Wróblewski, K., Wojtanowski, K., Mroczek, T., Juchnowicz, D., Karakuła-Juchnowicz, H. and Tuzimski, T. 2020. Comparison of various chromatographic systems for identification of vortioxetine in bulk drug substance, human serum, saliva, and urine samples by HPLC-DAD and LC-QTOF-MS. Molecules. https://doi.org/10.3390/molecules25112483

Pitt, J.J. 2009. Principles and applications of liquid chromatography-mass spectrometry in clinical biochemistry. Clin. Biochem. Rev.

Pluskal, T., Castillo, S., Villar-Briones, A. and Orešič, M. 2010. MZmine 2: Modular framework for processing, visualizing, and analyzing mass spectrometry-based molecular profile data. BMC Bioinformatics. https://doi.org/10.1186/1471-2105-11-395

Rabinowitz, J.D. and Kimball, E. 2007. Acidic acetonitrile for cellular metabolome extraction from Escherichia coli. Anal. Chem. https://doi.org/10.1021/ac070470c

Rashed, M.S., Bucknall, M.P., Little, D., Awad, A., Jacob, M., Alamoudi, M., Alwattar, M. and Ozand, P.T. 1997. Screening blood spots for inborn errors of metabolism by electrospray tandem mass spectrometry with a microplate batch process and a computer algorithm for automated flagging of abnormal profiles. Clin. Chem. https://doi.org/10.1093/clinchem/43.7.1129

Riyazuddin, M., Husain, A., Verma, S., Katekar, R., Garg, R., Kumar, S., Satish, S., Maurya, R., Narender, T., Chattopadhyay, N. and Gayen, J.R. 2020. Simultaneous quantification of five biomarkers in ethanolic extract of: Cassia occidentalis Linn. stem using liquid chromatography-tandem mass spectrometry: Application to its pharmacokinetic studies. RSC Adv. https://doi.org/10.1039/c9ra07482a

Salem, M.A., De Souza, L.P., Serag, A., Fernie, A.R., Farag, M.A., Ezzat, S.M. and Alseekh, S. 2020. Metabolomics in the context of plant natural products research: From sample preparation to metabolite analysis. Metabolites. https://doi.org/10.3390/metabo10010037

Salih, E.Y.A., Fyhrquist, P., Abdalla, A.M.A., Abdelgadir, A.Y., Kanninen, M., Sipi, M., Luukkanen, O., Fahmi, M.K.M., Elamin, M.H. and Ali, H.A. 2017. LC-MS/MS tandem mass spectrometry for analysis of phenolic compounds and pentacyclic triterpenes in antifungal extracts of Terminalia brownii (Fresen). Antibiotics. https://doi.org/10.3390/antibiotics6040037

Sánchez-Rabaneda, F., Jáuregui, O., Casals, I., Andrés-Lacueva, C., Izquierdo-Pulido, M. and Lamuela-Raventós, R.M. 2003. Liquid chromatographic/electrospray ionization tandem mass spectrometric study of the phenolic composition of cocoa (Theobroma cacao). J. Mass Spectrom. https://doi.org/10.1002/jms.395

Sapozhnikova, Y. 2014. Development of liquid chromatography-tandem mass spectrometry method for analysis of polyphenolic compounds in liquid samples of grape juice, green tea and coffee. Food Chem. https://doi.org/10.1016/j.foodchem.2013.10.131

Sawada, Y. and Yokota Hirai, M. 2013. Integrated LC-MS/MS system for plant metabolomics. Comput. Struct. Biotechnol. J. https://doi.org/10.5936/csbj.201301011

Senguttuvan, J., Paulsamy, S. and Karthika, K. 2014. Phytochemical analysis and evaluation of leaf and root parts of the medicinal herb, Hypochaeris radicata L. for in vitro antioxidant activities. Asian Pac. J. Trop. Biomed. https://doi.org/10.12980/APJTB.4.2014C1030

Shanmuga Sundaram, R., Ramanathan, M., Rajesh, R., Satheesh, B. and Saravanan, D. 2012. LC-MS quantification of rosmarinic acid and ursolic acid in the *Ocimum sanctum* Linn. leaf extract (Holy Basil, Tulsi). J. Liq. Chromatogr. Relat. Technol. https://doi.org/10.1080/10826076.2011.606583

Sharma, B., Yadav, A. and Dabur, R. 2019. Interactions of a medicinal climber Tinospora cordifolia with supportive interspecific plants trigger the modulation in its secondary metabolic profiles. Sci. Rep. https://doi.org/10.1038/s41598-019-50801-0

Smith, C.A., Want, E.J., O'Maille, G., Abagyan, R. and Siuzdak, G. 2006. XCMS: Processing mass spectrometry data for metabolite profiling using nonlinear peak alignment, matching, and identification. Anal. Chem. https://doi.org/10.1021/ac051437y

Smyth, W.F. and Brooks, P. 2004. A critical evaluation of high performance liquid chromatography-electrospray ionisation-mass spectrometry and capillary electrophoresis-electrospray-mass spectrometry for the detection and determination of small molecules of significance in clinical and f. Electrophoresis. https://doi.org/10.1002/elps.200305850

Song, L., Wang, J., Gao, Q., Ma, X., Wang, Y., Zhang, Y., Xun, H., Yao, X. and Tang, F. 2018. Simultaneous determination of five azadirachtins in the seed and leaf extracts of Azadirachta indica by automated online solid-phase extraction coupled with LC–Q-TOF–MS. Chem. Cent. J. https://doi.org/10.1186/s13065-018-0453-y

Song, X.Y., Li, Y.D., Shi, Y.P., Jin, L. and Chen, J. 2013. Quality control of traditional Chinese medicines: A review. Chin. J. Nat. Med. https://doi.org/10.1016/S1875-5364(13)60069-2

Springfield, E.P., Eagles, P.K.F. and Scott, G. 2005. Quality assessment of South African herbal medicines by means of HPLC fingerprinting. J. Ethnopharmacol. https://doi.org/10.1016/j.jep.2005.03.012

Stöckl, D., D'Hondt, H. and Thienpont, L.M. 2009. Method validation across the disciplines-Critical investigation of major validation criteria and associated experimental protocols. J. Chromatogr. B Anal. Technol. Biomed. Life Sci. https://doi.org/10.1016/j.jchromb.2008.12.056

Suleria, H.A.R., Barrow, C.J. and Dunshea, F.R. 2020. Screening and characterization of phenolic compounds and their antioxidant capacity in different fruit peels. Foods. https://doi.org/10.3390/foods9091206

Sumner, L.W., Amberg, A., Barrett, D., Beale, M.H., Beger, R., Daykin, C.A., Fan, T.W.M., Fiehn, O., Goodacre, R., Griffin, J.L., Hankemeier, T., Hardy, N., Harnly, J., Higashi, R., Kopka, J., Lane, A.N., Lindon, J.C., Marriott, P., Nicholls, A.W., Reily, M.D., Thaden, J.J. and Viant, M.R. 2007. Proposed minimum reporting standards for chemical analysis: Chemical Analysis Working Group (CAWG) Metabolomics Standards Initiative (MSI). Metabolomics. https://doi.org/10.1007/s11306-007-0082-2

Sun, Y., Li, W., Fitzloff, J.F. and Van Breemen, R.B. 2005. Liquid chromatography/electrospray tandem mass spectrometry of terpenoid lactones in Ginkgo biloba. J. Mass Spectrom. https://doi.org/10.1002/jms.795

Suryawanshi, S., Asthana, R.K. and Gupta, R.C. 2007. Simultaneous estimation of mangiferin and four secoiridoid glycosides in rat plasma using liquid chromatography-tandem mass spectrometry and its application to pharmacokinetic study of herbal preparation. J. Chromatogr. B Anal. Technol. Biomed. Life Sci. https://doi.org/10.1016/j.jchromb.2007.08.034

t'Kindt, R., Morreel, K., Deforce, D., Boerjan, W. and Van Bocxlaer, J. 2009. Joint GC-MS and LC-MS platforms for comprehensive plant metabolomics: Repeatability and sample pre-treatment. J. Chromatogr. B Anal. Technol. Biomed. Life Sci. https://doi.org/10.1016/j.jchromb.2009.08.041

Taamalli, A., Arráez-Román, D., Abaza, L., Iswaldi, I., Fernández-Gutiérrez, A., Zarrouk, M. and Segura-Carretero, A. 2015. LC-MS-based metabolite profiling of methanolic extracts from the medicinal and aromatic species Mentha pulegium and Origanum majorana. Phytochem. Anal. https://doi.org/10.1002/pca.2566

Tao, Y.I., Li, W., Liang, W. and Van Breemen, R.B. 2009. Identification and quantification of gingerols and related compounds in ginger dietary supplements using high-performance liquid chromatography-tandem mass spectrometry. J. Agric. Food Chem. https://doi.org/10.1021/jf9020224

Timsina, B., Kindlmann, P., Rokaya, M.B., Da Vrchotová, N., Tříska, J., Horník, T. and Sýkora, J. 2018. Xanthones content in swertia multicaulis D. Don from Nepal. Molecules. https://doi.org/10.3390/molecules23051067

Tistaert, C., Dejaegher, B. and Heyden, Y. Vander 2011. Chromatographic separation techniques and data handling methods for herbal fingerprints: A review. Anal. Chim. Acta. https://doi.org/10.1016/j.aca.2011.02.023

Tohge, T. and Fernie, A.R. 2010. Combining genetic diversity, informatics and metabolomics to facilitate annotation of plant gene function. Nat. Protoc. https://doi.org/10.1038/nprot.2010.82

Tozuka, Z., Kaneko, H., Shiraga, T., Mitani, Y., Beppu, M., Terashita, S., Kawamura, A. and Kagayama, A. 2003. Strategy for structural elucidation of drugs and drug metabolites using (MS)n fragmentation in an electrospray ion trap. J. Mass Spectrom. https://doi.org/10.1002/jms.511

Tuli, L. and Ressom, H.W. 2009. LC-MS based detection of differential protein expression. J. Proteomics Bioinforma. https://doi.org/10.4172/jpb.1000102

Ul Haq, F., Ali, A., Khan, M.N., Shah, S.M.Z., Kandel, R.C., Aziz, N., Adhikari, A., Choudhary, M.I., ur-Rahman, A., El-Seedi, H.R. and Musharraf, S.G. 2019. Metabolite Profiling and Quantitation of Cucurbitacins in Cucurbitaceae Plants by Liquid Chromatography coupled to Tandem Mass Spectrometry. Sci. Rep. https://doi.org/10.1038/s41598-019-52404-1

Vassiliadis, S., Elkins, A.C., Reddy, P., Guthridge, K.M., Spangenberg, G.C. and Rochfort, S.J. 2019. A simple Lc-Ms method for the quantitation of alkaloids in endophyte-infected perennial ryegrass. Toxins (Basel). https://doi.org/10.3390/toxins11110649

Vidal, R., Ma, Y. and Sastry, S.S. 2016. Principal component analysis, in: Interdisciplinary Applied Mathematics. https://doi.org/10.1007/978-0-387-87811-9_2

Wang, M., Park, C.H., Wu, Q. and Simon, J.E. 2005. Analysis of artemisinin in Artemisia annua L. by LC-MS with selected ion monitoring. J. Agric. Food Chem. https://doi.org/10.1021/jf051061p

Williamson, L.N. and Bartlett, M.G. 2007. Quantitative liquid chromatography/time-of-flight mass spectrometry. Biomed. Chromatogr. https://doi.org/10.1002/bmc.844

Yan, C., Liu, H. and Lin, L. 2013. Simultaneous determination of vitexin and isovitexin in rat plasma after oral administration of Santalum album L. leaves extract by liquid chromatography-tandem mass spectrometry. Biomed. Chromatogr. https://doi.org/10.1002/bmc.2780

Yasir, M., Sultana, B. and Amicucci, M. 2016. Biological activities of phenolic compounds extracted from Amaranthaceae plants and their LC/ESI-MS/MS profiling. J. Funct. Foods. https://doi.org/10.1016/j.jff.2016.08.029

Zhang, Q. and Ye, M. 2009. Chemical analysis of the Chinese herbal medicine Gan-Cao (licorice). J. Chromatogr. A.

Zhao, X., Wei, J. and Yang, M. 2018. Simultaneous analysis of iridoid glycosides and anthraquinones in morinda officinalis using UPLC-QqQ-MS/MS and UPLC-Q/TOF-MSE. Molecules. https://doi.org/10.3390/molecules23051070

Zhou, B., Xiao, J.F., Tuli, L. and Ressom, H.W. 2012. LC-MS-based metabolomics. Mol. Biosyst. https://doi.org/10.1039/c1mb05350g

Zunjar, V., Dash, R.P., Jivrajani, M., Trivedi, B. and Nivsarkar, M. 2016. Antithrombocytopenic activity of carpaine and alkaloidal extract of Carica papaya Linn. leaves in busulfan induced thrombocytopenic Wistar rats. J. Ethnopharmacol. https://doi.org/10.1016/j.jep.2016.01.035

10 Metabolomic Profiling of Some Medicinal Plants

Sahrish Aftab

CONTENTS

10.1 INTRODUCTION

Plants formed the basis of the traditional medicine system that continues to provide remedies to mankind for thousands of years. Natural products and their derivatives have been the source of potential chemotherapeutic agents. More than 50% of all the drugs in clinical use are represented by them. According to the WHO, 80% of the human population in developing countries rely on plant-derived medicines for their health. Plant-based formulations or their synthetic analogs represented a quarter of all the medical prescriptions (Pawar, Patil and Nagrik 2015). Almost 30% of new chemical substances released by the Food and Drug Administration (FDA) are either pure natural products, their semi-synthetic derivatives or botanical drugs (Newman and Cragg 2012).

Many indigenous plants in India are being used as herbal medicine in curing diseases and healing injuries. The important chemical substances found in plants are carbon-based compounds, alkaloids, glycosides, fatty acids, essential oils, resins, tannins, gums, mucilage, and so on (Pandey et al. 1978). Most of these are rich in therapeutic bioactive compounds or act as a precursor in the synthesis of useful drugs (Sofowora 1993). The use of plants as herbal remedies has been evident in history since the time of Neanderthals. A study of their dental plaques indicated the early consumption of surrounding vegetation, perhaps for self-medication purposes (Hardy et al. 2012). Historical records for the preparation of medicinal plant extracts and decoction were found in Ancient Egypt, Rome, Greece, the Middle East and China (Krikler 1985). In the modern era, more logical and evidence-based research began to be applied in studying herbal medicines. For example, *Digitalis purpurea* L. (Plantaginaceae) has therapeutic properties in treating edema and heart failure (Somberg, Greenfield and Tepper 1985). The beginning of the 19th century was a period of great development

DOI: 10.1201/9781003179139-10

in the pharmaceutical industry, particularly in the isolation of therapeutic and commercially beneficial compounds. In 1805, morphine, popularly used as an analgesic, was isolated and produced commercially in Europe and the United States (Brook, Bennett and Desai 2017), which paved the way for other compounds to be isolated and purified. Western medicine is based on a single compound with a particular biological activity. However, to study the synergistic effect and pro-drugs in an extract, a more holistic approach which is actually based on the traditional system of medicine is considered more nowadays (Verpoorte, Choi and Kim 2005).

All botanical drugs comprise of complex mixtures of plant secondary metabolites. Plants have evolved to produce them for defensive purposes, to attract pollinators and to disperse chemical signals in the surrounding environment. Secondary metabolites, though present in all higher plants, display a large structural diversity corresponding to their adaptation to various ecological niches (Wurtzel and Kutchan 2016). They are also difficult to isolate owing to their low amount and chemical complexity. In contrast, primary metabolites, which are related to highly conserved basic pathways in plants like growth, development and respiration, are well elucidated both at the biochemical and genetic level. (Scossa et al. 2018).

It is therefore necessary to study plant secondary metabolites as they comprise the natural products of medicinal value. Out of nearly 400,000 classified plants, only about 20,000 have been used for medicinal purposes (Wurtzel and Kutchan 2016), and very few among them have been studied for their metabolic composition and biological effects (Harvey, Edrada-Ebel and Quinn 2015). The Nobel Prize of 2015 in medicine was awarded for identifying an antimalarial compound from *Artemisia annua* L. The two principal drugs used to treat malaria are derived from plants (artemisinin and quinine from *Cinchona* L. barks). Furthermore, the last decade has seen several drugs based on plants being approved by the USFDA. For example, galanthamine hydrobromide; snowdrop (*Galanthus nivalis* L.), apomorphine hydrochloride; water lily (*Nymphaea* L. sp), tiotropium bromide; devil's snare (*Datura stramonium* L.), nitisinone, bottlebrush plant (*Callistemon citrinus* Curtis.) and Cytisine; golden rain acacia (*Cytisus laborinum* L.) (Auffret, Drapier and Vérin 2018). Thus, identifying biologically active potential medicinal plant sources by the use of high throughput analysis is a beneficial research area. Identifying the target bioactive compound from the plant material is a challenging task since the crude extract contains thousands of unknown compounds (Atanasov et al. 2015). It is done by comparing the structures of chemical compounds isolated to those already existing in compound libraries typically used in analytical processes (Dias et al. 2016).

10.2 METABOLOMICS: A POWERFUL TOOL FOR COMPREHENSIVE ANALYSIS OF SMALL MOLECULES

Metabolomics is a new addition to the "omics" field of genomics, transcriptomics and proteomics (Rochfort 2005). Metabolomics targets the quantitative and qualitative analysis of all the metabolites in an organism at a specific time under a particular set of conditions (Ulrich-Merzenich et al. 2007, Verpoorte, Choi and Kim 2007). A metabolome comprises the entire set of small molecules inside a cell, tissue or organism at a specific time under certain conditions (Ulrich-Merzenich et al. 2007). Metabolomics is emerging as a powerful tool for comprehensive analysis of small molecules and also for identifying biomarkers in a biological system (Wang et al. 2011). To detail out the primary and secondary metabolites in plants, animals and micro-organisms, bioinformatics tools like metabolomics are used (Sumner 2010). Metabolomics is considered as an important bioinformatics tool in understanding system biology as it involves the genotype. Some gene expression end products are considered as metabolomes, and, therefore, their study provides a comprehensive biochemical state of an organism and can also be used to monitor gene function (Sumner, Mendes and Dixon 2003). Studying gene function through proteomics and transcriptomics has several limitations as the changes in their profile do not always correlate with the biochemical phenotype. Also, the

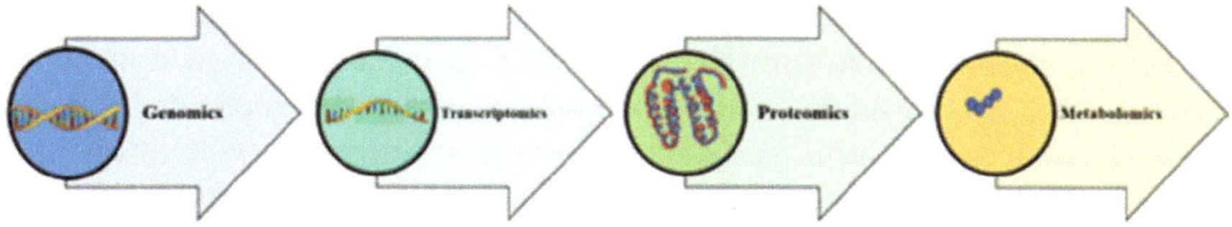

FIGURE 10.1 Flow of information from gene to metabolites.

protein after translation might not be enzymatically active (Rochfort 2005; Sumner, Mendes and Dixon 2003). The identification of mRNA and protein is made through sequence similarity in a database which can also result in limited information if the database source is missing. Therefore, metabolomics is considered as the most functional approach in all the omics technologies (Sumner, Mendes and Dixon 2003). The combined approach of genomics, proteomics and metabolomics is the best way to understand the functioning of a cell and organism (Sumner, Mendes and Dixon 2003; Ulrich-Merzenich et al. 2007) (Figure. 10.1). The quality and quantity of plant metabolites can provide a broad view of the biochemical status of any organism (Fiehn et al. 2000). Genomics and proteomics are helpful in studying the potential of a living system to adapt in response to external perturbations. The resultant metabolic changes are more useful in understanding biochemical reactions to stress (Bailey et al. 2003).

Metabolomics involves a comprehensive study of low molecular weight biological compounds to unfold modifications in metabolite profile. Over the last decade, metabolomics-based approaches have been used in medication or nutritional intervention by analyzing changes in metabolites between control and treatment sets (Kadurrah-Daouk, Weinshilboum and Pharmacometabolomics Research Network 2014 2015). These can be used as target and pathway biomarkers associated with toxicity, efficacy and side effects. Metabolomics also helps in investigating the pharmacological effects and mechanism of action of various medicinal herbs (Shi et al. 2016).

Metabolomics analysis is completed in three steps: (1) Preparation of the sample; (2) Data acquisition through analytical chemical methods; and (3) Data mining through suitable chemometric methods. The first two steps are strongly correlated since the property of the analyte depends on the sample preparation. A solvent should be chosen such that it can extract a diverse group of metabolites and meet the specific requirements of the analytical tool. As compared to proteins and RNA, secondary metabolites are broadly varied in physical and chemical properties such as polarity, solubility, volatility and molecular weight (Dunn and Ellis 2005). Thus, simultaneous profiling of all metabolites possesses technological difficulties. Therefore, there is a need to further refine the metabolomic experiment steps such as sampling, preparation of the sample, instrumental analysis, the interpretation of the enormous amount of data (Rochfort 2005), construction of combined metabolite libraries and data management system (Moco et al. 2006). Many analytical methods have been developed with better resolution, selectivity, reliability and robustness. To obtain an overview of the compound classes, multiple targets have to be identified and quantified instead of a few specific compounds (Fiehn 2002). The progress in the development of mass spectrometry and tandem mass spectrometry techniques has made it possible to detect a wide range of molecular weights and polarities (Verpoorte et al. 2008). Liquid chromatography is particularly used for non-volatile plant metabolites (Dunn and Ellis 2005). The sensitivity of mass spectrometry is best compared to other analytical methods. However, some limitation related to reproducibility is associated with it. Because of different detector responses, the absolute quantification requires a calibration curve for each individual compound. It poses a problem considering hundreds of peaks in the metabolomic profile. For this purpose, NMR spectrometry is a more suitable method as it allows relative quantification of the individual compounds in different experiments (Verpoorte et al. 2008).

Nature possesses the affluence of biologically important metabolites from both primary and secondary metabolic pathways in plants. All metabolic pathways produce metabolites as intermediates and end products. A total of about 500,000 metabolites is estimated to be produced, of which 100,000 are from secondary metabolism (Hadacek 2002). To evaluate this great diversity of structurally complex chemical compounds, various methods have been developed due to advancements in technology and data handling capacity. Combining mass spectrometry with in-line liquid or gas chromatography has led to the efficient separation of molecules. These multivariate spectrometric methods are useful for identifying and quantifying individual compounds in a mixture. Using monolithic silica capillary columns has proven to increase chromatographic resolution. For example, the metabolome of *Arabidopsis* showed reduced ionization suppression when a C-18 monolithic silica capillary column was used in HPLC combined with ion trap mass spectrometry (Tolstikov et al. 2003). NMR aims to look for compositional similarities to explore the overall natural variation in metabolic fingerprinting of plants (Fiehn 2002). Rapid screening of a large collection of biological samples like plant mutant population can be done by a technique called Fourier Transform Ion Cyclotron Mass Spectrometry (FTMS) (Cooper and Marshall 2001). As the chromatography methods were improved in the late 1960s and 1970s, identification of peaks in the complex matrix was made possible based totally on retention times.

The GC-MS-based profiling methods are not suitable for thermolabile, CoA adducts, lipids or large molecules such as bis and triphosphates. GC-TOF MS instruments can record sample spectra very quickly and allow much better deconvolution. Fourier transform ion cyclotron resonance (FT-ICR) mass spectrometry is a method with very high-resolution mass spectra. It separates and provides the empirical formula of many metabolites without chromatographic separation. The metabolomics analysis was first proposed by Jeremy Nicolson in the year 1999 (Fiehn et al. 2000). It includes the quantitative, qualitative and dynamic study of all the small molecules (typically < 1500 Da) present in a cell, tissue or organism at a specific time under certain environmental conditions (Feng et al. 2020). Advancements in technology have led to an increase in the number of metabolites that can be quantified simultaneously and the results of which are independent of the technology used for data acquisition (Lacalle-Bergeron et al. 2021). The study of metabolites is based on targeted and untargeted strategies. The targeted one is more hypothesis-based, which allows the absolute quantification of a set of compounds. It is limited as it focuses only on a small number of metabolites. It helps in studying the changes in the sample with respect to the known and well-characterized analyte (Leone et al. 2018; Pezzatti et al. 2020). The untargeted approach generates a hypothesis by investigating a wide range of metabolites. It is more common and can be used to identify new metabolites (Pezzatti et al. 2020). Through an untargeted approach, mass spectrometric features of unknown metabolites are known. In contrast, the targeted one is the more sensitive and discovery-based approach which measures the relative abundance of all detectable metabolites (Khakimov et al. 2014). It is divided into two main types: fingerprinting and profiling. Fingerprinting analyses the maximum number of compounds present, which includes their detection and statistical analysis without identifying and quantifying them. It is a fast and convenient way of classifying samples based on metabolite patterns in order to obtain sample fingerprinting. Profiling allows the analysis of a class of metabolites through identification and quantification with the help of databases and software (Castro-Puyana et al. 2017). To resolve the data complexity of untargeted metabolomics, statistical tools such as multivariate analysis are used (Lacalle-Bergeron et al. 2021). For metabolomics analysis, the two important platforms used are nuclear magnetic resonance (NMR) and high-resolution mass spectrometry (HRMS). NMR technique is more used for targeted approaches as it provides quantitative data. It detects molecular structure by observing the behavior of atomic nuclei when placed in the magnetic field. NMR is desirable as it is a non-invasive method that gives high quality and high repeatability with simplified extraction procedures and process automation. NMR can detect all organic chemical classes and identify compounds accurately. The solvent used in NMR is environmentally friendly such as deuterated water. Disadvantages associated with NMR are limited sensitivity, less spectral resolution, the high maintenance cost of a large instrument, low reproducibility and standardization of analysis protocols (Böhme et al.

2019; Feizi et al. 2021; Fraga-Corral et al. 2020; Mielko et al. 2021). Liquid or gas chromatography coupled with mass spectrometry LC or GC/MS-MS is a rapid and sensitive method for the detection of metabolites with the help of databases available. Techniques involving mass spectrometry can separate the metabolites very finely and are easily automated. But it also comes with several limitations, such as low resolution and selectivity, sample destruction and identification and the requirement of complicated derivatization (Feizi et al. 2021; Fraga-Corral et al. 2020; Mielko et al. 2021). The pre-processing of MS data includes challenges related to the data analysis software, such as correction of retention times shift and baseline variations (Leone et al. 2018). The huge amount of generated data is interpreted by mathematical tools in multivariate analysis such as clustering, regression analysis, principal component analysis (PCA) and discriminant analysis (DA) (Böhme et al. 2019).

The metabolome is of two types:

- Primary metabolites – These take part in the basic functioning of the living cell-like respiration and synthesis of compounds such as amino acids. The composition of primary metabolites is the same among a particular class of organism, though there may be differences between the classes.
- Secondary metabolites – They are produced as a result of the interaction of a cell with its environment, such as in defense mechanism and to attract pollinators. They are therefore species-specific.

The important criteria to be considered for selecting an appropriate metabolomics technology platform are reproducibility, ease of identification and quantitation, number of metabolites to be measured and the time required for analysis, including the preparation of a sample (Verpoorte, Choi and Kim 2007). The steps in the workflow of the metabolomic experiment include an appropriate experimental design, preparation and extraction of sample, detection of metabolites using suitable analytical techniques, processing and analysis of large data sets through bioinformatics (Ibarra-Estrada et al. 2016) (Figure 10.2). It is essential to carefully optimize the sample preparation, and extraction protocols as variations in metabolite concentration will interrupt the further processing. Integrated metabolomics can be used to identify the function of specialized metabolites by combining high throughput analysis like MS/MS, UV and isotope analysis, preparative LC-MS and LC–solid-phase extraction–nuclear magnetic resonance–MS (LC–SPE–NMR–MS) with computational approaches (Nakabayashi et al. 2015).

10.3 METABOLOME PROFILING OF SOME MEDICINAL PLANTS

10.3.1 *Catharanthus roseus*

Catharanthus roseus is an evergreen perennial plant of the Apocynaceae family. It is a well-known medicinal and ornamental plant around the world (Verma et al. 2015). Potentially bioactive chemical constituents of *C. roseus* are terpenoids, flavonoids, alkaloids, carbohydrates and proteins. It is an exceptional source of secondary metabolites producing more than 130 terpenoid indole alkaloids (TIAs) (Jacobs et al. 2004) that mediate plant adaptation to the environment, particularly during biotic interactions (St-Pierre et al. 2013) and are widely used to treat human diseases (Belal et al. 2009). Some of the widely used alkaloids are vinblastine, vincristine, ajmalicine, serpentine, catharanthine, tabersonine, lochnericine and horhammericine (Parr 1988). Of them, dimeric indole (bisindole) alkaloids vinblastine and vincristine are used as anticancer agents (O'Keefe et al. 1997) though they are produced in very low amounts (Tyler 1988) and monomeric alkaloids like ajmalicine are used as antihypertensive agent (Jacobs et al. 2004). The flowers possess an anthocyanin pigment, Rosindin (Bennouna 2008). It also possesses a broad spectrum of phenolic compounds besides TIAs (Mustafa and Verpoorte 2007) such as cinnamic acid derivatives, benzoic acid, caffeic acid,

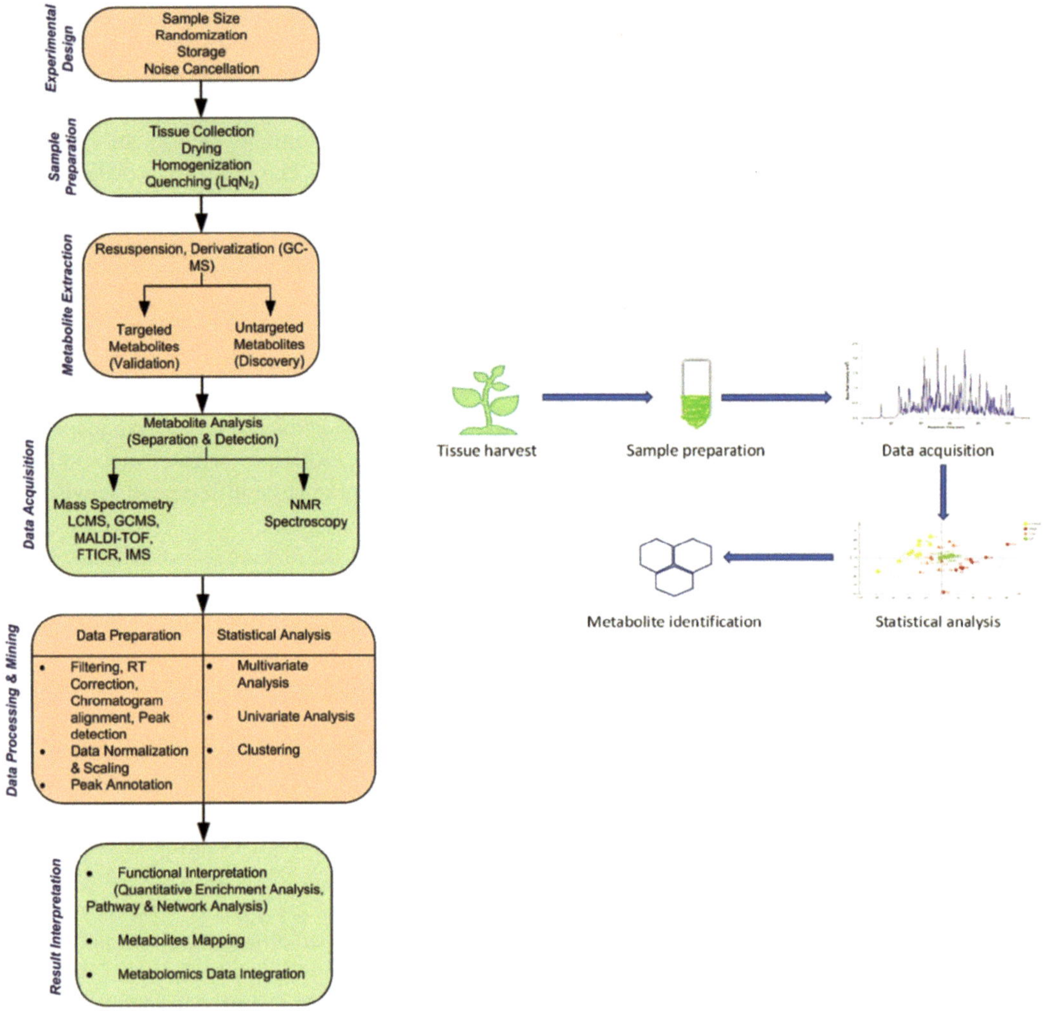

FIGURE 10.2 Workflow of metabolomic study.

flavonoids, and so on. In addition, they possess various biological activities like antibacterial, antioxidant and anticancer (Sousa et al. 2008).

The plant is widely studied due to its chemotherapeutic agents that are effective against several kinds of cancer (Schmeller 1998). Vinblastine is used in Hodgkin's disease and vincristine in leukemias. It also produces antihypertensive agents to overcome heart arrhythmias (Shanks et al. 1998) and improve blood circulation in the brain (Moreno, van der Heijden and Verpoorte 1994). Traditionally, *C. roseus* has been used to relieve muscular pain, depression and wasp stings. It has been used for bleeding gums, nose, mouth ulcers and sore throats. It has also been taken in case of memory loss, hypertension, gastritis, cystitis, diarrhea, enteritis and increased blood sugar level (Dessisa 2001). Its uses are reported in the treatment of malaria, fever, diabetes, chest complaints (The Wealth of India 1992) and as a wound healer, analgesic, vasodilatory (Nayak and Pinto Pereira 2006). The ethanolic extract of *C. roseus* shows significant antihelminthic activity (Agarwal 2011) and antioxidant potential (Bhutkar and Bhise 2011).

An untargeted metabolome analysis with PCA using 15^{N} labeling was done to characterize MIAs as having indolic skeleton in the leaves, flowers, stem, petiole and roots of *C. roseus* plant

(Nakabayashi et al. 2020). The effect of the phytohormones methyl jasmonate (MeJA) and ethylene in the shoots and roots of eight-day-old seedlings in two varieties of *C. roseus* was studied to examine the induction of TIAs biosynthesis and accumulation (Fraser, Philmus and Megraw 2020). Cell-specific TIAs localization and accumulation with 10μm spatial resolution were reported in the leaves of *C. roseus* using imaging mass spectrometry and live single-cell mass spectrometry (Yamamoto et al. 2019). By targeted metabolic analysis and RNA sequencing, it was demonstrated that folivory elicited leaves of *C. roseus* induce both local and systemic induction of MIAs, which results from the expression of MIAs biosynthesis specific genes (Duge De Bernonville et al. 2017). Comparable metabolic profiles in leaves of *C. roseus* and *V. minor* were investigated using non-targeted GC-MS for direct chemical screening of primary metabolites and targeted LC-MS to study species-specific TIAs levels (Chen et al. 2017). Phenolic compound profiling with combinational action of ethylene and MeJA was studied through non-targeted metabolomic using LC-ESI-QTOF/MS. A total of 34 phenolic compounds and seven intermediates for biosynthesis of indole terpenoid alkaloids TIAs and phenolic compounds were identified (Liu et al. 2016). Whole metabolome profiling of transgenic *C. roseus* plants overexpressing ORCA3 gene and G10H gene was performed by the NMR-based metabolomics (Pan et al. 2012). The effect of salicylic acid on the metabolite profile of *C. roseus* cell suspension culture was analyzed using NMR combined with PCA and PLS-DA during a period of 0–72 h (Mustafa et al. 2009). The cDNA-AFLP technique for genome-wide transcript profiling in combination with targeted and non-targeted LC-MS was used to study secondary metabolites in *C. roseus* elicited cells (Rischer et al. 2006). Broad metabolomic profiling of phytoplasma-infested *C. roseus* leaves was carried out using 1H NMR spectroscopy followed by multivariate data analysis. There was an increase in metabolites related to biosynthetic pathways of phenylpropanoid and terpenoid indole alkaloids. This acts as a major distinguishing factor between healthy and infected plants (Choi et al. 2004).

10.3.2 *Zingiber officinale*

Zingiber officinale Roscoe is an herbaceous perennial plant native to South Asia. It is used as a spice in foods and beverages due to its characteristic pungent and sharp flavor (Govindarajan and Connell 1982). Since antiquity, it has been used in the traditional medicine system (Ayurvedic, Chinese and Unani). It contains several bioactive phenolics, including non-volatile pungent compounds such as gingerols, shogoals, parasols and zingerones (Ali et al. 2008). Rhizome of *Z. officinale* Rosco, also called ginger, has been used in Asia for many years as a popular culinary spice and to treat rheumatism, arthritis, muscular aches, coughs, sinusitis, sore throat, cramps, colic, indigestion, diarrhea, fever, flu, chills, motion sickness, loss of appetite (College 1985). It has also been reported for its anti-inflammatory, antioxidative, anticancer, anti-platelet, anti-convulsive and analgesic effects (Parthasarathy, Chempakam and Zachariah 2008). It is used as a herbal medicine for catarrh, nervous diseases, toothache, gingivitis, asthma, strokes, diabetes and constipation (Ravindram and Babu 2004). Ginger also possesses the potential to prevent cardiovascular diseases (Akinyemi et al. 2015), respiratory diseases (Townsend et al. 2013), neurodegenerative diseases (Ho, Chang and Lin 2013), obesity (Suk et al. 2017), diabetes mellitus (Wei et al. 2017), chemotherapy-induced nausea and emesis (Walstab et al. 2013). It is known to enhance the motility of the gastrointestinal tract and has antiviral, antibacterial and antipyretic properties (O'Hara et al. 1998).

Rhizome of ginger contains 60–70% carbohydrates, 9% protein, 8% ash, 3–8% crude fiber, 3–6% fatty oil and 2–3% volatile oil (Ali et al. 2008). Flavoring bioactive compounds found abundantly in ginger are phenolic and terpenes. The main phenolics are gingerols, shogoals and parasols (Prasad and Tyagi 2015). Terpene compounds in ginger are the main constituents of its essential oil. They include zingiberene, α-curcumene, β-bisabolene, α-farnesene and β-sesquiphellandrene (Yeh et al. 2014). Principal pungency compounds in ginger were found to be diarylheptanoids and gingerols related compounds (Policegoudra and Aradhya 2007).

Untargeted metabolite profiling of Zingiberaceae plants, including *Zingiber officinale*, was conducted using GC-MS. It was done to compare the metabolic composition of leaves and rhizomes of common ginger plants. PCA revealed potential chemical markers between them (Barbosa et al. 2017). Two ginger varieties, namely *Guanogdong ginger* and *Chu ginger* from Taiwan, were compared based on their bioactive components and antioxidant properties. Extraction was carried out using HPLC, and volatile components were analyzed by GC-MS. Antioxidant properties were measured in terms of TEAC and FRAP methods (Yeh et al. 2014). To distinguish between two ginger varieties from two geographical locations Ghana in West Africa and China, an untargeted UPLC-QToF-MS-based metabolomics approach was used. Sixteen differential metabolites were identified, including six markers' components between them. The chemical properties of medicinal and edible ginger cultivars were investigated using the LC-MS approach. PCA results indicate that acetoxy derivates of gingerols were relatively higher in medicinal ginger than gingerols or shogoals (Tanaka et al. 2015).

Qualitative and quantitative analysis of constituents of steamed ginger was done by HPLC-ToF/MS and HPLC-Diode array detector (DAD), respectively. The anti-cancerous properties of fresh, dried and steamed ginger were evaluated using human Hela cancer cells. The study revealed that steaming the ginger increases its antiproliferative effects to approx. 1.5 to 2 folds (Cheng et al. 2011). For authentication of ginger species, phylogenetic and metabolic profiling was conducted using DNA sequence data, GC/MS profiling, HPLC quantitation of gingerols and anti-inflammatory assays (Jiang et al. 2006). LC/ESI-MS/MS coupled with diode array detection DAD led to the identification of 31 gingerol-related compounds in ginger rhizomes (Jiang et al. 2005). Metabolic profiling using GC/MS and LC-ESI-MS was done in in vitro micropropagated and traditionally greenhouse-grown ginger. The study revealed no significant chemical differences in properties indicating the in vitro micropropagation procedure efficient (Ma and Gang 2006).

According to traditional Chinese medicine (TCM), the cause of depression is mainly liver stagnancy. Xiaoyaosan [XYS] is a well-known TCM formula to treat depression by relieving liver stagnancy. Its medicinal composition includes *Z. officinale* Roscoe rhizome (Chen et al. 2020). A metabolomic study of UPLC-MS/MS and NMR was conducted to elucidate the anti-depressant mechanisms of XYS and to know its relationship between hepatoprotection and depression (Chen et al. 2019). GC/MS with multivariate statistical techniques was used to identify the effects of XYS potential biomarkers of the anti-depression effects using urine samples of CUMS rats (Dai et al. 2010).

Sini decoction (SND) is a traditional Chinese medicine formula that proved to be effective in treating doxorubicin (DOX)-induced cardiomyopathy. It is composed of three medicinal plants *Aconitum carmichaelii*, *Zingiber officinale* and *Glycyrrhiza uralensis*. GC/LC-MS-based metabolomics approaches have been used to investigate the therapeutic potential and action mechanisms of bioactive in SND (Chen et al. 2014).

10.3.3 *Rauvolfia Serpentina*

Rauvolfia is a plant genus of the family Apocynaceae that consists of 80 species of tropical distribution (Mabberly 2008). Species of *Rauvolfia* consists of medicinally important plants commonly used in traditional system of medicine (Monachino 1954). The importance of *Rauvolfia* genus attributes to its N-containing indole alkaloids, which are localized in roots (Hu et al. 2008), mainly in root bark (Mehrotra et al. 2013). *Rauvolfia serpentina*, commonly known as sarpagandha, is a medicinal undershrub belonging to South East Asian countries, including the Indian subcontinent (Kirtikar and Basu 1993). It is included in the 50 fundamental herbs in traditional Chinese medicine (Wu et al. 2016).

Several studies result in the identification of many heterocyclic alkaloids with monoterpene indole skeletons (Sahu 1983). The cell and tissue cultures of *R. serpentina* have reported more than 30 indole alkaloids belonging to different alkaloid groups. It has been used in India for about

3000 years to treat snake bites, fever and insanity (Stöckigt 1995). Major MIAs responsible for their therapeutic properties are reserpine, serpentine, ajmaline, ajmalicine and yohimbine (Lewis 1956). They have been used for the treatment of high blood pressure (Bhatia 1942), several mental disorders (Kline 1954), arrhythmia (Arora and Madan 1956), malaria fever (Zirihi et al. 2005), intestinal and spleen ailments (Yu et al. 2013), colic, fever, cholera and also as an antihelminthic (Chopra et al. 1956). In the Indian Ayurvedic system, its roots are used as a medication for treating hypertension, insomnia, gastrointestinal disorders, mental agitation, traumas, anxiety and epilepsy (Meena et al. 2009). Reserpine was the first major tranquilizer to be used, particularly for the treatment of paranoia and schizophrenia (Sahu 1983). It is also being used as adjunctive treatment in combination with other antipsychotics (Harisaranraj, Suresh and Babu 2009). Ajmaline from *R. serpentina* is used as an anti-arrhythmic drug (Kiesecker et al. 2004) and to diagnose irregular heartbeat conditions such as Brugada syndrome (Rolf et al. 2003). Recently these metabolites have shown promising results as targeted treatments for drug-resistant cancer lines (Abdelfatah and Efferth 2015). Due to its high demand for abundant medicinal properties, it suffered from habitat destruction (Mehrotra et al. 2015) and has been categorized as endangered species by the IUCN.

To identify high yielding reserpine sources, six *Rauvolfia* species were evaluated using HPLC-ESI-QToF-MS/MS and quantified by HPTLC. The study revealed that *R. micrantha* could replace endangered *Rauvolfia* species as a potential source of reserpine (Bindu et al. 2014). The effect of synergistic elicitation by methyl jasmonate and salicylic acid on the reserpine and ajmalicine production was studied in micropropagated *R. serpentina* roots by HPTLC method (Dey et al. 2020). An endophytic fungus strain *Cladosporium* sp. obtained from *R. serpentina* produces antibacterial and cytotoxic secondary metabolites anhydrofusarubin and methyl ether of fusarubin. They were isolated by chromatographic separation and re-crystallization. Their structures were further confirmed by 1H NMR, 13C NMR and mass spectroscopic data (Khan et al. 2016). An intergeneric somatic hybrid cell culture of *R. serpentina* and *Rhazya stricta* leads to the production of a monoterpenoid indole alkaloid 3-oxo-rhazinilam whose structure was determined by 1D and 2D NMR analysis (Gerasimenko, Sheludko and Stöckigt 2001). Extraction of indole alkaloids, particularly reserpine and ajmalicine from the in vitro raised *R. serpentina* plants was done using the monolithic reverse phase HPLC method (Goel et al. 2009).

Chemical fingerprinting of roots and leaves of six *Rauvolfia* species on the basis of MIA was developed by Direct analysis using real-time mass spectrometry (DARTMS). The data obtained were analyzed by PCA to identify distinguishing markers amongst the species. The method was also tested for the rapid identification, authentication and quality control of *Rauvolfia* species (Kumar et al. 2015). An ultra-high performance liquid chromatography coupled with hybrid triple quadrupole linear ion trap mass spectrophotometry [UHPLC-QqQlit-MS/MS] method in the multiple reaction monitoring modes MRM was performed to quantitate bioactive MIAs in the ethanolic extract of seven *Rauvolfia* species and herbal formulations (Kumar et al. 2016). A new mass spectrometric ionization technique, DART was used to rapidly and accurately profile the hairy root clones of *R. serpentina*. Two alkaloids were detected and identified as vomilenine and reserpine (Madhusudanan et al. 2008). Alkaloids from roots of *R. serpentina* and *R. vomitoria* were quantified using UHPLC-PDA method. The identification and characterization of the compounds were done by QToF-MS/MS (Sagi et al. 2016). Leaves of *Rauvolfia* species from south Western Ghats of India were screened for their antioxidant potential. *R. serpentina* showed the highest results in terms of total phenolic content, DPPH radical scavenging activity, pigment composition and vitamin E content (Nair, Panneerselvam and Gopi 2012).

10.3.4 *Ocimum sanctum*

Ocimum sanctum is distributed widely throughout the Indian subcontinent from Andaman and Nicobar Islands to the Himalayas up to 1800 m above sea level. Commonly known as holy basil in

English and Tulsi in Hindi, it belongs to the family Labiatae. The plant has a characteristic squared stem and specific aroma due to the presence of essential or volatile oil. It is concentrated in leaves and mainly comprises phenols, terpenes and aldehydes. The oil extracted from seeds is composed of fatty acids and is known as fixed oil (The Wealth of India 1991). Medicinal properties of Tulsi are valued not only in Ayurveda and Siddha but also in Roman, Greek and Unani system of medicines (Gupta, Prakash and Srivastava 2002).

Essential oil constituents of leaves are aromadendrene oxide, benzaldehyde, borneol, bornyl acetate, camphor, caryophyllene oxide, cis-α-terpineol, cubenol, cardinene, D-limonene, eicosane, eucalyptol, eugenol, farnesene, farnesol, furaldehyde, heptanol, germacrene, humulene, limonene, n-butyl benzoate, oleic acid, ocimene, sabinene, selinene, phytol, veridifloro, α-camphene, α-myrcene, α,β-pinene, α-thujene, β-guaiene, β-gurjunene, methyl chavicol and linalool (Naquvi et al. 2012). Fixed oil obtained from seeds contain linoleic acid, linolenic acid, oleic acid, palmitic acid, stearic acid. Mineral content includes vitamin C, vitamin A, calcium, phosphorous, chromium, copper, zinc and iron (Anbarasu and Vijayalakshmi 2007). Alcoholic extract of aerial parts and leaves include aesculectin, aesculin, apigenin, caffeic acid, chlorogenic acid, circineol, gallic acid, galuteolin, isorientin, isovitexin, luteolin, molludistin, orientin, procatechuic acid, stigmasterol, urosolic acid, vallinin, vicenin, vitexin (Mondal, Mirdha and Mahapatra 2009).

O. sanctum is known to promote longevity and is therefore referred to as "the elixir of life". Ayurveda and Siddha system of medicine mentioned several benefits of *O. sanctum* such as common cold, cough, flu, headache, earache, fever, sore throat, bronchitis, asthma, colic, hepatic disease, malaria fever.

Fixed oil and linolenic acid from *O. sanctum* seeds showed anti-inflammatory activity against PGE2, leukotrienes induce edema in rats by blocking arachidonic acid metabolism pathways (Singh and Majumdar 1994). Antioxidant effects were studied by Muralikrishnan, Pillai and Shakeel (2012) in streptozocin-induced diabetic rats. Fixed oil showed antihyperlipidemic and cardioprotective activity when tested with rats fed with a high-fat diet (Suanarunsawat et al. 2010). The essential oil showed potent antihelminthic activity against *Caenorhabditis elegans* (Asha et al. 2001). Fixed oil showed anticoagulant activity by prolonging the blood clotting time (Singh, Reban and Majumdar 2001). Aqueous, alcoholic, chloroform extract, as well as leaf oil, were studied against *E. coli*, *P. aeruginosa*, *S. typhimurium*, *S. aureus*. The extract was equally effective against gram+ve and gram-ve bacteria (Mishra and Mishra 2011). Oral administration of ethanolic extract of Tulsi in mice has been shown to improve the level of antistress hormones (Mondal 2009). Aqueous and acetone extracts of *O. sanctum* were found to have antifungal properties against *Alternaria tenuis*, *Helminthosporium* sps and *Curvalaria penniseli* (Sekhawat and Prasada 1971). Analgesic activity of its fixed oil was studied by Singh and Majumdar (1995) in mice and rats using acetic acid-induced writhing methods. The aqueous extract of *Ocimum sanctum* exhibited antiulcer activity in Wistar rats. It was also reported in the fixed oil due to its histamine antagonistic, lipoxygenase inhibitory and antisecretory effects (Ghangale et al. 2009). The fixed oil of seeds showed significant antiarthritic activity as reported by Singh and Majumdar (1996). Monga et al. (2011) studied the anti-melanoma activity of 50% alcoholic aqueous leaf extract of different *Ocimum* sps. Orally administered leaf extract (200 mg/kg p.o.) showed significant antitumor activity.

Variations in the profile of bioactive compounds were studied in the CC harvesting strategy in Sweet Basil using the LC-MS approach. Other quality attributes such as differences in biomass, nutrient use efficiency, physiology, antioxidant capacity and nitrate content were also analyzed. Antagonist anion-anion [NO3: Cl] effects were studied on the yield and leaf metabolome in Basil using an untargeted UHPLC-QTOF-MS approach (Corrado et al. 2020). The antifungal effect of essential oil of Basil (*Ocimum basilicum* L., BEO) on *Candida albicans* was studied by microbial metabolomics using GC-TOFMS (Miao et al. 2020). The effect of abiotic stress, cold, drought, flood and salinity were studied on the transcriptome and metabolome of the basil plant through de novo transcriptome sequencing and GC-MS, respectively (Rastogi et al. 2019). Leaf volatile profiling through GC-MS was performed in two basil cultivars, Napoletano and Genovese, to study water use efficiency in a saline environment. It was demonstrated that reduced stomatal density, high ascorbate and polyphenol oxidase (PPO) levels, together with different phenylpropanoids, contributed to their adaptation to salt stress (Barbieri et al. 2012). Volatile content, antioxidant capacity and mass in Sweet Basil showed significant differences under LED-based lighting (Carvalho et al. 2016). Total anthocyanins were isolated from the purple subtype of *O. sanctum* to develop a novel, cost-effective pH indicator strip. It was tested successfully for a wide range of buffers, soil solutions and irrigation water. The purple subtype showed higher expression of anthocyanin biosynthesis genes and regulating transcription factors (Suthar and Saran 2020).

Leaves of 30-day-old juvenile and 130-day-old pre-flowering holy basil plants were subjected to cold and drought stress for five days. GC-MS analysis revealed an increase in major essential oil components (Eugenol, methyl eugenol and β-caryophyllene). Gene expression analyses were done by qRT-PCR, which showed an increase in OtEOMT transcript level (Nguyen et al. 2020). Thymol-type essential oil of *Ocimum grassi* showed acaricidal activity (Hu et al. 2015). It was further evaluated to study the influence of season on oil activity (Silva-Lima et al. 2018). Anthocyanin extracts of *Ocimum tenuiflorum* were directly infused into an ESI-IT/MS/MS in positive ion mode. The method proved to be rapid and efficient in studying the exact mass, fragmentation patterns and molecular structures of anthocyanins (Sarkar, Kumar and Mukhopadhyay 2017). TiO2 nanoparticle bio-coronas *O. sanctum* leaves were subjected to untargeted MS analysis to identify various bioactive compounds (Kurepa and Smalle 2021). Leaf extract and quercetin separately were tested successfully as a reducing agent in the green synthesis of silver nanoparticles. The resultant AgNPs showed enhanced antibacterial properties against gram-negative *E. coli* (Jain and Mehata 2017).

10.3.5 *Nigella sativa*

Nigella sativa, commonly known as "Black cumin", is an annual herb from the family Ranunculaceae and is known for its wide range of medicinal properties (Chopra et al. 1982). The plant is distributed and cultivated throughout India, particularly in Punjab, Himachal Pradesh, Bihar, Bengal, Gangetic plains, Assam and Maharashtra. It is also grown in South Europe, Syria, Lebanon and Israel as well as in the Middle East, Turkey, Bangladesh and the Mediterranean basin (Naz 2011).

The most important active compounds found in *N. sativa* seeds are thymoquinone (30%–48%), thymohydroquinone, dithymoquinone, p-cymene (7%–15%), carvacrol (6%–12%), 4-terpineol (2%–7%), t-anethol (1%–4%), sesquiterpenelongifolene (1%–8%), α-pinene and thymol. Seeds also contain two different types of alkaloids isoquinoline alkaloids, for example nigellicimine and nigellicimine-N-oxide and pyrazol alkaloids or indazole bearing alkaloids,for example, nigellidine and nigellicine. Other component includes alpha-hederin, a water-soluble pentacyclic triterpene and saponin, a potential anticancer agent (Al-Jassir 1992).

The seeds and oil of *N. sativa* are considered an important drug in the Indian traditional system of medicine like Unani and Ayurveda (Sharma et al. 2005). The seeds are extensively used for the treatment of asthma, bronchitis, cough, headache, rheumatism, fever, liver and kidney disorders, influenza and eczema. The plant is also used as a diuretic, carminative, lactagogue and vermifuge (Khattak and Simpson 2008). According to recent scientific investigations bioactives present in *N. sativa* are anticarcinogenic (Rooney and Ryan 2005), antiulcer (Kanter et al. 2005), antibacterial and antifungal (Khan et al. 2003), antihypertensive (Najmi et al. 2013), hepatoprotective (Daba and Rahman 1998), anti-inflammatory, antipyretic and analgesic (AlGhamdi 2001), antioxidant (Thippeswamy and Naidu 2005), antiarthritic (Tekeoglu et al. 2007) and antihyperlipidemic (Ahmad and Beg 2013).

A multiplex metabolomics approach using UPLC-PDA-MS and GC-MS coupled with multivariate data analysis was used to study differences in primary and secondary metabolites profiles in 6 *Nigella* species (Farag et al. 2014). Callus cultures from root and sprout segments of *N. sativa* seedlings were elicitated with MeJA. A metabolomic approach based on UPLC-MS was used to investigate secondary metabolites composition and elicitated callus regulation (Farag et al. 2015). Different leukemic cancer cell lines were treated with thymoquinone (TQ) to study its various anticancer effects. An untargeted metabolic profile was performed using HPLC coupled to LTQ MS/MS to compare variations in major metabolic pathways (AlGhamdi et al. 2020). *N. sativa* and *N. damascena* were compared based on their polyphenolic composition by HPLC-MS, antioxidant activity through DPPH, TEAC, FRAP and CUPRAC and antidiuretic effects in rats by orally administered ethanolic extracts (Toma et al. 2015). Proton nuclear magnetic resonance 1HNMR spectrometric data was used to differentiate *N. sativa* seeds from four countries. Three assays, including the DPPH radical scavenging (DPPH-RS), nitric oxide inhibition (NOI) and α-glucosidase inhibitory (AGI) activities, were performed to relate their constituents and bioactivities (Maulidiani et al. 2015). Several phytoconstituents were identified from the seed extract of *N. sativa* using GC-MS and FT-IR methods (Nivetha and Prasanna 2016). *N sativa* oil (NSO) and its major bioactive constituent, Thymoquinone (TQ), were tested against the nephrotoxicity in rats induced by Cisplatin (CP) chemotherapy. Results showed the significant potential of NSO and TQ in suppressing CP-induced oxidative stress, thus protecting kidney structure and function (Farooqui et al. 2017). NSO at the dose of 5.0 mL/kg appeared to be effective against the binge ethanol-induced oxidative stress and hepatotoxicity in rats (Develi et al. 2014).

Black seeds (*N. sativa*) samples from Saudi Arabian, Indian and Pakistani markets were investigated for their quality and standardized according to the WHO guidelines. Instrumental analysis was performed by accelerated solvent extraction (ASE), IR, UHPLC and NMR. For pharmacological activities, chemical tests, ash values and antioxidant assays were performed (Ahmad et al. 2020). In vitro toxicity, genotoxicity and antigenotoxicity were investigated to study environmental variations in *N. sativa* seeds growing at three different locations in Morocco. Various tests like the neutral red uptake test, the bacterial Vitotox and Ames test, micronucleus and comet test were performed in human C3A cells (Nguyen et al. 2019).

10.3.6 *Crocus sativus*

Crocus sativus, a member of the Iridaceae family, is a geophyte herb. The dried stigmas of its flower are known as saffron; the most valuable and precious spice (Lim 2014) been used since antiquity for its aromatic property and golden-yellow hue it imparts to food and textiles (Negbi 1999). It is the most expensive spice in the world because of the requirement of direct labor for its cultivation, harvesting and handling (Melnyk, Wang and Marcone 2010). It is therefore known as "red gold" all over the world (Shahi, Assadpour and Jafari 2016). Iran is the world's largest producer and exporter of saffron. It is also successfully cultivated in countries like Greece, Spain, Italy, Morocco and India (Zeka et al. 2015).

The three main metabolites reported in phytochemical studies are crocin, picrocrocin and safranal (Baba et al. 2015). These metabolites determine the value of saffron, and their chemical composition is an important quality indicator (Abdullaev 2002). Saffron is an ancient spice used in the treatment of 90 medical indications for almost 4000 years (Bathaie and Mousavi 2010). In Ayurveda and folk medicine, the use of saffron is mentioned as a sedative, anti-asthmatic, expectorant, adaptogenic agent and as pain reliever (Schmidt, Betti and Hensel 2007). Various therapeutic effects of saffron, such as antihistamine, anti-convulsant and hypnotic, have been reported (Leone et al. 2018). Aqueous and ethanolic extracts of saffron showed significant anti-inflammatory properties (Hosseinzadeh and Younesi 2002). Moradzadeh, Kalani and Avan (2018) studied its anti-leukemic effects. Saffron, due to its antioxidant properties, prevents demyelination and reduces symptoms in multiple sclerosis (José Bagur et al. 2017). Ahmad et al. (2005) showed the neuromodulatory effects of crocetin in Wistar rats to prevent Parkinson's Disease. Saffron also has good efficacy in the treatment of mild to moderate depression with concomitant anxiety (Ghajar et al. 2016). When administered to patients with metabolic syndrome, Crocin supplements showed a significant increase in high-density lipoproteins (HDL) and a reduction in cholesterol (Ghaffari and Roshanravan 2019). Hypoglycemic effects of saffron were reported by Yaribeygi et al. (2019). Oral supplementation of saffron and crocin was tested as potential therapeutic in ocular diseases (Heitmar, Brown and Kyrou 2019). It also improves visual function in mild/moderate age-related macular degeneration (Broadhead et al. 2019).

Liquid chromatography coupled to high-resolution MS [LC-QTOF] was done to investigate the quality and authenticity of saffron by identifying possible markers between authentic and suspicious samples (Guijarro-Diez et al. 2015). Other methods reported to detect saffron adulteration include HPLC-PDA and ESI-MS (Sabatino et al. 2011), ^{1}HNMR metabolomics (Petrakis et al. 2015), electronic nose system coupled to multivariate statistical analysis (Heidarbeigi et al. 2015). A non-destructive and rapid detection method is NIR spectroscopy combined with multivariate data analysis (Shawky, El-Khair and Selim 2020). A combined approach of FT-IR and ^{1}HNMR metabolomics was implemented to trace back the age of commercial saffron samples of unknown history (Consonni et al. 2016).

The untargeted metabolomics approach through UHPLC-QTOF-MS facilitates the identification of phenolic markers to investigate the authenticity and traceability of saffron. Authenticity was investigated in the most difficult adulteration procedure of style mixing with different percentages of stamens and tepals. Chemical fingerprints also discriminate between PDO (Protected Designation of Origin) and non-PDO saffron samples (Senizza et al. 2019). NMR also provided very good discrimination between PDO Italian saffron and commercially Italian market saffron (Cagliani et al. 2015). To correlate the contents of bioactive compounds and the place of flower collection, a fast screening was done by HPTLC in saffron collected from different parts of Iran. Finer separation was further done by HPLC-DAD in samples with a high amount of crocins (Vahedi et al. 2018). The UPLC-HDMS combined with multivariate data analysis was done to analyze the bioactive compounds as potential markers between the stigmas, stamens and tepals of saffron (Xu et al. 2019).

UV-vis spectroscopy was done in four different sites of Italian saffron aqueous extracts to study potential geographical markers (D'Archivio and Maggi 2017). The two-step approach of applying both OPLS-DA and O2PLS-DA models to ^{1}HNMR data was done to detect plant adulterated saffron and to identify the adulterant plant (Petrakis et al. 2015). The NMR-based approach was also used to determine the adulteration of saffron with Sudan dyes (Petrakis and Polissiou 2017). Plant adulterants were also assessed through the application of diffuse reflectance infrared Fourier transform spectroscopy (DRIFTS). Major bioactive of saffron picrocrocin, crocin and safranal were extracted through microwave-assisted extraction (MAE). The effects of temperature, ethanol concentration and time on the contents during MAE were studied through response surface methodology (Sarfarazi et al. 2020).

GC-MS fingerprints of saffron were obtained with the help of a chemometric strategy based on multivariate curve resolution-alternating least squares (MCR-ALS) and multivariate pattern recognition methods. The technique greatly reduced the noise and resolved the chromatographic peaks (Aliakbarzadeh et al. 2016). The antioxidant and cytotoxic activities were evaluated in juices obtained from saffron cold-pressed floral by-products (Tuberoso et al. 2016). An integrated UHPLC-ESI-HRMS-based metabolomics approach was used to analyze plasma samples of mice after administration of TC4, a compound of saffron having potential activity against Alzheimer's Disease (Karkoula et al. 2020). Antioxidant activity of CSP (*Crocus sativus* petals) reduced the oxidative stress on APAP (Acetaminophen)-induced hepatoxicity in rats (Omidi et al. 2014). Crocin and safranal, the two main active components of saffron, were studied for their hMAO inhibitory activities. Human monooxidases are target enzymes for neuropsychiatric and neurodegenerative diseases. Safranal derivatives were synthesized, which displayed much more improved inhibitory activity (De Monte et al. 2014).

10.3.7 *Curcuma longa*

Curcuma longa, commonly known as turmeric, is a member of the Zingiberaceae family. The genus *Curcuma* constitutes approximately 80 species of rhizomatous plants that are primarily distributed in Asia, Africa and Australia (Kamazeri et al. 2012). They have traditionally been used as a spice, medicine and ceremonial dye (Wilson et al. 2005). It also has a long history in the Chinese and Ayurveda system of medicine. It is an important component in Indian medicines used for treating biliary disorders, diabetic wounds, hepatic disorders, anorexia, cough, rheumatism, sinusitis, sprain and swelling (Jurenka 2009).

Various pharmacological activities of *C. longa* include anti-inflammatory due to inhibiting effect on lipoxygenase and COX-2 inhibitor (Arora et al. 1971), anti-diabetic (Babu and Srinivasan 1997), anticancer (Aggarwal, Kumar and Bharti 2003), antioxidant by suppressing ROS and nitrite radicle generation (Menon and Sudheer 2007), anticoagulant by inhibiting collagen and adrenaline-induced platelet aggregation (Srivastava et al. 1986), antimicrobial and antiulcer. It also possesses wound healing properties by increasing the activity of nitric oxide synthase and the level of TGF-β1 (Mani et al. 2002).

Curcuma plants contain a varying number of phenolic pigments and essential oil as their main constituents (Jung et al. 2012). The phytochemical activities of *C. longa* is mainly attributed to the presence of diarylheptanoid compounds, curcuminoids such as curcumin, demethoxycurcumin and bisdemethoxycurcumin (Ireson et al. 2001). Curcumin, a yellow pigment, has reported anti-inflammatory, antioxidant and anticarcinogenic properties (Bamba et al. 2011). Biologically active ingredients of essential oils are found to be sesquiterpenoids, including β-element, curcumol, curdione, neocurdione and germacrone. Curdione is an important bioactive component known for its anticancer and antimicrobial activities (Yan et al. 2005).

^{1}HNMR-based metabolomics were used to profile the metabolites of four *Cucurma* species at three developmental stages (seven, eight and nine months). PCA data revealed the differences in the content of metabolites specific to each stage (Awin et al. 2019). Untargeted metabolite profiling using GC-MS was done to chemically differentiate the *Zingiber* species and to identify important markers in different plant tissues (Barbosa et al. 2017). The quality difference between commercial turmeric products was investigated by ^{1}HNMR spectroscopy coupled with PCA and HPTLC methods. About 72 turmeric samples in the form of starting materials, processed powders, extracted materials and end products were tested for phytochemical variation, adulteration, compositional diversity, and so on (Chatzinasiou et al. 2019).

^{1}HNMR spectroscopy coupled with multivariate analysis and HPTLC were used to understand the impact of value chains on the composition of products obtained from *C. longa*. This technique could help in understanding the composition and quality of herbal value chain products (Booker et al. 2014). Two Curcuma species *C. longa* and *C. aromatica* have grown in different regions in South Korea, and were subjected to primary metabolites profiling through NMR while secondary metabolites profiling using GC and LC-MS methods. NMR profiling proved to be a fast and efficient method to discriminate among species and origin. At the same time, GC and LC-MS is useful for analysis of their chemical composition. *Curcumae* Radix, a Chinese medicinal herb, was chemically discriminated by HHPLC/QTOF-MS method combined with multivariate statistical analysis from other *Curcuma* species (Liu et al. 2016). In vitro effect of the ethanolic, water and acetic acid extract of *C. longa* was studied in the microbiota of hypertensive patients to test dysbiosis (Vamanu et al. 2019).

NMR and MS-based metabolomics approaches were employed to study the antihyperlipidemic effect of curcumin in mice induced by a high-fat diet (Li et al. 2015). NMR-based metabolomics and GC-MS lipidomics were also used to study metabolic changes in rats fed with high fructose and saturated fatty acid diet (Tranchida et al. 2015). Extracts of *Curcuma longa* and *Cosmos caudatus* were analyzed by LC-HRMS study. The anticancer activity of the targeted compounds, that is,

Curcumin from *C. longa* and Lutein from *C. caudatus*, were determined through in-silico molecular docking on target protein caspase 8 (Safitri et al. 2020).

An endophytic bacterium, *Bacillus subtilis* was isolated from the dried rhizome of *Curcuma longa*. The mechanism of plant growth promotion and stress tolerance potential of *B. subtilis* was studied through phenomics and genomics approaches (Jayakumar, Nair and Radhakrishnan 2020). A unified metabolomic approach of NMR and MS was carried out in the urine samples of rats treated with *C. longa* extract. Variations in the urinary composition after treatment and the in-vivo antioxidant effects of curcuminoids were studied (Dall'Acqua et al. 2016). The work was further carried out in healthy adult humans to study the changes in 24-h urinary composition after the consumption of dried *C. longa* extracts for 28 days (Peron et al. 2020). Turmeric was subjected to microbial fermentation by applying *Monascus purpureus* and *Eurotium cristatum* fungi because of the reported benefits of fermented turmeric. Targeted metabolomics using LC-MS/MS was used to identify the variations in curcuminoids after fermentation (Xiang et al. 2020).

10.3.8 *Artimisia annua*

Three strains of a fungal endophyte, *Aspergillus terreus*, obtained from the leaves, stem and roots of *Artemisia annua*, were subjected to LC-MS-based metabolomics and multivariate analysis. About 682 compounds were detected, and the strains were classified based on their characteristic metabolites. The leaf-derived strain showed broad-spectrum in vitro antioxidant potential (Sayed et al. 2022). Antiplasmodial activity of *Artemisia annua* and *Artemisia afra* was tested against *Plasmodium falciparum* 3D7. To make a chemical differentiation, their polar metabolites were identified through 1D and 2D NMR, followed by multivariate data analysis (Liu et al. 2010). Transgenic *Artemisia annua* plant overexpressing the enzyme for artemisinin biosynthesis was subjected to GC and GC-MS to monitor the changes in the whole metabolic profile compared to control (Ma et al. 2008). It was also tested by another analytical method involving GC×GC-TOFMS followed by partial least square-discriminant analysis (PLS-DA) (Ma et al. 2009). qPCR was used to estimate the expression of genes of terpene metabolism in different parts of *Artemisia annua* plant (Olofsson et al. 2011). To investigate the fungal endophytes associated with antimalarial property in *A. annua*, an untargeted metabolomics approach based on LC-HRMS was used. About 11 endophytic fungi were isolated and screened for their in vitro antiplasmodial activity. The fungal extracts were chemically profiled through LC-HRMS. Multivariate statistical analyses (PCA, PLS-DA, OPLS-DA) were done to determine the variations in the metabolite composition. In-silico investigation using the neural-network-based prediction software PASS led to the selection of the most probable bioactive candidates (Alhadrami et al. 2021).

Scoparone, an important chemical substance in Herba *Artemisia scopariae*, is used as a therapy for liver injury in traditional Chinese medicine. To study its hepatoprotective effects, a metabolite profiling with UPLC/ESI-QToF/MS was done in CCl4-induced liver injury in rats. Differentiating metabolites were identified with pattern recognition approaches such as PCA and partial least square-discriminant analysis PLS-DA (Zhang et al. 2013). Similar approaches were used to investigate the hepatoprotective effect of *Artemisia annua* (Yinchenhaoren YCH) on acute liver injury induced by α-naphthylisothiocyanate in rats (Sun et al. 2014).

10.3.9 *Piper nigrum*

Piperaceae is one of the most ancient families of flowering plants in tropical regions consisting of 13 genera (Scott et al. 2007). It has a characteristic biting quality because of the presence of piperine and its isomers (Govindarajan 1977). It is used mainly for culinary purposes as a flavor enhancer and also as a preservative (Ravindran and Kallupurackal 2012). Besides being used as a food additive, it has its role in the treatment of dyspepsia, cholera, arthritic disorders and gastric ailments (Jung and Shin 1998). In the Asia Pacific islands, it is used to treat cold, chills, fever, flu, muscular aches and rheumatism (Mohammed, Omran and Hussein 2016). In traditional Chinese medicine, black pepper is used to treat coronary heart disease. Its essential oil is used to treat diarrhea, asthma, insomnia, chronic indigestion and as an immunity booster (Butt et al. 2013). Other pharmacological activities include anti-inflammatory, growth stimulatory, thermogenic action, anti-thyroid and chemopreventive (Panda and Kar 2003). Piperine has shown effective results against radiation in cancer patients undergoing radiotherapy (Subramanian et al. 2016). Effects of piperine are also reported on mood and cognitive disorders (Wattanathorn et al. 2008). *Piper* amides extracted from *P. nigrum* have insecticidal properties (Scott et al. 2005), while β-caryophyllene showed anesthetic activity (Santra et al. 2005). Pipeline, an important constituent of pepper oil, is a famous odorant. A secondary metabolite Nerolidol is used as a pesticide to control mites (Jayalekshmy et al. 2003). Black pepper is also reported to have antimutagenic (El-Hamss et al. 2003) antioxidant and free radical scavenging properties (Gülçin et al. 2005). Many valuable compounds have been isolated from the plant, including phenolics, derivatives of terpenes, lignans, chalcones, alkaloids, flavonoids and steroids (Parmar et al. 1997). Fruit extracts of *P. nigrum* (water, methanol and ethanol) revealed the presence of a wide variety of secondary metabolites, including glycosides, alkaloids, steroids, flavonoids, terpenoids, tannins and anthraquinones (Nahak and Sahu 2011). The main aromatic agents were α and β-pinene, limonene, α-phellandrene, linalool 2- and 3-methylbutanal, methylpropanal, butyric acid and 3-methylbutyric acid (Jagella and Grosch 1999).

To access the antioxidant potential and phenolic profile of three black spices namely *Piper nigrum* L., *Nigella sativa* L. and *Amomum subulatum*, a study based on LC-ESI-QTOF/MS and HPLC-PDA was conducted. Total phenolic content (TPC), total flavonoids, tannins as well as antioxidant activity was significantly higher in black pepper (Feng et al. 2020). A method involving ASE along with UHPLC-DAD was developed to evaluate the quality and quantity of piperine in black pepper samples. The method proved to be fast, reliable and accurate compared to previously reported methods (Ahmad et al. 2019). Secondary metabolites profiling in black pepper oleoresin extracts was done through LC-MS QToF using positive ESI mode. The mineral elements analysis was carried out using an Inductively Coupled Plasma/Mass spectrometer (Olalere et al. 2018). Chiral GC-MS analysis was carried out in *P. nigrum* oil to evaluate the enantiomeric distribution of compounds, mainly terpenes. It is used as a chiral marker to determine authenticity and quality control standards for pepper oil (Wang et al. 2019). The LC-MS/MS chromatogram profile compared the variations in the relative content of phenolic and polyhydroxy compounds in different pepper samples investigated. The total peak area was highest in fresh pepper berries while lowest in white pepper. One black pepper sample prepared by blanching and drying and the other by direct solarization showed moderate content and similar composition of compounds. This finding further verified the role of phenolic compounds in the blackening of pepper, contributing to color and flavor (Gu et al. 2018). GC-MS profiling was done to detect the adulteration of *Piper nigrum* with *Carica papaya*. On comparing the data 33 identified metabolites were unique to *C. papaya*, whereas 44 to *P. nigrum* (Gul et al. 2018). They are readily adulterated in the local markets of India due to their morphological similarity (Dhanya Syamkumar and Sasikumar 2009). This could lead to harmful consequences as *Carica* seeds have antifertility properties (Lohiya et al. 1999). The infection of *Fusarium solani* f. sp. piperis was studied on two cultivars *Bragantina* (susceptible) and *Cingapura* (tolerant). GC-MS analysis of volatile compounds revealed that *Bragantina* showed higher content of α-bisabolol and lower elemol. *Cingapura* responded to infection by decreasing δ-element production. HPLC-MS analysis of root extracts showed variation in the quality of alkamides during infection, indicating a possible relationship between plant tolerance against phytopathogens and its secondary metabolites (Da Luz et al. 2017). For rapid chemical profiling of *P. nigrum* from its inferior substitutes *P. longum* and *P. chaba*, DARTMS fingerprinting combined with a multivariate statistical approach was applied. It has been used for the first time in the interspecies classification of piper based on alkaloids and amides profiling. Significant differences in various plant parts used were also observed (Chandra et al. 2014). The technique is fast, reliable and non-invasive. It does not require sample preparation as the DART ion source ionize molecules directly from the surface (Bajpai et al. 2012).

REFERENCES

Abdelfatah, S. A. and Efferth, T. 2015. Cytotoxicity of the indole alkaloid reserpine from *Rauwolfia serpentina* against drug-resistant tumor cells. *Phytomedicine* 22: 308–318.

Abdullaev F. I. 2002. Cancer chemopreventive and tumoricidal properties of saffron (*Crocus sativus* L.). *Exp. Biol. Med.* 227: 20–25.

Agarwal, S., Jacob, S., Chettri, N., Bisoyi, S., Tazeen, A., Vedamurthy, A. B., ... Hoskeri, H. J. 2011. Evaluation of in-vitro anthelminthic activity of *Catharanthus roseus* extract. *Int. J. Pharm. Sci. Drug Res.* 3: 211–213.

Aggarwal, B. B., Kumar, A. and Bharti, A. C. 2003. Anticancer potential of curcumin: preclinical and clinical studies. *Anticancer Res.* 23: 363–398.

Ahmad, A. S., Ansari, M. A., Ahmad, M., Saleem, S., Yousuf, S., Hoda, M. N. and Islam, F. 2005. Neuroprotection by crocetin in a hemi-parkinsonian rat model. *Pharmacol. Biochem. Behavior.* 81(4): 805–813.

Ahmad, R., Ahmad, N. and Shehzad, A. 2019. Solvent and temperature effects of accelerated solvent extraction (ASE) with Ultra-high pressure liquid chromatography (UHPLC-PDA) technique for determination of Piperine and its ICP-MS analysis. *Industrial Crops Prod.* 136: 37–49.

Ahmad, R., Ahmad, N., Mohd, A., AlJHISI, F., Alamer, M. H., Al-Shaban, H. R., ... Aslam, K. 2020. Variation in *Nigella sativa* quality and its standardization via instrumental analysis: A study based on geographical origin. *Notulae Botanicae Horti Agrobotanici Cluj-Napoca* 48: 1141–1154.

Ahmad, S. and Beg, Z. H. 2013. Elucidation of mechanisms of actions of thymoquinone-enriched methanolic and volatile oil extracts from *Nigella sativa* against cardiovascular risk parameters in experimental hyperlipidemia. *Lipids Health Dis.* 12: 1–12.

Akinyemi, A. J., Thome, G. R., Morsch, V. M., Stefanello, N., Goularte, J. F., Belló-Klein, A., ... Schetinger, M. R. C. 2015. Effect of dietary supplementation of ginger and turmeric rhizomes on angiotensin-1 converting enzyme (ACE) and arginase activities in L-NAME induced hypertensive rats. *J. Func. Foods* 17: 792–801.

AlGhamdi, A. A., Mohammed, M. R. S., Zamzami, M. A., Al-Malki, A. L., Qari, M. H., Khan, M. I. and Choudhry, H. 2020. Untargeted metabolomics identifies key metabolic pathways altered by thymoquinone in leukemic cancer cells. *Nutrients* 12: 1792.

Al-Ghamdi, M. S. 2001. The anti-inflammatory, analgesic and antipyretic activity of *Nigella sativa*. *J. Ethnopharmacol.* 76: 45–48.

Alhadrami, H. A., Sayed, A. M., El-Gendy, A. O., Shamikh, Y. I., Gaber, Y., Bakeer, W., ... Hozzein, W. N. 2021. A metabolomic approach to target antimalarial metabolites in the *Artemisia annua* fungal endophytes. *Sci. Rep.* 11: 1–11.

Ali, B. H., Blunden, G., Tanira, M. O. and Nemmar, A. 2008. Some phytochemical, pharmacological and toxicological properties of ginger (*Zingiber officinale* Roscoe): A review of recent research. *Food Chem. Toxicol.* 46: 409–420.

Aliakbarzadeh, G., Sereshti, H. and Parastar, H. 2016. Pattern recognition analysis of chromatographic fingerprints of *Crocus sativus* L. secondary metabolites towards source identification and quality control. *Anal. Bioanal. Chem.* 408: 3295–3307.

Al-Jassir, M. S. 1992. Chemical composition and microflora of black cumin (*Nigella sativa* L.) seeds growing in Saudi Arabia. *Food Chem.* 45: 239–242.

Anbarasu, K., Vijayalakshmi, G. 2007. Improved shelf life of protein-rich tofu using *Ocimum sanctum* (tulsi) extracts to benefit Indian rural population. *J. Food Sci.* 72: M300–M305.

Anonymous. Wealth of India 1991; Vol. 7. Publication and Information Directorate, CSIR, New Delhi, pp. 79–89.

Arora, R. B. and Madan, B. R. 1956. Antiarrhythmics. VI. Ajmaline and serpentine in experimental cardiac arrhythmias. *J. Pharmacol. Exp. Therapeutics* 117: 62–67.

Arora, R. B., Kapoor, V., Basu, N. and Jain, A. P. 1971. Anti-inflammatory studies on *Curcuma longa* (turmeric). *Indian J. Med. Res.* 59: 1289–1295.

Asha, M. K., Prashanth, D., Murali, B., Padmaja, R. and Amit, A. 2001. Anthelmintic activity of essential oil of *Ocimum sanctum* and eugenol. *Fitoterapia* 72: 669–670.

Atanasov, A. G., Waltenberger, B., Pferschy-Wenzig, E. M., Linder, T., Wawrosch, C., Uhrin, P., et al. 2015. Discovery and resupply of pharmacologically active plant-derived natural products: A review. *Biotechnol. Adv.* 33: 1582–1614

Auffret, M., Drapier, S. and Vérin, M. 2018. The many faces of apomorphine: Lessons from the past and challenges for the future. *Drugs in R & D* 18: 91–107.

Awin, T., Mediani, A., Leong, S. W., Faudzi, S. M. M., Shaari, K. and Abas, F. 2019. Phytochemical and bioactivity alterations of *Curcuma* species harvested at different growth stages by NMR-based metabolomics. *J. Food Composition Anal.* 77: 66–76.

Baba, S. A., Malik, A. H., Wani, Z. A., Mohiuddin, T., Shah, Z., Abbas, N. and Ashraf, N. (2015). Phytochemical analysis and antioxidant activity of different tissue types of Crocus sativus and oxidative stress alleviating potential of saffron extract in plants, bacteria, and yeast. *South African Journal of Botany*, 99: 80–87.

Babu, P. S. and Srinivasan, K. 1997. Hypolipidemic action of curcumin, the active principle of turmeric (*Curcuma longa*) in streptozotocin induced diabetic rats. *Mol. Cell. Bioche.* 166: 169–175.

Bailey, N. J., Oven, M., Holmes, E., Nicholson, J. K. and Zenk, M. H. (2003). Metabolomic analysis of the consequences of cadmium exposure in Silene cucubalus cell cultures via 1H NMR spectroscopy and chemometrics. Phytochemistry, 62(6): 851–858.

Bajpai, V., Pandey, R., Negi, M. P. S., Kumar, N. and Kumar, B. 2012. DART MS based chemical profiling for therapeutic potential of *Piper* betle landraces. *Nat. Prod. Comm.* 7: 1934578X1200701222.

Bamba, Y., Yun, Y. S., Kunugi, A. and Inoue, H. 2011. Compounds isolated from *Curcuma aromatica* Salisb. inhibit human P450 enzymes. *J. Nat. Med.* 65: 583–587.

Barbieri, G., Vallone, S., Orsini, F., Paradiso, R., De Pascale, S., Negre-Zakharov, F. and Maggio, A. 2012. Stomatal density and metabolic determinants mediate salt stress adaptation and water use efficiency in basil (*Ocimum basilicum* L.). *J. Plant Physiol.* 169: 1737–1746.

Barbosa, G. B., Jayasinghe, N. S., Natera, S. H., Inutan, E. D., Peteros, N. P. and Roessner, U. 2017. From common to rare *Zingiberaceae* plants-a metabolomics study using GC-MS. *Phytochemistry* 140: 141–150.

Bathaie, S. Z. and Mousavi, S. Z. 2010. New applications and mechanisms of action of saffron and its important ingredients. *Critical Rev. Food Sci. Nut.* 50: 761–786.

Belal, T. S., Barary, M. H., Sabry, S. M., Elsayed, M. and Ibrahim, A. L. 2009. Kinetic spectrophotometric analysis of naftidrofuryl oxalate and vincamine in pharmaceutical preparations using alkaline potassium permanganate. *J. Food Drug Anal.* 17: 415–423.

Bennouna, J., Delord, J. P., Campone, M. and Nguyen, L. 2008. Vinflunine: A new microtubule inhibitor agent. *Clinical Cancer Res.* 14: 1625–1632.

Bhatia, B. B. 1942. On the use of *Rauwolfia serpentina* in high blood pressure. *J. Indian Med. Assoc.* 11: 262–265.

Bhutkar, M. A. and Bhise, S. B. 2011. Studies on antioxidant properties of *Catharanthus rosea* and *Catharanthus alba*. *J. Curr. Pharma Res.* 1: 337.

Bindu, S., Rameshkumar, K. B., Kumar, B., Singh, A. and Anilkumar, C. 2014. Distribution of reserpine in *Rauvolfia* species from India–HPTLC and LC–MS studies. *Indus. Crops Prod.* 62: 430–436.

Böhme, K., Calo-Mata, P., Barros-Velázquez, J. and Ortea, I. 2019. Recent applications of omics-based technologies to main topics in food authentication. *Trends Anal. Chem.* 110: 221–232.

Booker, A., Frommenwiler, D., Johnston, D., Umealajekwu, C., Reich, E. and Heinrich, M. 2014. Chemical variability along the value chains of turmeric (*Curcuma longa*): a comparison of nuclear magnetic resonance spectroscopy and high performance thin layer chromatography. *J. Ethnopharmacol.* 152: 292–301.

Broadhead, G. K., Grigg, J. R., McCluskey, P., Hong, T., Schlub, T. E. and Chang, A. A. 2019. Saffron therapy for the treatment of mild/moderate age-related macular degeneration: a randomised clinical trial. *Graefe's Archive Clin. Exp. Ophthalmology* 257: 31–40.

Brook, K., Bennett, J. and Desai, S. P. 2017. The chemical history of morphine: An 8000-year journey, from resin to de-novo synthesis. *J. Anesthesia History* 3: 50–55.

Butt, M. S., Pasha, I., Sultan, M. T., Randhawa, M. A., Saeed, F. and Ahmed, W. 2013. Black pepper and health claims: A comprehensive treatise. *Critical Rev. Food Sci. Nut.* 53: 875–886.

Cagliani, L. R., Culeddu, N., Chessa, M. and Consonni, R. 2015. NMR investigations for a quality assessment of Italian PDO saffron (*Crocus sativus* L.). *Food Control* 50: 342–348.

Carvalho, S. D., Schwieterman, M. L., Abrahan, C. E., Colquhoun, T. A. and Folta, K. M. 2016. Light quality dependent changes in morphology, antioxidant capacity, and volatile production in sweet basil (*Ocimum basilicum*). *Front. Plant Sci.* 7: 1328.

Castro-Puyana, M., Pérez-Míguez, R., Montero, L. and Herrero, M. 2017. Reprint of: Application of mass spectrometry-based metabolomics approaches for food safety, quality and traceability. *Trends Anal. Chem.* 96: 62–78.

Chandra, P., Bajpai, V., Srivastva, M., Kumar, K. R. and Kumar, B. 2014. Metabolic profiling of *Piper* species by direct analysis using real time mass spectrometry combined with principal component analysis. *Anal. Methods* 6: 4234–4239.

Chatzinasiou, L., Booker, A., MacLennan, E., Mackonochie, M. and Heinrich, M. 2019. Turmeric (*Curcuma longa* L.) products: What quality differences exist?. *J. Herbal Med.* 17: 100281.

Chen, S., Wu, S., Li, W., Chen, X., Dong, X., Tan, G., ... and Chai, Y. 2014. Investigation of the therapeutic effectiveness of active components in Sini decoction by a comprehensive GC/LC-MS based metabolomics and network pharmacology approaches. *Mol. BioSystems* 10: 3310–3321.

Chen, Q., Lu, X., Guo, X., Guo, Q. and Li, D. 2017. Metabolomics characterization of two Apocynaceae plants, *Catharanthus roseus* and *Vinca minor*, using GC-MS and LC-MS methods in combination. *Molecules* 22: 997.

Chen, C., Yin, Q., Tian, J., Gao, X., Qin, X., Du, G. and Zhou, Y. 2020. Studies on the potential link between antidepressant effect of Xiaoyao San and its pharmacological activity of hepatoprotection based on multi-platform metabolomics. *Journal of ethnopharmacology* 249: 112432.

Chen, C., Yin, Q., Tian, J., Gao, X., Qin, X., Du, G. and Zhou, Y. 2020. Studies on the potential link between antidepressant effect of Xiaoyao San and its pharmacological activity of hepatoprotection based on multi-platform metabolomics. *J. Ethnopharmacol.* 249: 112432.

Cheng, X. L., Liu, Q., Peng, Y. B., Qi, L. W. and Li, P. 2011. Steamed ginger (*Zingiber officinale*): Changed chemical profile and increased anticancer potential. *Food Chem.* 129: 1785–1792.

Choi, Y. H., Tapias, E. C., Kim, H. K., Lefeber, A. W., Erkelens, C., Verhoeven, J. T. J. ... and Verpoorte, R. 2004. Metabolic discrimination of *Catharanthus roseus* leaves infected by phytoplasma using 1H-NMR spectroscopy and multivariate data analysis. *Plant Physiol.* 135: 2398–2410.

Chopra, R. N., Chopra, I. C., Handa, K. L. and Kapur, L. D.1982. *Indigenous Drugs of India*, Academic Publishers: Calcutta, India.

Chopra, R. N., Nayar, S. L. and Chopra, I. C. 1956. *Glossary of Indian Medicinal Plants Council of Scientific and Industrial Research*. New Delhi, 89.

College, J. N. M. 1985. *The dictionary of traditional Chinese medicine*. Shanghai: Shanghai Sci-Tech Press.

Consonni, R., Ordoudi, S. A., Cagliani, L. R., Tsiangali, M. and Tsimidou, M. Z. 2016. On the traceability of commercial saffron samples using 1H-NMR and FT-IR metabolomics. *Molecules* 21: 286.

Cooper, H. J. and Marshall, A. G. 2001 Electrospray ionization Fourier transform mass spectrometric analysis of wine. *J Agri Food Chem* 49: 5710–5718.

Corrado, G., Chiaiese, P., Lucini, L., Miras-Moreno, B., Colla, G. and Rouphael, Y. 2020. Successive harvests affect yield, quality and metabolic profile of sweet basil (*Ocimum basilicum* L.). *Agronomy* 10: 830.

Dall'Acqua, S., Stocchero, M., Boschiero, I., Schiavon, M., Golob, S., Uddin, J., ... & Schievano, E. 2016. New findings on the in vivo antioxidant activity of Curcuma longa extract by an integrated 1H NMR and HPLC–MS metabolomic approach. *Fitoterapia*, 109, 125–131.

D'Archivio, A. A. and Maggi, M. A. 2017. Geographical identification of saffron (*Crocus sativus* L.) by linear discriminant analysis applied to the UV-visible spectra of aqueous extracts. *Food Chem.* 219: 408–413.

Da Luz, S. F., Yamaguchi, L. F., Kato, M. J., De Lemos, O. F., Xavier, L. P., Maia, J. G. S., ... and Da Silva, J. K. D. R. 2017. Secondary metabolic profiles of two cultivars of *Piper nigrum* (black pepper) resulting from infection by *Fusarium solani* f. sp. piperis. *International J. Mol. Sci.* 18: 2434.

Daba, M. H. and Abdel-Rahman, M. S. 1998. Hepatoprotective activity of thymoquinone in isolated rat hepatocytes. *Toxicology Lett.* 95: 23–29.

Dai, Y., Li, Z., Xue, L., Dou, C., Zhou, Y., Zhang, L. and Qin, X. 2010. Metabolomics study on the anti-depression effect of xiaoyaosan on rat model of chronic unpredictable mild stress. *J. Ethnopharmacology* 128: 482–489.

Dall'Acqua, S., Stocchero, M., Boschiero, I., Schiavon, M., Golob, S., Uddin, J. ... and Schievano, E. 2016. New findings on the in vivo antioxidant activity of *Curcuma longa* extract by an integrated 1H NMR and HPLC–MS metabolomic approach. *Fitoterapia* 109: 125–131.

De Monte, C., Carradori, S., Chimenti, P., Secci, D., Mannina, L., Alcaro, F. ... and Petzer, J. P. 2014. New insights into the biological properties of *Crocus sativus* L.: Chemical modifications, human monoamine oxidases inhibition and molecular modeling studies. *European J. Med. Chem.* 82: 164–171.
Desisa, D. 2001. A preliminary economic evaluation of medicinal plants in Ethiopia: trade, volume and price. In *Proceedings of the National Workshop on Biodiversity Conservation and Sustainable Use of Medicinal Plants in Ethiopia*, Addis Abeba (Ethiopia), 28 Apr–1 May 1998. pp. 176–187.
Develi, S., Evran, B., Kalaz, E. B., Koçak-Toker, N. and Erata, G. Ö. 2014. Protective effect of *Nigella sativa* oil against binge ethanol-induced oxidative stress and liver injury in rats. *Chinese J. Nat. Med.* 12: 495–499.
Dey, A., Nandy, S., Nongdam, P., Tikendra, L., Mukherjee, A., Mukherjee, S. and Pandey, D. K. 2020. Methyl jasmonate and salicylic acid elicit indole alkaloid production and modulate antioxidant defence and biocidal properties in *Rauvolfia serpentina* Benth. ex Kurz. in vitro cultures. *South African J. Bot.* 135: 1–17.
Dhanya, K., Syamkumar, S. and Sasikumar, B. 2009. Development and application of SCAR marker for the detection of papaya seed adulteration in traded black pepper powder. *Food Biotechnol.* 23: 97–106.
Dias, A. D., Jones, A. O., Beale, J. D., Boughton, A. B., Benheim, D., Kouremenos, A. K. et al. 2016. Current and future perspectives on the structural identification of small molecules in biological systems. *Metabolites* 6: E46.
Dugé de Bernonville, T., Carqueijeiro, I., Lanoue, A., Lafontaine, F., Sánchez Bel, P., Liesecke, F. ... and Courdavault, V. 2017. Folivory elicits a strong defense reaction in *Catharanthus roseus*: Metabolomic and transcriptomic analyses reveal distinct local and systemic responses. *Sci. Rep.* 7: 1–14.
Dunn, W. B. and Ellis, D. I. 2005. Metabolomics: Current analytical platforms and methodologies. *Trends Anal. Chem.* 24: 285–294.
El Hamss, R., Idaomar, M., Alonso-Moraga, A. and Serrano, A. M. 2003. Antimutagenic properties of bell and black peppers. *Food Chem. Toxicol.* 41: 41–47.
Farag, M. A., El Sayed, A. M., El Banna, A. and Ruehmann, S. 2015. Metabolomics reveals distinct methylation reaction in MeJA elicited *Nigella sativa* callus via UPLC–MS and chemometrics. *Plant Cell Tissue Org. Cult.* 122: 453–463.
Farag, M. A., Gad, H. A., Heiss, A. G. and Wessjohann, L. A. 2014. Metabolomics driven analysis of six *Nigella* species seeds via UPLC-qTOF-MS and GC–MS coupled to chemometrics. *Food Chem.* 151: 333–342.
Farooqui, Z., Shahid, F., Khan, A. A. and Khan, F. 2017. Oral administration of *Nigella sativa* oil and thymoquinone attenuates long term cisplatin treatment induced toxicity and oxidative damage in rat kidney. *Biomed. Pharmacotherapy* 96: 912–923.
Feizi, N., Hashemi-Nasab, F. S., Golpelichi, F., Saburouh, N. and Parastar, H. 2021. Recent trends in application of chemometric methods for GC-MS and GC× GC-MS-based metabolomic studies. *Trends Anal. Chem.* 138: 116239.
Feng, Z., Ding, C., Li, W., Wang, D. and Cui, D. 2020. Applications of metabolomics in the research of soybean plant under abiotic stress. *Food Chem.* 310: 125914.
Fiehn, O. 2002. Metabolomics – the link between genotypes and phenotypes. *Func. Genomics* 155–171.
Fiehn, O., Kopka, J., Dörmann, P., Altmann, T., Trethewey, R. N. and Willmitzer, L. 2000. Metabolite profiling for plant functional genomics. *Nat. Biotechnol.* 18: 1157–1161.
Fraga-Corral, M., Carpena, M., Garcia-Oliveira, P., Pereira, A. G., Prieto, M. A. and Simal-Gandara, J. 2020. Analytical metabolomics and applications in health, environmental and food science. *Crit. Rev. Anal. Chem.* 52(4): 712–734.
Fraser, V. N., Philmus, B. and Megraw, M. 2020. Metabolomics analysis reveals both plant variety and choice of hormone treatment modulate vinca alkaloid production in *Catharanthus roseus*. *Plant Direct* 4: e00267.
Gerasimenko, I., Sheludko, Y. and Stöckigt, J. 2001. 3-oxo-rhazinilam: A new indole alkaloid from *Rauvolfia serpentina*× *Rhazya stricta* hybrid plant cell cultures. *J. Nat. Prod.* 64: 114–116.
Ghaffari, S. and Roshanravan, N. 2019. Saffron: An updated review on biological properties with special focus on cardiovascular effects. *Biomed. Pharmacotherapy* 109: 21–27.
Ghajar, A., Neishabouri, S. M., Velayati, N., Jahangard, L., Matinnia, N., Haghighi, M., ... and Akhondzadeh, S. 2017. *Crocus sativus* L. versus citalopram in the treatment of major depressive disorder with anxious distress: a double-blind, controlled clinical trial. *Pharmacopsychiatry* 50: 152–160.
Ghangale, G. R., Tushar, M. and Jadhav, N. D. 2009. Evaluation of antiulcer activity of *Ocimum sanctum* in rats. *Veterinary World* 2: 465–466.

Goel, M. K., Mehrotra, S., Kukreja, A. K., Shanker, K. and Khanuja, S. P. S. 2009. In vitro propagation of *Rauwolfia serpentina* using liquid medium, assessment of genetic fidelity of micropropagated plants, and simultaneous quantitation of reserpine, ajmaline, and ajmalicine. In *Protocols for In Vitro Cultures and Secondary Metabolite Analysis of Aromatic and Medicinal Plants* (pp. 17–33). Humana Press, Totowa, NJ.

Govindarajan, V. S. and Stahl, W. H. 1977. Pepper – chemistry, technology, and quality evaluation. *Crit. Rev. Food Sci. Nut.* 9: 115–225.

Govindarajan, V. S. and Connell, D. W. (1983). Ginger – chemistry, technology, and quality evaluation: part 1. *Critical Reviews in Food Science & Nutrition*, 17(1): 1–96.

Gu, F., Huang, F., Wu, G. and Zhu, H. 2018. Contribution of polyphenol oxidation, chlorophyll and vitamin C degradation to the blackening of *Piper nigrum* L. *Molecules* 23: 370.

Guijarro-Díez, M., Nozal, L., Marina, M. L. and Crego, A. L. 2015. Metabolomic fingerprinting of saffron by LC/MS: novel authenticity markers. *Anal. Bioanal. Chem.* 407: 7197–7213.

Gul, I., Nasrullah, N., Nissar, U., Saifi, M. and Abdin, M. Z. 2018. Development of DNA and GC-MS Fingerprints for authentication and quality control of *Piper nigrum* L. and its adulterant *Carica papaya* L. *Food Anal. Meth.* 11:1209–1222.

Gülçin, İ. 2005. The antioxidant and radical scavenging activities of black pepper (*Piper nigrum*) seeds. *International J. Food Sci. Nut.* 56: 491–499.

Gupta, S. K., Prakash, J. and Srivastava, S. 2002. Validation of traditional claim of Tulsi, *Ocimum sanctum* Linn. as a medicinal plant. *Indian J. Exp. Biol.* 40(7): 765–773.

Hadacek, F. 2002. Secondary metabolites as plant traits: Current assessment and future perspectives. *Critical Re. Plant Sci.* 21(4): 273–322.

Hardy, K., Buckley, S., Collins, M. J., Estalrrich, A., Brothwell, D., Copeland, L., ... and Rosas, A. 2012. Neanderthal medics? Evidence for food, cooking, and medicinal plants entrapped in dental calculus. *Naturwissenschaften* 99(8): 617–626.

Harisaranraj, R., Suresh, K. and Babu, S. S. 2009. Production of reserpine in somatic embryos of Rauwolfia serpentina cultured in bioreactors by the induction of elicitor (Methyl Jasmonate). *Glob. J. Biotechnol. Biochem.* 4(2): 143–147.

Harvey, A. L., Edrada-Ebel, R. and Quinn, R. J. 2015. The re-emergence of natural products for drug discovery in the genomics era. *Nat. Rev. Drug Discovery* 14: 111–129.

Heidarbeigi, K., Mohtasebi, S. S., Foroughirad, A., Ghasemi-Varnamkhasti, M., Rafiee, S. and Rezaei, K. 2015. Detection of adulteration in saffron samples using electronic nose. *Int. J. Food Properties* 18(7): 1391–1401.

Heitmar, R., Brown, J. and Kyrou, I. 2019. Saffron (*Crocus sativus* L.) in ocular diseases: A narrative review of the existing evidence from clinical studies. *Nutrients* 11(3): 649.

Ho, S. C., Chang, K. S. and Lin, C. C. 2013. Anti-neuroinflammatory capacity of fresh ginger is attributed mainly to 10-gingerol. *Food Chem.* 141(3): 3183–3191.

Hosseinzadeh, H. and Younesi, H. M. 2002. Antinociceptive and anti-inflammatory effects of *Crocus sativus* L. stigma and petal extracts in mice. *BMC Pharmacol.* 2: 1–8.

Hu, X. J., He, H. P., Zhou, H., Di, Y. T., Yang, X. W., Hao, X. J. and Kong, L. Y. 2008. New indole alkaloids from *Rauvolfia verticillatae*. *Helv. Chim. Acta*, 89: 1344–1350.

Hüe, T., Cauquil, L., Fokou, J. B., Dongmo, P. M., Bakarnga-Via, I. and Menut, C. (2015). Acaricidal activity of five essential oils of Ocimum species on Rhipicephalus (Boophilus) microplus larvae. Parasitology research, 114(1): 91–99.

Ibarra-Estrada, E., Soto-Hernández, R. M. and Palma-Tenango, M. 2016. Metabolomics as a Tool in Agriculture. In *Metabolomics: Fundamentals and Applications*. London: IntechOpen, pp. 148–168.

Ireson, C., Orr, S., Jones, D. J., Verschoyle, R., Lim, C. K., Luo, J. L., ... and Gescher, A. 2001. Characterization of metabolites of the chemopreventive agent curcumin in human and rat hepatocytes and in the rat in vivo, and evaluation of their ability to inhibit phorbol ester-induced prostaglandin E2 production. *Cancer Res.* 61(3): 1058–1064.

Jacobs, D. I., Snoeijer, W., Hallard, D. and Verpoorte, R. 2004. The Catharanthus alkaloids: Pharmacognosy and biotechnology. *Curr. Med. Chem.* 11(5): 607–628.

Jagella, T. and Grosch, W. 1999. Flavour and off-flavour compounds of black and white pepper (*Piper nigrum* L.) I. Evaluation of potent odorants of black pepper by dilution and concentration techniques. *European Food Res. Tech.* 209: 16–21.

Jain, S. and Mehata, M.S. 2017. Medicinal plant leaf extract and pure flavonoid mediated green synthesis of silver nanoparticles and their enhanced antibacterial property. *Sci. Rep.* 7: 1–13.

Jayakumar, A., Nair, I. C. and Radhakrishnan, E. K. 2021. Environmental adaptations of an extremely plant beneficial Bacillus subtilis Dcl1 identified through the genomic and metabolomic analysis. *Microbial Eco.* 81(3): 687–702.

Jayalekshmy, A., Menon, A. N. and Padmakumari, K. P. 2003. Essential oil composition of four major cultivars of black pepper (*Piper nigrum* L.). *J. Essential Oil Res* 15: 155–157.

Jiang, H., Sólyom, A. M., Timmermann, B. N. and Gang, D. R. 2005. Characterization of gingerol-related compounds in ginger rhizome (*Zingiber officinale* Rosc.) by high-performance liquid chromatography/electrospray ionization mass spectrometry. *Rapid Communications in Mass Spectrometry* 19: 2957–2964.

Jiang, H., Xie, Z., Koo, H. J., McLaughlin, S. P., Timmermann, B. N. and Gang, D. R. 2006. Metabolic profiling and phylogenetic analysis of medicinal Zingiber species: Tools for authentication of ginger (*Zingiber officinale* Rosc.). *Phytochemistry*, 67: 1673–1685.

José Bagur, M., Alonso Salinas, G. L., Jiménez-Monreal, A. M., Chaouqi, S., Llorens, S., Martínez-Tomé, M. and Alonso, G. L. 2017. Saffron: An old medicinal plant and a potential novel functional food. *Molecules* 23(1): 30.

Jung, B. S. and Shin, M. K. 1998. Encyclopedia of illustrated Korean natural drugs, *Araliaceae. Seoul: Young Lim Sa*, pp. 439–443.

Jung, Y., Lee, J., Kim, H. K., Moon, B. C., Ji, Y. and Hwang, G. S. (2012). Metabolite profiling of Curcuma species grown in different regions using 1 H NMR spectroscopy and multivariate analysis. *Analyst* 137(23): 5597–5606.

Jurenka, J. S. 2009. Anti-inflammatory properties of curcumin, a major constituent of *Curcuma longa*: a review of preclinical and clinical research. *Alternative Med. Rev.* 14(2): 141–153.

Kaddurah-Daouk, R., Weinshilboum, R. M. and Pharmacometabolomics Research Network. 2014. Pharmacometabolomics: implications for clinical pharmacology and systems pharmacology. *Clinical Pharmacology & Therapeutics* 95(2): 154–167.

Kaddurah-Daouk, R., Weinshilboum, R. and Pharmacometabolomics Research Network. 2015. Metabolomic signatures for drug response phenotypes: pharmacometabolomics enables precision medicine. *Clinical Pharmacology & Therapeutics* 98(1): 71–75.

Kamazeri, T. S. A. T., Abd Samah, O., Taher, M., Susanti, D. and Qaralleh, H. 2012. Antimicrobial activity and essential oils of Curcuma aeruginosa, Curcuma mangga, and Zingiber cassumunar from Malaysia. *Asian Pacific J. Trop. Med.* 5(3): 202–209.

Kanter, M., Demir, H., Karakaya, C. and Ozbek, H. 2005. Gastroprotective activity of Nigella sativa L oil and its constituent, thymoquinone against acute alcohol-induced gastric mucosal injury in rats. *World J. Gastroenterology* 11(42): 6662.

Karkoula, E., Dagla, I. V., Baira, E., Kokras, N., Dalla, C., Skaltsounis, A. L., ... and Tsarbopoulos, A. 2020. A novel UHPLC-HRMS-based metabolomics strategy enables the discovery of potential neuroactive metabolites in mice plasma, following ip administration of the main *Crocus sativus* L. bioactive component. *J. Pharmaceu. Biomed. Anal.* 177: 112878.

Khakimov, B., Bak, S. and Engelsen, S. B. 2014. High-throughput cereal metabolomics: Current analytical technologies, challenges and perspectives. *J. Cereal Sci.* 59(3): 393–418.

Khan, M. A. U., Ashfaq, M. K., Zuberi, H. S., Mahmood, M. S. and Gilani, A. H. 2003. The in vivo antifungal activity of the aqueous extract from Nigella sativa seeds. *Phytotherapy Res.* 17: 183–186.

Khan, M. I. H., Sohrab, M. H., Rony, S. R., Tareq, F. S., Hasan, C. M. and Mazid, M. A. 2016. Cytotoxic and antibacterial naphthoquinones from an endophytic fungus, *Cladosporium* sp. *Toxicology Reports*, 3: 861–865.

Khattak, K. F. and Simpson, T. J. 2008. Effect of gamma irradiation on the extraction yield, total phenolic content and free radical-scavenging activity of *Nigella staiva* seed. *Food Chem.* 110(4): 967–972.

Kiesecker, C., Zitron, E., Lück, S., Bloehs, R., Scholz, E. P., Kathöfer, S., ... and Kiehn, J. 2004. Class Ia antiarrhythmic drug ajmaline blocks HERG potassium channels: mode of action. *Naunyn-Schmiedeberg's Arch. Pharmacol.* 370(6): 423–435.

Kirtikar, K. R. and Basu, B. D. 1933. Indian medicinal plants. *Indian Medicinal Plants.* Allahabad, India.

Kline, N. S. 1954. Use of *Rauwolfia serpentina* Benth. in neuropsychiatric conditions. *Ann. New York Acad. Sci.* 59: 107–132.

Krikler, D. M. 1985. The foxglove, "The old woman from Shropshire" and William Withering. *J. Am. College Cardiology* 5(5): 3A–9A.

Kumar, S., Bajpai, V., Singh, A., Bindu, S., Srivastava, M., Rameshkumar, K. B. and Kumar, B. 2015. Rapid fingerprinting of *Rauwolfia* species using direct analysis in real time mass spectrometry combined with principal component analysis for their discrimination. *Anal. Meth.* 7: 6021–6026.

Kumar, S., Singh, A., Bajpai, V., Srivastava, M., Singh, B. P., Ojha, S. and Kumar, B. 2016. Simultaneous determination of bioactive monoterpene indole alkaloids in ethanolic extract of seven *Rauvolfia* species using UHPLC with hybrid triple quadrupole linear ion trap mass spectrometry. *Phytochem. Anal.* 27(5): 296–303.

Kurepa, J. and Smalle, J. A. 2021. Composition of the metabolomic bio-coronas isolated from *Ocimum sanctum* and *Rubia tinctorum*. *BMC Res. Notes* 14: 1–6.

Lacalle-Bergeron, L., Izquierdo-Sandoval, D., Sancho, J. V., López, F. J., Hernandez, F. and Portolés, T. 2021. Chromatography hyphenated to high resolution mass spectrometry in untargeted metabolomics for investigation of food (bio) markers. *Trends Anal. Chem.* 135: 116.

Leone, S., Recinella, L., Chiavaroli, A., Orlando, G., Ferrante, C., Leporini, L., ... and Menghini, L. 2018. Phytotherapic use of the *Crocus sativus* L Saffron) and its potential applications: A brief overview. *Phytotherapy Res.* 32: 2364–2375.

Lewis, J. J. 1956. The *Rauwolfia* alkaloids. *J. Pharmacy Pharmacol.* 8: 465–494.

Li, Z. Y., Ding, L. L., Li, J. M., Xu, B. L., Yang, L., Bi, K. S. and Wang, Z. T. 2015. 1H-NMR and MS based metabolomics study of the intervention effect of curcumin on hyperlipidemia mice induced by high-fat diet. *Plos One* 10(3): e0120950.

Lim, T. K. 2014. *Crocus sativus*. In *Edible Medicinal and Non Medicinal Plants* (pp. 77–136. Springer, Dordrecht.

Liu, J., Liu, Y., Wang, Y., Zhang, Z. H., Zu, Y. G., Efferth, T. and Tang, Z. H. 2016. The combined effects of ethylene and MeJA on metabolic profiling of phenolic compounds in *Catharanthus roseus* revealed by metabolomics analysis. *Front. Physiol.* 7: 217.

Liu, N. Q., Cao, M., Frédérich, M., Choi, Y. H., Verpoorte, R. and van der Kooy, F. 2010. Metabolomic investigation of the ethnopharmacological use of *Artemisia afra* with NMR spectroscopy and multivariate data analysis. *J. Ethnopharmacol.* 128: 230–235.

Lohiya, N. K., Mishra, P. K., Pathak, N., Manivannan, B. and Jain, S. C. 1999. Reversible azoospermia by oral administration of the benzene chromatographic fraction of the chloroform extract of the seeds of *Carica papaya* in rabbits. *Adv. Contraception* 15: 141–161.

Ma, C., Wang, H., Lu, X., Wang, H., Xu, G. and Liu, B. 2009. Terpenoid metabolic profiling analysis of transgenic *Artemisia annua* L. by comprehensive two-dimensional gas chromatography time-of-flight mass spectrometry. *Metabolomics* 5(4): 497–506.

Ma, X. and Gang, D. R. 2006. Metabolic profiling of in vitro micropropagated and conventionally greenhouse grown ginger (*Zingiber officinale*). *Phytochemistry* 67: 2239–2255.

Ma, C., Wang, H., Lu, X., Xu, G. and Liu, B. 2008. Metabolic fingerprinting investigation of *Artemisia annua* L. in different stages of development by gas chromatography and gas chromatography–mass spectrometry. *J. Chromatography A.* 1186: 412–419.

Mabberley, D. J. 2008. *Mabberley's plant-book: A portable dictionary of plants, their classifications and uses* (3rd ed.). Cambridge: Cambridge University Press.

Madhusudanan, K. P., Banerjee, S., Khanuja, S. P. and Chattopadhyay, S. K. 2008. Analysis of hairy root culture of *Rauvolfia serpentina* using direct analysis in real time mass spectrometric technique. *Biomed. Chromatography* 22: 596–600.

Mani, H., Sidhu, G. S., Kumari, R., Gaddipati, J. P., Seth, P. and Maheshwari, R. K. 2002. Curcumin differentially regulates TGF-β1, its receptors and nitric oxide synthase during impaired wound healing. *Biofactors* 16: 29–43.

Maulidiani, M., Sheikh, B. Y., Mediani, A., Wei, L. S., Ismail, I. S., Abas, F. and Lajis, N. H. 2015. Differentiation of *Nigella sativa* seeds from four different origins and their bioactivity correlations based on NMR-metabolomics approach. *Phytochem. Lett.* 13: 308–318.

Meena, A. K., Bansal, P. and Kumar, S. 2009. Plants-herbal wealth as a potential source of ayurvedic drugs. *Asian J. Tradit. Med.* 4(4): 152–170.

Mehrotra, S., Goel, M. K., Srivastava, V. and Rahman, L. U. 2015. Hairy root biotechnology of *Rauwolfia serpentina*: A potent approach for the production of pharmaceutically important terpenoid indole alkaloids. *Biotechnol. Lett.* 37: 253–263.

Mehrotra, S., Srivastava, V., Rahman, L. U. and Kukreja, A. K. 2013. Overexpression of a *Catharanthus* tryptophan decarboxylase (tdc) gene leads to enhanced terpenoid indole alkaloid (TIA) production in transgenic hairy root lines of *Rauwolfia serpentina*. Plant Cell Tiss. *Organ Cult.* 115(3): 377–384.

Melnyk, J. P., Wang, S. and Marcone, M. F. 2010. Chemical and biological properties of the world's most expensive spice: Saffron. *Food Res. Int.* 43(8): 1981–1989.

Menon, V. P. and Sudheer, A.R. (2007). Antioxidant and anti-inflammatory properties of curcumin. In: Aggarwal, B.B., Surh, YJ., Shishodia, S. (eds) *The Molecular Targets and Therapeutic Uses of Curcumin in Health and Disease. Advances in Experimental Medicine and Biology*, vol 595. Springer, Boston, MA.

Miao, Q., Zhao, L., Wang, Y., Hao, F., Sun, P., He, P., ... and Jia, W. 2020. Microbial metabolomics and network analysis reveal fungistatic effect of basil (*Ocimum basilicum*) oil on *Candida albicans*. *J. Ethnopharmacol.* 260: 113002.

Mielko, K. A., Pudełko-Malik, N., Tarczewska, A. and Młynarz, P. 2021. NMR spectroscopy as a "green analytical method" in metabolomics and proteomics studies. *Sustainable Chem. Pharmacy* 22: 100474.

Mishra, P. and Mishra, S. 2011. Study of antibacterial activity of Ocimum sanctum extract against gram positive and gram negative bacteria (No. RESEARCH). *American Journal of Food Technology* 6(4): 336–341.

Moco, S., Bino, R. J., Vorst, O., Verhoeven, H. A., de Groot, J., van Beek, T. A., ... and De Vos, C. R. 2006. A liquid chromatography-mass spectrometry-based metabolome database for tomato. *Plant Physiol.* 141: 1205–1218.

Mohammed, G. J., Omran, A. M. and Hussein, H. M. 2016. Antibacterial and phytochemical analysis of *Piper nigrum* using gas chromatography-mass Spectrum and Fourier-transform infrared spectroscopy. *Int. J. Pharmacog. Phytochem. Res.* 8: 977–996.

Monachino, J. 1954. *Rauvolfia serpentina* – its history, botany and medical use. *Econ. Bot.* 8: 349–365.

Mondal, S., Mirdha, B. R. and Mahapatra, S. C. 2009. The science behind sacredness of Tulsi (*Ocimum sanctum* Linn.). *Indian J. Physiol. Pharmacol.* 53: 291–306.

Monga, J., Sharma, M., Tailor, N. and Ganesh, N. 2011. Antimelanoma and radioprotective activity of alcoholic aqueous extract of different species of *Ocimum* in C57BL mice. *Pharmaceu. Biol.* 49: 428–436.

Moradzadeh, M., Kalani, M. R. and Avan, A. 2018. The antileukemic effects of saffron (*Crocus sativus* L.) and its related molecular targets: A mini review. *J. Cellular Biochem.* 120: 4732–4738.

Moreno, P. R., van der Heijden, R. and Verpoorte, R. 1994. Elicitor-mediated induction of isochorismate synthase and accumulation of 2, 3-dihydroxy benzoic acid in *Catharanthus roseus* cell suspension and shoot cultures. *Plant Cell Rep.* 14: 188–191.

Muralikrishnan, G., Pillai, S. K. and Shakeel, F. 2012. Protective effects of *Ocimum sanctum* on lipid peroxidation and antioxidant status in streptozocin-induced diabetic rats. *Natural Prod. Res.* 26: 474–478.

Mustafa, N. R. and Verpoorte, R. 2007. Phenolic compounds in *Catharanthus roseus*. *Phytochem. Rev.* 6: 243–258.

Mustafa, N. R., Kim, H. K., Choi, Y. H. and Verpoorte, R. 2009. Metabolic changes of salicylic acid-elicited *Catharanthus roseus* cell suspension cultures monitored by NMR-based metabolomics. *Biotech. Lett.* 31: 1967–1974.

Nahak, G. and Sahu, R. K. 2011. Phytochemical evaluation and antioxidant activity of *Piper cubeba* and *Piper nigrum*. *J. Appl. Pharmaceu. Sci.* 01(08): 153–157.

Nair, V. D., Panneerselvam, R. and Gopi, R. 2012. Studies on methanolic extract of *Rauvolfia* species from Southern Western Ghats of India – In vitro antioxidant properties, characterisation of nutrients and phytochemicals. *Industrial Crops Prod.* 39: 17–25.

Najmi, A., Nasiruddin, M., Khan, R. A. and Haque, S. F. 2013. Indigenous herbal product *Nigella sativa* proved effective as an antihypertensive in metabolic syndrome. *Asian J. Pharm. Clin. Res.* 6: 61–64.

Nakabayashi, R. and Saito, K. 2015. Integrated metabolomics for abiotic stress responses in plants. *Curr. Opi. Plant Biol.* 24: 10–16.

Nakabayashi, R., Mori, T., Takeda, N., Toyooka, K., Sudo, H., Tsugawa, H. and Saito, K. 2020. Metabolomics with 15N labeling for characterizing missing monoterpene indole alkaloids in plants. *Anal. Chem.* 92: 5670–5675.

Naquvi, J. K., Dohare, L. S., Shuaib, M. and Ahmad, I. M. 2012. Chemical composition of volatile oil of *Ocimum sanctum* Linn. *Int. J. Biomed. Adv. Res.* 3: 129–131.

Nayak, B. S. and Pinto Pereira, L. M. 2006. Catharanthus roseus flower extract has wound-healing activity in Sprague Dawley rats. *BMC Compl. Alt. Med.* 6: 1–6.

Naz, H. 2011. *Nigella sativa*: the miraculous herb. *Pak. J. Biochem. Mol. Biol.* 44: 44–48.

Negbi, M. 1999. Saffron cultivation: past, present and future prospects. In: Saffron: *Crocus sativus* L. (Ed. Negbi, M.), London: Harwood Academic Publisher, 1–17.

Newman, D. J. and Cragg, G. M. 2012. Natural products as sources of new drugs over the 30 years from 1981 to 2010. *J. Nat. Prod.* 75: 311–335.

Nguyen, C. T. T., Nguyen, N. H., Choi, W. S., Lee, J. H. and Cheong, J. J. 2020. Biosynthesis of essential oil compounds in *Ocimum tenuiflorum* is induced by abiotic stresses. *Plant Biosyst.* 156(2): 1–5.

Nguyen, T., Talbi, H., Hilali, A., Anthonissen, R., Maes, A. and Verschaeve, L. 2019. In vitro toxicity, genotoxicity and antigenotoxicity of *Nigella sativa* extracts from different geographic locations. *South Afri. J. Bot.* 126: 132–141.

Nivetha, K. and Prasanna, G. 2016. GC-MS and FT-IR analysis of *Nigella sativa* L. seeds. *Int. J. Adv. Res. Biol. Sci.* 3: 45–54.

O'Hara, M., Kiefer, D., Farrell, K. and Kemper, K. 1998. A review of 12 commonly used medicinal herbs. *Arch. Family Med.* 7: 523.

O'Keefe, B. R., Mahady, G. B., Gills, J. J., Beecher, C. W. and Schilling, A. B. 1997. Stable vindoline production in transformed cell cultures of *Catharanthus roseus*. *J. Nat. Prod.* 60: 261–264.

Olalere, O. A., Abdurahman, N. H., bin Mohd Yunus, R. and Alara, O. R. 2018. Data article on elemental and metabolomic-based alkaloidal composition in black pepper oleoresin using a positive ESI-mode LC-QToF and ICP-mass spectroscopy. *Data in Brief* 19: 1627–1630.

Olofsson, L., Engström, A., Lundgren, A. and Brodelius, P. E. 2011. Relative expression of genes of terpene metabolism in different tissues of *Artemisia annua* L. *BMC Plant Biol.* 11: 1–12.

Omidi, A., Riahinia, N., Torbati, M. B. M. and Behdani, M. A. 2014. Hepatoprotective effect of *Crocus sativus* (saffron) petals extract against acetaminophen toxicity in male Wistar rats. *Avicenna J. Phytomed.* 4: 330.

Pan, Q., Wang, Q., Yuan, F., Xing, S., Zhao, J., Choi, Y. H., ... and Tang, K. 2012. Overexpression of ORCA3 and G10H in Catharanthus roseus plants regulated alkaloid biosynthesis and metabolism revealed by NMR-metabolomics. *PLoS One* 7(8): e43038.

Panda, S. and Kar, A. 2003. Piperine lowers the serum concentrations of thyroid hormones, glucose and hepatic 5′ D activity in adult male mice. *Hormone Metabolic Res.* 35: 523–526.

Pandey, B. P. 1978. *Economic Botany: For Degree, Honours and Post-graduate Students*. New Delhi: S. Chand and Company.

Parmar, V. S., Jain, S. C., Bisht, K. S., Jain, R., Taneja, P., Jha, A. ... and Boll, P. M. 1997. Phytochemistry of the genus *Piper*. *Phytochem.* 46: 597–673.

Parr, A. J., Peerless, A. C. J., Hamill, J. D., Walton, N. J., Robins, R. J. and Rhodes, M. J. C. 1988. Alkaloid production by transformed root cultures of *Catharanthus roseus*. *Plant Cell Rep.* 7: 309–312.

Parthasarathy, V. A., Chempakam, B. and Zachariah, T. J. 2008. Chemistry of Spices. CAB International, Wallingford.

Pawar, M. A., Patil, S. S. and Nagrik, D. M. 2015. Phytochemical and physicochemical investigation of *Curcuma longa* Linn. rhizome. *Int. J. Chem. Phys. Sci.* 4: 458–463.

Peron, G., Sut, S., Dal Ben, S., Voinovich, D. and Dall'Acqua, S. 2020. Untargeted UPLC-MS metabolomics reveals multiple changes of urine composition in healthy adult volunteers after consumption of *Curcuma longa* L. extract. *Food Res. Int.* 127: 108730.

Petrakis, E. A. and Polissiou, M. G. 2017. Assessing saffron (*Crocus sativus* L.) adulteration with plant-derived adulterants by diffuse reflectance infrared Fourier transform spectroscopy coupled with chemometrics. *Talanta* 162: 558–566.

Petrakis, E. A., Cagliani, L. R., Polissiou, M. G. and Consonni, R. 2015. Evaluation of saffron (Crocus sativus L.) adulteration with plant adulterants by 1H NMR metabolite fingerprinting. *Food Chem.* 173: 890–896.

Pezzatti, J., Boccard, J., Codesido, S., Gagnebin, Y., Joshi, A., Picard, D. ... and Rudaz, S. 2020. Implementation of liquid chromatography – high resolution mass spectrometry methods for untargeted metabolomic analyses of biological samples: A tutorial. *Anal. Chimica Acta* 1105: 28–44.

Phondke, G. P. 1992. *The Wealth of India, Raw Materials*, Revised Edition, Vol. VIII, NISCAIR, CSIR, New Delhi, 162.

Policegoudra, R. S. and Aradhya, M. S. 2007. Accumulation of bioactive compounds during growth and development of mango ginger (*Curcuma amada* Roxb.) rhizomes. *J. Agric. Food Chem.* 55: 8105–8111.

Prasad, S. and Tyagi, A. K. (2015). Ginger and its constituents: role in prevention and treatment of gastrointestinal cancer. *Gastroenterology Research and Practice*, 2015(11): 142979.

Rastogi, S., Shah, S., Kumar, R., Vashisth, D., Akhtar, M. Q., Kumar, A. ... and Shasany, A. K. 2019. Ocimum metabolomics in response to abiotic stresses: Cold, flood, drought and salinity. *PloS one* 14: e0210903.

Ravindran, P. N. and Kallupurackal, J. A. 2012. Black pepper. In *Handbook of Herbs and Spices* (pp. 86–115). Cambridge, UK: Woodhead Publishing.

Ravindran, P. N. and Babu, K. N. (Eds.). 2016. *Ginger: the genus Zingiber*. Boca Raton: CRC Press.

Rischer, H., Orešič, M., Seppänen-Laakso, T., Katajamaa, M., Lammertyn, F., Ardiles-Diaz, W., ... and Goossens, A. 2006. Gene-to-metabolite networks for terpenoid indole alkaloid biosynthesis in *Catharanthus roseus* cells. *Proc. Nat. Acad. Sci.* 103: 5614–5619.

Rochfort, S. 2005. Metabolomics reviewed: A new "omics" platform technology for systems biology and implications for natural products research. *J. Nat. Prod.* 68: 1813–1820.

Rolf, S., Bruns, H. J., Wichter, T., Kirchhof, P., Ribbing, M., Wasmer, K. ... and Eckardt, L. 2003. The ajmaline challenge in Brugada syndrome: Diagnostic impact, safety, and recommended protocol. *European Heart J.* 24*:* 1104–1112.

Rooney, S. and Ryan, M. F. 2005. Effects of alpha-hederin and thymoquinone, constituents of *Nigella sativa*, on human cancer cell lines. *Anticancer Res.* 25(3B): 2199–2204.

Sabatino, L., Scordino, M., Gargano, M., Belligno, A., Traulo, P. and Gagliano, G. 2011. HPLC/PDA/ESI-MS evaluation of saffron (*Crocus sativus* L.) adulteration. *Nat. Prod. Commun.* 6: 1934578X1100601220.

Safitri, A., Putri, A. S., Octavianty, T. D. and Sari, D. R. T. 2020, October). Metabolomic profiles of *Curcuma longa* L. and *Cosmos caudatus* extracts and their in-silico anti-cancer activity. *Journal of Physics: Conference Series* 1665(1): 012022.

Sagi, S., Avula, B., Wang, Y. H. and Khan, I. A. 2016. Quantification and characterization of alkaloids from roots of *Rauwolfia serpentina* using ultra-high performance liquid chromatography-photo diode array-mass spectrometry. *Anal. Bioanal. Chem.* 408: 177–190.

Sahu, B. N. 1983. *Rauvolfias: Vol. II: Chemistry and Pharmacology*. New Delhi: Today and Tomorrow's Printers and Publishers.

Santra, M., Santra, D. K., Rao, V. S., Taware, S. P. and Tamhankar, S. A. 2005. Inheritance of β-carotene concentration in durum wheat (*Triticum turgidum* L. ssp. durum). *Euphytica* 144: 215–221.

Sarfarazi, M., Jafari, S. M., Rajabzadeh, G. and Galanakis, C. M. 2020. Evaluation of microwave-assisted extraction technology for separation of bioactive components of saffron (*Crocus sativus* L.). *Industrial Crops Prod.* 145: 111978.

Sarkar, B., Kumar, M. and Mukhopadhyay, K. 2017. A rapid method for detection and characterization of anthocyanins from *Hibiscus*, *Ocimum* and *Syzygium* species and evaluation of their antioxidant potential. In *Applications of Biotechnology for Sustainable Development* (pp. 129–137). Springer, Singapore.

Sayed, A. M., Sherif, N. H., El-Gendy, A. O., Shamikh, Y. I., Ali, A. T., Attia, E. Z. ... and Abdelmohsen, U. R. 2022. Metabolomic profiling and antioxidant potential of three fungal endophytes derived from *Artemisia annua* and *Medicago sativa*. *Nat. Prod. Res.* 36: 2404–2408.

Schmeller, T. and Wink, M. 1998. Utilization of alkaloids in modern medicine. In *Alkaloids* (pp. 435–459). Springer, Boston, MA.

Schmidt, M., Betti, G. and Hensel, A. 2007. Saffron in phytotherapy: Pharmacology and clinical uses. *Wiener Medizinische Wochenschrift* 157(13): 315–319.

Scossa, F., Benina, M., Alseekh, S., Zhang, Y. and Fernie, A. R. 2018. The integration of metabolomics and next-generation sequencing data to elucidate the pathways of natural product metabolism in medicinal plants. *Planta Med.* 84: 855–873.

Scott, I. M., Gagnon, N., Lesage, L., Philogene, B. J. R. and Arnason, J. T. 2005. Efficacy of botanical insecticides from *Piper* species (Piperaceae) extracts for control of *European chafer* (Coleoptera: Scarabaeidae). *J. Eco. Entomol.* 98: 845–855.

Scott, I. M., Helson, B. V., Strunz, G. M., Finlay, H., Sánchez-Vindas, P. E., Poveda, L. ... and Arnason, J. T. 2007. Efficacy of *Piper nigrum* (Piperaceae) extract for control of insect defoliators of forest and ornamental trees. *Canadian Entomol.* 139*:* 513–522.

Sekhawat, P. S. and Prasada, R. 1971. Antifungal properties of some plant extracts II: Growth inhibition studies. *Sci. Cult.* 37: 40–41.

Senizza, B., Rocchetti, G., Ghisoni, S., Busconi, M., Pascual, M. D. L. M., Fernandez, J. A. ... and Trevisan, M. 2019. Identification of phenolic markers for saffron authenticity and origin: An untargeted metabolomics approach. *Food Res. Int.* 126: 108584.

Shahi, T., Assadpour, E. and Jafari, S. M. 2016. Main chemical compounds and pharmacological activities of stigmas and tepals of "red gold"; saffron. *Trends Food Sci. Tech.* 58: 69–78.

Shanks, J. V., Bhadra, R., Morgan, J., Rijhwani, S. and Vani, S. 1998. Quantification of metabolites in the indole alkaloid pathways of *Catharanthus roseus*: Implications for metabolic engineering. *Biotech. Bioeng.* 58: 333–338.

Sharma, Y., Bashir, S., Irshad, M., Gupta, S. D. and Dogra, T. D. 2005. Effects of acute dimethoate administration on antioxidant status of liver and brain of experimental rats. *Toxicology* 206: 49–57.

Shawky, E., El-Khair, R. M. A. and Selim, D. A. 2020. NIR spectroscopy-multivariate analysis for rapid authentication, detection and quantification of common plant adulterants in saffron (*Crocus sativus* L.) stigmas. *LWT* 122: 109032.

Shi, J., Cao, B., Wang, X. W., Aa, J. Y., Duan, J. A., Zhu, X. X. ... and Liu, C. X. 2016. Metabolomics and its application to the evaluation of the efficacy and toxicity of traditional Chinese herb medicines. *J. Chromatography B* 1026: 204–216.

Silva Lima, A., Milhomem, M. N., Santos Monteiro, O., Arruda, A. C. P., de Castro, J. A. M., Fernandes, Y. M. L. ... and Costa-Junior, L. M. 2018. Seasonal analysis and acaricidal activity of the thymol-type essential oil of *Ocimum gratissimum* and its major constituents against *Rhipicephalus microplus* (Acari: Ixodidae). *Parasitol. Res.* 117: 59–65.

Silva Lima, A., Milhomem, M. N., Santos Monteiro, O., Arruda, A. C. P., de Castro, J. A. M., Fernandes, Y. M. L., ... & Costa-Junior, L. M. 2018. Seasonal analysis and acaricidal activity of the thymol-type essential oil of Ocimum gratissimum and its major constituents against Rhipicephalus microplus (Acari: Ixodidae). *Parasitology research* 117(1), 59–65.

Singh, S. and Majumdar, D. K. 1994. Toxicological studies of the fixed oil of *Ocimum sanctum* Linn. (Tulsi). *New Bot.* 21: 139–146.

Singh, S. and Majumdar, D. K. 1996. Effect of fixed oil of Ocimum sanctum against experimentally induced arthritis and joint edema in laboratory animals. International Journal of Pharmacognosy, 34(3), 218–222.

Singh, S., Rehan, H. M. S. and Majumdar, D. K. 2001. Effect of Ocimum sanctum fixed oil on blood pressure, blood clotting time and pentobarbitone-induced sleeping time. J Ethnopharmacol. 2001; 78: 13943.

Sofowora, A. 1993. Medicinal plants and medicine in Africa. *John Willey Spectrum, Ibadan Nigeria*, 281–285.

Somberg, J., Greenfield, B. and Tepper, D. 1985. Digitalis: Historical development in clinical medicine. *J. Clin. Pharmacol.* 25(7): 484–489.

Sousa, C., Valentão, P., Ferreres, F., Seabra, R. M. and Andrade, P. B. 2008. Tronchuda cabbage (*Brassica oleracea* L. var. costata DC): Scavenger of reactive nitrogen species. *J. Agric. Food Chem.* 56: 4205–4211.

Srivastava, R., Puri, V., Srimal, R. C. and Dhawan, B. N. 1986. Effect of curcumin on platelet aggregation and vascular prostacyclin synthesis. *Arzneimittel-forschung* 36(4), 715–717.

Stöckigt, J. 1995. Biosynthesis in Rauwolfia serpentina modern aspects of an old medicinal plant. *The Alkaloids: Chemistry and Pharmacology* 47: 115–172.

Stöckigt, J., Sheludko, Y., Unger, M., Gerasimenko, I., Warzecha, H. and Stöckigt, D. 2002. High-performance liquid chromatographic, capillary electrophoretic and capillary electrophoretic – electrospray ionisation mass spectrometric analysis of selected alkaloid groups. *J. Chromatography A* 967: 85–113.

St-Pierre, B., Besseau, S., Clastre, M., Courdavault, V., Courtois, M., Creche, J., ... and Giglioli-Guivarc'h, N. 2013. Deciphering the evolution, cell biology and regulation of monoterpene indole alkaloids. *Advances in Botanical Research* 68: 73–109.

Suanarunsawat, T., Boonnak, T., Ayutthaya, W. N. and Thirawarapan, S. 2010. Anti-hyperlipidemic and cardioprotective effects of *Ocimum sanctum* L. fixed oil in rats fed a high fat diet. *J. Basic Clin. Physiol. Pharmacol.* 21(4): 387–400.

Subramanian, R., Subbramaniyan, P., Ameen, J. N. and Raj, V. 2016. Double bypasses soxhlet apparatus for extraction of piperine from Piper nigrum. *Arabian J. Chem.* 9: S537–S540.

Suk, S., Kwon, G. T., Lee, E., Jang, W. J., Yang, H., Kim, J. H., ... and Lee, K. W. 2017. Gingerenone A, a polyphenol present in ginger, suppresses obesity and adipose tissue inflammation in high-fat diet-fed mice. *Mol. Nutrition Food Res.* 61: 1700139.

Sumner, L. W. 2010. Recent advances in plant metabolomics and greener pastures. *F1000 Biol. Rep.* 2.

Sumner, L. W., Mendes, P. and Dixon, R. A. 2003. Plant metabolomics: large-scale phytochemistry in the functional genomics era. *Phytochemistry* 62: 817–836.

Sun, H., Zhang, A. H., Zou, D. X., Sun, W. J., Wu, X. H. and Wang, X. J. 2014. Metabolomics coupled with pattern recognition and pathway analysis on potential biomarkers in liver injury and hepatoprotective effects of yinchenhao. *Appl. Biochem. Biotech.* 173: 857–869.

Suthar, M. K. and Saran, P. L. 2020. Anthocyanins from Ocimum sanctum L., a promising biomolecule for development of cost-effective and widely applicable pH indicator. *3 Biotech* 10: 1–11.

Tanaka, K., Arita, M., Sakurai, H., Ono, N. and Tezuka, Y. 2015. Analysis of chemical properties of edible and medicinal ginger by metabolomics approach. *BioMed Res. Int.* 2015(Article ID 671058):1–7.

Tekeoglu, I., Dogan, A., Ediz, L., Budancamanak, M. and Demirel, A. 2007. Effects of thymoquinone (volatile oil of black cumin) on rheumatoid arthritis in rat models. *Phytotherapy Res.* 21: 895–897.

Thippeswamy, N. B. and Naidu, K. A. 2005. Antioxidant potency of cumin varieties – cumin, black cumin and bitter cumin – on antioxidant systems. *European Food Res. Tech.* 220: 472–476.

Tolstikov, V. V., Lommen, A., Nakanishi, K., Tanaka, N. and Fiehn, O. 2003. Monolithic silica-based capillary reversed-phase liquid chromatography/electrospray mass spectrometry for plant metabolomics. *Anal. Chem.* 75: 6737–6740.

Toma, C. C., Olah, N. K., Vlase, L., Mogoşan, C. and Mocan, A. 2015. Comparative studies on polyphenolic composition, antioxidant and diuretic effects of *Nigella sativa* L. (black cumin) and *Nigella damascena* L.(lady-in-a-mist) seeds. *Molecules* 20: 9560–9574.

Townsend, E. A., Siviski, M. E., Zhang, Y., Xu, C., Hoonjan, B. and Emala, C. W. 2013. Effects of ginger and its constituents on airway smooth muscle relaxation and calcium regulation. *Am. J. Respiratory Cell Mol. Biol.* 48: 157–163.

Tranchida, F., Shintu, L., Rakotoniaina, Z., Tchiakpe, L., Deyris, V., Hiol, A. and Caldarelli, S. 2015. Metabolomic and lipidomic analysis of serum samples following *Curcuma longa* extract supplementation in high-fructose and saturated fat fed rats. *PLoS One* 10: e0135948.

Tuberoso, C. I., Rosa, A., Montoro, P., Fenu, M. A. and Pizza, C. 2016. Antioxidant activity, cytotoxic activity and metabolic profiling of juices obtained from saffron (*Crocus sativus* L.) floral by-products. *Food Chem.* 199: 18–27.

Tyler, V. E. 1988. Medicinal plant research: 1953–1987. *Planta Med.* 54: 95–100.

Ulrich-Merzenich, G., Zeitler, H., Jobst, D., Panek, D., Vetter, H. and Wagner, H. 2007. Application of the "-Omic-" technologies in phytomedicine. *Phytomed.* 14: 70–82.

Vahedi, M., Kabiri, M., Salami, S. A., Rezadoost, H., Mirzaie, M. and Kanani, M. R. 2018. Quantitative HPLC-based metabolomics of some Iranian saffron (*Crocus sativus* L.) accessions. *Industrial Crops Prod.* 118: 26–29.

Vamanu, E., Gatea, F., Sârbu, I. and Pelinescu, D. 2019. An in vitro study of the influence of Curcuma longa extracts on the microbiota modulation process, in patients with hypertension. *Pharmaceu.* 11: 191.

Verma, P., Sharma, A., Khan, S. A., Shanker, K. and Mathur, A. K. 2015. Over-expression of Catharanthus roseus tryptophan decarboxylase and strictosidine synthase in rol gene integrated transgenic cell suspensions of *Vinca minor*. *Protoplasma* 252: 373–381.

Verpoorte, R., Choi, Y. H. and Kim, H. K. 2005. Ethnopharmacology and systems biology: A perfect holistic match. *J. Ethnopharmacol.* 100: 53–56.

Verpoorte, R., Choi, Y. H. and Kim, H. K. 2007. NMR-based metabolomics at work in phytochemistry. *Phytochem. Rev.* 6: 3–14.

Verpoorte, R., Choi, Y. H., Mustafa, N. R. and Kim, H. K. 2008. Metabolomics: back to basics. *Phytochem. Rev.* 7: 525–537.

Walstab, J., Krüger, D., Stark, T., Hofmann, T., Demir, I. E., Ceyhan, G. O., ... and Niesler, B. 2013. Ginger and its pungent constituents non-competitively inhibit activation of human recombinant and native 5-HT3 receptors of enteric neurons. *Neurogastroenterol. Motility* 25: 439–e302.

Wang, X., Sun, H., Zhang, A., Sun, W., Wang, P. and Wang, Z. 2011. Potential role of metabolomics apporoaches in the area of traditional Chinese medicine: as pillars of the bridge between Chinese and Western medicine. *Journal of pharmaceutical and biomedical analysis* 55(5), 859–868.

Wang, M., Chittiboyina, A. G., Parcher, J. F., Ali, Z., Ford, P., Zhao, J. ... and Khan, I. A. 2019. *Piper nigrum* oil – determination of selected terpenes for quality evaluation. *Planta Med.* 85: 185–194.

Wattanathorn, J., Chonpathompikunlert, P., Muchimapura, S., Priprem, A. and Tankamnerdthai, O. 2008. Piperine, the potential functional food for mood and cognitive disorders. *Food Chem. Toxicol.* 46: 3106–3110.

Wei, C. K., Tsai, Y. H., Korinek, M., Hung, P. H., El-Shazly, M., Cheng, Y. B. ... and Chang, F. R. 2017. 6-paradol and 6-shogaol, the pungent compounds of ginger, promote glucose utilization in adipocytes and myotubes, and 6-paradol reduces blood glucose in high-fat diet-fed mice. *Int. J. Mol. Sci.* 18: 168.

Wilson, B., Abraham, G., Manju, V. S., Mathew, M., Vimala, B., Sundaresan, S. and Nambisan, B. 2005. Antimicrobial activity of *Curcuma zedoaria* and *Curcuma malabarica* tubers. *J. Ethnopharmacol.* 99: 147–151.

Wu, F., Kerčmar, P., Zhang, C. and Stöckigt, J. 2016. Sarpagan-Ajmalan-Type Indoles: Biosynthesis, structural biology, and chemo-enzymatic significance. *Alkaloids: Chem. Biol.* 76: 1–61.

Wurtzel, E. T. and Kutchan, T. M. 2016. Plant metabolism, the diverse chemistry set of the future. *Science* 353: 1232–1236.

Xiang, X., Song, C., Shi, Q., Tian, J., Chen, C., Huang, J. ... and Jin, S. 2020. A novel predict-verify strategy for targeted metabolomics: Comparison of the curcuminoids between crude and fermented turmeric. *Food Chem.* 331: 127281.

Xu, S., Ge, X., Li, S., Guo, X., Dai, D. and Yang, T. 2019. Discrimination of different parts of saffron by metabolomic-based ultra-performance liquid chromatography coupled with high-definition mass spectrometry. *Chem. Biodiversity* 16: e1900363.

Yamamoto, K., Takahashi, K., Caputi, L., Mizuno, H., Rodriguez-Lopez, C. E., Iwasaki, T. ... and Mimura, T. 2019. The complexity of intercellular localisation of alkaloids revealed by single-cell metabolomics. *New Phytol.* 224: 848–859.

Yan, J., Chen, G., Tong, S., Feng, Y., Sheng, L. and Lou, J. 2005. Preparative isolation and purification of germacrone and curdione from the essential oil of the rhizomes of *Curcuma wenyujin* by high-speed counter-current chromatography. *J. Chromatography A* 1070: 207–210.

Yaribeygi, H., Zare, V., Butler, A. E., Barreto, G. E. and Sahebkar, A. 2019. Antidiabetic potential of saffron and its active constituents. *J. Cell. Physiol.* 234: 8610–8617.

Yeh, H. Y., Chuang, C. H., Chen, H. C., Wan, C. J., Chen, T. L. and Lin, L. Y. 2014. Bioactive components analysis of two various gingers (*Zingiber officinale* Roscoe) and antioxidant effect of ginger extracts. *Food Sci. Tech.* 55: 329–334.

Yu, J., Ma, Y., Drisko, J. and Chen, Q. 2013. Antitumor activities of Rauwolfia vomitoria extract and potentiation of carboplatin effects against ovarian cancer. *Curr. Therapeutic Res.* 75: 8–14.

Zeka, K., Ruparelia, K. C., Continenza, M. A., Stagos, D., Vegliò, F. and Arroo, R. R. 2015. Petals of Crocus sativus L. as a potential source of the antioxidants crocin and kaempferol. *Fitoterapia*, 107, 128–134.

Zhang, A., Sun, H., Dou, S., Sun, W., Wu, X., Wang, P. and Wang, X. 2013. Metabolomics study on the hepatoprotective effect of scoparone using ultra-performance liquid chromatography/electrospray ionization quadruple time-of-flight mass spectrometry. *Analyst* 138: 353–361.

Zirihi, G. N., Mambu, L., Guédé-Guina, F., Bodo, B. and Grellier, P. 2005. In vitro antiplasmodial activity and cytotoxicity of 33 West African plants used for treatment of malaria. *J. Ethnopharmacol.* 98: 281–285.

11 Mutagenesis in Medicinal Plants

Ambreen Asif

CONTENTS

11.1 INTRODUCTION

Genetic variability is an asset of domestication of crops. It is also used for crop improvement and genetic research. Until the 20th century, spontaneous mutation was the source of novel genetic variability, which is used in selecting plants and animals for breeding and domestication purposes. With the discovery of ionizing radiation as a source of modification of the genetic make-up of organisms, a leap in plant breeding programs came. In the late 1920s, Stadler started the work of plant mutation breeding. The past decade has witnessed the rapid emergence of new tools for plant breeders. Induced mutagenesis offers a valuable tool for plant breeders for the production of useful mutations in crop plants and to induce desirable mutated alleles, which were previously absent in the existing germplasm. The process of mutation and various mutational experiments can be utilized to overcome the limits of variability and to produce variability in a brief time period (Akgun and Tosun 2004). Plant mutagenesis has proven a practical tool for increasing genetic variability. With the fast-accruing knowledge of plant mutagenesis and with the new tools, plant mutagenesis is now

DOI: 10.1201/9781003179139-11

fully capitalizing on advances in molecular and biotechnologies. New species are emerging as being important in agriculture, but also to the environment, medicine and energy production. There was a resurgence in mutation technologies in the early years of the 21st century. This is due to abundant familiarity to mutagenesis and related disciplines, leading to its further application. On account of the rapid understanding of molecular mutagenesis, chance-based mutation induction has been transformed successfully into science-based techniques. The use of molecular and genomic tools for mutant screening and characterization also enabled mutation breeding to embrace and utilize the very recent findings and technological innovations in plant genomics and molecular biology research.

11.2 MUTAGENS AS SOURCE OF MUTAGENESIS IN PLANTS

From 1950s onwards, mutagenesis became very popular as a breeding tool for increasing trait variation of crops and ornamental plants through the treatment of irradiation (Leitao 2012). Not only was mutagenesis used for breeding, but also for analysis of nature and gene functioning as the genes are the basis of the growth and development of plants. Identified genes have become the source for genetic improvement of economic crops. Irradiation (X-rays, gamma rays, ion beam, etc.) treatment with chemical mutagens, site-directed mutagenesis and insertion mutagenesis are used for creating mutations (Kharkwal and Shu 2009, Forster and Shu 2012). In induced mutagenesis, mutagens play a pivotal role for creating novel changes and for improving the traits. It has been suggested that mutagens may induce genetic modifications in organisms, linkage break and introduce new characters for crop improvement (Shah et al. 2008). Mutagens have extraordinary potential for crop improvement for their qualitative and quantitative traits like height, growth, seed size, maturity time, improvement in yield, and so on (Asif et al. 2016, Asif and Ansari 2019) and several other characters have been found (Sigurbjornson and Micke 1969).

11.2.1 Chemical Mutagens for Mutagenesis in Plants

The discovery of chemical mutagens has revolutionized the plant breeding programs. Chemical agents are useful because of high mutation rates, especially point mutations (Jain 2005). Rapoport (1966) discovered most of the strong chemical mutagens that are broadly utilized in research in the field of genetic and breeding. Three groups of chemical mutagens are available, which are alkylating agents, base analogues and others. Alkylating agents are ethyl methane sulphonate (EMS), ethyl ethane sulphonate (EES), methyl methane sulphonate (MMS), diethyl sulphate (DES), ethyl imines (IE) and nitroso compounds. Base analogues are 5-bromo uracil and 5-bromo deoxyuridine. Azides (sodium azide), acridines, nitrous acid and hydroxylamine are in another category. Alkylating agents have been described to be more active than the other chemical mutagens (Swaminathan 1966). Alkylating agents are categorized into monofunctional and bifunctional or polyfunctional on the basis of the number of alkyl groups present in the compound (Natarajan 2005). In alkylation, an alkyl group (e.g., C_2H_5 of EMS) is substituted for hydrogen in the nitrogenous bases (Sharma and Chopra 1994). As per the work of Siddiqui (1999) and Natarajan (2005) DNA alkylation causes the alkylation of phosphate group and base of DNA and depurination. The mutagenic action of different chemicals has been studied in various crop plants and since it is observed that these chemicals are efficient in inducing wide spectrum of useful mutations, a number of mutagens have been used by several workers in crop plants to induce genetic variability (Shah et al. 2008, Mostafa 2011).

EMS is a known potent mutagen, in comparison to other chemical mutagens. The Environmental Mutagen Information Center cited more than 3400 references in which EMS was taken as a source of generating mutations. The EMS is now an established very powerful and effective mutagen (Hajra 1979), due to its capability to produce a wide spectrum and high frequency of mutations through induced mutagenesis approach (Hohmann, Jacobs and Jung 2005). Fahmy and Fahmy (1957) studied earlier the mutagenic effect of EMS in *Drosophila*. Effect of EMS mutagenesis has also been studied in various crop plants including barley and wheat (Swaminathan, Chopra and Bhaskaran 1962), barley (Khalatkar, Gopal-Ayengar and Bhatia 1977), tomato (Saba and Mirza 2002), *Vigna* spp. (Kozgar, Goyal and Khan 2011).

11.2.2 Physical Mutagens for Mutagenesis in Plants

Physical mutagens or radiations (ionizing and non- ionizing) have also been frequently used for inducing mutations. Radiations are high-energy levels of electromagnetic spectrum that are able to dislodge electrons from the orbits to the nucleus of the impacted atoms, and thereby transform the atoms to ions. Therefore, the term ionizing radiation is given. Cosmic rays, gamma rays and X-rays are the ionizing components of the electromagnetic spectrum (Mba 2013). Further advancement in the field of induced mutagenesis for causing genetic variations in plants was seen after the discovery of the X-rays mutagenic action, and with its demonstration in *Zea mays*, *Hordeum* and *Triticum* by Stadler (1930). In the past 80 years, hereditary aberrations have been induced through physical mutagens as over 70% of mutant varieties were established by means of these mutagens (Mba 2013, Wani et al. 2014). Various other particles like protons, neutrons, alpha and beta particles have also been produced by nuclear reactors (Acquaah 2007). Gamma (γ) rays are commonly generated from radioactive ^{60}Co (Cobalt-60) as these are of shorter wavelength and having more energy than proton. X-rays penetrate deeper into the tissues (Amano 2006). Whole plants and delicate materials are irradiated by these radiations. Ionizing radiations usually cause chromosomal rearrangements and deletions whereas chemicals induce mainly point mutations. Chemical mutagens are found to be more potent over physical ones (Bhat, Khan and Parveen 2005a). Therefore, chemical mutagenesis has become popular for genetic studies, even with the advent of new technologies (Greene et al. 2003). In contrast, Jain (2005) stated that use of physical mutagens is more advantageous due to accurate dosimetry, sufficient reproducibility and accuracy. Gamma rays are particularly used as these possess uniform and high penetrating power in the tissues of the plant than the chemical mutagens (Yamaguchi et al. 2003). According to Jain (2005), developing direct mutant varieties through radiation-induced mutagenesis is the most common method, as evident from the development of 90% varieties by irradiation (64% with gamma-rays, 22% with X-rays). Among the physical mutagens, gamma rays stand at the top in efficiency for mutation induction. They interact with atoms or molecules to produce free radicals which act directly on the DNA (Kovacs and Keresztes 2002). During early 1950s the effect of various physical mutagens on cytogenetics was evaluated; studies and comparisons in numerous plants have also been carried out in *V. rosea* (Sudhakaran 1971), Sorghum (Ramulu 1971), *V. faba* (Amer and Mikheal 1972), *O. sativa* (Siddiq 1973) and barley (Srivastava 1974).

11.2.3 Combination of Physical and Chemical Mutagens for Mutagenesis in Plants

In order to increase the mutation frequencies in crop plants, a combination of both the chemical and physical mutagens was also used by the breeders. A wider spectrum of mutation is induced by using combinations of physical and chemical mutagens (Mehandjiev, Kosturkova and Mihov 2001). It has been further suggested that both the quantitative and qualitative variations in the mutation spectrum are enhanced with the combination of chemical and physical mutagens, thus the frequency of the particular mutation types is increased or decreased. Effects of gamma rays, EMS and DES singly and in combination upon seedlings of three rice varieties have been discussed by Bhan and Kaul (1976). The induced mutagenic effect of gamma rays and sodium azide singly and in combination have also been described by Raina et al. (2020) in *Vigna unguiculata* to induce the genetic variability. Goyal et al. (2021) have also revealed a wide extent of genetic variability in *Vigna mungo* through individual and combination treatments of gamma rays and EMS. Gamma rays and EMS in their different single and combination doses have also been utilized to induce wide genetic variability in two Sorghum genotypes (Kalpande et al. 2022). Many reports have evidenced the advantage of combined doses of physical and chemical mutagens over a single dose for improving crop varieties in lentil (Reddy and Annadurai 1992), tomato (Zeerak 1992), soybean (Khan and Tyagi 2009), blackgram (Goyal and Khan 2010) and rapeseed (Siddiqui, Khan and Khatri 2009).

Several attributes in the context of development of different types of mutants after combination treatment have been put forward. Gecheva (1983) suggested that combination treatment of gamma rays and EMS resulted into the early maturing mutant variety "Boriana" at M_2 generation level in soybean which has a 30 days lesser vegetation period than parents. Combined treatment of EMS and gamma rays has also developed a highly productive and resistant to lodging mutant variety of soybean "Biser" (Mehandjiev 2005). Privalov (1986) reported high yielding mutant with increased medicinal oil content in *Hippophea rhamnoides* by the treatment in combination of gamma rays (150 Gy) and Nitroso methyl urea (0.01%). Seeds after treating with EMS (0.1% and 0.2%) for 2 h and then SA treatment for 1 h have developed a semi-dwarf mutant of *Avena sativa* (Konzak 1993). A mutant in rice with different genetic lesions has been observed by Wu et al. (2005) by using various physical and chemical mutagens. Various grain shape mutants in rice have also been recovered by Reddy and Reddy (1973) after treatment with EMS and DES.

11.3 CHARACTER-SPECIFIC MUTANTS

11.3.1 Stem Mutants

Bifurcation, trifurcation, twisting, unbranched and twining nature of stem in M_1 generation of *Nigella sativa* due to X-ray irradiations and EMS treatments were reported by Datta and Biswas (1985). On the other hand, mutations in the stem for height have also been shown by many workers. Gnanamurthy and Dhanavel (2014) obtained viable morphological mutants in cowpea. Tall mutants showed increased height with more branches and clusters at 20 mM of EMS. Contrastingly, dwarf mutants showed reduced height with lesser branches and clusters at 30 mM of EMS. Dwarf mutants of lentil and blackgram have also been described (Arulbalachandran and Mullainathan 2009). The stimulatory effect of lower doses of caffeine on growth and seedling height was shown in *Helianthus annuus* L. by Khursheed, Ansari and Shahab (2009). Gaur and Gaur (1999) developed a fasciated mutant in *Crcer arietinum* which has been characterized by its broadened and flattened stem with irregular leaf arrangement and clustering of pods at the tip of stem. EMS induced stem mutation in the form of fasciation has also been reported in sunflower (Lyakh, Soroka and Vasin 2005). The EMS and gamma rays treatments also produced similar mutation in *Cicer arietinum* var. ILC-3279 (Haq et al. 1994). A male sterile mutant has been isolated from M_2 generation in *Cajanus cajan*, which possessed a slender stem in comparison to normal plants (Narasimha and Bhalla 1988), whereas giant bushy mutants isolated by Khan, Ansari and Gupta (2012) in 0.8% 6-AP possessed long leafy branches and thick stem. Narrow-leaf mutant was developed in *Vigna mungo* at 0.02% sodium azide which had a stout stem (Goyal and Khan 2010). Prostrate habit, dwarfing, thick and glabrous stems have been noticed in chickpea (*Cicer arietinum*) following EMS treatments (Van Rheenen, Pundir and Miranda 1993). A wide spectrum of macromutations affecting all plant parts in chickpea have been reported by Laskar et al. (2015) following physical and chemical mutagenesis. In terms of stem mutations, they reported dwarf mutant, bushy habit, bushy compact mutant with reduced internode length, increased branching, dark pigmentation of stem, simple single stem with no branches and no secondary branching with unique primary branching pattern. Asif and Ansari (2019) reported significant changes in the height of *Nigella sativa* that was treated with chemical mutagen (EMS) and physical mutagen (γ-rays) (Figure 11.1).

FIGURE 11.1 Variations in the stem height of mutants of Nigella sativa L.

11.3.2 Leaf Mutants

Induced mutations-generated leaf variants have been reported in chickpea (Salimath et al. 2007, Toker et al. 2012). Fawole (2001) reported altered leaflet shape and arrangements in *Vigna unguiculata*. Combination treatment of 1% EMS + 2% DMSO for 6 h displayed a significant increase in total leaf area of plant (20.31%), with the increase in number of leaflets in *Nigella sativa* L. (Tantray et al. 2017). Khan, Parveen and Goyal (2011) isolated morphological mutants in *Cicer arietinum* L. where a mutant contained narrow leaves with pointed tips. Similarly, a leaf mutant in *Sesamum indicum* L. has also been induced (Sengupta and Datta 2005). A number of mutants including some of leaf mutations like feathery, crumpled, needle and crinkle leaves were developed in *Nigella* (Datta and Biswas 1985). Lobed lamina, altered orientation and several other morphological mutants have been induced by base purine analog in *Vigna radiata* (Mahna and Garg 1994). Bifurcated/bilobed leaves have been observed in high yielding mutant of lentil using 0.5% dose of caffeine (Shahwar et al. 2017). Simple leaf, multipinnate leaf and broad leaflets mutants were also reported (Toker and Cagirgan 2004). Gamma rays and EMS-induced tetrafoliate and pentafoliate mutants of *Vigna radiata* have been isolated by Singh and Sharma (1993). Various leaf sizes and shapes including tetrafoliate, pentafoliate crumpled, ovate, lanceolate and lobed leaves have been observed in mungbean after gamma ray irradiation (Tah 2008). Fast neutron irradiation in *Capsicum* has also caused leaf abnormalities which essentially affected productivity negatively (Daudu and Falusi 2011). The leaf mutant with pentafoliate leaves was developed using EMS in *Vigna unguiculata* (Gnanamurthy and Dhanavel 2014). Induced mutagenesis with EMS and gamma rays have also resulted in various leaf mutants namely unifoliate, bifoliate, tetrafoliate, wrinkled, narrow rugose, biforked and triforked leaves in *Lablab purpureus* (Monica and Seetharaman 2017). Wani (2011) isolated various leaf shape mutants from M_2 generation of *Cicer arietinum* following mutagenic treatment of EMS, gamma rays and their combinations. Gamma rays generated a large number of leaf mutations including narrow-rugose, lanceolate, multiple, round-cuneate, unifoliate and wrinkled type in mungbean (Sangsiri, Sorajjapinun and Srinives 2005). A chickpea mutant due to gamma irradiation with enlarged leaves, pods, seeds and flowers has been reported by Davis, Foster and Phillips (1985). In *Nigella sativa*, a male sterile mutant was isolated possessing dark green, thick and leather-like pinnae of the leaves (Datta and Biswas 1984). Some mutants of leaf color variation have been isolated in black cumin following the

FIGURE 11.2 Leaf colours and shapes of mutants of Nigella sativa L.

treatment of EMS and gamma rays (Asif and Ansari 2019) (Figure 11.2). Similarly, a male sterile mutant with lax pinnae along with yellowish green pinnae was isolated by Rang and Datta (1998). EMS treated chilli have shown different leaf mutations like small-sized, spathulate, obcordate, single and double notched (Jabeen and Mirza 2004). In chickpeas various leaf mutations using gamma rays and hydrazine hydrates have been induced. These mutants have broad leaflets with deep serrations, opposite leaflets on long rachis, low-pigmented leaf, reduced rachis with small leaves and overlapping ovate leaflets (Laskar et al. 2015). Various leaf derivatives were observed in chickpea (Rao and Laxmi 1980, Toker and Ceylan 2013). Male sterile mutants containing long and narrow leaves have been reported in pigeon pea using EMS in M_2 generation (Narasimha and Bhalla 1988). Van Rheenen, Pundir and Miranda (1993) noticed different mutations after EMS treatment in *Cicer arietinum*. A six-hour treatment of 0.4% EMS induced variegated leaf mutant (JGM-3) in chickpea (Gaur et al. 2004).

11.3.3 Chlorophyll Mutants

In basic and applied research, chlorophyll mutations have been given an immense attention as the chloromutations have been used as genetic markers and mainly used as dependable measure for the evaluation of genetic effect of mutagenic treatment in several crops (Gustaffson 1947). Chlorophyll mutations are useful in finding threshold level of a mutagen for increasing the genetic variability.

In several crops, chlorophyll mutations namely xantha, albina, striata, viridis, and so on have been induced using chemical and physical mutagens (Das and Kundagrami 2000). Many other researchers (Makeen et al. 2013, Singh et al. 2006, Singh et al. 2007) have also reported chlorophyll mutations in several crops. Four kinds of chlorophyll mutants (striata, chlorina, xantha, albino) in M_2 generation of *Allium cepa* L. by gamma radiation and sodium azide individually and in combinations have been reported by Kirtane (2014). Srinivas and Veerabadhiran (2010) have induced four kinds of chlorophyll mutants (albina, chlorine, xantha and viridis) in Lablab by the treatments of gamma rays, sodium azide and combinations of both. Chlorophyll mutations as induced by combination treatments of mutagens have also been described in black gram (Lal, Toms and Smith 2009a) and urd bean (Makeen et al. 2013). Bhattacharya (2003) reported a wider spectrum of chlorophyll mutants in *Dianthus caryophyllus* L. with high frequency by using EMS. More chlorophyll mutations in EMS treated plants than gamma irradiated ones in M_2 generation of cowpea were observed by Girija and Dhanavel (2009). Similar observations were also reported by Solanki (2005) in lentil. Several types of chlorophyll mutants have been produced in rice using EMS and MMS (Reddi and Suneetha 1992). Chemical mutagens were observed as more effective for inducing chloromutations (Kharakwal 2000). On the other hand, Geetha and Vaidyanathan (2000) reported in two cultivars of *Glycine max* that physical mutagen was more effective than chemical mutagens in generating chlorophyll mutations.

Several types of morphological and chlorophyll mutants in *Pennisetum typhoides* were successfully observed in M_2 generation after mutagenic treatment of gamma rays and EMS (Ambli and Mullainathan 2015). Mahna and Singh (1982) observed different types of chlorophyll mutations by gamma rays in M_2 generation of *Physalis ixocarpa*. Jain, Ranwah and Bordia (2005) displayed dose-dependent chlorophyll mutations by gamma rays. Five categories of chlorophyll mutants from their highest to lowest frequency of occurrence (Viridis > Xantha > Albina > Albo-Viridis> Chlorina) were isolated in M_3 generation of rice and bean with gamma rays (150, 300, 450 and 600 Gy) treatment (Devi et al. 2002). Frequency of chlorophyll mutation enhanced with increasing doses up to 450 Gy and then decreased at 600 Gy. However, the most effective dose in generating chlorophyll mutation was 450 Gy. Rate of chlorophyll mutations in *Setaria italica* was influenced by dry and wet irradiated conditions (Ichitani et al. 2003). The trends of increasing chlorophyll mutation frequency with increased concentrations of mutagens have also been recorded by different workers in several crops (Reddy and Annadurai 1992, Mahamune and Kothekar 2012).

11.3.4 Flower Mutants

Nigella sativa generally has pentamerous flowers , however, hexapetalous mutant at 100 Gy gamma rays was observed in M_1 generation (Kumar and Gupta 2007). Similarly, several interesting floral variations in black cumin were found to occur in X-ray and EMS treatments. Some of the alterations included adnation of sepals and petals, triforked and biforked sepals, shield shaped sepals, smaller sized sepals, elongated and strap shaped petals, two gynoecia in the same flower and presence of bract like structures (incompletely forked) similar to that of the petalloid sepals (Datta and Biswas 1985). In colchicine treated population of *Impatiens balsamina* L. Raut and Thombre (1977) obtained autotetraploids, which possessed bigger flowers and longer blooming period.

Mutants with change in color like white, blue and pink colored flowers in EMS treatment have been reported by Girija, Dhanavel and Gnanamurthy (2013a) in *Vigna unguiculata* L. Walp. Flower mutants having pink and light pink colors were observed in cowpea at 25 mM a 20 mM of EMS treatments, respectively (Gnanamurthy and Dhanavel 2014). Flower color modification (white flowered mutant) in *Spathoglottis plicata* Blume, a terrestrial orchid was reported using sodium azide (Roy and Biswas 2005). Laskar et al. (2015) observed light/dark pink, blue and white colored flowers in M_2 population of chickpea by the treatments of mutagens. EMS, gamma rays and their combination successfully induced a broad range of flower color mutants (white, purple, pink and yellow against blue in control) in M_2, M_3 and M_4 generations of *Lathyrus sativus* (Ramezani et al. 2017). They recorded maximum flower color frequency in EMS (20 mM) in 4th generation and lowest in gamma rays (100 Gy) in second generation. Induced flower color mutation has also been recorded in *Vigna unguiculata* (Oluwatosin 2001) and *Lathyrus sativus* (Tripathy, Lenka and Ranjan 2011).

Different flower mutants induced by caffeine in *Capsicum annuum* L. were investigated by Aslam et al. (2017). They observed a broad spectrum of flower oddities, that is, tetramerous, fused petals, hexamerous, heptamerous, flowers with different anther colors, presence and absence of petals, sepals, anthers and so on. Extra standard petal and extra wing petal in terms of flower mutations have been developed in *Lablab purpureus* (Monica and Seetharaman 2017). Zeerak (1998) suggested change in carpel number and morphology of *Solanum melongena*. In chickpea, mutants having double flower, white flower and absence of flower were reported after mutagenic treatments (Wani 2011).

Gamma ray irradiation of 20 kR in *Vigna mungo* displayed mutant plants from M_2 progeny showing flowers similar to control but the upper part of the style and stigma was curved (Singh 1996).) Gamma ray treatment at 500 Gy induced a flower mutant as cock's comb raceme with pollen sterility in mungbean (Sangsiri, Sorajjapinun and Srinives 2005). Gamma ray treatment followed by EMS in cotton lead to male sterile mutants in M_2. The mutant displayed smaller flowers with rudimentary, indehiscent anthers, short filaments and characteristic feature of curved stigma (Badigannavar, Katageri and Khadi 2003). Significant reduction in flower characters like petal size and number, anther size, non-dehiscent anthers was noticed in *Gossypium* (Mehetre and Patil 2004). Geddam et al. (2013) also observed significant reduction in different flower morphological traits. These traits are lengths of flower pedicel, staminal column, style and filament in sterile genotypes compared to their fertile ones. Altered flower arrangements from single to triple flowers at each peduncle were reported in *Lens culinaris* L. by chemical mutagenesis, which was against two flowers per peduncle in the control plant (Amin, Laskar and Khan 2015). They also noticed reduced flower numbers in high doses of mutagen.

11.3.5 Seed Mutants

Diverse seed peculiarities, that is, altered shape, size and coat color were displayed by Mahamune and Kothekar (2011). They obtained bold and wrinkled seeds with shiny, dark red and cream seed coat in M_2 generation of *Phaseolus vulgaris* L. after gamma ray treatment. EMS-induced seed coat color mutants in French bean have been found by Barbosa, de Souza Carneiro and Vieira (1988). Chemical mutagen (EMS) has been successfully employed to generate early maturing mutants of *Trigonella* plants with high seed yield and quality (Basu, Acharya and Thomas 2008). Brown, white, bold size seeds and several other mutations have been informed as viable mutants in cowpea using EMS (Gnanamurthy and Dhanavel 2014).

Gamma rays have been efficiently used in beans to induce alterations in flower color and seed characteristics (Al-Rubeai 1982). Mutants with high seed yield have been obtained in chickpea (Wani and Anis 2008). Singh and Raghuvanshi (1991) have obtained bold seeded mutant in blackgram by gamma ray treatment with slightly increased protein content over control. Gamma ray (20 kR) treatment has also induced a bold seeded mutant in M_2 progeny of *Vigna mungo* (Singh 1996). Bold seeded mutants have also been developed earlier in other legumes (Gottschalk and Wolff 1983). The various types of morphological mutants including plant height, leaf, pod and seed mutants have been reported by Wani (2011). These mutants were isolated from individual and combined doses of gamma rays and EMS in chickpea. Gamma rays and EMS subjected seed mutants (dark brown, light brown, white, bold and wrinkled) in *Vigna unguiculata* L. were induced by Girija, Dhanavel and Gnanamurthy (2013a). Moh (1972) reported that mutagenesis with EMS and gamma rays in *Phaseolus vulgaris* resulted in some seed coat color mutants varied from white, yellow to various degrees of brown. The seed coat color was associated with a change in hypocotyl color from red to green by employing special screening technique for the isolation of potential mutants at a very early stage of seedling development. Rang and Datta (2001) spotted dark reddish brown, yellowish brown and peach color seeded plants in M_2, by different treatments of gamma irradiations and EMS in *Nigella sativa*. Alka and Ansari (2013a) have induced polyploid mutant by the combined action of

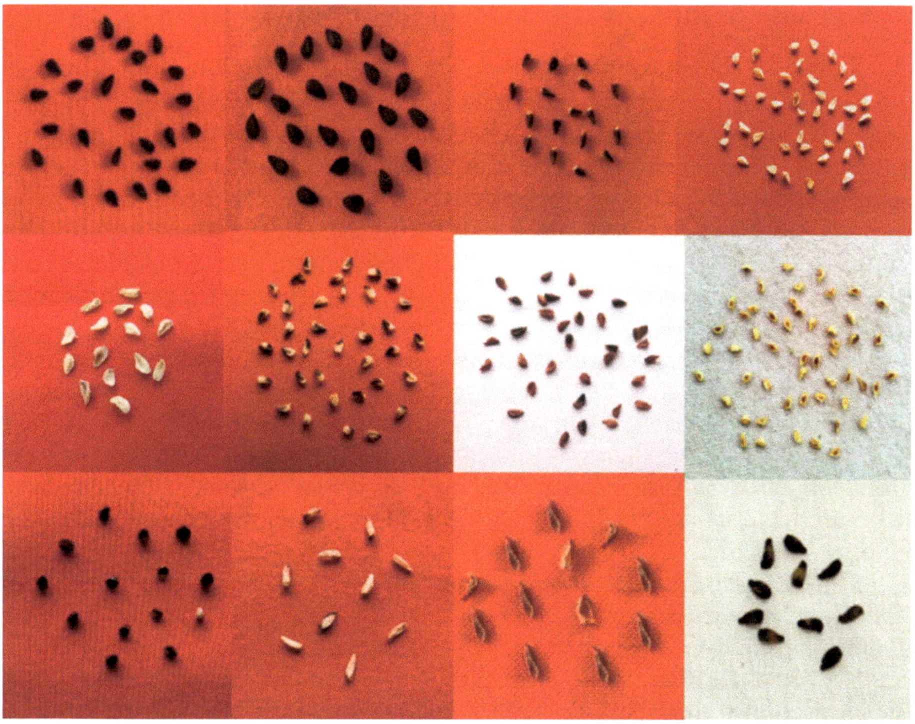

FIGURE 11.3 Seed mutants of Nigella sativa L.

physical and chemical mutagen (5 kR gamma rays + 0.10% sodium azide) in *Linum usitatissimum* L. The mutant showed different seed characters like light brown, seed tip dark brown and beaked, broad, thick, bigger size and shining surface. Girhe and Choudhary (2002) isolated mutants with dark coat color, light color and red seeds in sweet pea. Asif and Ansari (2019) have also reported various seed mutants of *Nigella sativa*, following the treatment of chemical and physical mutagens (Figure 11.3).

Early maturing mutant at 0.25% caffeine was induced by Asif et al. (2016) in black cumin. The mutant showed different altered quantitative and qualitative characters and possessed bigger sized capsules and seeds in comparison to control. The 6-AP induced mutant in flaxseed was observed to have big sized capsules and seeds in comparison to control and seed color varied from brown to dark brown in mutant (Ahmad et al. 2017).

11.4 ASSESSMENT OF MUTANTS INDUCED BY PHYSICAL AND CHEMICAL MUTAGENS

The different mutagens effect can be accessed through variations in different biological parameters in M_1 generation (Edwin and Reddy 1993). These variations at M_1 generation level can be utilized as a measure of efficiency of mutagen treatment (Plesnik 1993), though are less significant for procuring stable mutations in gene. These M_1 parameters may be informative in deciding the mutagenic doses or rejection of less mutagenized populations with less efforts and minimum time in any plant breeding program (Singh and Singh 2003). Mutagenic treatments have been reported to affect different biological and morphological parameters including seed germination, plant survival, pollen fertility, plant height, yield and other morphological features.

11.4.1 Assessment of Mutants Through Biological Parameters

11.4.1.1 Biological Damage and Germination

Biological damage may be used as an important index for determining sensitivity of germinating seeds to mutagens and it is an important aspect in mutation breeding. Mutagen-induced plant injury and lethality in M_1 generation as a result of physiological, chromosomal and factor mutations have been reported by Singh and Chaturvedi (1980). Biological damage in terms of seed germination, seedling height and pollen fertility in *Lens culinaris* by means of chemical mutagens have also been reported by Wani (2021). Two inbreds of *Vigna radiata* were treated with different mutagens (EMS, NMU and gamma-rays). EMS was found to be most effective for germination, NMU for pollen fertility and seedling height in M_1 generation and for chlorophyll mutations in M_2 generation (Singh and Chaturvedi 1980). Lower doses of mutagens have a promoting effect on germination in several crops (Kumar and Dubey 1998). However, the effect of mutagens on germination mostly showed an inhibitory effect on variety of crop plants. Effect of gamma rays (10kR-15kR) and EMS (0.15–0.90%) on *Vigna unguiculata* was studied by Manju, Mercy and Nair (1983). Gamma rays had a stimulatory effect on germination, while EMS decreased the same. General reduction was observed in survival, plant height and pollen fertility. The toxic consequences of mutagens on germination of seeds has been observed previously in many plants such as *Nigella sativa* (Mitra and Bhowmik 1998), *Plantago ovata* (Lal and Sharma 2000), *Trigonella foenum graecum* (Siddiqui et al. 2008). Variation in the percentage of germination due to the effect of sodium azide in *Nigella sativa*, *Trigonella foenum-graecum* and *Plantago ovata* L. has also been reported by Prabha, Dixit and Chaudhary (2011). A dose-dependent change in seed germination and pollen fertility due to mutagenic impact of two alkylating agents, EMS and MMS was studied by Jahan, Ansari and Asif (2016) in *Lycopersicon esculentum* Mill.

Reduced germination percentage by chemical mutagens has been discussed in *Vigna mungo* L. (Sagade and Apparao 2011), *Solanum melongena* L. (Shahab, Jaiswal and Khan 2007), *Zea mays* L. (Gnanamurthy et al. 2012), *Moluccella laevis* (Minisi et al. 2013), *Psoralea corylifolia* (Bhat et al. 2015) and *Nigella sativa* (Kumbhakar et al. 2016). Many researchers stated reduced germination percentage with the increasing dose of gamma radiations in *Secale montanum* (Akgun and Tosun 2004), *Pinus* (Thapa 2004), *Psoralea corylifolia* (Jan, Parween and Siddiqi 2012). Mokobia and Anomohanran (2005) in *Arachis hypogaea*, Minisi et al. (2013) in *Moluccella laevis* and Bhosle and More (2014) in *Withania somnifera* have discussed reduced germination percentage and growth rate with increasing doses of gamma rays. When seeds of rice varieties were treated with gamma rays and ethyl methane sulfonate, the germination pattern following gamma ray and EMS treatment showed gradual reduction (Prakash and Shambhulingappa 2000). Increasing levels of gamma rays and EMS generally decreased the germination percentage in *Abelmoschus esculentus* (Kumar and Mishra 2004). Seeds of mustard, when treated with five mutagens (gamma-rays, ethyl methane sulfonate, ethidium bromide, acriflavine and malic hydrazide), showed the increased damaging effect on germination with their increasing doses (Chauhan, Kumar and Krishna 2001). In the laboratory germination test, it has been reported in *Jatropha curcas* that increasing levels of gamma rays and EMS had a negative effect (Dhakshanamoorthy, Selvaraj and Chidambaram 2010). Similarly, gamma irradiation and EMS have inhibited the germination percentage in two soybean varieties (Co1and Co2) (Karthika and Lakshmi 2006). Among all treatments in M_1 generation, Jabeen and Mirza (2002) have shown lowest germination percentage (84%) in *Capsicum annuum* after treating the seeds with 1.5% EMS. Colchicine (0.2 %) pretreated seeds of *Phaseolus aureus* Roxb. showed decreased germination with increasing dose of gamma radiations (Raghuvanshi, Pathak and Singh 1978). Irradiation with gamma rays in *Phaseolus vulgaris* showed stimulatory effect on germination of seed, height of seedling and primary leaves length (Mujeeb and Greig 1976). Doses of gamma rays (3 krad and 5 krad) increased the percentage germination in *Sorghum vulgare* (Sharon and Muralidharan 1978). Mutagens in combination may have adverse effect on seed germination in comparison to their single or individual treatments. Chopra and Singh (1978) studied the effect of gamma rays, 2, 4-D and their combined treatment in *Guizotia abyssinica* and reported that combined treatments have a deleterious effect on germination as compared to individual ones. The negative effect of combination of both chemical and physical mutagens on germination and survival in *Cicer arietinum* has also been reported by Bhatnagar (1984). Mutagenic sensitivity of single doses of gamma rays, sodium azide and combinations of both in different doses in M_1 generation was studied by Lal, Toms and Lal (2009) in black gram and concluded that increased level of azide doses decreased the seed germination at M_1 generation. However, a radio-protective effect of EMS in *Abelmoschus esculentus* was investigated by Gupta and Yashvir (1975). Higher germination percentage was found in combinations of gamma rays and EMS treatments, compared to corresponding EMS treatments.

11.4.1.2 Plant Survival, Pollen Fertility and Sterility

Pollen fertility is significantly correlated with meiotic behavior of chromosomes as the fertile pollen grains indicate the normal chromosomal behavior. A significant inverse correlation of pollen fertility with chromosomal irregularities has been recorded by various workers (Kumari et al. 2011, Choudhary et al. 2012). Investigation of the MMS effect on pollen fertility in urdbean was carried out by Khan and Siddiqui (1987). However, Maldhure (1990) has correlated pollen sterility with chromosomal irregularities in urdbean. Due to the effect of mutagen at molecular level in *Gloriosa superba* the decrease in pollen fertility led to a rise in chromosomal anomalies generated at meiosis (Chandra 1988). Plant survival and pollen fertility generally decreased, when the doses of mutagens were increased. The increasing doses of gamma rays (100, 200, 300 and 400 Gy) and EMS (0.1, 0.2, 0.3 and 0.4 %) decreased the seed germination, survival of plant and fertility of pollen in two cultivars of *Vigna mungo* (Sharma, Yelne and Dennis 2005). The effect of gamma rays and EMS was also studied earlier by Prem Sekar and Appadurai (1982). They observed that higher doses gave less pollen and seed fertility in *Cajanus cajan*. Size and shape of pollen grains in the irradiated and control pollens were not significantly different in *Gossypium hirsutum* L. (Savaskan 2002). A direct relationship of pollen and ovule sterility has been observed in *Vigna mungo* at higher doses of gamma rays and EMS. Mutagenicity of gamma-rays and EMS and combinations of both was also studied by Vadivelu and Rathinam (1980) in *Cicer arietinum* L., where it has been observed that EMS produced greater reduction than gamma rays. With increasing concentration of mutagens, the germination percentage, survival, seedling growth, dry matter production and seed fertility decreased. Due to the effect of caffeine, a dose-dependent decrease in survival of plant and pollen fertility has been described in chilli by Aslam et al. (2017). The effect of gamma-radiations in M_1, M_2 and M_3 generations was studied on three species of *Vigna*. Germination of seeds, survival of seedlings and growth rate decreased when level of dose increased (Subramanian 1980). EMS and MMS treatments in *Vigna sinensis* resulted in reduced survival percentage, average pod number, seed yield per plant and average pollen fertility in M_1 and M_2 generations. The reduction was less in M_2 (Narsinghani and Kumar 1976). Seeds of sweet pepper were treated with several chemical mutagens. In M_1 generation, a decreased germination and survival rate with the increased mutagenic treatments was noticed by Skripnikova (1976). Oommen and Gopimany (1986) reported the reduced survival of plants, growth of plants and pollen fertility in M_1 generation of cowpea after irradiating the seeds with six treatments of gamma rays (5–30 kR) and seven doses of EMS (0.1–0.7%). The EMS treated population of chili exhibited a gradual decrease in growth of seedling and pollen fertility (Sharma and Anis 1995). Germination and survival rate declined linearly with the enhancement of radiation dose in gamma rays and fast neutrons (Din, Qasim and Ahmad 2004).

Reduced pollen fertility in *Oryza sativa* following the mutagenic treatment (physical and chemical mutagens) has been noticed by Reddi (1977). Several other workers have also recorded reduced pollen fertility by gamma rays in *Gossypium hirsutum* L. (Muthuswamy and Jayabalan 2002); EMS, MMS and gamma rays in *Vicia faba* (Bhat et al. 2005a, b); EMS in *Brassica oleracea* (Vesselina et al. 2006), $PbNO_3$ in *Capsicum annuum* L. (Kumar and Gupta 2008), Pb and Zn in *Lathyrus sativus* (Tripathy and Kumar 2010).

M_1 parameters were studied in barley after treatment with gamma rays, EMS and their combined treatment (Khalatkar and Bhatia 1975). Injury of seedlings, aberrations of chromosomes and sterility of pollens and seeds were lower in combined treatments as compared to individual treatments. The percentage of pollen sterility increased due to the greater meiotic abnormalities or cumulative outcome of various meiotic anomalies. Reddy and Rao (1982) considered different aberrations like univalents, micronuclei, stickiness and laggards that were closely associated with sterility of pollens and was the outcome of segmental interchanges of heterologous chromosomes. A study by Jahan et al. (2016) in *Lycopersicon esculentum* revealed that the lower doses of the mutagens (EMS and MMS) cause less pollen sterility and chromosomal damage as compared to the higher doses of these mutagens. Rao and Laxmi (1980) found that there was increased pollen sterility and decreased germination when doses of gamma rays in *Capsicum annuum* were increased. Sharma and Sharma (1981) observed in microsperma and macrosperma lentils that NMU had 1.5–2.0 times higher effect on plant survival and sterility than gamma rays.

11.4.2 Assessment of Mutants Through Morphological and Yield Parameters

11.4.2.1 Plant Height

The height of seedlings is a simple and quick method which is considered as an index for the determination of the biological effects of mutagens in M_1 generation (Kumar and Mani 1997). The plant height usually decreases at higher mutagenic doses but lower concentrations have proved stimulatory in showing a slight increase in plant height (Aruldoss and Mullainathan 2015). The stimulatory effect of lower concentration and inhibitory effect of high concentration of caffeine has been reported in *Trigonella foenum-graecum* (Anis and Wani 1997), *Capsicum annuum* L. (Kumar and Tripathi 2004). The stimulating effect of low doses of gamma radiations on seedling height of *Leucocephala* has been studied by Cepero et al. (2001). Gamma rays and EMS at their lower doses showed stimulatory effect on root and shoot lengths of the seedlings of *Psoralea corylifolia* (Jan, Parween and Siddiqi 2012, Bhat et al. 2015). Badigannavar and Mondal (2010) developed a tall mutant at 10kR gamma irradiation in *Arachis hypogaea*. Reduction in plant height has been reported after gamma ray treatment by Das et al. (2004) in *Vigna radiata*. Shakoor, Ahsan-Ul-Haq and Sadiq (1978) also noticed the adverse effect on plant height in *Vigna radiata* in M_1 generation. According to Srivastava et al. (2011) sodium azide reduced the root depth and shoot height in *Triticum aestivum*. In *Plantago*, a dose-dependent decrease in height and dry weight of seedling was found with increment in the doses of gamma rays (Jain, Ranwah and Bordia 2005).

The increasing EMS concentrations have been informed to reduce root length in *Cicer arietinum* (Haq et al. 1992) and *Vicia hirsuta* (Kumari 1994). With the increasing concentrations of EMS, plant height was also found to decrease in *Jatropa curcas* L. (Dhakshanamoorthy, Selvaraj and Chidambaram 2010). Amin et al. (2015) also noticed the same results in lentil, where plant length reduced with increasing doses of chemical mutagens. The mutagenic effect of EMS at 0.25% and 0.50% doses in T-9 variety of black gram was studied earlier by Kundu and Singh (1981). They observed that mean plant height and yield per plant decreased at 0.50% EMS treatment.

Gamma radiation-induced variability in *Phaseolus aureus* in the M_2 generation was studied by Rajput (1974). A positive as well as a negative shift in mean values for different polygenic traits including plant height, flowering time, pods number per plant, number of grains per pod, 100-seed weight and seed yield was determined. A negative shift in the mean values for plant height was also recorded in *Trifolium* using X-rays by Brock (1965), in *Triticale* using EMS by Sinha and Joshi (1986).

The influence of different individual and combinations of gamma rays (10–30 kR) and MMS (0.2%) was studied by Das and Prasad (1978) on some varieties of *Lathyrus sativus* L. They observed a reduction in height of plant and number of branches per plant in all the varieties in M_1 generation, which was dose-dependent. But, in M_2 both increase and decrease was reported when compared with control. Mohammed, Rounds and Tarpley (2007) have also noticed linear reduction in plant height after exposure of low ultraviolet (UV-B) radiations in rice. Gamma rays, EMS and their combinations have proven successful for inducing desirable height mutants in three *Triticale* varieties (Reddy 2001).

A remarkable variation in height followed by mutagenic treatment has been observed in different plants. Variability in plant height in *Cicer arietinum* has been induced by Athwal et al. (1970) through gamma radiations. Early maturing mutant with increased height in *Nigella sativa* L. has been induced using caffeine (Asif et al. 2016), dwarf mutant in *Vigna mungo* has been isolated using EMS (Goyal and Khan 2010), dwarf plants in *Hevea brasiliensis* have been identified from 30 Gy gamma rays exposed population (Amma, Namboodiri, Panikkar and Sethuraj 1990). Forster (2001) reported gamma rays induced semi-dwarf mutant in *Hordeum vulgare* where the semi-dwarfing genes showed positive response to salt stress.

11.4.2.2 Yield

In recent years, the genetic manipulations have clearly displayed the great possibility of improving yield and other related characters. Many workers have investigated the effect of the number of mutagens on yield and yield attributing traits like 1000-seed weight, number of capsules per plant, number of seeds per capsules, yield, and so on, in different plants and concluded either increase or decrease in these parameters. There have been reports showing an improvement in polygenic traits including grain yield, grain quality, early maturity, quality characters, abiotic stress and biotic resistance by mutagenesis (Kharakwal 1996). Mutations have developed several crop varieties with higher yield and quality, thereby refining agronomical inputs and user acceptance (Shehata, Allah and Zayed 2009, Aruna, Prakash and Sunil Kumar 2010). Variation of yield contributing characters by various doses of mutagens has been reported in *Lens culinaris* (Ali and Shaikh 2007).

Basu, Acharya and Thomas (2008) successfully used EMS in *Trigonella foenum-graecum* for developing mutants for improved early seed maturity, improved seed yield and quality. Seed mutagenesis through EMS treatment has also been displayed earlier by Krishna and Vasudevan (1984) in *Avena sativa* L. for quantitative variations in different yield traits. Several viable mutants including earliness, erect plants, profuse flowering and high yielding mutants in *Phaseolus lunatus* L. have been developed by gamma rays (Kumar, Kesarwani and Sharma 2003). Dry seeds after exposure to low gamma ray doses have shown increased yield in sunflower and *Ammi visnaga* (El-Shafie et al. 1993). Seeds of *Sesame indicum* L. were treated with varying concentrations (0–0.25%) of sodium azide and colchicine solutions. Early maturing and high yield variants were obtained by low doses of both mutagens (<0.125 %) (Mensah et al. 2007). Alka and Ansari (2013b) studied morphology of *Linum usitatissimum* by applying combined doses of gamma rays and SA. Fasciated plants in both M_2 and M_3 generation were obtained that had increased the circumference of stem and number of leaves, buds, flowers, fruits and seeds. Higher yielding mutants of mung bean in 40 kR gamma rays treatments were isolated in M_2 generation (Chaturvedi and Singh 1980). Effect of different doses of gamma rays (100 Gy to 500 Gy) was studied by Kumar and Rai (2006) in soybean. Seed setting was reported to be adversely affected by the increased doses of γ-rays. Seed and pod number per plant also decreased as compared to control when doses of γ-rays were increased, except for 100 Gy dose at which number of pods per plant exhibited a minor improvement over control. Panigrahi et al. (2015) suggested the increased yield and variations in quantitative parameters at significant levels that showed the stable gene mutations in subsequent generations. Higher concentrations of EMS were found to be effective in improving soybean oil through enhancement in the variability in the content of the fatty acid (Patil et al. 2007). Different types of macro and micro mutants after gamma rays, EMS and their combination treatments have been recorded in M_2 generation of *Vigna mungo*. All viable mutants bred true in the M_3 generation and some of them were out yielded over control (Gautam and Mittal 1998). EMS, gamma rays and their combination treatments induced six high yielding mutants in chickpea at M_2 generation level. A positive correlation between total Fe and

Mn contents to yield at lower dose and a negative correlation between Fe and Mn contents in these mutants were obtained (Kozgar, Goyal and Khan 2012). A positive seed yield after gamma irradiation was reported by Sharma and Mishra (2007).

Contrary to this, the negative effects of different mutagens on yield were also reported. The decrease in yield and yield ascribing qualities were noticed by Waghmare and Mehra (2001) by gamma rays in *Lathyrus sativus*, Sundaravadivelu et al. (2006) by EMS in *Gossypium hirsutum* and Amin, Laskar and Khan (2015) by MMS and DMSO in *Lens culinaris*. In *Trigonella foenum-graecum* gamma rays-induced mutations were obtained where, in mature plants, the increasing radiation doses resulted in late maturity, flowering, reduced plant height and yield components (Rokadia et al. 1991). A fasciated mutant was developed through EMS mutagenesis in *Cicer arietinum* in M_2 generation, which showed delayed maturity, larger seed size but lesser yield as compared to its parental cultivar (Gaur and Gaur 1999).

11.4.3 Assessment of Mutants Through Cytological Aberrations

Cytological aberrations in plants as a consequence of mutagenic treatment serve as an excellent monitoring system for the elucidation of variations and genotoxiciy induced by mutagens. Cytological studies constitute a dependable means to validate the DNA damage due to mutagenesis. According to Sharma, Khan and Bhat (2009) meiotic lesion analysis and the genetic consequences of these lesions are commonly used in most mutation studies. Induced chromosomal variations either during mitosis or meiosis are valuable cytogenetical tools for identification of potency of mutagen and for estimation of mutagenic sensitivity of species.

Different mutagens have been used time to time for the induction of chromosomal alterations such as chemical agents (Ganai et al. 2005), heavy metals (Kumar and Gupta 2008) and radiations (Ahmad and Godward 1993). EMS and SA treated *Nigella sativa* L. cultivars showed various chromosomal disorders in meiosis. These disorders included the formation of laggards, bridges, stray chromosomes, micronuclei, stickiness, disturbed polarity and cytomixis (Dixit, Prabha and Chaudhary 2013). Cytomixis creates variation in the chromosome number of the gametes by transferring the nuclear material through cytomictic channels. Cytomixis takes place in all phases of meiotic cycle at equal or different intensity (Shamina et al. 2000). Cytomixis phenomenon and its related meiotic abnormalities like pycnotic chromatin, chromatin stickiness, late disjunction of bivalents, out of plate bivalents, inter bivalent connections, bridges and laggards have caused partial pollen sterility and variable sizes of pollen grains in *Vicia faba*, *Caltha palustris*, *Meconopsis aculeata* and *Hippophae rhamnoides* (Kumar and Singhal 2008, Singhal, Kaur and Kumar 2008). Datta and Biswas (1981) studied different types of EMS induced mitotic anomalies in *N. sativa* and *N. damascena*. Numerous aberrations in chromosomes in EMS treated *Vicia faba* L. have also been observed by Bhat et al. (2005a). Various meiotic anomalies namely stickiness, univalents, multivalents, unequal separation, precocious separation, unorientation of chromosome, laggards, bridges have been reported in *Cicer arietinum* L. using EMS by Sharma and Kumar (2004). Different chromosomal lesions as sticky and disturbed metaphase with two stray chromosomes, disturbed metaphase II non-synchronization, disturbed polarity at anaphase II, laggard at telophase I were documented by Shahwar et al. (2017). They isolated a mutant of lentil with high seed yield at 0.25% caffeine with all these cytological anomalies. Sharifi-sirchi et al. (2012) observed partial desynapsis in *Hordeum vulgare* using EMS, SA and gamma rays. EMS and DES induced meiotic anomalies have also been reported in *Vicia faba* by Vandana and Dubey (1996).

A spectrum of chromosomal abnormalities as a result of irradiation at different stages of cell division has also been shown in different plants. In *Psoralea corylifolia*, laggards, stray chromosomes, micronuclei and bridges were found at varying gamma rays doses (Bhat et al. 2015). Gamma rays have also been reported to induce three translocation heterozygotes, exhibiting the formation of either a ring or a chain of four chromosomes in meiocytes of *Nigella damascena* L. (Saha and Datta

2000). Lakshmi et al. (1997) recorded a mutant exhibiting "multiploid sporocyte" in *Capsicum annuum* after gamma ray treatment. Gamma ray irradiation of pollens and thermal neutron irradiation of seeds resulted in monosomic *Gossypium hirsutum* L. (Sanamyan et al. 2000). Jayabalan and Rao (1987) observed clumping, stickiness and irregular grouping of chromosomes in *Lycopersicon esculentum* at metaphase-I after gamma rays irradiations. In gamma ray-treated *Vigna mungo* Shah et al. (1992) reported various mitotic abnormalities at metaphase, bridges at anaphase and multinucleate condition at telophase. Kumar and Srivastava (2010) studied radio-cytological outcome of gamma and laser rays in *Carthamus tinctorius* L. They encountered a wide spectrum of chromosomal lesions in both laser and gamma rays treatments but stickiness of chromosomes was most frequently noticed. It was also recorded that gamma rays caused a higher percentage of chromosomal aberrations than laser rays, suggesting that gamma rays could be successfully employed for creating additional genetic variability in Safflower. Colchicine and gamma-rays treatment in pollens resulted in translocation mutations in cotton (Sanamyan and Rakhmatullina 2003). Kumar et al. (2003) found the greater frequency of meiotic irregularities, when combination of gamma rays and EMS was given as compared to the individual ones in *Lens culinaris*.

Ganai et al. (2005) observed a prominent increase in the frequency of meiotic abnormalities in *Cicer arietinum* L.-var (GNG-469 and BG-372) by MMS treatment. Reduction in chiasma frequency with increase in the mutagenic concentration of sodium azide has been displayed by Kumar (2001) in *Psyllium* husk. A wide spectrum of meiotic anomalies in the flower mutants of *Capsicum annuum* after treating with caffeine were recorded by Aslam et al. (2017). They also observed that meiotic irregularities were dose dependent that is, as the concentration increased anomalies also increased. Similar observations were also made earlier in various crops by many workers (Kumar and Tripathi 2004, Khan et al. 2009).

11.4.4 Assessment of Mutants Through Phytochemical Analysis

The usage of synthetic chemicals and their side effects increased the interest in natural products from plant sources. The bioactive compounds or phytochemicals obtained from medicinal plants are highly useful to protect the mankind from various diseases. Phytochemicals are biologically active plant chemicals which have health-related effects on humans. These compounds are produced during the normal metabolic processes of plants and include alkaloids, phenols, coumarins, glycosides, flavonoids, polysaccharides, terpenes and terpenoids (Okwu 2004). Among the wide variety of medicinal plants, black cumin is of specific attention because of rich pool of phytochemicals with important biological assets and health benefits (Salem 2005). Most of these properties are attributed to volatile and fixed oils. Pharmacologically active constituents of volatile oil are thymoquinone (TQ), dithymoquinone (DTH), thymohydroquinone (THQ) and thymol (THY) (Ghosheh, Houdi and Crooks 1999). These are major quinonic phenol compounds that have been identified in the seeds of *Nigella* (Morikawa et al. 2004). Thymoquinone (27–57%) is the most abundant component of the quinone compounds (Gali-Muhtasib Roessner and Schneider-Stock 2006). Other literatures also showed that the valuable constituents which have beneficial chemical effects were principally thymoquinone (Nader, El-Agamy and Suddek 2010), p-cymene (Kisko and Roller 2005), carvacrol (Mastelic et al. 2008) and 4-terpineol (Burits and Bucar 2000). Thymoquinone is a major bioactive and pharmacologically active quinone of *Nigella* seeds containing anti-inflammatory and analgesic properties (Abdel-Fattah, Matsumoto and Watanabe 2000). For many types of human tumor cells, thymoquinone and dithymoquinone were found to be cytotoxic (Worthen, Ghosheh and Crooks 1998).

Many researchers are of the opinion that preparation of sample is important during any analytical study (Majors 1999). Extraction is the first step in the phytochemical studies. Therefore, selection of a proper extraction method is crucial for the study of active compounds of plants (Sasidharan et al. 2011). Karna (2013) showed various secondary metabolites including steroids, tannins flavonoids, coumarins, cardiac glycosides, saponins and diterpenes in methanolic seed extract of *N. sativa*.

Hydro-distillation extraction of the essential oil of *Nigella sativa* showed the occurrence of para-cymene, alpha thujene, beta pinene and alpha pinene as the major components using GC and GC/MS technique (Singh and Jhunjhunwalla 2014). GC and GC/MS techniques have been used to determine fixed and volatile oils from black seeds. Burits and Bucar (2000) reported thymoquinone as the major component of essential oil whereas Singh et al. (2005) represented *p*-cymene a major essential oil component. GC/MS analysis of seed oil of *Nigella* showed a variety of compounds having antimicrobial properties (Benkaci-Ali et al. 2007). These compounds are carvacrol (Ultee, Bennik and Moezelaar 2002), limonene, carvone (Oumzil 2002), thymol (Rasooli and Mirmostafa 2003), *p*-cymene and γ-terpinene (Güllüce et al. 2003) and thymoquinone (Benkaci-Ali et al. 2007). Agbaria et al. (2015) reported the anticancer activity of *Nigella sativa*. Ainane et al. (2014) analyzed monoterpene hydrocarbons and phenolic monoterpenes as major essential oil constituents of the *N. sativa* using GC/MS. The antibacterial activity of the essential oil was also determined against a panel of strains of bacteria using a broth microdilution method. In *Leidium sativum*, the qualitative and quantitative analysis was done by Nayak, Upadhyaya and Upadhyaya (2009). Sinapic acid was quantified using HPTLC. Atta (2003) revealed that TLC and GLC analysis of black seed oil contained significant amounts of sterols. It has been confirmed that the oil was rich in β-sitosterol that inhibits the absorption of dietary cholesterol. Ghosheh, Houdi and Crooks (1999) used a simple and sensitive isocratic high performance liquid chromatographic method to examine TQ, THQ, DTQ and THY in commercial *N. sativa* seed oil using a reversed-phase column and a mobile phase of water, methanol and propanol. El Dakhakhny (1963) has also isolated thymoquinone (TQ) from the essential oil of black cumin seed and determined that the nigellone isolated earlier was a dimer of thymoquinone, which was named as dithymoquinone (DTQ). The content of TQ in *N. sativa* seed oil samples from different origins have been analyzed by Houghton et al. (1995) through gas chromatography and found it to be in the range of 0.13–0.17% w/v of the oil. Thin layer chromatographic assay of TQ and identification of DTQ and THY have been done in *N. sativa* seed by Basha, Rashed and Aboul-Enein (1995). They have also developed a normal phase HPLC method for the determination of TQ in the seed oil and found significant variations of TQ content between different commercial samples of *Nigella* seed oil.

Thymoquinone being the main component of volatile oil of *Nigella sativa*, its relative content in control versus mutant plants may be considered a significant criterion for assessment of the superiority of genotypes. Dixit, Prabha and Chaudhary (2013) treated the seeds of two cultivars (NRCSSAN-1 and BHUVN-1) of *Nigella sativa* with EMS and SA. Besides reporting genotoxic effect in terms of meiotic aberrations, they have also reported a mutant line (NSM 8) with significant increase in total oil and thymoquinone (THQ) content over its respective control. Asif and Ansari (2019) have also reported significant variations in TQ content in all the nine mutant lines that were selected at M_3 generation. Khattak and Simpson (2008) revealed that gamma rays (2, 4, 8, 10, 12 and 16 kGy) irradiated seeds of *Nigella sativa* enhanced yield of oil and total phenolic content. The extraction yield was found to be increased with increasing doses of radiation.

11.4.5 Assessment of Mutants Through Molecular Marker

The success of a plant breeding experiment depends on genetic variability in a population. Induced mutations have been used to generate genetic variability. For the evaluation of genetic variation, classical methods have been complemented by molecular techniques. DNA-based marker development has significantly enabled research in a variety of many biological branches such as taxonomy, phylogenetic relationships and genetics (Halward et al. 1992). DNA-based markers are now prevalently known for the identification of medicinal plants (Yip et al. 2007). Various techniques based on molecular biology tools are used for authentication of medicinal plants.

Species-specific variations in the DNA sequences that are randomly spread over the whole genome produce characteristic "fingerprints" of genomic DNA. Various plant species have been authenticated using different molecular techniques such as RAPD (Randomly Amplified Polymorphic DNA) (Khan et al. 2009a), RFLP (Restriction Fragment Length Polymorphism) (Biswas, Kapoor and Biswas 2013), AFLP (Amplified Fragment Length Polymorphism) (Ghosh, Majumder and Mandi 2011), microsatellite markers (Balloux and Lugon-Moulin 2002), microchip electrophoresis (Qin et al. 2005), sequencing of rRNA and chloroplast mat K gene (Yan et al. 2008), sequencing of 5S-rRNA spacer domain (Zhao et al. 2003), ISSR (Inter-Simple Sequence Repeats) and SSR (Simple Sequence Repeats) polymorphism. Sequence-Characterized Amplified Regions (SCAR) (Paran and Michelmore 1993), Amplified Refractory Mutation System (ARMS) and Multiplex Amplification Refractory Mutation System (MARMS) and DNA microarray are also used for authentication of medicinal plants.

Molecular analysis using DNA markers viz., RAPD, RFLP, AFLP, SSR, SNPs, and so on allow the assessment of genetic variability and can be used for diversity studies. Costa et al. (2016) employed RAPD, ISSR and AFLP markers for fingerprinting the genotypes of orchard grass (*Dactylis glomerata*) and detected genetic variation between the three different sub-species. In this study, RAPD showed 97 bands (40 polymorphic), ISSR primers amplified 91 bands (54 polymorphic) and AFLP showed 100 bands (92 polymorphic). It was concluded that AFLP was the best for molecular assay among genotypes of *Dactylis glomerata* for evaluating genetic relationship and fingerprinting. While RAPD is used for finding genetic diversity among individuals of same species or different species (Sen, Skaria and Muneer 2010), the ISSR is employed for assessing the relationships between very closely related individuals (Mao and Fang 2014). The AFLP technique was developed by Vos et al. (1995). It has been widely utilized in studies related to analysis of genetic diversity in numerous plant species (Sharma et al. 2018). The AFLP technique is favored over other markers such as RAPD, RFLP and ISSR due to their high multiplex ratio and reproducibility (Powell et al. 1996). The AFLP technique is based on the selective PCR amplification of restriction fragments from a total digest of genomic DNA. Three steps are involved in this technique: (i) restriction of the DNA and ligation of oligonucleotide adapters; (ii) selective amplification of sets of restriction fragments; and (iii) gel analysis of the amplified fragments. Restriction fragment amplification through PCR is attained by using the adapter and restriction site sequence as target sites for primer annealing. The selective amplification is achieved by the use of primers that extend into the restriction fragments, amplifying only those fragments in which the primer extensions match the nucleotides flanking the restriction sites. With this method, sets of restriction fragments may be visualized by polymerase chain reaction without the information of sequence of nucleotides. The method consents the specific co-amplification of greater numbers of restriction fragments. The number of fragments that can be analyzed simultaneously, however, is dependent on the resolution of the detection system. Typically, 50–100 restriction fragments are amplified and detected on denaturing polyacrylamide gels (Vos et al. 1995). The AFLP technique provides a novel and very powerful DNA fingerprinting technique for DNAs of any origin or complexity. In contrast to other molecular markers, the AFLP have been very effective for the identification of variations in closely linked cultivars (Patzak 2001). The AFLP offers a perfect device for evaluating genetic variability in both natural and cultivated populations considering the quantity of information generated, replicability, resolution, ease of use and cost efficiency (Mueller and Wolfenbarger 1999). Analysis of inter-specific and intra-specific genetic variation in medicinal plant has also been carried out through AFLP (Varma and Shrivastava 2018).

11.4.6 Assessment of Mutants Through Proteomics

Proteomics is an indispensable tool for measuring large number of proteins qualitatively and quantitatively that directly influence cellular biochemistry. Various stress stimuli or gene mutations causing alteration of protein can also be examined with this technique. Proteome technology paves the way to monitor protein profiles and to identify and characterize different proteomes (Binita et al.

2014, Kumari, Tajmul and Yadav 2015). From the last few years, proteomics has emerged very fast and appealed to the attention of the scientists working on various aspects. An increasing trend of using proteomic analysis has been seen in fields of life science in the last decade.

The application of a proteomics-based approach is most promising to boost agricultural production (Dhand 2000), analyzing plant metabolic pathways involved in the synthesizing spectrum of bioactive compounds, identification of induced, repressed or post-transcriptionally modified proteins, estimation of genetic variability, characterization of individuals or lines or mutants (Thiellement et al. 1999). Plant growth and different biological alterations in response to environmental stresses have been identified using proteomic techniques in major crop species (Hashiguchi and Komatsu 2017, Ghatak, Chaturvedi and Weckwerth 2017). Development of mutants through mutagenesis is of agricultural importance. Proteome profiling of isolated mutants is of great importance and the change in their proteins confirms the genetic diversity. Protein profiling has been reported in number of isolated mutants by several workers in different crops like *Lycopersicon esculentum* (Mahmoud and Al-Twaty 2006), *Vigna radiata* (Auti and Apparao 2009), *Cicer arietinum* (Barshile and Apparao 2009), *Glycine max* (Nakagawa 2009) and *Vigna aconitifolia* (Khadke and Kothekar 2011). Alanazi et al. (2016) have analyzed the proteome profile of *Nigella sativa* for the first time. They have identified a total of 277 proteins which were characterized on the basis of their functions and locations. They have also related to the proteins with the known metabolic processes of significance. Asif et al. (2019) have also analyzed the leaf proteome of macro-mutant lines of *N. sativa*. They observed 23 proteins were differentially expressed which were categorized into groups on the basis of their function. A comparative proteomic analysis in *Ginkgo biloba* under ultraviolet-B (UV-B) treatment has been performed by Zheng et al. (2015) to determine the differential expression of proteins under UV-B radiation treatment. A total of 27 differential proteins have been identified and annotated.

During the past decade proteomic technologies have considerably improved. Protein analysis due to the availability of standardized procedures like two-dimensional electrophoresis (2-DE) made large-scale investigations possible. The 2-DE approach supports genetic determination of the quantity of protein for numerous gene products at a time, whether their function is known or not. Yousuf et al. (2017) analyzed leaf proteins in two wheat cultivars to identify the functional proteins produced in response to high temperature and low nitrogen stress. The proteome maps of wheat leaf (Donnelly et al. 2005, Bahrman et al. 2004), flour (Mamone et al. 2000) and amyloplast (Andon et al. 2002) were developed using 2-DE technique. The 2-DE technique is also an important procedure for evaluating alterations of protein expression caused by a mutation (Damerval et al. 1988). Santoni, Bellini and Caboche (1994) analyzed the proteome profile of developmental mutants of *Arabidopsis thaliana*. It has been shown that wild plant proteome patterns were distinguishable from that of the mutants which had their specific proteome pattern.

REFERENCES

Abdel-Fattah, A.F.M., Matsumoto, K. and Watanabe, H. 2000. Antinociceptive effects of *Nigella sativa* oil and its major component, thymoquinone, in mice. *Eur. J. Pharmaco.* 400(1): 89–97.

Acquaah, G. 2007. *Principles of Plant Genetics and Breeding*. Blackwell, Malden, UK.

Agbaria, R., Gabarin, A., Dahan, A. and Ben-Shabat, S. 2015. Anticancer activity of *Nigella sativa* (black seed) and its relationship with the thermal processing and quinone composition of the seed. *Drug Des. Dev. Ther.* 9: 3119–3124.

Ahmad, S. and Godward, M.B.E. 1993. Gamma radiation induced mutations in *Cicer arietinum* L. *Acta Bot. Indica.* 21: 1–8.

Ahmad. T., Ansari. M.Y.K. and Asif, A. 2017. Mutagenic effect of 6-aminopurine (base analogue) on the quantitative and qualitative traits of *Linum usitatissimum* L. *Int. J. Sci. Res.* 6(9): 484–485.

Ainane, T., Askaoui, Z., Elkouali, M., Talbi, M., Lahsasni, S., Warad, I. and Hadda, T.B. 2014. Chemical composition and antibacterial activity of essential oil of *Nigella sativa* seeds from Beni Mellal (Morocco): What is the most important part, essential oil or the rest of seeds? *J. Mater. Environ. Sci.* 5(6): 2017–2020.

Akgun, I. and Tosun, M. 2004. Agricultural and cytological characteristics of M_1 perennial rye (*Secale montanum* Guss.) as effected by the application of different doses of gamma rays. *Pak. J. Biol. Sci.* 7(5): 827–833.

Alanazi, I.O., Benabdelkamel, H., Alfadda, A.A., AlYahya, S.A., Alghamdi, W.M., Aljohi, H.A., Almalik, A. and Masood, A. 2016. Proteomic analysis of the protein expression profile in the mature *Nigella sativa* (black seed). *Appl. Biochem. Biotechnol.* 179(7): 1184–1201.

Ali, F. and Shaikh, N.A. 2007. Genetic exploitation of lentil through induced mutations. *Pak. J. Bot.* 39(7): 2379–2388.

Alka and Ansari, M.Y.K. 2013a. Induction of polyploid mutant in *Linum usitatissimum* L. by the combined action of 5 KR gamma rays + 0.10% sodium azide. *Indian streams res. j.* 3(4): 1–6.

Alka and Ansari, M.Y.K. 2013b. Fasciated mutant induced by the combined treatment of gamma rays and sodium azide (20 KR + 0.10% SA) in *Linum usitatissimum* L. *Int. J. Pharma Bio Sci.* 4(2): 1196–1200.

Al-Rubeai, M.A.F. 1982. Radiation induced mutations in *Phaseolus vulgaris* L. *Rev Bras Genet.* 5(3): 503–515.

Amano, E. 2006. Use of induced mutants in rice breeding in Japan. *Plant Mutation Reports.* 1(1): 21–24.

Ambli, K. and Mullainathan, L. 2015. Chlorophyll and morphological mutants of Pearl millet (*Pennisetum tphoides* (Burn.) stapf. var. CO (cu) 9. *Eur. J. Exp. Biol.* 5(3): 72–77.

Amer, S. M. and Mikheal, E. 1972. Cytogenetic studies on the effect of Co^{60} γ-irradiation on *Vicia faba. Cytologia.* 37: 169–174.

Amin, R., Laskar, R.A. and Khan, S. 2015. Assessment of genetic response and character association for yield and yield components in Lentil (*Lens culinaris* L.) population developed through chemical mutagenesis. *Cogent Food Agric.* 1(1): 1–15.

Amma, C.K.S., Namboodiri, A.N., Panikkar, A.O.N. and Sethuraj, M.R. 1990. Radiation-induced male sterility in *Hevea brasiliensis* (Wild ex Adr. de Juss) Muell., *Agricultural Cytologia.* 55: 547–551.

Andon, N.L., Hollingworth, S., Koller, A., Greenland, A.J., Yates, J.R. and Haynes, P.A. 2002. Proteomic characterization of wheat amyloplasts using identification of proteins by tandem mass spectrometry. *Proteomics.* 2(9): 1156–1168.

Anis, M. and Wani, A.A. 1997. Caffeine induced morpho-cytological variability in Fenugreek, *Trigonella foenum-graecum* L. *Cytologia.* 62: 343–349.

Arulbalachandran, D. and Mullainathan, L. 2009. Chlorophyll and morphological mutants of blackgram (*Vigna mungo* (L.) Hepper) derived by gamma rays and EMS. *J. Phytol.* 1(4): 236–241.

Aruldoss, T. and Mullainathan, L. 2015. Studies on effect of induced mutagenesis on *Capsicum annuum* (L.) var K1 in M_1 generation. *J. Chem. Biol. Phys.* 5(1): 347–352.

Aruna, J., Prakash, M. and Sunil Kumar, B. 2010. Studies on effect of physical and chemical mutagens on seedling characters in Brinjal (*Solanum melongena* L.). *Int. J. Curr. Res.* 3: 38–41.

Asif, A., Ansari, M.Y.K., Jahan, N. and Ahmad, T. 2016. Mutational bioassay and ameliorative action of caffeine for induction of early maturing mutant in *Nigella sativa* L. *Int. Res. J. Pharm.* 7(5): 33–37.

Asif, A., Ansari, M.Y.K., Hashem, A., Tabassum, B., Abd_Allah, E.F. and Ahmad, A. 2019. Proteome Profiling of the Mutagen-Induced Morphological and Yield Macro-Mutant Lines of *Nigella sativa* L. *Plants.* 8(9): 321.

Asif, A. and Ansari, M.Y.K. 2019. Generation of mutant lines of *Nigella sativa* L. by induced mutagenesis for improved seed yield. *Ind Crops Prod.* 139: 111552.

Aslam, R., Bhat, T.M., Choudhary, S., Ansari, M.Y.K. and Shahwar, D. 2017. Estimation of genetic variability, mutagenic effectiveness and efficiency in M_2 flower mutant lines of *Capsicum annuum* L. treated with caffeine and their analysis through RAPD markers. *J. King Saud Univ. Sci.* 29(3): 274–283.

Athwal, D.S., Bhalla, S.K., Sandhu, S.S. and Brar, H.S. 1970. A fertile dwarf and three other mutants in *Cicer. Indian J Genet Plant Breed.* 30(1): 261–266.

Atta, M.B. 2003. Some characteristics of nigella (*Nigella sativa* L.) seed cultivated in Egypt and its lipid profile. *Food Chem.* 83(1): 63–68.

Auti, S.G. and Apparao, B.J. 2009. Induced mutagenesis in mungbean (*Vigna radiata* (L.) Wilczek). In: *Induced Plant Mutations in the Genomics Era.* Ed. Q.Y. Shu, Food and Agriculture Organization of the United Nation, Rome, Italy, pp. 97–100.

Badigannavar, A.M., Katageri, I.S. and Khadi, B.M. 2003. Induced pleiotropy for curved stigma and genetic male sterility in tetraploid cotton (*Gossypium hirsutum* L.). *Indian J Genet Plant Breed.* 63(3):271–272.

Badigannavar, A.M. and Mondal, S. 2010. Induction of mutations for plant height and inheritance of dwarf mutant in groundnut (*Arachis hypogaea* L.) through gamma ray irradiation. *Electron. J. Plant Breed.* 1(2): 156–161.

Bahrman, N., Negroni, L., Jaminon, O. and Le Gouis, J. 2004. Wheat leaf proteome analysis using sequence data of proteins separated by two-dimensional electrophoresis. *Proteomics.* 4(9): 2672–2684.

Balloux, F. and Lugon-Moulin, N. 2002. The estimation of population differentiation with microsatellite markers. *Mol. Ecol.* 11(2):155–165.

Barbosa, H.M., de Souza Carneiro, J.E. and Vieira, C. 1988. Ethyl methansulfonate-induced seed coat color mutants in *Phaseolus vulgaris* L. *Rev Bras Genet.* 11(1): 123–128.

Barshile, J.D. and Apparao, B.J. 2009. Genetic improvement of chickpea (*Cicer arietinum* L.) using induced mutations. In: *Induced Plant Mutations in Genomic Era.* Food and Agriculture Organization of the United Nations, Rome, pp. 91–94.

Basha, L.A., Rashed, M.S. and Aboul-Enein, H.Y. 1995. TLC assay of thymoquinone in black seed oil (*Nigella sativa* Linn) and identification of dithymoquinone and thymol. *J. Liq. Chromatogr. Relat. Technol.* 18(1): 105–115.

Basu, S. K., Acharya, S.N. and Thomas, J.E. 2008. Genetic improvement of Fenugreek (*Trigonella foenum-graecum* L.) through EMS induced mutation breeding for higher seed yield under western Canada prairie conditions. *Euphytica.* 160(2): 249–258.

Benkaci-Ali, F., Baaliouamer, A., Meklati, B.Y. and Chemat, F. 2007. Chemical composition of seed essential oils from Algerian *Nigella sativa* extracted by microwave and hydrodistillation. *Flavour Fragr J.* 22(2):148–153.

Bhan, A.K. and Kaul, M.L.H. 1976. Frequency and spectrum of chlorophyll-deficient mutations in rice after treatment with radiation and alkylating agents. *Mutat. Res. – Fundam. Mol. Mech. Mutagen.* 36(3): 311–317.

Bhat, T.A., Khan, A.H. and Parveen, S. 2005a. Comparative analysis of meiotic abnormalities induced by gamma rays, EMS and MMS in *Vicia faba* L. *J. Indian Bot. Soc.* 84: 45–48.

Bhat, T.A., Khan, A.H., Parveen, S. and Ganai, F.A. 2005b. Clastogenic effect of EMS and MMS in *Vicia faba* L. *Cytol. Genet.* 6: 117–122.

Bhat, T.M., Ansari, M.Y.K., Choudhary, S., Aslam, R. and Bhat, W.F. 2015. Alteration in anti-oxidant defense system and protein expression in response to varied concentrations of EMS in *Psoralea corylifolia. Acta Physiol. Plant.* 37(1): 1–13.

Bhatnagar, S.M. 1984. Interaction of physical and chemical mutagens in Kabuli Chickpea. *Int. Chickpea Newsl.* 11: 17–17.

Bhattacharya, C. 2003. Effect of ethyl methane sulphonate on carnation (*Dianthus caryophyllus* L.). *Environ. Ecol.* 21(2): 301–305.

Bhosale, R.S. and More, A.D. 2014. Effect of gamma radiation on seed germination, seedling height and seedling injury in *Withania somnifera* (L.) Dunal. *Int. J. Life Sci.* 2(3): 226–228.

Binita, K., Kumar, S., Sharma, V.K., Sharma, V. and Yadav, S. 2014. Proteomic identification of *Syzygium cumini* seed extracts by MALDI-TOF/MS. *Appl. Biochem. Biotechnol.*172(4): 2091–2105.

Biswas, K., Kapoor, A. and Biswas, R. 2013. Authentication of herbal medicinal plant-*Boerhavia diffusa* L. using PCR-RFLP. *Curr Trends Biotechnol Pharm.* 7(3): 725–731.

Brock, R.D. 1965. Response of *Trifolium subterraneum* to X-rays and thermal neutrons. *Radiat Bot.* 5(6): 543–555.

Burits, M. and Bucar, F. 2000. Antioxidant activity of *Nigella sativa* essential oil. *Phytother Res.* 14(5): 323–328.

Cepero, L., Mesa, R.A., Lajonchere, G., Prieto, M. and Forrajes, P.Y. 2001. Stimulating growth of *Leucaena leucocephala* cv. Cunningham with gamma rays from 60Co. *Pastos y. Forrajes.* 24(3): 235–240.

Chandra, N. 1988. Cytological and mutational studies on *Gloriosa superba* Linn. Ph.D. thesis, Nagpur University, Nagpur, India.

Chaturvedi, S.N. and Singh, V.P. 1980. Gamma-ray induced quantitative variation in mung bean. *Cytol. Genet.* 15(1): 66–67.

Chauhan, Y.S., Kumar, K. and Krishna, H. 2001. Effectiveness and efficiency of physical and chemical mutagens on germination in *Brassica juncea* L. *Crucif. Newsl.* 23: 41–42.

Chopra, S. and Singh, R.P. 1978. Effect of gamma rays and 2, 4-D on germination, growth and morphogenetic responses in *Guizotia abyssinica*. *Phytomorphology.* 28: 82–87.

Choudhary, S., Ansari, M.Y.K., Khan, Z. and Gupta, H. 2012. Cytotoxic action of lead nitrate on cytomorphology of *Trigonella foenum-graecum* L. *Turk. J. Biol.* 36(3): 267–273.

Costa, R., Pereira, G., Garrido, I., Tavares-de-Sousa, M.M. and Espinosa, F. 2016. Comparison of RAPD, ISSR, and AFLP molecular markers to reveal and classify orchardgrass (*Dactylis glomerata* L.) germplasm variations. *PloS one.* 11(4): 1–15.

Damerval, C., Zivy, M., Granier, F. and de Vienne, D. 1988. Two-dimensional electrophoresis in plant biology. In: *Advances in Electrophoresis.* Eds. A. Chrambach, M.J. Dunn, B.J. Radola. VCH Verlagsgesellschaft, Weinheim, Germany, pp. 263–340.

Das, A.K. and Prasad, A.B. 1978. Variation induced in quantitative characters of some varieties of *Lathyrus sativus* L. *J. Indian Bot. Soc.* 57(suppl.): 73–74.

Das, P.K. and Kundagrami, S. 2000. Frequency and spectrum of chlorophyll mutations in grasspea induced by gamma rays. *Indian J Genet Plant Breed.* 60(2): 239–241.

Das, M.L., Kabir, S.M.R., Begum, S. and Ahmed, Z.U. 2004. Radio-siensitivity and selection of elite mutants of mungbean (*Vigna radiata* [L.] Wilczek). *Bangladesh J. Bot.* 33(1): 41–45.

Datta, A.K. and Biswas, A.K. 1981. EMS induced mitotic aberrations in *Nigella sativa* L. and *N. damascena* L. *Cell Chrom. Newsl.* 4: 1–2.

Datta, A.K. and Biswas, A.K. 1984. Induced male sterile mutant in *Nigella sativa* L. *Cell Chrom. Res.* 7: 24–25.

Datta, A.K. and Biswas, A.K. 1985. Induced mutagenesis in *Nigella sativa* L. *Cytologia.* 50(3): 545–562.

Daudu, O.Y. and Falusi, O.A. 2011. Induced Genetic variability for morphological and yield parameters in *Capsicum annum* and *Capsicum frutescens. Waste Manag. Bioresour. Technol.* 2011(1): 31–37.

Davis, T.M., Foster, K.W. and Phillips, D.A. 1985. Nodulation mutants in chickpea. *Crop Sci.* 25(2): 345–348.

Devi, T.R., Prodhan, H.S., Brajendra, N. and Rohinikumar, M. 2002. Effectiveness and efficiency of gamma rays in inducing chlorophyll mutations in M_2 generation of three rice bean genotypes (*Vigna umbellata* (L.) Thunb. Ohwi & Ohashi). *Environ. Ecol.* 20(4): 890–893.

Dhakshanamoorthy, D., Selvaraj, R. and Chidambaram, A. 2010. Physical and chemical mutagenesis in *Jatropha curcas* L. to induce variability in seed germination, growth and yield traits. *Rom. J. Biol.* 55(2): 113–125.

Dhand, R. 2000. Nature insight functional genomics. *Nature.* 405(6788): 819.

Din, R., Qasim, M. and Ahmad, K. 2004. Radiosensitivity of various wheat genotypes in M_1 generation. *Int J Agric Biol.* 6: 898–900.

Dixit, V., Prabha, R. and Chaudhary, B.R. 2013. Effects of EMS and SA on meiotic cells and thymoquinone content of *Nigella sativa* L. cultivars. *Caryologia.* 66(2): 178–185.

Donnelly, B.E., Madden, R.D., Ayoubi, P., Porter, D.R. and Dillwith, J.W. 2005. The wheat (*Triticum aestivium* L.) leaf proteome. *Proteomics.* 5(6): 1624–1633.

Edwin, R. and Reddy, V.R.K. 1993. Effects of gamma rays and EMS on certain biological parameters in triticale. *J. Indian Bot. Soc.* 72(1–2): 97–101.

El–Dakhakhny, M. 1963. Studies on the chemical constitution of Egyptian *Nigella sativa* L. seeds. II1) the essential oil. *Planta Med.* 11(4): 465–470.

El-Shafie, S.A., Mazrou, M.M., El Sayed, S.A. and Aly, H.F.M. 1993. Physiological influence of pre-sowing gamma irradiation on the growth, drug yield and some chemical constituents of *Ammi visnage* L. Plants. *Minufiya J Agric Res.*18: 2565–2578.

Fahmy, O.G. and Fahmy, M.J. 1957. Mutagenic response to the alkyl-methanesulphonates during spermatogenesis in *Drosophila melanogaster. Nature.* 180(4575): 31–34.

Fawole, I. 2001. Maternal inheritance of plant variegation in cowpea, *Vigna unguiculata* (L.) Walp. *Theor. Appl. Genet.*102(2–3): 458–462.

Forster, B.P. 2001. Mutation genetics of salt tolerance in barley: An assessment of Golden Promise and other semi-dwarf mutants. *Euphytica.* 120(3): 317–328.

Forster, B.P. and Shu, Q.Y. 2012. Plant mutagenesis in crop improvement: basic terms and applications. In: *Plant Mut. Breed. Biotechnol.* Eds. Q.Y. Shu, B.P. Forster and H. Nakagawa. Wallingford, CABI. UK, pp. 9–20.

Gali-Muhtasib, H., Roessner, A. and Schneider-Stock, R. 2006. Thymoquinone: A promising anti-cancer drug from natural sources. *Int. J. Biochem. Cell Biol.* 38(8): 1249–1253.

Ganai, F.A., Khan, A.H., Bhat, T.A., Parveen, S. and Wani, N.A. 2005. Cytogenetic effects of methylmethane sulphonate (MMS) on two varieties of chickpea (*Cicer arietinum* L.). *Cytol. Genet.* 6(2): 97–102.

Gaur, P.M. and Gaur, V.K. 1999. An induced fasciated mutant of chickpea (*Crcer arietinum* L.). *Indian J Genet Plant Breed.* 59(3): 325–330.

Gaur, P.M., Gour, V.K. and Singh, K. 2004. Induction and inheritance of a variegated leaf and an apical chlorosis mutant in chickpea (*Cicer arietinum* L.). *Indian J Genet Plant Breed.* 64(3): 208–211.

Gautam, A.S. and Mittal, R.K. 1998. Induced mutation in black gram [*Vigna mungo* (L.) Hepper.]. *Crop Research Hisar.* 16(3): 344–348.

Gecheva, A.M.P. 1983. New mutant soybean variety "Boriana". *Mut. Breed. Newsl.* 23: 3–4.

Geddam, S.B., Khadi, B.M., Mogali, S.C., Nagappa, H., Patil, R.S. and Katageri, I.S. 2013. Genetics and morphological studies of genetic male sterile lines in diploid cotton (*Gossypium arboreum* L.). *J. Cotton Res. Dev.* 27(1): 32–36.

Geetha, K. and Vaidyanathan, P. 2000. Studies on induction of chlorophyll mutations in soybean through physical and chemical mutagens. *Agric. Sci. Digest.* 20(1): 33–35.

Ghatak, A., Chaturvedi, P. and Weckwerth, W. 2017. Cereal crop proteomics: Systemic analysis of crop drought stress responses towards marker-assisted selection breeding. *Front. Plant Sci.* 8: 1–25.

Ghosh, S., Majumder, P.B. and Mandi, S.S. 2011. Species-specific AFLP markers for identification of *Zingiber officinale, Z. montanumand and Z. zerumbet* (Zingiberaceae). *Genet. Mol. Res.* 10(1): 218–229.

Ghosheh, O.A., Houdi, A.A. and Crooks, P.A. 1999. High performance liquid chromatographic analysis of the pharmacologically active quinones and related compounds in the oil of the black seed (*Nigella sativa* L.). *J. Pharm. Biomed. Anal.* 19(5): 757–762.

Girhe, S. and Choudhary, A.D. 2002. Induced morphological mutants in *Lathyrus sativus. J. Cytol. Genet.* 3: 1–6.

Girija, M. and Dhanavel, D. 2009. Mutagenic effectiveness and efficiency of gamma rays, ethyl methylmethane sulphonate and their combined treatment in cowpea (Vigna unguiculata L. Walp.). *Global J. Mol. Sci.* 4: 68–75.

Girija, M., Dhanavel, D. and Gnanamurthy, S. 2013a. Gamma rays and EMS induced flower color and seed mutants in cowpea (*Vigna unguiculata* L. Walp). *Adv. Appl. Sci. Res.* 4(2): 134–139.

Gnanamurthy, S., Dhanavel, D., Girija, M., Pavadai, P. and Bharathi, T. 2012. Effect of chemical mutagenesis on quantitative traits of maize (*Zea mays* L.). *Int. J. Res. Botany.* 2(4): 34–36.

Gnanamurthy, S. and Dhanavel, D. 2014. Effect of EMS on induced morphological mutants and chromosomal variation in Cowpea (*Vigna unguiculata* (L.) Walp). *Int. Lett. Nat. Sci.* 22: 33–43.

Gottschalk, W. and Wolff, G. 1983. Induced mutations in plant breeding. Monograph on theoretical and Applied Genetics 7. Berlin, Springer-Verlag.

Goyal, S. and Khan, S. 2010. Cytology of induced morphological mutants in *Vigna mungo* (L.) Hepper. *Egypt. J. Biol.* 12 (2): 81–85.

Goyal, S., Wani, M.R., Raina, A., Laskar, R.A. and Khan, S. 2021. Phenotypic diversity in mutagenized population of urdbean (Vigna mungo (L.) Hepper). *Heliyon.* 7(5): e06356.

Greene, E.A., Codomo, C.A., Taylor, N.E., Henikoff, J.G., Till, B.J., Reynolds, S.H., Enns, L.C., Burtner, C., Johnson, J.E., Odden, A.R. and Comai, L. 2003. Spectrum of chemically induced mutations from a large-scale reverse-genetic screen in *Arabidopsis. Genetics.* 164(2): 731–740.

Güllüce, M., Sökmen, M., Daferera, D., Ağar, G., Özkan, H., Kartal, N., Polissiou, M., Sökmen, A. and Şahin, F. 2003. In vitro antibacterial, antifungal, and antioxidant activities of the essential oil and methanol extracts of herbal parts and callus cultures of *Satureja hortensis* L. *J. Agric. Food Chem.* 51(14): 3958–3965.

Gupta, P.K. and Yashvir. 1975. Induced mutation in foxtail millet (*Setaria italic* Beauv). I: Chlorophyll mutations induced by gamma rays, EMS and DES. *Theor. Appl. Genet.* 45(6): 242–249.

Gustaffson, A. 1947. Mutations in agricultural plants. *Hereditas.* 33(1–2): 1–100.

Hajra, N.G. 1979. Induction of mutations by chemical mutagen in tall indica rice [India]. *Indian Agric.* 23: 67–72.

Halward, T., Stalker, T., LaRue, E. and Kochert, G. 1992. Use of single-primer DNA amplifications in genetic studies of peanut (*Arachis hypogaea* L.). *Plant Mol. Biol.* 18(2): 315–325.

Haq, M.A., Singh, K.B., Abidin, Z. and Ahmad, M.S. 1992. Mutation studiesin chickpea (*Cicer arietinum* L.). I. Mutagen sensitivity. *Pak. J. Agric. Sci.* 29(4): 429–438.

Haq, M.A., Singh, K.B., Abidin, Z. and Ahmad, M.S. 1994. Mutation studies in chickpea (*Cicer arietinum* L.) III. Selection of mutants in M_2 generation. *Pak. J. Agric. Sci.* 31: 13–22.

Hashiguchi, A. and Komatsu, S. 2017. Proteomics of soybean plant. In: *Proteomics in Food Science: From Farm to Fork.* Ed. M. Colgrave. Elsevier, Amsterdam, The Netherlands, pp. 89–105.

Hohmann, U., Jacobs, G. and Jung, C. 2005. An EMS mutagenesis protocol for sugar beet and isolation of non-bolting mutants. *Plant Breed.* 124(4): 317–321.

Houghton, P.J., Zarka, R., de las Heras, B. and Hoult, J.R.S. 1995. Fixed oil of *Nigella sativa* and derived thymoquinone inhibit eicosanoid generation in leukocytes and membrane lipid peroxidation. *Planta Med.* 61(1): 33–36.

Ichitani, K., Takayama, Y., Sato, M., Hashimoto, F., Ishiguro, E. and Sakata, Y. 2003. Effect of irradiation conditions on mutation rate induced by gamma-ray irradiation in foxtail millet (*Setaria italica* (L.) P. Beauv.). *Memoirs of the Faculty of Agriculture-Kagoshima University (Japan)* 38: 1–8.

Jabeen, N. and Mirza, B. 2002. Ethyl methane sulphonate genetic variability in *Capsicum annuum. Asian J. Plant Sci.* 1(4): 25–428.

Jabeen, N. and Mirza, B. 2004. Ethyl methane sulfonate induces morphological mutations in *Capsicum annuum. Int J Agric Biol.* 6(2): 340–345.

Jahan, N., Ansari, M.Y.K. and Asif, A. 2016. Comparative spectrum of cellular damage caused by two major alkylating agents (EMS and MMS) in *Lycopersicon esculentum* Mill. *Int. J. Plant Animal Env. Sci.* 6(3): 18–26.

Jain, D.K., Ranwah, B.R. and Bordia, P.C. 2005. Mutagenic effectiveness and efficiency of gamma rays in Isabgol (*Plantago ovata* Forsk). *Crop. Improv. India.* 32(1): 71–77.

Jain, S. 2005. Major mutation-assisted plant breeding programs supported by FAO/IAEA. *Plant Cell Tissue Organ Culture.* 82(1): 113–123.

Jan, S., Parween, T. and Siddiqi, T.O. 2012. Anti-oxidant modulation in response to gamma radiation induced oxidative stress in developing seedlings of *Psoralea corylifolia* L. *J. Environ. Radioact.* 113: 142–149.

Jayabalan, N. and Rao, G.R. 1987. Gamma radiation induced cytological abnormalities in Lycopersicon esculentum Mill. var. pusa ruby. *Cytologia*, 52: 1–4.

Kalpande, H.V., Surashe, S.M., Badigannavar, A., More, A. and Ganapathi, T.R. 2022. Induced variability and assessment of mutagenic effectiveness and efficiency in sorghum genotypes [Sorghum bicolor (L.) Moench]. *Int. J. Radiat. Biol.* 98(2): 230–243.

Karna, S.K.L. 2013. Phytochemical screening and gas chromatography – Mass spectrometry and analysis of seed extract of *Nigella sativa* Linn. *Int. J. Chem. Stud.* 1(4):183–187.

Karthika, R. and Lakshmi, B.S. 2006. Effect of gamma rays and EMS on two varieties of soybean. *Asian J. Plant Sci.* 5(4): 721–724.

Khadke, S.G. and Kothekar, V.S. 2011. Genetic improvement of moth bean (*Vigna acconitifolia* (Jacq.) Marechal) through mutation breeding. In: *Breeding of Pulse Crops*. Eds. S. Khan and M.I. Kozgar. Kalyani Publishers, Ludhiana, India, pp. 34–54.

Khalatkar, A.S. and Bhatia, C.R. 1975. Synergistic effect of combined treatments of gamma radiation and ethyl methanesulfonate in barley. *Radiat. Bot.* 15(3): 223–229.

Khalatkar, A.S., Gopal-Ayengar, A.R. and Bhatia, C.R. 1977. Correlation between EMS uptake by barley embryos under different treatment conditions and mutation frequency. *Mutat. Res. – Fundam. Mol. Mech. Mutagen.* 43(1): 45–55.

Khan, M.H. and Tyagi, S.D. 2009. Cytological effects of different mutagens in soybean [*Glycine max* (L.) Merrill]. *Front. Agric. China.* 3(4): 397–401.

Khan, S., Mirza, K.J., Tayaab, M. and Abdin, M.Z. 2009. RAPD profile for authentication of medicinal plant *Glycyrrhiza glabra* Linn. *Internet J. Food Saf.* 11: 24–28.

Khan, S., Parveen, K. and Goyal, S. 2011. Induced mutations in chickpea-morphological mutants. *Front. Agric. China.* 5(1): 35–39.

Khan, S. and Siddiqui, B.A. 1987. Effect of methyl methane sulphonate (MMS) on pollen fertility in urdbean (*Vigna mungo* L.). *Indian J. Biol. Res.* 3: 216–217.

Khan, Z., Ansari, M.Y.K. and Gupta, H. 2012. Induction of mutations in *Cichorium intybus* L. by base analogue 6-aminopurine (6-AP) and their detection with random amplified polymorphic DNA (RAPD) analysis. *Afr. J. Biotechnol.* 11(56): 11901–11906.

Kharakwal, M.C. 1996. Accomplishments of mutation breeding in crop plants in India. In: *Isotopes and Radiation in Agriculture and Environmental Research.* Eds. M.S. Sachdev and D.L. Deb. Indian Society for Nuclear Techniques in Agriculture and Biology, New Delhi, India, pp. 196–218.

Kharkwal, M.C. 2000. Induced mutations in chickpea (*Cicer arietinum* L.) IV. Types of macromutations induced. *Indian J. Genet. Plant Breed.* 60(3): 305–320.

Kharkwal, M.C. and Shu, Q.Y. 2009. The role of induced mutations in world food security. In: *Induced Plant Mutations in the Genomics Era*. Ed. Q.Y. Shu. Food and Agriculture Organization of the United Nations, Rome, pp. 33–38.

Khattak, K.F. and Simpson, T.J. 2008. Effect of gamma irradiation on the extraction yield, total phenolic content and free radical-scavenging activity of *Nigella staiva* seed. *Food Chem.* 110(4): 967–972.

Khursheed, T., Ansari, M.Y.K. and Shahab, D. 2009. Studies on the effect of caffeine on growth and yield parameters in *Helianthus annuus* L. variety Modern. *Biol. Med.* 1(2): 56–60.

Kirtane, S.A. 2014. Studies on induced mutations in onion: frequency and spectrum of chlorophyll mutations. *Biol. Forum.* 6(2): 146–149.

Kiskó, G. and Roller, S. 2005. Carvacrol and p-cymene inactivate *Escherichia coli* O157: H7 in apple juice. *BMC Microbiol.* 5(1): 36–44.

Konzak, C.F. 1993. Induction of mutations in oats. *Mut. Breed. Newsl.* 40: 6.

Kovacs, E. and Keresztes, A. 2002. Effect of gamma and UV-B/C radiation on plant cells. *Micron.* 33(2): 199–210.

Kozgar, M.I., Goyal, S. and Khan, S. 2011. EMS induced mutational variability in *Vigna radiata* and *Vigna mungo*. *Res.J. Bot.* 6(1): 31–37.

Kozgar, M.I., Khan, S. and Wani, M.R. 2012. Variability and correlations studies for total iron and manganese contents of chickpea (*Cicer arietinum* L.) high yielding mutants. *Am. J. Food Technol.* 7: 437–444.

Krishna-Murthy, C.S. and Vasudevan, K. 1984. Induced polygenic variation following single and combination treatment with azide, EMS and gamma rays in Oats. *Crop Improv.* 11: 128–131.

Kumar, A. and Mishra, M.N. 2004. Effect of gamma-rays, EMS and NMU on germination, seedling vigour, pollen viability and plant survival in M_1 and M_2 generation of Okra (*Abelmoschus esculentus* (L.) Moench). *Adv. Plant Sci.* 17(1): 295–297.

Kumar, D.S., Nepolean, T. and Gopalan, A. 2003. Effectiveness and Efficiency of the mutagens gamma rays and ethylmethane sulphonate on umabean (*Phaseolus lunatus* L.). *Indian J. Agric. Res.* 37(2): 115–119.

Kumar, G. 2001. Cytomictic variation in Isabgol (*Plantago ovata* forsk). *Nucleus.* 44: 180–182.

Kumar, S. and Dubey, D.K. 1998. Effects of separate and simultaneous application of gamma rays and EMS on germination, growth, fertility and yield in cultivars Nirmal and LSD-3 of khesari (*Lathyrus sativus* L.). *J. Phytol. Res.* 11(2): 165–170.

Kumar, G. and Gupta, P. 2007. Mutagenic efficiency of lower doses of gamma rays in black cumin (*Nigella sativa* L.). *Cytologia.* 72(4): 435–440.

Kumar, G. and Gupta, P. 2008. Lead induced differential genotoxicity in *Capsicum annuum* L. *Curr. Biot.* 3(2): 261–273.

Kumar, R. and Mani, S.C. 1997. Chemical mutagenesis in Manhar variety of rice (*Oryza sativa* L.). *Indian J. Genet. Plant Breed.* 57(2): 120–126.

Kumar, G. and Rai, P. 2006. Partial genome elimination through micronuclei in soybean (*Glycine max*). *Natl Acad Sci Lett.* 29(11–12): 417–421.

Kumar, P. and Singhal, V.K. 2008. Cytology of *Caltha palustris* L. (Ranunculaceae) from cold regions of Western Himalayas. *Cytologia.* 73(2): 137–143.

Kumari, M., Sinhal, V.K., Srivastava, A. and Singh, V.P. 2011. Cytogenetic effects of individual and combined treatment of Cd^{2+}, Cu^{2+} and Zn^{2+} in *Vigna radiata* (L.) Wilczeck. *J. Phytol.* 3(8): 38–42.

Kumar, G. and Srivastava, P. 2010. Comparative radiocytological effect of gamma rays and laser rays on safflower. *Rom. J. Biol. Plant Biol.* 55(2): 105–111.

Kumari, N., Tajmul, M. and Yadav, S. 2015. Proteomic analysis of mature *Lagenaria siceraria* seed. *Appl. Biochem. Biotechnol.* 175(8): 3643–3656.

Kumari, S. 1994. Effect of gamma rays on germination and seedling growth in *Vicia*. *Neo Bot.* 2(1): 65–66.

Kumbhakar, D.V., Datta, A.K., Mandal, A., Das, D., Gupta, S., Ghosh, B., Halder, S. and Dey, S. 2016. Effectivity of copper and cadmium sulphide nanoparticles in mitotic and meiotic cells of *Nigella sativa* L. (black cumin)–can nanoparticles act as mutagenic agents? *J. Exp. Nanosci.* 11(11): 823–839.

Kundu, S.K. and Singh, D.P. 1981. EMS induced variability in black gram. *Crop Improv.* 8: 71–72.

Lakshmi, N., Gupta, C.G. and Srivalli, T. 1997. Morphology and cytogenetics of a multiploid sporocyte mutant in *Capsicum*. *J. Indian Bot. Soc.* 76: 281–282.

Lal, G.M., Toms, B. and Smith, S. 2009a. Induced chlorophyll mutations in black gram. *Asian J. Agric. Sci.* 1(1): 1–3.

Lal, G. M., Toms, B. and Lal, S.S. 2009b. Mutagenic sensitivity in early generation in blackgram. *Asian J. Agric. Sci.* 1(1): 9–11.

Lal, R.K. and Sharma, J.R. 2000. Effects of gamma irradiation (Co60) on economic traits in isabgol (*Plantago ovata*). *J. Med. Aromat. Plants Sci.* 22(2/3): 251–255.

Laskar, R.A., Khan, S., Khursheed, S., Raina, A. and Amin, R. 2015. Quantitative analysis of induced phenotypic diversity in chickpea using physical and chemical mutagenesis. *J. Agron.* 14(3): 102–111.

Leitao, J.M. 2012. Chemical mutagenesis. In: *Plant Mutation Breeding and Biotechnology*. Eds. Q.Y. Shu, B.P. Forster and H. Nakagawa. CABI, Wallingford, UK, pp. 135–158.

Lyakh, V., Soroka, A. and Vasin, V. 2005. Influence of mature and immature sunflower seed treatment with ethylmethanesulphonate on mutation spectrum and frequency. *Helia.* 28(43): 87–98.

Mahamune, S.E. and Kothekar, V.S. 2011. Gamma ray induced flower colour and seed mutants in French bean (*Phaseolus vulgaris* L.). *Recent Res. Sci. Technol.* 3(5): 33–35.

Mahamune, S.E. and Kothekar, V.S. 2012. Induced mutagenic frequency and spectrum of chlorophyll mutants in French bean. *Int. Multidiscip. Res. J.* 2(3): 30–32.

Mahmoud, A.A. and AL-Twaty, N. 2006. Effect of gamma irradiation and sodium azide on some economic traits in tomato. *Saudi J. Biol. Sci.* 13(1): 44–49.

Mahna, S.K. and Garg, R. 1994. Induction of mutations in *Vigna radiata* L. Wilczek using hypoxanthine, a purine base analogue. *J. Indian Bot. Soc.* 73(1–2): 51–54.

Mahna, S.K. and Singh, D. 1982. Induced mutation in tomatillo (*Physalis ixocarpa* Brot.) Biol. *Plant.* 24(4): 307–310.

Majors, R.E. 1999. An overview of sample preparation methods for solids. *LC-GC Europe.* 17(6): 8–13.

Makeen, K., Suresh, B.G., Lavanya, G.R. and Kumari, A. 2013. Study of chlorophyll and macromutations induced by gamma rays and sodium azide in urd bean (*Vigna mungo* L. Hepper). *Afr. J. Agric. Res.* 8(47): 5958–5961.

Maldhure, B.B. 1990. Cytogenetic studies in *Vigna mungo* (L.) Hepper. Ph.D. Thesis. Nagpur University, Nagpur, India.

Mamone, G., Ferranti, P., Chianese, L., Scafuri, L. and Addeo, F. 2000. Qualitative and quantitative analysis of wheat gluten proteins by liquid chromatography and electrospray mass spectrometry. *Rapid Commun. Mass Spectrom.* 14(10): 897–904.

Manju, P.S., Mercy, T. and Nair, V.G. 1983. Induction of variability in horse-gram *(Vigna unguiculata)* with EMS and γ -rays. *Legume Res.* 6: 21–28.

Mao, L.H. and Fang, Y.M. 2014. ISSR primer screening and preliminary evaluation of genetic diversity in *Haplocladium microphyllum. Biochem. Syst. Ecol.* 55:107–111.

Mastelic, J., Jerkovic, I., Blažević, I., Poljak-Blaži, M., Borović, S., Ivančić-Baće, I., Smrečki, V., Žarković, N., Brčić-Kostic, K., Vikić-Topić, D. and Müller, N. 2008. Comparative study on the antioxidant and biological activities of carvacrol, thymol, and eugenol derivatives. *J. Agric. Food Chem.* 56(11): 3989–3996.

Mba, C. 2013. Induced mutations unleash the potentials of plant genetic resources for food and agriculture. *Agronomy.* 3(1): 200–231.

Mehandjiev, A. 2005. Role of experimental mutagenesis for genetic improvement of pea and soybean. In: *Role of Classical Mutation Breeding in Crop Improvement*. Ed. S.K. Datta. Daya Publishing House, Delhi, India, pp. 73–98.

Mehandjiev, A., Kosturkova, G. and Mihov, M. 2001. Enrichment of *Pisum sativum* gene resources through combined use of physical and chemical mutagens. *Isr. J. Plant Sci.* 49(4): 280–284.

Mehetre, S.S. and Patil, V.R. 2004. Genetics and morphological studies of new genetic male sterile line of cotton (*Gossypium arboreum* L.). *Indian J. Genet. Plant Breed.* 64(2): 161–162.

Mensah, J.K., Obadoni, B.O., Akomeah, P.A., Ikhajiagbe, B. and Ajibolu, J. 2007. The effects of sodium azide and colchicine treatments on morphological and yield traits of sesame seed (*Sesame indicum* L.). *Afr. J. Biotechnol.* 6(5): 534–538.

Minisi, F.A., El-mahrouk, M.E., Rida, M.E.F. and Nasr, M.N. 2013. Effects of gamma radiation on germination, growth characteristics and morphological variations of *Moluccella laevis* L. *Am.-Eurasian J. Agric. Environ. Sci.* 13: 696–704.

Mitra, P.K. and Bhowmik, G. 1998. Effect of mutagens on some biological parameters of *Nigella sativa* L. *Adv. Plant Sci.* 11: 155–162.

Moh, C.C. 1972. Induced seed coat colour mutations in beans and their significance for bean improvement. In: *Induced Mutations and Plant Improvement*, Panel Proceedings Series – International Atomic Energy Agency No., IAEA, Vienna, Austria, pp. 67–72.

Mohammed, A.R., Rounds, E.W. and Tarpley, L. 2007. Response of rice (*Oryza sativa* L.) tillering to sub-ambient levels of ultraviolet-B radiation. *J. Agron. Crop Sci.* 193(5): 324–335.

Mokobia, C.E. and Anomohanran, O. 2005. The effect of gamma irradiation on the germination and growth of certain Nigerian agricultural crops. *J. Radiol. Prot.* 25(2): 181–188.

Monica, S. and Seetharaman, N. 2017. Induced chlorophyll and viable mutations in *Lablab purpureus* (L.) sweet var. *Typicus* through gamma rays and ethyl methane sulphonate. *Int. J. Pharma Bio Sci.* 8(2): 240–249.

Morikawa, T., Xu, F., Kashima, Y., Matsuda, H., Ninomiya, K. and Yoshikawa, M. 2004. Novel dolabellane-type diterpene alkaloids with lipid metabolism promoting activities from the Seeds of *Nigella sativa*. *Org. Lett.* 6(6): 869–872.

Mostafa, G.G. 2011. Effect of sodium azide on the growth and variability induction in *Helianthus annuus* L. *Int. J. Plant Breed. Genet.* 5(1): 76–85.

Mueller, U.G. and Wolfenbarger, L.L. 1999. AFLP genotyping and fingerprinting. *Trends Ecol. Evol.* 14(10): 389–394.

Mujeeb, K.A. and Greig, J.K. 1976. Growth stimulation in *Phaseolus vulgaris* L. induced by gamma irradiation of seeds. *Biol. Plant.* 18(4): 301–303.

Muthuswamy, A. and Jayabalan, N. 2002. Effect of mutagens on pollen fertility of cotton (*Gossypium hirsutum* L.). *Indian J. Genet.* 62(2): 187.

Nader, M.A., El-Agamy, D.S. and Suddek, G.M. 2010. Protective effects of propolis and thymoquinone on development of atherosclerosis in cholesterol-fed rabbits. *Arch. Pharm. Res.* 33(4): 637–643.

Nakagawa, H. 2009. Induced mutations in plant breeding and biological researches in Japan. *Crops.* 242(188): 48.

Narasimha, C.S. and Bhalla, J.K. 1988. EMS induced male sterile mutant in pigeonpea (*Cajanus cajan* (L.) Millsp.). *Indian J. Genet. Plant Breed.* 48(3): 303–304.

Narsinghani, V.G. and Kumar, S. 1976. Mutation studies in cowpea. *Indian J. Agric. Sci.* 46(2): 61–64.

Natarajan, A.T. 2005. Chemical mutagenesis: From plants to human. *Curr. Sci.* 89(2): 312–317.

Nayak, P.S., Upadhyaya, S.D. and Upadhyaya, A. 2009. A HPTLC densitometer determination of sinapic acid in chandrasur (*Lepidium sativum*). *J. Sci. Res.* 1(1): 121–127.

Okwu, D.E. 2004. Phytochemicals and vitamin content of indigenous spices of South Eastern Nigeria. *J. Sustain. Agric. Env.* 6(1): 30–37.

Oluwatosin, O.B. 2001. Inheritance of genes controlling flower color in cowpea (*Vigna unguiculata* L. Walp). *J. Genet. Plant Breed.* 1(4): 259–262.

Oommen, S.K. and Gopimany, R. 1986. Efficient mutagenesis in cowpea. *Agric. Res. J. Kerala.* 22: 57–62.

Oumzil, H., Ghoulami, S., Rhajaoui, M., Ilidrissi, A., Fkih-Tetouani, S., Faid, M. and Benjouad, A. 2002. Antibacterial and antifungal activity of essential oils of *Mentha suaveolens*. *Phytotherapy Research: An International Journal Devoted to Pharmacological and Toxicological Evaluation of Natural Product Derivatives*. 16(8): 727–731.

Panigrahi, K.K., Mohanty, A. and Jyotshnarani, B.B. 2015. Mutagenic efficiency and effectiveness of gamma rays, ethyl methane sulphonate (ems), nitrosoguanidine (ng) and their synergistic effect for different polygenic traits in black gram (*Vigna mungo* (l.) hepper) through induced mutagenesis. *Int. J. Plant Animal Env. Sci.* 5(1): 292–299.

Paran, I. and Michelmore, R.W. 1993. Development of reliable PCR-based markers linked to downy mildew resistance genes in lettuce. *Theor. Appl. Genet.* 85(8): 985–993.

Patil, A., Taware, S.P., Oak, M.D., Tamhankar, S.A. and Rao, V.S. 2007. Improvement of oil quality in soybean [*Glycine max* (L.) Merrill] by mutation breeding. *J. Am. Oil Chem. Soc.* 84(12): 1117–1124.

Patzak, J. 2001. Comparison of RAPD, STS, ISSR and AFLP molecular methods used for assessment of genetic diversity in hop (*Humulus lupulus* L.). *Euphytica.* 121(1): 9–18.

Plesnik, S. 1993. The evaluation of some quantitative traits in M1 generation in soybean after laser emission and ethyleneimine treatment. *Acta facultatis rerum naturalium universitiatis comenianae, Genetica et biologia molecularis*. 24(25): 105–113.

Powell, W., Morgante, M., Andre, C., Hanafey, M., Vogel, J., Tingey, S. and Rafalski, A. 1996. The comparison of RFLP, RAPD, AFLP and SSR (microsatellite) markers for germplasm analysis. *Mol. Breed.* 2(3): 225–238.

Prabha, R., Dixit, V. and Chaudhary, B.R. 2011. Comparative spectrum of sodium azide responsiveness in plants. *World J. Agric. Res.* 7(1): 104–108.

Prakash, B.G. and Shambhulingappa, K.G. 2000. Effect of gamma rays and EMS on biological end points and estimation of LD_{50} value in rice bean. *Karnataka J. Agric. Sci.* 13(1): 155–157.

Prem Sekar, S. and Appadurai, R. 1982. Effect of doses of γ -rays and EMS on the germination and survival of induced mutations in' pigeon pea (*Cajanus cajan*). *Indian J. Agric. Sci.* 51: 381–386.

Privalov, G.F. 1986. "Zyrianka": A mutant variety of sea buckthorn. *Mut. Breed. Newsl.* 28: 4–5.

Qin, J., Leung, F.C., Fung, Y., Zhu, D. and Lin, B. 2005. Rapid authentication of ginseng species using microchip electrophoresis with laser-induced fluorescence detection. *Anal. Bioanal. Chem.* 381(4): 812–819.

Raghuvanshi, S.S., Pathak, C.S. and Singh, A.K. 1978. Effect of preirradiation colchicine treatment on mutation spectrum of *Phaseolus aureus* Roxb. *Cytologia*. 43(1): 143–151.

Raina, A., Laskar, R.A., Tantray, Y.R., Khursheed, S., Wani, M. and Khan, S. 2020. Characterization of induced high yielding cowpea mutant lines using physiological, biochemical and molecular markers. *Sci. Rep.* 10(1): 1–22.

Rajput, M.A. 1974. Increased variability in the M_2 of gamma-irradiated mung beans (*Phaseolus aureus* Roxb.). *Radiat. Bot.* 14(2): 85–89.

Ramezani, P., Siavoshi, M., More, A.D., Ebrahimi, M. and Dastan, S. 2017. Gamma rays and EMS induced flower color mutation in grasspea (*Lathyrus sativus* Linn.). *J. Agric. Sci.* 23(4): 423–427.

Ramulu, K. S. 1971. Induced structural changes and meiotic aberrations in *Sorghum*. *Cytologia*. 36(2): 229–236.

Rang, S.K. and Datta, A.K. 1998. A male sterile mutant with desynaptic behaviour of chromosome in *Nigella sativa* L. *J. Phytol. Res.* 11: 91–94.

Rang, S.K. and Datta, A.K. 2001. Mutation in seed-coat colour in black cumin (*Nigella sativa* L.). *Indian J. Genet. Plant Breed.* 61(1): 80–81.

Rao, N.B. and Laxmi, N. 1980. Gamma ray induced meiotic abnormalities in *Capsicum annuum* L. *Caryologia*. 33(4): 509–518.

Rapoport, I.A. 1966. *Peculiarities and Mechanism of Action of Supermutagens*. Supermutagens, Nauka, Moscow, pp. 9–23.

Rasooli, I. and Mirmostafa, S.A. 2003. Bacterial susceptibility to and chemical composition of essential oils from *Thymus kotschyanus* and *Thymus persicus*. *J. Agric. Food Chem.* 51(8): 2200–2205.

Raut, R.S. and Thombre, P.G. 1977. Induced autotetraploids of *Impatiens balsamina* L. [India]. *J. Maharashtra Agric. Univ.* 2(2): 141–144.

Reddy, G.M. and Reddy, T.P. 1973. Induction of some grain shape and morphological mutations in rice, variety IR-8. *Radiat. Bot.* 13(3): 181–184.

Reddy, S.S. and Rao, G.M. 1982. Cytogenetic effects of agricultural chemicals II. Effects of herbicides "Lasso and basagran" on chromosomal mechanism in relation to yield and yield components in chilli (*Capsicum annuum* L.). *Cytologia*. 47(2): 257–267.

Reddi, T.V.V.S. 1977. *Cytogenetics of mutants induced by physical and chemical mutagens in three varieties of rice* (*Oryza sativa* L.). Ph.D. Thesis. Andhra University. Waltair. India. p. 145.

Reddi, T.V.V.S. and Suneetha, J. 1992. Chlorophyll-deficient mutations induced in rice by alkylating agents and azide. *Cytologia*. 57(2): 285–288.

Reddy, V.R. and Annadurai, M. 1992. Induction of chlorophyll mutants in lentil. *Res. JPI Env.* 8(1): 59–69.

Reddy, V.R.K. 2001. Mutation breeding in hexapliod triticlaes-desirable mutants. *Adv. Plant Sci.* 14(1): 255–258.

Rokadia, P., Mathur, V.L., Pandya, R. and Kothari, P.R. 1991. Note on mutagenic efficiency of gamma rays in fenugreek. *Indian J. Hortic.* 48(3): 269–272.

Roy, S. and Biswas, A.K. 2005. Isolation of a white flowered mutant through seed culture in *Spathoglottis plicata* Blume. *Cytologia*. 70(1): 1–6.

Saba, N. and Mirza, B. 2002. Ethyl methane sulfonate induced genetic variability in *Lycopersicon esculentum*. *Int. J. Agric. Biol.* 4(1): 89–92.

Sagade, A.B. and Apparao, B.J. 2011. M_1 generation studies in urdbean (*Vigna mungo* (L.) Hepper). *Asian J. Exp. Biol. Sci.* 2(2): 372–375.

Saha, A. and Datta, A.K. 2000. Gamma-rays induced reciprocal translocations in *Nigella damascena* L. *Biol. Sci.* 70(2): 159–163.

Salem, M.L. 2005. Immunomodulatory and therapeutic properties of the *Nigella sativa* L. seed. *Int. Immunopharmacol.* 5(13–14): 1749–1770.

Salimath, P.M., Toker, C., Sandhu, J.S., Kumar, J., Suma, B., Yadav, S.S. and Bahl, P.N. 2007. Conventional Breeding Methods. In: *Chickpea Breeding and Management*. Eds. S.S. Yadav, R. Redden, W. Chen and B. Sharma. CAB International, London, pp. 369–390.

Sanamyan, M.F., Petlyakova, J.E. and Musaev, D.A. 2000. The development and a cytogenetic study of monosomics of *Gossypium hirsutum* L. *Biol. Plant.* 43(2): 193–197.

Sanamyan, M.F. and Rakhmatullina, E.M. 2003. Cytogenetic analysis of translocations in cotton. *Plant Breed.* 122(6): 511–516.

Sangsiri, C., Sorajjapinun, W. and Srinives, P. 2005. Gamma radiation induced mutations in mungbean. *Sci. Asia.* 31: 251–255.

Santoni, V., Bellini, C. and Caboche, M. 1994. Use of two-dimensional protein-pattern analysis for the characterization of *Arabidopsis thaliana* mutants. *Planta.* 192(4): 557–566.

Sasidharan, S., Chen, Y., Saravanan, D., Sundram, K.M. and Latha, Y.L. 2011. Extraction, isolation and characterization of bioactive compounds from plants' extracts. *Afr. J. Tradit. Complement. Altern. Med.* 8(1): 1–10.

Savaskan, C. 2002. The effects of gamma irradiation on the pollen size of *Gossypium hirsutum* L. *Turk. J. Bot.* 26(6): 477–480.

Sen, S., Skaria, R. and Muneer, P.A. 2010. Genetic diversity analysis in *Piper* species (Piperaceae) using RAPD markers. *Mol. Biotechnol.* 46(1): 72–79.

Sengupta, S. and Datta, A.K. 2005. Induced narrow leaf mutant of sesame (*Sesamum indicum* L.). *Indian J. Genet. Plant Breed.* 65(1): 59–60.

Shah, D., Singh, R.R. and Kaur, A. 1992. Gamma ray induced nucleolar behaviour and mitotic anomalies in *Vigna mungo. Acta Bot. Indica.* 20: 300–302.

Shah, T.M., Mirza, J.I., Haq, M.A. and Atta, B.M. 2008. Induced genetic variability in chickpea *(Cicer arietinum* L.). II. Comparative mutagenic effectiveness and efficiency of physical and chemical mutagens. *Pak. J. Bot.* 40(2): 605–613.

Shahab, D., Jaiswal, K. and Khan, M.A. 2007. Effect of diethyl sulphate (DES) on seed germination and seedling growth of *Solanum melongena* L. *Indian J. App. Pure Biol.* 22(2): 227–229.

Shahwar, D., Ansari, M.Y.K., Bhat, T.M., Choudhary, S. and Aslam, R. 2017. Evaluation of high yielding mutant of lentil developed through caffeine of an exotic germplasm. *Int. J. Plant Breed. Genet.* 11(2): 55–62.

Shakoor, A., Ahsan-Ul-Haq, M. and Sadiq, M. 1978. Induced variation in mung bean (*Vigna radiata* (L.) Wilczek). *Environ. Exp. Bot.* 18(3):169–175.

Shamina, N.V., Dorogova, N.V., Zagorskaia, A.A., Deĭneko, E.V. and Shumnyĭ, V.K. 2000. Disruption of male meiosis in transgenic line tobacco res91. *Tsitologiia.* 42(12): 1159–1164.

Sharifi-sirchi, G.R., Baghizadeh, A., Golestani, M. and Saffari, V.R. 2012. The effects of chemical and physical mutagens on morphological and cytological characters of barley (Iranian cv. Nosrat). *Afr. J. Biotechnol.* 11(16): 3842–3848.

Sharma, B.K. and Mishra, M.N. 2007. Micro-mutations for fruit number, fruit length and fruit yield characters in gamma-irradiated generation of ANKUR-40 variety of okra [*Abelmoschus esculentus* (L.) Monech]. Int. *J. Plant Sci. Muzaffarnagar.* 2(2): 208–211.

Sharma, M., Khan, A.H. and Bhat, T.A. 2009. Assessment of mutagenicity of individual and combination treatments of gamma rays and MMS in broad bean (*Vicia faba* L.). *Cytologia.* 74(2): 235–241.

Sharma, N. and Anis, M. 1995. Effect of MMS on seed germination and quantitative characters in chili. *Adv. Plant Sci.* 8(1): 100–105.

Sharma, P.C., Yelne, M.B. and Dennis, T.J. 2005. *Database on Medicinal Plants used in Ayurveda*, Vol. 6, CCRAS, New Delhi. 420–440.

Sharma, R.P. and Chopra, V.L. 1994. Mutation breeding. In: *Plant Breeding: Theory and Practice.* Ed. V.L. Chopra. Oxford and IBH Pub. Co, New Delhi, India, pp. 409–466.

Sharma, S., Kumar, P., Gambhir, G., Kumar, R. and Srivastava, D.K. 2018. Assessment of genetic diversity in lettuce (*Lactuca sativa* L.) germplasm using RAPD markers. *3 Biotechnology.* 8(1): 9.

Sharma, S.K. and Sharma, B. 1981. Induced chlorophyll mutations in lentil. *Indian J. Genet. Plant Breed.* 41(3): 328–333.

Sharma, V. and Kumar, G. 2004. Meiotic studies in two cultivars of *Cicer arietinum* L. after EMS treatment. *Cytologia.* 69(3): 243–248.

Sharon, M. and Muralidharan, K. 1978. Effect of -irradiation on the growth of *Sorghum vulgare. Indian J. Plant Physiol.* 21: 156–161.

Shehata, S.M., Allah, A.A. and Zayed, B.A. 2009. Development of salt tolerant rice lines through mutation breeding. *J. Agric. Res. Kafrelsheikh Uni.* 35(4): 954–963.

Siddiq, E.A. 1973. Cytogenetical effects of physical and chemical mutagens in rice. Indian J. Genet. Pl. Breed. 33: 162–171.

Siddiqui, B.A. 1999. Mutagenesis – Tools and techniques: A practical view. In: *Breeding in Crop Plants: Mutations and in vitro Mutation Breeding*. Eds. Bahar A. Siddiqui and Samiullah Khan, Kalyani Publishers, Ludhiana, India, pp. 20–34.

Siddiqui, M.A., Khan, I.A. and Khatri, A. 2009. Induced quantitative variability by gamma rays and ethylmethane sulphonate alone and in combination in rapeseed (*Brassica napus* L.). *Pak. J. Bot.* 41(3): 1189–1195.

Siddiqui, S., Meghvansi, M.K., Khan, S.S. and Aali, N.S. 2008. Mutagenic effect of herbicide (Maleic hydrazide) on seed germination and radicle length on *Trigonella foenum-graecum* L. *Ind. J. App. Pure Biol.* 23: 103–106.

Sigurbjornson, B. and Micke, A. 1969. *Progress in Mutation Breeding*. In: *Induced Mutations and Plant Improvement*, Panel Proceedings Series – International Atomic Energy Agency No., Vienna, pp. 673–697.

Singh, G., Marimuthu, P., de Heluani, C.S. and Catalan, C. 2005. Chemical constituents and antimicrobial and antioxidant potentials of essential oil and acetone extract of *Nigella sativa* seeds. *J. Sci. Food Agric.* 85(13): 2297–2306.

Singh, R.H. and Raghuvanshi, S.S. 1991. Bold seeded mutant in blackgram. *Mut. Breed. Newsl.* 38: 5–6.

Singh, R.K. 1996. Gamma-ray induced bold seeded mutant in *Vigna mungo* (L.) Hepper. *Indian J. Genet. Plant Breed.* 56(1): 104–108.

Singh, R.K. and Jhunjhunwalla, K.N. 2014. Chemical composition of volatile oil of *Nigella sativa* seeds. *World J. Pharm. Pharm. Sci.* 3(10): 1588–1594.

Sinha, R.P. and Joshi, M.G. 1986. Induced polygenic variation in hexaploid triticale in M_3 generation. *J. Nucl. Agric. Biol.* 15(3): 170–174.

Singh, S.P., Singh, N.K., Singh, R.P. and Prasad, J.P. 2006. Mutagenic effect of gamma rays and EMS on nodulation, yield and yield traits on lentil. *Indian J. Pul. Res.* 19(1): 53–55.

Singh, S.P., Singh, R.P., Singh, N.K., Prasad, J.P. and Shahi, J.P. 2007. Mutagenic efficiency of gamma-rays, ethyl methane sulphonate and its combination on micro-sperma lentil (*Lens culinaris* Medik). *Int. J. Agric. Sci.* 3(1): 113–118.

Singh, V.P. and Chaturvedi, S.N. 1980. Mutagenic efficiency of EMS, NMU, gamma-rays and their combined treatments in *Vigna radiata* (L.) Wilczek. *Genet. Agrar.* 34(3/4): 331–348.

Singh, V.P. and Sharma, R. 1993. γ-Rays- and EMS-induced leaf mutants in mung bean (*Vigna radiata* (L) Wilczek). *Curr. Sci.* 65(8): 636–638.

Singhal, V.K., Kaur, D. and Kumar, P. 2008. Effect of cytomixis on the pollen size in "Seabuckthorn" (*Hippophae rhamnoides* L., Elaeagnaceae). *Cytologia.* 73(2): 167–172.

Skripnikova, Z.V. 1976. Mutants of sweet pepper induced by chemical mutagens. (Russian). *Genetika.* 12: 37–44.

Solanki, I.S. 2005. Isolation of macromutations and mutagenic effectiveness and efficiency in lentil (*Lens culinaris* Medik.). *Indian J. Genet. Plant Breed.* 65(4): 264–268.

Srinivas, T.R. and Veerabadhiran, P. 2010. Efficiency and effectiveness of physical and chemical mutagens and their combination in inducing chlorophyll mutations in M_2 generation of Lablab [*Lablab purpureus* (l.) Sweet var. Typicus]. *Electron. J. Plant Breed.* 1(4): 752–757.

Srivastava, M. 1974. Radiation induced desynaptic mutant in barley. *Cytologia.* 39(1): 63–68.

Srivastava, P., Marker, S., Pandey, P. and Tiwari, D.K. 2011. Mutagenic effects of sodium azide on the growth and yield characteristics in wheat (*Triticum aestivum* L em. Thell.). *Asian J. Plant Sci.* 10(3): 190–201.

Stadler, L.J. 1930. Some genetic effects of X-rays in plants. *Journal of Heredity.* 21: 3–19.

Subramanian, D. 1980. Effect of gamma-radiation in *Vigna*. *Indian J. Genet. Plant Breed.* 40(1): 187–194.

Sudhakaran, I.V. 1971. Meiotic abnormalities induced by gamma rays in *Vinca rosea* Linn. *Cytologia.* 36(1): 67–79.

Sundaravadivelu, K., Ranjithselvi, P. and Reddy, V.R.K. 2006. Induced genetic variability in cotton (*Gossypium hirsutum* L.) for yield and its components. *Crop Res. Hisar.* 32(3): 442–446.

Swaminathan, M.S., Chopra, V.L. and Bhaskaran, S. 1962. Chromosome aberrations and the frequency and spectrum of mutations induced by ethylmethane sulphonate in barley and wheat. *Indian J. Genet. Plant Breed.* 22(3): 192–207.

Swaminathan, M.S. 1966. Use of induced mutation. *Indian Farming.* 34: 125–127.

Tah, P.R. 2008. Studies of leaflet mutants in Mungbean (*Vigna radiata* (L.) Wilczek). *Int. J. Plant Breed. Genet.* 2(2): 75–84.

Tantray, A.Y., Raina, A., Khursheed, S., Amin, R. and Khan, S. 2017. Chemical mutagen affects pollination and locule formation in capsules of black cumin (*Nigella sativa* L.). *Int. J. Agric. Sci.* 8(1): 108–118.

Thapa, C.B. 2004. Effect of acute exposure of gamma rays on seed germination and seedling growth of *Pinus kesiya* Gord and *P. wallichiana* AB Jacks. *Our Nature.* 2(1):13–17.

Thiellement, H., Bahrman, N., Damerval, C., Plomion, C., Rossignol, M., Santoni, V., de Vienne, D. and Zivy, M. 1999. Proteomics for genetic and physiological studies in plants. *Electrophoresis.* 20(10): 2013–2026.

Toker, C. and Cagirgan, M.I. 2004. Spectrum and frequency of induced mutations in chickpea. *International and Chickpea Pigeonpea Newsletter.* 11: 8–10.

Toker, C., Ceylan, F.O., Inci, N.E., Yildirim, T. and Cagirgan, M.I. 2012. Inheritance of leaf shape in the cultivated chickpea (*Cicer arietinum* L.). *Turkish J. Field Crop.* 17(1): 16–18.

Toker, C. and Ceylan, F.O. 2013. Induction and inheritance of compound leaf and cone stipule in the cultivated chickpea (*Cicer arietinum* L.). *Turkish J. Field Crop.* 18(2): 211–214.

Tripathi, R. and Kumar, G. 2010. Genetic loss through heavy metal induced chromosomal stickiness in grasspea. *Caryologia.* 63(3): 223–228.

Tripathy, S. K., Lenka, D. and Ranjan, R. 2011. Maximization of mutation of mutation frequency in grasspea (*Lathyrus sativus* L.). *Legume Res.* 34(4): 296–299.

Ultee, A., Bennik, M.H.J. and Moezelaar, R. 2002. The phenolic hydroxyl group of carvacrol is essential for action against the food-borne pathogen *Bacillus cereus*. *Appl. Environ. Microbiol.* 68(4): 1561–1568.

Vadivelu, K.K. and Rathinam, M. 1980. Mutagenicity of gamma rays and EMS and their combination in Bengalgram (*Cicer arietinum* L.). *Madras Agric. J.* 67: 360–365.

Van Rheenen, H.A., Pundir, R.P.S. and Miranda, J.H. 1993. How to accelerate the genetic improvement of a recalcitrant crop species such as chickpea. *Curr. Sci.* 65(5): 414–417.

Vandana, D.D. and Dubey, D.K. 1996. Meiotic anomalies induced by EMS and DES in faba bean (*Vicia faba* L.). *J. Indian Bot. Soc.* 75: 237–240.

Varma, A. and Shrivastava, N. 2018. Genetic structuring in wild populations of two important medicinal plant species as inferred from AFLP markers. *Plant Biosyst. – Int. J. Plant Biol.* 152(5): 1088–1100.

Vesselina, N., Galina, A., Nasya, T. and Valentina, P. 2006. Chemical mutagenesis in head cabbage (*Brassica oleracea* var. capitata L.): III Effect of ethyl methanesulphonate on micro sporogenesis and pollen fertility. *Selekcijai semenarstvo.* 12(1–2): 89–93.

Vos, P., Hogers, R., Bleeker, M., Reijans, M., Lee, T.V.D., Hornes, M., Friters, A., Pot, J., Paleman, J., Kuiper, M. and Zabeau, M. 1995. AFLP: a new technique for DNA fingerprinting. *Nucleic Acids Res.* 23(21): 4407–4414.

Waghmare, V.N. and Mehra, R.B. 2001. Induced chlorophyll mutations, mutagenic effectiveness and efficiency in *Lathyrus sativus* L. *Indian J. Genet. Plant Breed.* 61(1): 53–56.

Wani, A.A. 2011. Spectrum and frequency of macromutations induced in chickpea (*Cicer arietinum* L.). *Turk. J. Biol.* 35(2): 221–231.

Wani, A.A. and Anis, M. 2008. Gamma ray- and EMS-induced bold-seeded high-yielding mutants in chickpea (*Cicer arietinum* L.). *Turk. J. Biol.* 32(3): 161–166.

Wani, M.R., Kozgar, M.I., Tomlekova, N., Khan, S., Kazi, A.G., Sheikh, S.A. and Ahmad, P. 2014. Mutation breeding: A novel technique for genetic improvement of pulse crops particularly Chickpea (*Cicer arietinum* L.). In: *Improvement of Crops in the Era of Climatic Changes*. Eds. P. Ahmad, M.R. Wani, M.M. Azooz and L-S.P. Tran. Springer, New York, USA, pp. 217–248.

Wani, M.R. 2021. Comparative biological sensitivity and mutability of chemo-mutagens in lentil (*Lens culinaris* Medik). *Legume Res.* 44: 26–30.

Worthen, D.R., Ghosheh, O.A. and Crooks, P.A. 1998. The in vitro anti-tumor activity of some crude and purified components of blackseed, *Nigella sativa* L. *Anticancer Res.* 18(3A): 1527–1532.

Wu, J.L., Wu, C., Lei, C., Baraoidan, M., Bordeos, A., Madamba, M.R.S. et al. 2005. Chemical-and irradiation-induced mutants of indica rice IR64 for forward and reverse genetics. Plant Molecular Biology. 59:85–97.

Yamaguchi, H., Nagatomi, S., Morishita, T., Degi, K., Tanaka, A., Shikazono, N. and Hase, Y. 2003. Mutation induced with ion beam irradiation in rose. *Nuclear Instruments and Methods in Physics Research Section B: Beam Interactions with Materials and Atoms.* 206: 561–564.

Yan, P., Pang, Q.H., Jiao, X.W., Zhao, X., Shen, Y.J. and Zhao, S.J. 2008. Genetic variation and identification of cultivated *Fallopia multiflora* and its wild relatives by using chloroplast matK and 18S rRNA gene sequences. *Planta Med.* 74(12): 1504–1509.

Yip, P.Y., Chau, C.F., Mak, C.Y. and Kwan, H.S. 2007. DNA methods for identification of Chinese medicinal materials. *Chin. Med.* 2(1): 1–19.

Yousuf, P.Y., Abd_Allah, E.F., Nauman, M., Asif, A., Hashem, A., Alqarawi, A.A. and Ahmad, A. 2017. Responsive proteins in wheat cultivars with contrasting nitrogen efficiencies under the combined stress of high temperature and low nitrogen. *Genes.* 8(12): 1–23.

Zeerak, N.A. 1992. Cytogenetical effects of gamma rays and ethyl methane sulphonate in tomato (*Lycopersicon esculentum* var. Cerasiforme). *Phytomorphology.* 42: 81–86.

Zeerak, N.A. 1998. Some induced mutations of phylogenetic interest in *Solanum melongena. J. Indian Bot. Soc.* 77(1–4): 253–254.

Zhao, K.J., Dong, T.T.X., Cui, X.M., Tu, P.F. and Tsim, K.W.K. 2003. Genetic distinction of radix adenophorae from its adulterants by the DNA sequence of 5S-rRNA spacer domains. *Am. J. Chin. Med.* 31(6): 919–926.

Zheng, W., Li, X., Zhang, L., Zhang, Y., Lu, X. and Tian, J. 2015. Improved metabolites of pharmaceutical ingredient grade *Ginkgo biloba* and the correlated proteomics analysis. *Proteomics.* 15(11): 1868–1883.

12 Nanoparticle Fabrication from Herbal Plants and Their Role in Antimicrobial Activities

Fekadu Ashine C., Rakesh Kumar Bachheti, Archana Bachheti, Limenew Abate and Azamal Husen

CONTENTS

12.1 INTRODUCTION

Nanotechnology is a recently emerged field of modern science and technology which involves the production of enormous nanomaterials at the molecular level (nanoscale) and their dimension of less than 100 nm that exhibit distinct material properties compared to bulk counterparts (Sinha et al. 2009, Khodakovskaya et al. 2012, Patel et al. 2015, Siddiqui, Al-Whaibi and Mohammad 2015, Husen 2017, Albrahim et al. 2021a, b). Currently, the synthesis of NPs and their characterization are rapidly gaining importance because of huge applications in the different fields of study such as biology, chemistry, physics, medicine, engineering and technology and photocatalytic degradation (Eyasu, Yadav and Bachheti 2013, Patel et al. 2015, Siddiqui, Al-Whaibi and Mohammad 2015, Bachheti et al. 2019, Joshi et al. 2019). Thus, the fabrication of nanomaterials is initiated via chemical, physical and biological approaches globally. More recently, rather than designing new nanomaterials, developing its novel synthesis methods is gaining more attraction, thus attaining special focus from many researchers. Top-down and bottom-up (Figure 12.1) are the most common approaches to the synthesis of NPs via the conversion of bulk materials and atoms into the nanoscale level (Demirbas et al. 2020, Hyeon 2003).

The most common bottom-up approaches are chemical and biological methods, but chemical methods have their own limitations. Hence, biological approaches to synthesizing NPS is a very promising research thematic area for many researchers due to limitation of chemical and physical approaches, such as toxicity, radiation, expensive processes and high energy requirements (Sinha et al. 2009, Saratale et al. 2018, Fardsadegh et al. 2019, Kannan et al. 2020, Albrahim et al. 2021b). In the last few years, bottom-up approaches, especially the biofabrications of metal and metal oxide NPs from biological resources, have become popular in chemistry and

DOI: 10.1201/9781003179139-12

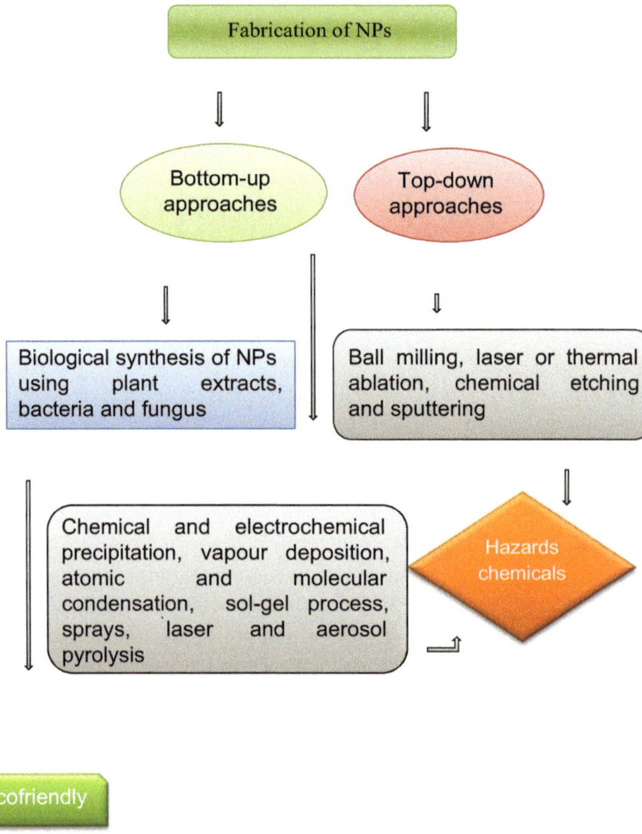

FIGURE 12.1 Approaches for NPs fabrications.

chemical technologies and now become a major focus of many researchers (Sharma, Yngard and Lin 2009, Mubayi et al. 2012, Rauwel et al. 2015, Hussain et al. 2016, Parveen. Banse and Ledwani 2016, Singh et al. 2018, Gour and Jain 2019). Hence, biological synthesis methods have been proposed as a cost-effective, available, ecofriendly and inexpensive process alternatives to traditional methods. These methods encompass biological entities such as bacteria, fungi, yeast, peptides, algae and plant extracts for NPs fabrications (Figure 12.2) and are becoming more popular (Thakkar, Mhatre and Parikh 2010, Durán et al. 2011, Zikalala et al. 2018). Compared to bacterial and fungi, NPs synthesis from plants and plant-related products need less time to produce Ag-NPs (Rauwel et al. 2015).

In general, biological approaches of NPs synthesis have several benefits such as simple, rapid, non-toxic, avoiding toxic chemicals, cleaner production, use of ecofriendly benign solvents, cost-effective, ease of preparation, control of size and shape, low energy demand and performs under moderate operations (Parveen, Banse and Ledwani 2016, Ahmed and Mustafa 2020). The fabricated NPs have high surfaces, high reactions, tunable, pore sizes and particularly morphology of engineered NPs (Siddiqui, Al-Whaibi and Mohammad 2015, Albrahim et al. 2021b). The effectiveness of nanomaterials depends on their chemical structure, surface area, particle pores size, shape and their reactivity (Khodakovskaya et al. 2012, Albrahim et al. 2021b). Utilization of nanomaterials (dimension sized from 1 nm to 100 nm) is a developing area of science and technology and provides potential solutions to technological and environmental challenges more frequently found in medical, consumer products, water purifications, industrial as well as agricultural fields (Sharma, Yngard and Lin 2009, Hussain et al. 2016, Teimouri et al. 2018, Ahmed and

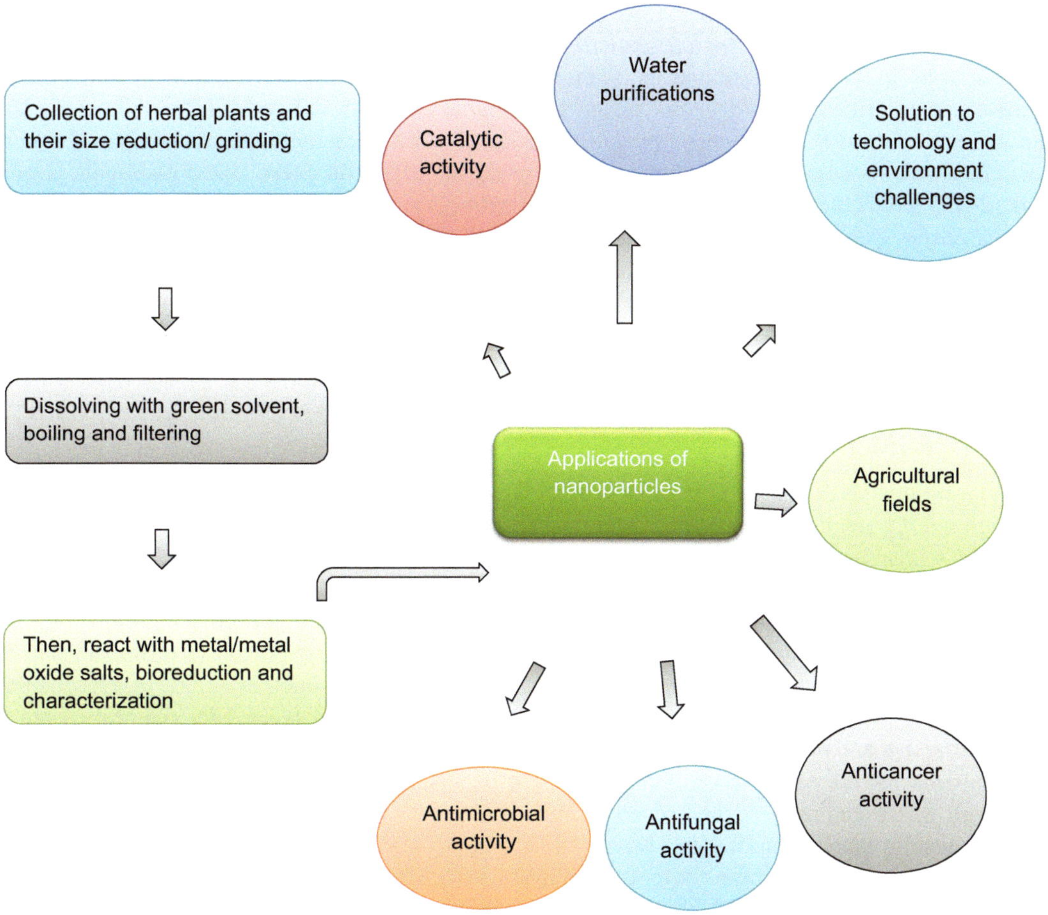

FIGURE 12.2 NPs synthesis using herbal plants and its applications.

Mustafa 2020). Furthermore, NPs have also attracted high attention in their potential applications, such as antimicrobial activity, antifungal activity and anticancer activities (Jayandran, Haneefa and Balasubramanian 2015, Ibrahim 2015, Wang et al. 2016, Jayarambabu et al. 2020). The fabricated NPs have emerged as sole antimicrobial agents due to their high surface area to volume ratios (Varier et al. 2019). Many researchers studied the fabrication of metal/metal oxide NPs using plants and every part of plant extract and their role in antimicrobial activities (Cremonini et al. 2016, Wang et al. 2017, Jayarambabu et al. 2020,). The main factors that can affect the effectiveness of NPs in the antimicrobial activities include size, shape, chemical structure, surface area, reactivity as well as the materials used for their fabrications (Mohan, Shanmugharaj and Sung Hun 2011, Hussen 2017, Jayarambabu et al. 2020, Albrahim et al. 2021b). The efficiency of Ag-NPs in antimicrobial agents is highly dependent on the shape of the NPs. Spherical shape Ag-NPs have a significant effect on antimicrobial activity compared to triangular NPs since antimicrobial activities depend mainly on the shape of fabricated NPs (Acharya et al. 2018). Fabrication of NPs using plant seed extracts was also well established and has been extensively utilized, particularly for Ag-NPs in antimicrobial activities(Suresh 2008). Currently, several research works reported on the fabrication of NPs using the extract from each part of plants, such as flowers (Bachheti et al. 2020), roots (Sre et al. 2015, Pethakamsetty et al. 2017, Bachheti et al. 2020), leaves (Kumar et al. 2014, Obeizi et al. 2020, Kannan et al. 2020), seeds (Xu et al. 2015, Sumitha et al. 2018), fruits (Gavade et al. 2015), bark (Sasikala et al. 2015), and so on, for antimicrobial applications. Miri et al. (2015) studied the fabrication of Ag-NPs using *Prosopis farcta* extract, spherical in shape and with a particle size of 10.8 nm at normal temperature and its potential in antimicrobial activity tested against the gram-positive bacteria like *Staphylococcus aureus* and *Bacillus subtilis* and gram-negative bacteria like *Escherichia coli* and *Pseudomonas aeruginosa* using disc diffusion methods and compared to control. The result showed the fabricated Ag-NPs significant effect on every tested pathogen. Kumar et al. (2015) biosynthesized Ag-NPs using tea leaf extracts and tested them against bacteria activity such as *Staphylococcus aureus* and *Escherichia coli*. The result showed that Ag-NPs are more effective in inhibition against *Staphylococcus aureus* and lead to impairment bacterial cell-cell adhesion. Therefore, this review chapter summarizes the fabrication of NPs from various herbal plants, their characterization and their potential antimicrobial activities.

12.2 FABRICATION AND CHARACTERIZATION OF NANOPARTICLES FROM HERBAL PLANTS

The conventional NPs fabrication approaches involve hazardous chemicals, expensive processes and high energy requirements. In recent times, many scientists have focused on the utilization of biological resources for nanoparticle fabrication in the field of nanotechnology for a wide range of applications in different sectors (Rao et al. 2015). Almost all biological entities, including yeast, algae, bacteria, fungi and plant extracts, used possess high bioreduction ability in controlling the size and shape of metal/metal oxide NPs (Gopinath et al. 2012). They evolved as a promising substitute for the physical and chemical procedures of NPs synthesis (Bachheti et al. 2021a, Husen et al. 2019). Among these biological resources, plants and their part extracts for NPs synthesis seem to be very effective and environmentally sound (Akhtar, Panwar and Yun 2013). Moreover, employing plant extracts for the fabrication of NPs also can be valuable over bio entities such as bacteria and fungi, which can resolve high reaction times and preserve their culture which can limit their potential for the fabrication of NPs (Kavitha et al. 2013, Roy and Das 2015, Ahmed et al. 2016). Metal/metal oxide NPs are non-toxic, have a high surface-to-volume ratio and are easy and eco-friendly (Saratale et al. 2018). Plant extracts are rich in phytoconstituents in which many researchers exploited each part of the plant for the NPs fabrications (Bachheti et al. 2020). Herbal plants have been potentially used for NPs synthesis due to being rich in phytochemical constituents alkaloids, saponins, phenolics, terpenoids, tannins, etc. (Rao et al. 2015, Bachheti et al. 2020). They play a vital role in metal/metal oxide NPs fabrication as stabilizing, reducing and capping agents for their potential applications in antimicrobial activities, catalytic activity, antioxidant, anti-inflammatory, antibiofilm, and so on. Thus, many researchers have utilized herbal plants such as *Aloe vera* (Maensiri et al. 2008, Nasrollahzadeh and Sajadi 2015, Rao et al. 2015), *Echinacea angustifolia* (Ranjbar et al. 2021), *Tanacetum parthenium* (Ranjbar, Dastani and Khakdan 2021), *Allium satiuum* (El-El-Refai et al. 2018, Vijaya et al. 2019, Kumar et al. 2019), *Zingiber officinale* (Iravani 2011, Raj and Jayalakshmy 2015, Vijaya et al. 2017), *Ginkgo biloba* (Nasrollahzadeh and Sajadi 2015), *Panax ginseng* (Singh et al. 2016) and *Hydrastis canadensis* (Singh et al. 2016, Mandala et al. 2020) for fabrication of NPs.

12.3 METAL NANOPARTICLES FROM HERBAL PLANTS AND THEIR CHARACTERIZATION

Plants extracts have been of interest to investigators because of their advantages over other biological entities in metallic NPs fabrications. They are inexpensive, easy and stable in harsh conditions, easily prepared within a limited time, ecofriendly and have various applications in different sectors (Demirbas et al. 2020). Herbal plants are an ideal choice medicinal plant for metallic NPs fabrication because they are regarded as safe, one-step, fast, easily available and ecofriendly. Furthermore, the bioactive molecules present in each part of herbal plants are used as capping and reducing agents for metallic NPs fabrication (Demirbas et al. 2020). Metallic NPs are of great interest at the nanoscale with their dimension 1 nm-100, which enhance unique physical and chemical properties compared to bulk counterparts (Iravani et al. 2014). Thus, this enables them to find their potential applications in different sectors such as catalysis, environmental remediation, chemical sensing, drug delivery, antimicrobial, antioxidant, biomedical, chemical industries, agricultural fields, and so on (Vijayaraghavan and Kumar 2017, Bachheti et al. 2020 2020a). Due to the rich plant biodiversity in the world, their potential utilization for the fabrication of NPs is yet to be explored (Akhtar, Panwar and Yun 2013). Among all plant biodiversity, herbal plants are the ideal choice plants for metal NPs synthesis because they are rich in biomolecules such as alkaloids, saponins, phenols, terpenoids, steroids, tannins and polysaccharides. Some of the herbal plants that are frequently used for NPs fabrication are shown in Figure 12.3.

Gopinath et al. (2012) explored the biofabrication of silver NPs using extract of dried *Tribulus terrestris* fruit body. The bioactive molecules present in dried fruit bodies of *Tribulus terrestris* plant played a vital role in the bioreduction of Ag+ to size range of 16–28 nm and spherical Ag-NPs were obtained. Characterization of fabricated Ag-NPs was done by TEM, XRD, FTIR, AFM and UV-Vis spectroscopy. Kumar et al. (2011) studied the fabrication of Au-NPs of size 5–15nm using *Zingiber officinale* extract. They were characterized by TEM, XRD, FTIR and UV-Vis. *Zingiber officinale* extract contains bioactive molecules such as alkaloids, phenolics and flavonoids that also has a role of bio-reducing Ag^+ into Ag^0. Velmurugan et al. (2014) studied biofabricated Ag-NPs and Au-NPs via using root extract of *Zingiber officinale*as reducing agent. The optimized parameters for the synthesis of Ag-NPs and Au-NPs were pH of 8, metal ion concentration of 9.1 mM, temperature of 50°C and reaction time of 2:30–3hrs. UV-Vis analysis indicated surface Plasmon absorption peak at 436 -531nm of Ag-NPs and Au-NPs, respectively, depending on the size of NPs. TEM characterization has shown a particle size of 10–20 nm of Ag-NPs and Au-NPs of 5–20 nm with FCC crystal structure has been confirmed by XRD. FTIR showed potential peaks that have a role in the bioreduction of Ag and Au-metal ions for their nanoparticle fabrications. Moreover, Ag-NPs have been shown to moderate antimicrobial activity against bacterial food pathogens. Yang et al. (2017) biosynthesized Ag-NPs using the ginger rhizomes extract. It was characterized by TEM, UV-Vis, XRD and FTIR. It has higher antibacterial activity and stability compared with chemically synthesized Ag-NPs. Recent publications studied the microwave-assisted green synthesis of Ag-NPs and CuO-NPs using fresh ginger rhizomes extract. Characterizations were done by UV-Vis, FTIR, XRD, FTIR and XRD. The particle size of Ag-NPs and Cu-NPs were 2.68 nm–42.49 nm and 11.32 nm–33.70 nm. These small particles' size is due to the fast and homogeneous heating system in microwave irradiation. Hashoosh Fadhil and Al-Ani (2014) carried out the synthesis of Ag-NPs using *Aloe vera* extract as a reducing agent. UV-Vis and SEM characterization showed peaks at 400 nm and rectangular morphology of Ag-NPs with a particle size of 500 nm. *Aloe vera* leaf extracts are also used for the fabrication of Ag-NPs and Au-NPs. They were characterized by UV-Vis, and TEM showed single-crystalline gold nano triangles with sizes ranging from 50–350 nm, by which only adjusting *Aloe vera* extract content was used in the Au^+ bioreduction. And spherical Ag-NPs were fabricated with single-step process with a particle size of 15.2 ±4.2nm (Chandran et al. 2006). Logaranjan et al. (2017) studied controlling the size and shape of Ag-NPs by manipulating the amount of *Aloe vera* extract and evaluated their antimicrobial activity. It was characterized by FTIR, UV-Vis, TEM and XRD. The particle size of 5–50 nm and octahedron shape of fabricated Ag-NPs indicated that *Aloe vera* extract played a vital role in directing the shape and size of Ag-NPs.

Karimi and Mohsenzadeh (2015) fabricated Cu-NPs using flower extract of *Aloe vera* played a role in reduction of Cu^{++} to Cu^o, and they were characterized by UV-Vis, FTIR and FESEM. FTIR analysis showed absorption bands at 578 nm, which confirmed the fabrication of Cu-NPs. FESEM analysis showed to be spherical in shape and had a particle size of 40 nm. Aerial parts of *Echinacea purpurea* (L.) were used for the fabrication of Ag-NPs. UV-Vis, FTIR, XRD and SEM techniques

FIGURE 12.3 Some of herbal plants frequently used for NPs fabrication (A) *Aloe vera*; (B) *Echinacea purpurea* flower; (C) *Allium sativum* cloves; (D) *Zingiber officinale*; (E) *Ginkgo biloba* leaf; and (F) *Panax ginseng* root.

were used for characterization. The particle size of fabricated Ag-NPs is 68.24nm and spherical in shape (Gecer et al. 2021). Anu et al. (2017) biofabricated Se-NPs using *Allium satiuum* extract as metal ion reducing and capping agent. Synthesized NPs were characterized by UV-Vis, SEM, TEM, EDAX, XRD and FTIR. Au-NPs were fabricated by using an aqueous extract of *Allium satiuum* and were characterized using TEM and FTIR (Rastogi and Arunachalam 2013). In an experiment, *Zingiber officinale* and *Allium satiuum* plant extract was used for the fabrication of Ag, Au, Fe and Zn-NPs. Zn-NPs fabricated from ginger plant extract as a reducing and capping agent has shown the best antioxidant activity. Results have shown average particle size of Ag (13.22–22.69 nm) with spherical and tetragonal in shape, Cu (14.62–22.80 nm) with spherical and tetragonal in shape, Fe (60.30–82.63 nm) tetragonal in shape and Zn (79.88–100.62 nm) tetragonal and hexagonal in shape using garlic extract as reducing agent while average particle size of Ag (10.10–18.33 nm) with spherical and tetragonal in shape, Cu (14.94–27.4 nm) with spherical and linear in shape, Fe (14.08–21.57 nm) with spherical in shape and Zn (99.34–134.57nm with spherical and tetragonal in shape) for *Zingiber officinale* extracts. Fabricated Ag-NPs using *Allium satiuum* and *Zingiber officinale* water extract, Ag-NPs and Cu-NPs using *Allium satiuum* and *Zingiber officinale* ethanolic extract have shown significant effect on antimicrobial activity against different microorganisms (El-Refai et al. 2018). Fardsadegh and Malmiri et al. (2019), Se-NPs were fabricated by hydrothermal synthesis using *Aloe vera* leaf extract. Results have shown excellent antibacterial and antifungal activities against pathogenic and spoilage fungi strains. Joseph, Prakash and Narvi (2016) studied phyto-fabrications of Cu-NPs using *Allium satiuum*. Fabricated Cu-NPs were examined via UV-Vis, FTIR, SEM, TEM and XRD. The average particle size was 100 nm and spherical in shape. The synthesized Cu-NPs have antimicrobial activity against gram-positive bacteria like *Bacillus subtilis*. Zha et al. (2017) also biofabricated Au-NPs by employing *Ginkgo biloba* leaf extract as stabilizing and reducing agents, which were highly crystalline in nature with FCC structure, spherical and particle size ranging from 10–40 nm. They were SEM, TEM, XRD, UV-Vis and DTS. Cu-NPs synthesized using *Ginkgo biloba* leaf extract as a catalyst in Huisgen [3+2] cycloaddition of azides and alkynes (Nasrollahzadeh and Sajadi 2015). Biosynthesis of Ag-NPs using *Ginkgo biloba* leaf extract acts as a reducing agent. The NPs were characterized via TEM, UV-Vis, XRD and FTIR. TEM analysis indicated a particle size of 10–16 nm (Ren et al. 2016). Sreekanth et al. (2018) studied biofabricated Ag-NPs using aqueous extract *Panax ginseng* root via an ultra-sonication method. UV-Vis, XRD, FTIR, HRTEM and EDX were used for characterization. According to HRTEM analysis, fabricated Ag-NPs were particle sizes ranging from 5–15nm and quasi-spherical in shape. Ag-NPs and Au-NPs were also fabricated using aqueous extract of fresh roots of *Panax ginseng*, and their formation was confirmed via UV-Vis, EDX, TEM and XRD. The fabricated Ag-NPs and Au-NPs with particle sizes of 10–30 nm and 10–40 nm, respectively, were spherical in shape. There has been effective antibacterial activity against *Bacillus anthracis*, *Vibrio parahaemolyticus*, *Staphylococcus aureus*, *Escherichia coli* and *Bacillus cereus* (Singh et al. 2016). Akhtar, Panwar and Yun (2013) also used red ginseng root extract for the fabrication of Ag-NPs and Au-NPs. UV-Vis results showed absorption peaks at 410 nm and 560 nm for Ag-NPs and Au-NPs. Owaid et al. (2019) biofabricated Zn-NPs using red ginseng roots aqueous extract. AFM, SEM, FTIR, UV-Vis, EDS and SPM were used for characterization. UV-Vis confirmed the formation of Zn-NPs at an absorption peak at 340 nm. The obtained Zn-NPs had a particle size of 59.76 nm with spherical to irregular shape via AFM, SEM and EDS analysis. Tagad et al. (2017) studied one-spot green fabrication of Ag-NPs using aqueous *Panax ginseng* root extract as a capping and reducing agent for the detection of Hg^{2+}. Their morphological properties are checked by UV-Vis, FTIR, TEM and AFM. Ag-NPs have shown a linear response at a concentration ranging from 10μM–1mM Hg^{+2} with a minimum detection limit of 5 μM, thus sensing properties were examined with heavy metals and found to be linear. Behravan et al. (2019) fabricated Ag-NPs using *Berberis vulgaris* leaf and root extract. TEM and XRD analysis of fabricated Ag-NPs showed particle size ranging from 30–70 nm and spherical in shape. Ag-NPs synthesis process parameters such as $AgNO_3$ concentration (10, 3, 1, 0.5mM), aqueous plant extract (30, 15, 10, 5, 3ml) and reaction time (24, 12, 6 2, 1h) were examined.

It has been known that different herbal plant extracts are composed of various phytochemical constituents, such as alkaloids, flavonoids, terpenoids, tannins, and so on. All these phytochemicals play a vital role in NPs fabrication. They can act as reducing, capping and stabilizing agents for NPs fabrications (Das et al. 2013, Andra et al. 2019, Bachheti et al. 2021b). Thus, NPs fabrication mechanisms using herbal plant extracts vary as they obtain from different herbal plant species. But, the possible mechanism of nanoparticle fabrication is indicated here below in Figure 12.4.

12.4 METAL OXIDE NANOPARTICLES FROM HERBAL PLANTS AND THEIR CHARACTERIZATION

Due to increasing demand, biofabricated metal oxide NPs using all plant extracts have been studied in recent years. It was mainly focused on CuO, ZnO, Fe_2O_3, TiO_2, CeO_2, SeO_2 and other NPS because of their versatile industrial applications (Ishak et al. 2019). Metallic oxide NPs using plant and related products are also the promising areas in modern science and technology, especially

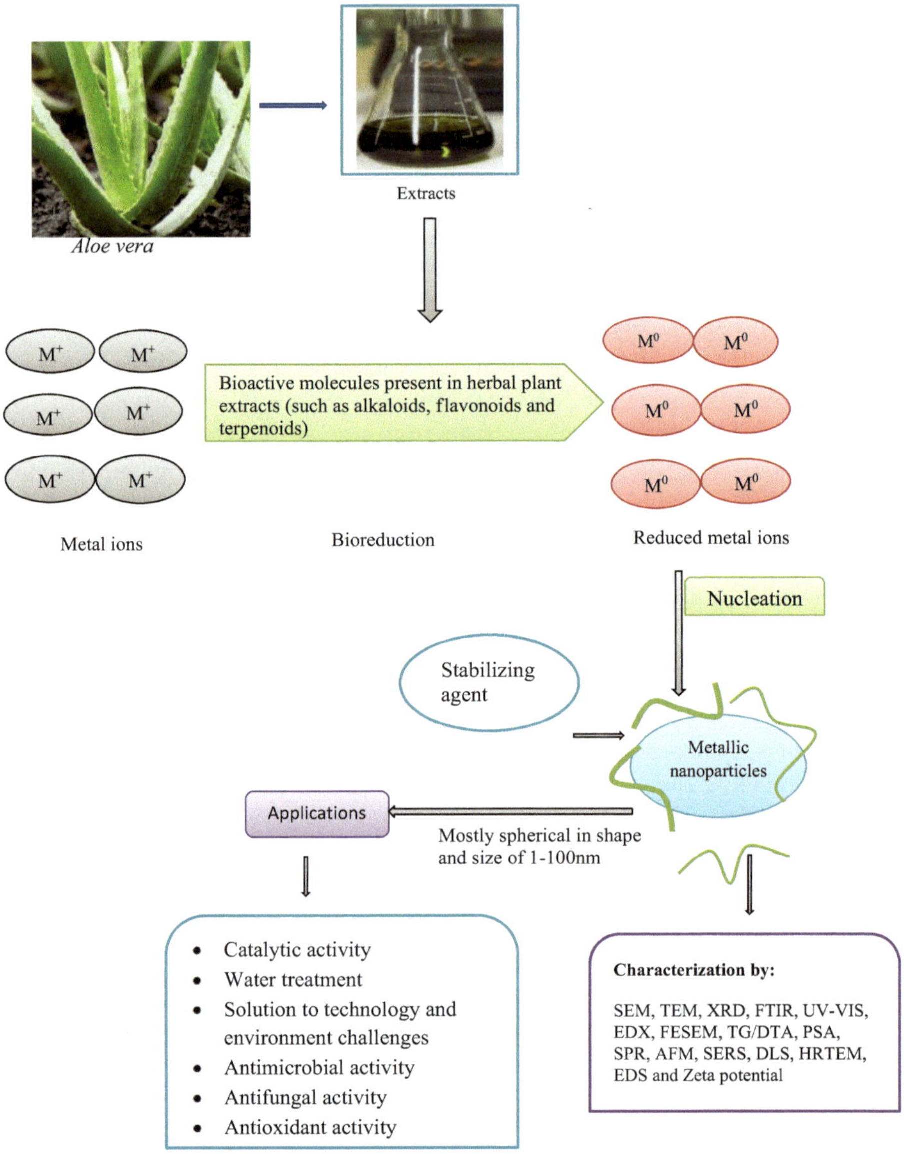

FIGURE 12.4 Possible mechanism, characterization and application of NPs fabricated from herbal plants.

nanotechnology (Andra et al. 2019, Kannan et al. 2020, El-Shafey 2020). Kumar et al. (2015) studied the biosynthesis of CuO-NPs using *Aloe vera* leaf extract with an average size of 20nm and spherical shape. They were characterized by XRD and TEM.

TiO_2-NPs were fabricated from *Aloe vera* leaf extract and characterized by XRD, SEM, TG/DTA, PSA and TEM. They were used in products for hair and skin due to the presence of high vitamins, minerals, amino acids and fatty acids. XRD results showed a particle size of 20 nm and tetragonal in shape (Rao et al. 2015). Nasrollahzadeh and Sajadi (2015) used an aqueous extract of *Euphorbia heteradena* Jaub root for the fabrication of TiO_2-NPs and characterized by UV-Vis, SPR and XRD. Attar and Yapaoz (2018) used *Aloe vera* plant extract, biomimetically fabricated ZnO-NPs and Au-NPs using *Echinacea angustifolia* flower extract as a reducing and stabilizing agent. SEM, UV-Vis,and FTIR techniques were used for NPs characterizations. SEM analysis showed fabricated ZnO-NPs and Au-NPs were spherical in shape and average particle size ranging from 90–170 nm for ZnO-NPs and 80–120nm for Au-NPs. CuO-NPs were fabricated using *Tanacetum parthenium* extract, and their effects were evaluated for their antimicrobial activity, catalytic activity and anticancer activity; and they were characterized by XRD, FTIR, UV-Vis, FESEM and EDX. UV-Vis showed a surface plasmatic absorption band at 298nm for CuO-NPs. Due to the presence of bioactive molecules in feverfew (*Tanacetum parthenium*), CuO-NPs were reduced and stabilized, which was confirmed by FTIR analysis (Ranjbar, Dastani and Khakdan 2021). Thandapani et al. (2018) reported TiO_2-NPs fabricated using the aqueous leaf extracts of *Parthenium hysterophorus* by microwave irradiation. FTIR analysis confirmed that the presence of bioactive molecules such as fluoroalkanes, phenols, alkanes and alcohols in *Parthenium hysterophorus* leaf extract played a vital role in bioreduction from TiO_4 into TiO_2-NPs. They were characterized by FETEM, UV-Vis, FTIR, XRD and EDX. FESEM analysis showed average particle size ranging between 20–50 nm and spherical in shape. Phumying et al. (2013) studied magnetite Fe_3O_4-NPs by novel hydrothermal method and *Aloe vera* extracted solution. TEM analysis had shown that fabricated NPS have crystalline and particle sizes in the range of 6–30 nm. Slman, Jalill and Abd (2018) also reported biosynthesized ZnO-NPs using hot aqueous *Allium sativum* extract and were confirmed by AFM, XRD, FTIR, UV-Vis. ZnO-NPs are fabricated when caped with *Allium* sativum and *Hydrastis canadensis* root extract. It was obtained using zinc nitrate hexahydrate and aqueous plant extracts at 65°C. Their formations were checked via TGA-MS, FTIR, XRD, UV-Vis and SEM. The phytochemical constituents of garlic extract (contains tannin, quinones, terpenoids and steroids) and goldenseal root extract (contains steroids, terpenoids, saponins and phenols) play a basic role as reducing and capping agents for NPs formation. XRD analysis indicated average particle sizes ZnO-NPs were 2.6(9) nm for garlic extract and 5(2) nm *Hydrastis canadensis* root extract. The plant goldenseal roots and rhizomes mainly contain three alkaloids such as berberine, hydrastine and canadine. Berberine is a major bioactive alkaloid present in different parts of goldenseal extract, which play a fundamental role in antimicrobial activity, antiviral, anti-inflammatory, hypoglycemic and antioxidant activities (Mandal et al. 2020).

12.5 METAL/METAL OXIDE NANOPARTICLES FROM HERBAL PLANTS AND THEIR ANTIMICROBIAL ACTIVITIES

Numerous research reports were presented in the literature about fabricated metal/metal oxide NPs using herbal plant extracts and their antimicrobial activities, as shown in Table 12.1. Metal/Metal oxide NPs possess unique physical and chemical properties linked to a particle size at nanoscale compared to their bulk size material, thus offering their potential applications in antimicrobial activities (Bouqellah, Mohamed and Ibrahim 2019, Hussain et al. 2019, Raghunath and Perumal 2017, Tippayawat et al. 2016). However, their antimicrobial activity effectiveness depends on the size and shape of particle as well as plant extracts used for NPs fabrications (Anu et al. 2017, Attar and Yapaoz 2018, Bachheti et al. 2020, Bouqellah, Mohamed and Ibrahim 2019). The small metal/metal oxide NPs size leads to the higher surface area to volume ratios of NPs, thus allowing better contact with the microbe's cell membranes. Hence, metal/metal oxide NPs that have smaller particle size and higher surface area enhance better contact with microbes, thereby affecting the DNA replication process, formation of cell membranes and cell division of fungi, bacteria and virus via releasing their metal ions (Singh et al. 2018, Some 2018). Herbal plant extracts contain phytochemical constituents such as terpenoids, flavonoids, phenolic acid and proteins, thus, they have played roles in biofabrication of various metal/metal oxide NPs that can be utilized for antimicrobial activities (Hashoosh, Fadhil and Al-Ani 2014, El-Refai et al. 2018, Hussain et al. 2019). For instance, it is a well-known fact that Ag-NPs are highly toxic to microorganisms, mainly to different species of bacteria, via releasing Ag^+. Thus, Ag-NPs are the best choices for antimicrobial

TABLE 12.1
Metal/Metal Oxide NPs, Characterization Techniques, and Their Antimicrobial Activities

Plant name	NPs	Size (nm)	Shape	Characterization techniques	Activity	Reference(s)
Allium sativa extract	Au	6	-	UV-Vis, TEM, and DSL	High antiviral activity.	Villanueva et al. (2019)
Allium sativum clove extract	Ag	9.42	Cubic	XRD, AFM, TEM, SEM, FTIR and UV-Vis	Effective antibacterial activity.	Lakhan et al. (2020)
Allium sativum clove extract	Ag	10 – 50	Spherical	UV-VIS, FTIR, XRD and TEM	High antibacterial, antibiofilm, antihelminthic, anti-inflammatory, and anti-breast cancer.	Vijaya et al. (2019), Kumar et al. (2019)
Allium sativum clove extract	Ag	12	Spherical shape	DLS, FETEM, XRD, EDX and UV-Vis	-	Ahamed et al. (2011)
Allium sativum cloves aqueous extracts	Se	40–110	Trigonal	UV-vis, TEM, SEM, EDX, FTIR and XRD	High cytotoxicity on Vero cells.	Anu et al. (2017)
*Allium sativum*plant extract	Ag	13.13 -22.69	Spherical and Tetragonal	TEM, SEM, UV-Vis and EDX	Fe and Zn-NPs high antioxidant activity, Ag and Cu-NPs-antibacterial against bacterial pathogens.	El-Refai et al. (2018)
	Cu	14.62–22.80	Tetragonal			
	Fe	60.30–82.63	Tetragonal			
	Zn	79.88–100.6	Hexagonal			
Aloe barbadensis miller plant extract	CeO_2	63.60	FCC cubic	XRD, FTIR and TEM	-	Priya et al. (2014)
Aloe vera plant aqueous extracts	Ag, Fe	36.61, 34.93	Spherical Spherical	UV-Vis, FTIR and TEM	High inhibition effect against *proteus mirabilis* and *Klebsiella pneumonia*. Ag-NPs has high antibacterial effect than Fe-NPs and is even better against a wide range of microbes.	Yadav et al. (2016)
Aloe vera plant Leaf extract	Au	50–350	Triangular	UV-Vis, XRD, TEM, SEM, TEM, FTIR and SERS	Au-NPs-anticancer activity.	Chandran et al. (2006)
	Ag	5–50	Spherical		Ag-NPs- antibacterial activity.	Logaranjan et al. (2016)

Aloe vera plant leaf extract	In_2O_3	5–50	Cubic	XRD and TEM	-	Maensiri et al. (2008)
Aloe vera plant leaf extract	CuO	20	Spherical	XRD and TEM	High antibacterial activity against pathogenic bacteria such as *Aeromonas hydrophila* and *Pseudomonas fluorescens* at 20ppm.	Kumar et al. (2015)
Aloe vera plant leaf extract	TiO_2	20–32	Tetragonal	XRD, PSA, SEM, TEM and TG/DTA	-	Rao et al. (2015)
Aloe vera leaf extract	Se	121–3243	Spherical	FTIR, UV-Vis, DLS and TEM	High antibacterial activity against such as *Staphylococcus aureus* and *Escherichia coli. High* antifungal activity against *Colletotrichum coccodes* and *Penicillium digitatum.*	Fardsadegh and Malmiri. (2019)
Echinacea angustifolia flower extract	ZnO Au	90–170 80–120	Spherical Spherical	SEM, FTIR and UV-Vis	Significant effect against both gram-positive and gram-negative bacteria.	Dobrucka (2017)
Emblica officialis fruit extracts	Se	15–40	Spherical in shape	UV-Vis, EDX, TEM, SEM, FTIR, XRD, DLS and HRTEM	High antifungal activity, high antibacterial activity.	Gunti et al. (2019)
Ginkgo biloba leaf extract	Ag	10–16	Spherical	UV-Vis, XRD, TEM and FTIR	Effective antibacterial activities against gram-positive and gram-negative bacteria.	Ren et al. (2016)
Ginkgo biloba leaf extract	Au	10–40	Spherical	TEM, SEM, UV-Vis, EDS, DLS, XRD and FTIR	-	Zha et al. (2017)
Ginkgo biloba leaf extract	Cu	15–20	Spherical	TEM, EDS, FTIR, UV-Vis, and HPLC	-	Nasrollahzadeh and Sajadi (2015)
Hydrasis canadensis extract	Ag	111.3	Spherical	Zeta, TEM, UV-Vis, DSL, FTIR and XRD	Effective antibacterial activity at 120ppm.	Das et al. (2013)
Panax ginseng (red) root extract	Ag Au	83 183	Spherical	FESEM, EDX, and DLS	High antibacterial activity against *Staphylococcus aureus* and *Pseudomonas aeruginosa* even at 4μg.	Singh et al. (2016)
Panax ginseng fresh leaves	Ag Au	5–15 10–20	Spherical Spherical	UV-Vis, FE-TEM, EDX and XRD	An efficient antimicrobial activity.	Singh et al.(2016b)
Panax ginseng fresh roots	Ag Au	10–30, 10–40	Spherical Spherical	XRD, EDX, TEM and UV-VIS	Ag-NPs- antibacterial activity.	Singh et al.(2016a)

(*continued*)

TABLE 12.1 (Continued)
Metal/metal Oxide NPs, Characterization Techniques, and Their Antimicrobial Activities

Plant name	NPs	Size (nm)	Shape	Characterization techniques	Activity	Reference(s)
Panax ginseng root extract	Ag	4–20	Spherical	UV-Vis, FTIR, TEM and AFM	-	Tagad et al. (2017)
Panax ginseng root extracts (red)	Ag AU	83 183	Spherical	FETEM, EDX and DLS	Antibacterial activity against *Staphylococcus aureus* and *Pseudomonas aeruginosa* even at 4µg.	Singh et al. (2016)
Tanacetum parthenium extract	CuO	16	Cubic	XRD, FETEM, EDX, EDX and UV-VIS	High anticancer and antibacterial activities.	Ranjbar et al. (2021)
Tribulus terrestris herbal methanol extracts	Ag ZnO Cu	40 30 70		UV-Vis, XRD and FTIR	Excellent antibacterial activity against *Escherichia coli, Enterococcus faecaliss, proteus spmorganella sp, Staphylococcus aureus* and *Candida albicans.*	Albalawi et al. (2017)
Zingiber officinale aqueous root extracts	ZnO	30–50	Spherical	SEM, FTIR, FESEM and EDX	-	Raj and Jayalkshmy (2015)
Zingiber officinale dried root extract	Ag	10	Spherical	UV-Vis, HRTEM, FTIR, XRD, HRSEM and XRD	Antibacterial activity.	Vijaya et al. (2017)
Zingiber officinale extracts	Ag Cu Fe Zn	10.10–18.33 14.92–27.4 14.08–21.57 99.34–134.57	Spherical Tetragonal Spherical Linear	TEM and UV-Vis	Ag-NPs, and Cu-NPs high antibacterial activity.	El-Refai et al. (2018)
Zingiber officinale rhizomes extract	Ag	10	Spherical	UV-Vis, TEM, XRD and FTIR	High stability and antimicrobial activity against aquatic pathogens.	Yang et al. (2017)
Zingiber officinale rhizome extract	Ag	4–15	Spherical	UV-Vis, HR-TEM, XRD and FTIR	High antibacterial activity against *Staphylococcus aureus* and *Escherichia coli.*	Mathew et al. (2018)
Zingiber officinale root extract	Ag Au	10–20 5–20	Spherical cubic	TEM and SEM-EDS	Moderate antibacterial activity against food pathogenic bacteria.	Velmurugan et al. (2014)

activities, when Ag-NPs are used, they have a high surface area to volume ratios available for microbes to be exposed to, making them excellent for antimicrobial activities in a wide range of bacterial strains (Prabhu and Poulose 2012). In an experiment, Gopinath et al. (2012) explored the biofabrication of silver NPs using an aqueous extract of dried *Tribulus terrestris* fruit body. The bioactive molecules present in the dried fruit bodies of *Tribulus terrestris* plant played a vital role in quickly reducing Ag^+ to spherical shaped Ag-NPs. It has been more effective in antimicrobial activity against *Escherichia coli*, *Streptococcus pyogenes*, *Bacillus subtilis*, *Pseudomonas aeruginosa* and *Staphylococcus aureus*. Hashoosh et al. (2014) carried out the production of Ag-NPs using *Aloe vera* extract. Synthesized Ag-NPs showed a significant effect on *Escherichia coli* and *Staphylococcus aureus* even at low concentration. However, it has no significant impact on antifungal. Sreekanth et al. (2018) fabricated Ag-NPs using aqueous extract *Panax ginseng* root via ultra-sonication method. Antiviral activity was tested against *Influenza virus*. Ag-NPs and Au-NPs were also fabricated using fresh roots of *Panax ginseng* and synthesized NPs exhibited effective antibacterial activity against *Vibrio parahaemolyticus*, *Staphylococcus aureus*, *Bacillus anthracis*, *Escherichia coli* and *Bacillus cereus* (Singh et al. 2016a). In another experiment, Kumar et al. (2015) studied biofabrication of CuO-NPs using *Aloe vera* leaf extract and tested for antibacterial activity against pathogenic bacteria such as *Aeromonas hydrophila*, *Pseudomonas fluorescens* and *Flavobacterium branchiophilum*. Results showed that synthesized CuO-NPs have effective antimicrobial activity against *Aeromonas hydrophila*, *Pseudomonas fluorescens* and *Flavobacterium branchiophilum*. Attar and Yapaoz (2018) biomimetically fabricated ZnO-NPs and Au-NPs using *Echinacea angustifolia* flower aqueous extract. Both ZnO-NPs and Au-NPs have significant activity against both *Staphylococcus aureus* (gram-positive) and *Escherichia coli* (gram-negative) bacterial strains. CuO-NPs were fabricated using *Tanacetum parthenium* extract and were evaluated for their antimicrobial activity, anticancer activity and catalytic activity. It has been shown that CuO-NPs have excellent antimicrobial activity against *Escherichia coli* and *Candida albicans* (Ranjbar, Dastani and Khakdan 2021). Akhtar, Panwar and Yun (2013) also used red ginseng root extract for the fabrication of Ag-NPs and Au-NPs. They were examined against pathogenic microorganisms including *Vibrio parahaemolyticus*, *Staphylococcus aureus*, *Bacillus cereus* and *Candida albicans* and found that they were also effective on biofilm degrading activity against *Staphylococcus aureus* and *Pseudomonas aeruginosa*.

12.6 CONCLUSION AND FUTURE PERSPECTIVES

Biofabrications of NPs using bacteria, fungi, algae, seaweeds and plants are becoming a hot issue for several ecofriendly applications. The fabrication of NPs using plants and related products is a promising area and has extraordinary applications in antimicrobial activity, catalytic activity, medical fields, consumer products, water purifications and industrial as well as agricultural fields. The herbal plants and their parts, such as roots, stems, stalks, barks, leaves, flowers and seeds, are known for their different activities, including antimicrobial activity. The efficacy of these herbal can be enhanced by converting them into NPs. Different metal and metal oxide NPs such as silver, gold, copper, zinc, titanium, iron and selenium synthesized from herbal plants have already proven their antimicrobial activity. In the future, the herbal plants have a wide range of perspectives for the fabrication of metal/metal oxide NPs for various new applications in medical fields, consumer products, degradation of organic pollutants by optimizing different parameters that could affect the NPs synthesis process, and, concurrently, degradation efficiency such as pH, volume ratios between templates, solvent effect, calcination temperature and the precursor salt and concentration of the precursor.

REFERENCES

Acharya, D., K. M. Singha, P. Pandey, B. Mohanta, J. Rajkumari and L. P. Singha. 2018. Shape dependent physical mutilation and lethal effects of silver nanoparticles on bacteria. *Scientific Reports* 8(1): 1–11.

Ahamed, M., M. M. Khan, M. Siddiqui, M. S. AlSalhi and S. A. Alrokayan. 2011. Green synthesis, characterization and evaluation of biocompatibility of silver nanoparticles. *Physica E: Low-Dimensional Systems and Nanostructures* 43(6): 1266–1271.

Ahmed, M. J., G. Murtaza, A. Mehmood and T. M. Bhatti. 2016. Process parameters for green synthesis of silver Nanoparticles using leaves extract of *Aloe vera* plant. *International Journal of Multidisciplinary and Current Research* 3: 966–975.

Ahmed, R. H. and D. E. Mustafa. 2020. Green synthesis of silver nanoparticles mediated by traditionally used medicinal plants in Sudan. *International Nano Letters* 10(1): 1–14.

Ahmed, S., M. Ahmad, B. L. Swami and S. Ikram. 2016. A review on plants extract mediated synthesis of silver nanoparticles for antimicrobial applications: A green expertise. *Journal of Advanced Research* 7(1): 17–28.

Akhtar, M. S., J. Panwar and Y. S. Yun. 2013. Biogenic synthesis of metallic nanoparticles by plant Extracts. *ACS Sustainable Chemistry and Engineering* 1(6): 591–602.

Albalawi, M. A., J. Eldiasty, S. Khasim and N. Badi. 2017. Effect of Ag,Cu, and ZnO nanoparticle suspensions on the antimicrobial activity of *Tribulus terrestris* herbal extracts. Paper presented at the Journal of Nano Research.

Albrahim, J. S., J. S. Alosaimi, A. M. Altaher, R. N. Almulayfi and N. F. Alharbi. 2021a. Employment of*Cassia angustifolia* leaf extract for zinc nanoparticles fabrication and their antibacterial and cytotoxicity. *Saudi Journal of Biological Sciences* 28(6): 3303–3308.

Albrahim, J. S., J. S. Alosaimi, A. M. Altaher, R. N. Almulayfi and N. F. Alharbi. 2021b. Employment of *Cassia angustifolia* leaf extract for Zinc Nanoparticles fabrication and their antibacterial and cytotoxicity. *Saudi Journal of Biological Sciences* 28(6): 3303–3308.

Andra, S., S. K. Balu, J. Jeevanandham, M. Muthalagu, M. Vidyavathy, Y. San Chan and M. K. Danquah. 2019. Phytosynthesizedmetal oxide nanoparticles for pharmaceutical applications. *Naunyn- Schmiedeberg's Archives of Pharmacology* 392(7): 755–771.

Anu, K., G. Singaravelu, K. Murugan and G. Benelli. 2017. Green-synthesis of selenium nanoparticles using garlic cloves (*Alliumsativum*): biophysical characterization and cytotoxicity on Vero cells. *Journal of Cluster Science* 28(1): 551–563.

Attar, A. and M. A. Yapaoz. 2018. Biomimetic synthesis, characterization and antibacterial efficacy of ZnO and Au nanoparticles using Echinacea flower extract precursor. *Materials Research Express* 5(5): 055403.

Bachheti, A., R. K. Bachheti, L. Abate and A. Husen. 2021b. Current status of Aloe-based nanoparticle fabrication, characterization and their application in some cutting-edge areas. *South African Journal of Botany*. https://doi.org/10.1016/j.sajb.2021.08.021

Bachheti, R. K., L. Abate, A. Bachheti, A. Madhusudhan and A. Husen. 2021a. Algae-, fungi-, and yeast-mediated biological synthesis of nanoparticles and their various biomedical applications. In Boris Kharisov and Oxana Kharissova, eds., *Handbook of Greener Synthesis of Nanomaterials and Compounds* (pp. 701–734). Elsevier, The Netherlands.

Bachheti, R. K., R. Konwarh, V. Gupta, A. Husen and A. Joshi. 2019. Green synthesis of iron oxide nanoparticles: cutting edge technology and multifaceted applications. In *Nanomaterials and Plant Potential* (pp. 239–259). Springer, Cham.

Bachheti, R., Y. Godebo, A. Bachheti, M. O. Yassin and A. Husen. 2020. Root-based fabrication of metal/metal-oxide nanomaterials and their various applications. *Nanomaterials for Agriculture and Forestry Applications*, https://doi.org/10.1016/B978-0-12-817852-2.00006-8

Bachheti, R. K., A. Fikadu, A. Bachheti, A. and A. Husen 2020. Biogenic fabrication of nanomaterials from flower-based chemical compounds, characterization and their various applications: A review. *Saudi Journal of Biological Sciences* 27(10): 551–2562.

Bachheti, R. K., Sharma, A., Bachheti, A., Husen, A., Shanka, G. M. and D. P. Pandey. 2020a. Nanomaterials from various forest tree species and their biomedical applications. In Azamal Husen and Mohammad Jawaid, eds., *Nanomaterials for Agriculture and Forestry Applications* (pp. 81–106). Elsevier, Cambridge, MA.

Behravan, M., A. H. Panahi, A. Naghizadeh, M. Ziaee, R. Mahdavi and A. Mirzapour. 2019. Facile green synthesis of silver nanoparticles using Berberis vulgaris leaf and root aqueous extract and its antibacterial activity. *International Journal of Biological Macromolecules* 124: 148–154.

Bouqellah, N. A., M. M. Mohamed and Y. Ibrahim. 2019. Synthesis of eco-friendly silver nanoparticles using *Allium* sp. and their antimicrobial potential selected vaginal bacteria. *Saudi Journal of Biological Sciences* 26(7): 1789–1794.

Chandran, S. P., M. Chaudhary, R. Pasricha, A. Ahmad and M. Sastry. 2006. Synthesis of gold nano triangles and silver nanoparticles using Aloe vera plant extract. *Biotechnology Progress* 22(2): 577–583.

Cremonini, E., E. Zonaro, M. Donini, S. Lampis, M. Boaretti, S. Dusi, P. Melotti, M. M. Lleo and G. J. M. B. Vallini. 2016. Biogenic selenium nanoparticles: characterization, antimicrobial activity and effects on human dendritic cells and fibroblasts. *Microbial Biotechnology* 9(6): 758–771.

Das, S., J. Das, A. Samadder, S. S. Bhattacharyya, D. Das, A. R. Khuda-Bukhsh and S. B. Biointerfaces. 2013. Biosynthesized silver nanoparticles by ethanolic extracts of *Phytolacca decandra*, Gelsemium sempervirens, *Hydrastis canadensis* and *Thuja occidentalis* induce differential cytotoxicity through G2/M arrest in A375 cells. *Colloids and Surfaces B: Biointerfaces* 101: 325–336.

Demirbas, A., T. Karaytuğ, N. Arabaci, E. S. Yilmaz and I. Ocsoy 2020. Synthesis of Metallic and Metal Oxide nanomaterials. In: Neha Srivastava, Manish Srivastava, P. K. Mishra and Vijai Kumar Gupta, eds., *Green Synthesis of Nanomaterials for Bioenergy Applications* (pp. 99–123), https://doi.org/10.1002/9781119576785.ch4

Dobrucka, R. 2017. Synthesis of titanium dioxide nanoparticles using *Echinacea purpurea* herba. *Iranian Journal of Pharmaceutical Research: International Journal of Pharmaceutical Research* 16(2): 756.

Durán, N., P. D. Marcato, M. Durán, A. Yadav, A. Gade and M. Rai. 2011. Mechanistic aspects in the biogenic synthesis of extracellular metal nanoparticles by peptides, bacteria, fungi, and plants. *Applied Microbiology and Biotechnology* 90(5): 1609–1624.

El-Refai, A. A., G. A. Ghoniem, A. Y. El-Khateeb and M. M. Hassaan 2018. Eco-friendly synthesis of metal nanoparticles using ginger and garlic extracts as biocompatible novel antioxidant and antimicrobial agents. *Journal of Nanostructure in Chemistry* 8(1): 71–81.

El-Shafey A. M. 2020. Green synthesis of metal and metal oxide nanoparticles from plant leaf extracts and their applications: A review. *GreenProcessing and Synthesis* 9(1): 304–339.

Eyasu, A., O. Yadav and R. J. Bachheti. 2013. Photocatalytic degradation of methyl orange dye using Cr-doped ZnS nanoparticles under visible radiation. *International Journal of Chemtech Research* 5(4): 1452–1461.

Fardsadegh, B. and H. Jafarizadeh-Malmiri. 2019. *Aloe vera* leaf extract mediated green synthesis of selenium nanoparticles and assessment of their in vitro antimicrobial activity against spoilage fungi and pathogenic bacteria strains. *Green Processing and Synthesis* 8(1): 399–407.

Fardsadegh, B., H. Vaghari, R. Mohammad-Jafari, Y. Najian and H. Jafarizadeh-Malmiri. 2019. Biosynthesis, characterization and antimicrobial activities assessment of fabricated selenium Nanoparticles using *Pelargonium zonale* leaf extract. *Green Processing and Synthesis* 8(1): 191–198.

Gavade, S. M., G. Nikam, R. Dhabbe, S. Sabale, B. Tamhankar and G. Mulik. 2015. Green synthesis of silver nanoparticles by using carambola fruit extract and their antibacterial activity. *Advances in Natural Sciences: Nanoscience and Nanotechnology* 6(4): 045015.

Gecer, E. N., R. Erenler, C. Temiz, N. Genc and I. Yildiz. 2021. Green synthesis of silver nanoparticles From *Echinacea purpurea* L. *Particulate Science and Technology* https://doi.org/10.1080/02726351.2021.1904309

Gopinath, V., D. MubarakAli, S. Priyadarshini, N. M. Priyadharsshini, N. Thajuddin and P. Velusamy. 2012. Biosynthesis of silver nanoparticles from *Tribulus terrestris* and its antimicrobial activity: a novel biological approach. *Colloids and Surfaces B: Biointerfaces* 96: 69–74.

Gour, A. and N. K. Jain. 2019. Advances in green synthesis of nanoparticles. *Artificial cells, Nanomedicine and Biotechnology* 47(1): 844–851.

Gunti, L., R. S. Dass and N. Kalagatur. 2019. Phytofabrication of selenium nanoparticles from *Emblica officinalis* fruit extract and exploring its biopotential applications: antioxidant, antimicrobial, and biocompatibility. *Frontiers in Microbiology*, 10: 931.

Hashoosh, S. I., A. M. Fadhil and N. K. Al-Ani. 2014. Production of Ag nanoparticles using *Aloe vera* extract and its antimicrobial activity. *Al-Nahrain Journal of Science* 17(2): 165–171.

Husen, A. 2017. Gold nanoparticles from plant system: Synthesis, characterization and their application. *Nano-science and plant–soil systems*, https://doi.org/10.1007/978-3-319-46835-8-17

Husen, A., Q. I. Rahman, M. Iqbal, M. O. Yassin and R. K. Bachheti. 2019. Plant-mediated fabrication of gold nanoparticles and their applications. In: A. Husen and M. Iqbal, eds., Nanomaterials and *Plant Potential* (pp. 71–110). Springer International Publishing AG, Cham.

Hussain, A., M. Oves, M. F. Alajmi, I. Hussain, S. Amir, J. Ahmed, M. T. Rehman, H. R. El-Seedi and I. J. R. A. Ali. 2019. Biogenesis of ZnO nanoparticles using *Pandanus odorifer* leaf extract: anticancer and anti-microbial activities. *RSC Advances* 9(27): 15357–15369.

Hussain, I., N. Singh, A. Singh, H. Singh and S. J. Singh. 2016. Green synthesis of nanoparticles and its potential application. *Biotechnology Letters* 38(4): 545–560.

Hyeon, T. 2003. Chemical synthesis of magnetic nanoparticles. *Chemical Communications* (8): 927–934.

Ibrahim, H. M. 2015. Green synthesis and characterization of silver nanoparticles using banana peel extract and their antimicrobial activity against representative microorganisms. *Journal of Radiation Research and Applied Sciences* 8(3): 265–275.

Iravani S. 2011. Green synthesis of metal nanoparticles using plants. *Green Chemistry* 13(10): 2638–2650.

Iravani, S., H. Korbekandi, S. V. Mirmohammadi and B. J. Zolfaghari. 2014. Synthesis of silver nanoparticles: chemical, physical and biological methods. *Research in Pharmaceutical Sciences* 9(6): 385.

Ishak, N. M., S. Kamarudin and S. J. Timmiati. 2019. Green synthesis of metal and metal oxide nanoparticles via plant extracts: an overview. *Materials Research Express* 6(11): 112004.

Jayandran, M., M. M. Haneefa and V. J. Balasubramanian. 2015. Green synthesis and characterization of Manganese nanoparticles using natural plant extracts and its evaluation of antimicrobial activity. *Journal of Applied Pharmaceutical Science* 5(12): 105–110.

Jayarambabu, N., A. Akshaykranth, T. V. Rao, K. V. Rao and R. R. Kumar. 2020. Green synthesis of Cu nanoparticles using *Curcuma longa* extract and their application in antimicrobial activity. *Materials Letters* 259: 126813.

Joseph, A. T., P. Prakash and S. J. Narvi. 2016. Phytofabrication and characterization of copper Nanoparticles using *Allium sativum* and its antibacterial activity. *International Journal of Engineering, Science and Technology* 4: 463–472.

Joshi A, A. Sharma, R.K. Bachheti, A. Husen, V.K. Mishra. 2019. Plant-mediated synthesis of copper oxide nanoparticles and their biological applications. In: Husen A, Iqbal M, eds., *Nanomaterials and Plant Potential* (pp. 221–237). Springer International Publishing AG, Cham.

Kannan, K., D. Radhika, S. Vijayalakshmi, K. K. Sadasivuni, A. A. Ojiaku and U. J. Verma. 2020. Facile fabrication of CuO nanoparticles via microwave-assisted method: photocatalytic, antimicrobial and anticancer enhancing performance. *International Journal of Environmental Analytical Chemistry*, https://doi.org/10.1080/03067319.2020.1733543

Karimi, J. and S. Mohsenzadeh. 2015. Rapid, green, and eco-friendly biosynthesis of copper Nanoparticles using flower extract of *Aloe vera*. Synthesis and reactivity in inorganic. *Metal-organic, and Nano-Metal Chemistry* 45(6): 895–898.

Kavitha, K., S. Baker, D. Rakshith, H. Kavitha, H. Yashwantha Rao, B. Harini and S. J. Satish. 2013. Plants as green source towards synthesis of nanoparticles. *International Research Journal* of *Biological Sciences* 2(6): 66–76.

Khodakovskaya, M. V., K. De Silva, A. S. Biris, E. Dervishi and H. J. Villagarcia. 2012. Carbon nanotubes induce growth enhancement of tobacco cells. *ACS Nano* 6(3): 2128–2135.

Kumar, K. P., W. Paul and C. Sharma. 2011. Green synthesis of gold nanoparticles with *Zingiber Officinale* extract: Characterization and blood compatibility. *Process Biochemistry* 46(10): 2007–2013.

Kumar, P. V., S. M. J. Kala and K. Prakash. 2019. Green synthesis of gold nanoparticles using *Croton caudatus* Geisel leaf extract and their biological studies. *Materials Letters* 236: 19–22.

Kumar, P. V., S. Pammi, P. Kollu, K. Satyanarayana and U. Shameem. 2014. Green synthesis and characterization of silver nanoparticles using *Boerhaavia diffusa* plant extract and their antibacterial activity. *Industrial Crops and Products* 52: 562–566.

Kumar, P. V., U. Shameem, P. Kollu, R. Kalyani and S. J. B. Pammi. 2015. Green synthesis of copper oxide nanoparticles using *Aloe vera* leaf extract and its antibacterial activity against fish bacterial pathogens. *Bio NanoScience* 5(3): 135–139.

Lakhan, M. N., R. Chen, A. H. Shar, K. Chand, A. H. Shah, M. Ahmed, I. Ali, R. Ahmed, J. Liu and K. J. Takahashi. 2020. Eco-friendly green synthesis of clove buds extract functionalized silver nanoparticles and evaluation of antibacterial and antidiatom activity. *Journal of Microbiological Methods* 173: 105934.

Logaranjan, K., A. J. Raiza, S. C. Gopinath, Y. Chen, K. Pandian. 2017. Shape-and size-controlled Synthesis of silver nanoparticles using *Aloe vera* plant extract and their antimicrobial activity *Nanoscale research letters* 11(1): 1–9.

Maensiri, S., P. Laokul, J. Klinkaewnarong, S. Phokha, V. Promarak and S. J. Seraphin. 2008. Indium oxide (In_2O_3) nanoparticles using *Aloe vera* plant extract: Synthesis and optical properties. *Journal* of *Optoelectronics and Advanced Materials* 10(3): 161–165.

Mandala, S. K., A. K. Maji, S. K. Mishra, P. M. Ishfaq, H. P. Devkota, A. S. Silva and N. J. Das. 2020. Goldenseal (*Hydrastis canadensis* L.) and its active constituents: A critical review of their efficacy and toxicological issues. *Pharmacological Research*, https://doi.org/ 10.1016/ j.phrs. 2020.105085

Mathew, S., A. Prakash and E. Radhakrishnan. 2018. Sunlight mediated rapid synthesis of small size range silver nanoparticles using *Zingiber officinale* rhizome extract and its antibacterial activity analysis. *Inorganic and Nano-Metal Chemistry* 48(2): 139–145.

Villanueva, M. A., K. Morán-Santibañez, J. J. Martínez-Sanmiguel, R. Rangel-López, M. A. Garza-Navarro, C. Rodríguez-Padilla, D. G. Zarate-Triviño and L. M. Trejo-Ávila. 2019. Virucidal activity of gold nanoparticles synthesized by green chemistry using garlic extract. *Viruses* 11(12): 1111.

Miri, A., M. Sarani, M. R. Bazaz and M. Darroudi. 2015. Plant-mediated biosynthesis of silver nanoparticles using Prosopis farcta extract and its antibacterial properties. *Spectrochimica Acta Part a: Molecular and Biomolecular Spectroscopy* 141: 287–291.

Mohan, R., A. Shanmugharaj and R. Sung Hun. 2011. An efficient growth of silver and copper nanoparticles on multiwalled carbon nanotube with enhanced antimicrobial activity. *Journal of Biomedical Materials Research Part B: Applied Biomaterials* 96(1): 119–126.

Mubayi, A., S. Chatterji, P. M. Rai and G. Watal. 2012. Evidence-based green synthesis of nanoparticles. *Advanced Materials Letter* 3(6): 519–525.

Nasrollahzadeh M., S. M. Sajadi. 2015. Synthesis and characterization of titanium dioxide Nanoparticles using *Euphorbia heteradena* Jaub root extract and evaluation of their stability. *Ceramics International* 41(10): 14435–14439.

Nasrollahzadeh, M. and S. M. Sajadi. 2015. Green synthesis of copper nanoparticles using *Ginkgo biloba* L. leaf extract and their catalytic activity for the Huisgen [3+ 2] cycloaddition of azides and alkynes at room temperature. *Journal of Colloid and Interface Science* 457: 141–147.

Obeizi, Z., H. Benbouzid, S. Ouchenane, D. Yılmaz, M. Culha and M. Bououdina. 2020. Biosynthesis of Zinc oxide nanoparticles from essential oil of *Eucalyptus globulus* with antimicrobial and anti-biofilm activities. *Materials Today Communications* 25: 101553.

Owaid, M. N., T. A. Zaidan, R. F. Muslim and M. A. Hammood. 2019. Biosynthesis, characterization and cytotoxicity of zinc nanoparticles using *Panax ginseng* roots, Araliaceae. *Acta Pharmaceutica Sciencia* 57(1): 19–32.

Parveen, K., V. Banse and L. Ledwani. 2016. Green synthesis of nanoparticles: Their advantages and disadvantages. Paper presented at the AIP conference proceedings. Vol. 1724, No. 1, p. 020048). AIP Publishing LLC.

Patel, P., P. Agarwal, S. Kanawaria, S. Kachhwaha and S. Kothari. 2015. Plant-based synthesis of silver nanoparticles and their characterization. In H. M. Siddiqui, M. H. Al-Whaibi and F. Mohammad, eds., *Nanotechnology and Plant Sciences* (pp. 271–288). Springer, Cham.

Pethakamsetty, L., K. Kothapenta, H. R. Nammi, L. K. Ruddaraju, P. Kollu, S. G. Yoon and S. V. N. Pammi. 2017. Green synthesis, characterization and antimicrobial activity of silver nanoparticles using methanolic root extracts of *Diospyros sylvatica. Journal of Environmental Sciences* 55: 157–163.

Phumying, S., S. Labuayai, C. Thomas, V. Amornkitbamrung, E. Swatsitang and S. Maensiri. 2013. *Aloe vera* plant-extracted solution hydrothermal synthesis and magnetic properties of magnetite (Fe_3O_4) nanoparticles. *Applied Physics A* 111(4): 1187–1193.

Prabhu, S. and E. K. Poulose. 2012. Silver nanoparticles: mechanism of antimicrobial action, synthesis, medical applications, and toxicity effects. *International Nano Letters* 2(1): 1–10.

Priya, G. S., A. Kanneganti, K. A. Kumar, K. V. Rao and S. Bykkam. 2014. Biosynthesis of cerium oxide nanoparticles using *Aloe barbadensis* miller gel. *International Journal of Scientific and Research Publication* 4(6): 199–224.

Raghunath, A. and E. Perumal. 2017. Metal oxide nanoparticles as antimicrobial agents: A promise for the future. *International Journal of Antimicrobial Agents* 49(2): 137–152.

Raj, L. and E. Jayalakshmy. 2015. Biosynthesis and characterization of zinc oxide nanoparticles using root extract of *Zingiber officinale*. *Oriental Journal of Chemistry* 31(1): 51–56.

Ranjbar, M., M. Dastani and F. Khakdan. 2021. Evaluation of antimicrobial, cytotoxicity and catalytic activities of CuO-NPs synthesized by *Tanacetum Parthenium* extract, https://doi.org/10.21203/rs.3.rs-154772/v1

Rao, K. G., C. Ashok, K. V. Rao, C. Chakra and P. Tambur. 2015. Green synthesis of TiO_2 nanoparticles using *Aloe vera* extract. *International Journal of Advanced Research in Physical Science* 2(1A): 28–34.

Rastogi, L. and J. Arunachalam. 2013. Green synthesis route for the size-controlled synthesis of biocompatible gold nanoparticles using aqueous extract of garlic (*Allium sativum*). *Advanced Materials Letter* 4: 548–555.

Rauwel, P., S. Küünal, S. Ferdov and E. Rauwel. 2015. A review on the green synthesis of silver nanoparticles and their morphologies studied via TEM. *Advances in Materials Science and Engineering* 2015, https://doi.org/10.1155/2015/682749

Ren, Y.Y., H. Yang, T. Wang and C. Wang. 2016. Green synthesis and antimicrobial activity of monodisperse silver nanoparticles synthesized using *Ginkgo biloba* leaf extract. *Physics Letters A* 380(45): 3773–3777.

Roy, S. and T. K. Das. 2015. Plant mediated green synthesis of silver nanoparticles-A. *International Journal of Plant Biology* 3(3): 1044.

Saratale, R. G., G. Benelli, G. Kumar, D. S. Kim and G. D. Saratale. 2018. Bio-fabrication of silver nanoparticles using the leaf extract of an ancient herbal medicine, dandelion (*Taraxacum officinale*), evaluation of their antioxidant, anticancer potential, and antimicrobial activity against phytopathogens. *Environmental Science and Pollution Research* 25(11): 10392–10406.

Sasikala, A., M. L. Rao, N. Savithramma and T. Prasad. 2015. Synthesis of silver nanoparticles from stem bark of *Cochlospermumreligiosum* (L.) Alston: an important medicinal plant and evaluation of their antimicrobial efficacy. *Applied Nanoscience* 5(7): 827–835.

Sharma, V. K., R. A. Yngard and Y. Lin. 2009. Silver nanoparticles: Green synthesis and their antimicrobial activities. *Advances in Colloid and Interface Science* 145(1–2): 83–96.

Siddiqui, M. H., M. H. Al-Whaibi and F. Mohammad. 2015. Nanotechnology and plant sciences. *International Publishing Switzerland* 10: 978–973.

Singh, J., T. Dutta, K.H. Kim, M. Rawat, P. Samddar and P. Kumar. 2018. Green synthesis of metals and their oxide nanoparticles: Applications for environmental remediation. *Journal of Nano-Biotechnology* 16(1): 1–24.

Singh, P., Y. J. Kim and D. C. Yang. 2016b. A strategic approach for rapid synthesis of gold and silver nanoparticles by *Panax ginseng* leaves. *Artificial Cells, Nano-Medicine, and Biotechnology* 44(8): 1949–1957.

Singh, P., Y. J. Kim, C. Wang, R. Mathiyalagan and D. C. Yang. 2016a. The development of a green approach for the biosynthesis of silver and gold nanoparticles by using *Panax ginseng* root extract, and their biological applications. *Artificial Cells, Nano-Medicine, and Biotechnology* 44(4): 1150–1157.

Singh, P., Y. J. Kim, C. Wang, R. Mathiyalagan, M. El-Agamy Farh and D. C. Yang. 2016. Biogenic silver and gold nanoparticles synthesized using red ginseng root extract, and their applications. *Artificial Cells, NanoMedicine, and Biotechnology* 44(3): 811–816.

Sinha, S., I. Pan, P. Chanda and S. K. Sen. 2009. Nanoparticle's fabrication using ambient biological resources. *Journal of Applied Biosciences* 19: 1113–1130.

Slman, D. K., R. D. A. Jalill and A. N. Abd. 2018. Biosynthesis of zinc oxide nanoparticles by hot aqueous extract of *Allium sativum* plants. *Journal of Pharmaceutical Sciences and Research* 10(6): 1590–1596.

Some, S., I. K. Sen, A. Mandal, T. Aslan, Y. Ustun, E. Ş. Yilmaz, A. Katı, A. Demirbas, A. K. Mandal and I. Ocsoy. 2018. Biosynthesis of silver nanoparticles and their versatile antimicrobial properties. *Materials Research Express* 6(1): 012001.

Sre, P. R., M. Reka, R. Poovazhagi, M. A. Kumar and K. Murugesan. 2015. Antibacterial and cytotoxic effect of biologically synthesized silver nanoparticles using aqueous root extract of *Erythrina indica* lam. *Spectrochimica Acta Part a: Molecular and Biomolecular Spectroscopy* 135: 1137–1144.

Sreekanth, T., P. Nagajyothi, P. Muthuraman, G. Enkhtaivan, S. Vattikuti, C. Tettey, D. H. Kim, J. Shim and K. Yoo. 2018. Ultra-sonication-assisted silver nanoparticles using Panax ginseng root extract and their anti-cancer and antiviral activities. *Journal of Photochemistry and Photobiology B: Biology* 188 : 6–11.

Sumitha, S., S. Vasanthi, S. Shalini, S. V. Chinni, S. C. Gopinath, P. Anbu, M. B. Bahari, R. Harish, S. Kathiresan and V. Ravichandran. 2018. Phyto-mediated photo catalyzed green synthesis of silver nanoparticles using *Durio zibethinus* seed extract: antimicrobial and cytotoxic activity and photocatalytic applications. *Molecules* 23(12): 3311.

Suresh, K. 2008. Antimicrobial and phytochemical investigation of the leaves of *Carica papaya* L., *Cynodondactylon* L., Pers., *Euphorbia hirta* L., *Melia azedarach* L., and *Psidium guajava* L. *Ethnobotanical Leaflets* 2008(1): 157.

Tagad, C., H. H. Seo, R. Tongaonkar, Y. W. Yu, J. H. Lee, M. Dingre, A. Kulkarni, H. Fouad, S. A. Ansari and S. H. Moh. 2017. Green Synthesis of silver nanoparticles using *Panax ginseng* root extract for the detection of Hg^{2+}. *Sensors and Materials* 29(2): 205–215.

Teimouri, M., F. Khosravi-Nejad, F. Attar, A. A. Saboury, I. Kostova, G. Benelli and M. Falahati. 2018. Gold nanoparticles fabrication by plant extracts: synthesis, characterization, degradation of 4- nitrophenol from industrial wastewater, and insecticidal activity–a review. *Journal of Cleaner Production* 184: 740–753.

Thakkar, K. N., S. S. Mhatre and R. Y. Parikh. 2010. Biological synthesis of metallic nanoparticles. *Nanomedicine: Nanotechnology, Biology and Medicine* 6(2): 257–262.

Thandapani, K., M. Kathiravan, E. Namasivayam, I. A. Padiksan, G. Natesan, M. Tiwari, B. Giovanni and V. Perumal. 2018. Enhanced larvicidal, antibacterial, and photocatalytic efficacy of TiO2 nanohybrids green synthesized using the aqueous leaf extract of *Parthenium hysterophorus*. *Environmental Science and Pollution Research* 25(11): 10328–10339.

Tippayawat, P., N. Phromviyo, P. Boueroy and A. Chompoosor. 2016. Green synthesis of silver nanoparticles in *Aloe vera* plant extract prepared by a hydrothermal method and their synergistic antibacterial activity. *Peer Journal* 4: e2589.

Varier, K.M., M. Gudeppu, A. Chinnasamy, S. Thangarajan, J. Balasubramanian, Y. Li and B. Gajendran. 2019. Nanoparticles: antimicrobial applications and its prospects. In: M. Naushad, S. Rajendran and F. Gracia, eds., *Advanced nanostructured materials for environmental remediation* (pp. 321–355). Springer, Cham.

Velmurugan, P., K. Anbalagan, M. Manosathyadevan, K. J. Lee, M. Cho, S. M. Lee, J. H. Park, S. G. Oh, K. S. Bang and B. T. Oh. 2014. Green synthesis of silver and gold nanoparticles using *Zingiber officinale* root extract and antibacterial activity of silver nanoparticles against food pathogens. *Bioprocess and BioSystems Engineering* 37(10): 1935–1943.

Vijaya, J. J., N. Jayaprakash, K. Kombaiah, K. Kaviyarasu, L. J. Kennedy, R. J. Ramalingam, H. A. Al-Lohedan, V. Mansoor-Ali and M. Maaza. 2017. Bioreduction potentials of dried root of *Zingiber officinale* for a simple green synthesis of silver nanoparticles: antibacterial studies. *Journal of Photochemistry and Photobiology B: Biology* 177: 62–68.

Vijaya, J. J., N. Jayaprakash, K. Kombaiah, K. Kaviyarasu, L. J. Kennedy, R. J. Ramalingam, H. A. Al-Lohedan, V. Mansoor-Ali and M. Maaza. 2019. Garlic clove extract assisted silver nanoparticle – Antibacterial, antibiofilm, antihelminthic, anti-inflammatory, anticancer and ecotoxicity assessment. *Journal of Photochemistry and Photobiology B: Biology* 198: 111558.

Vijayaraghavan, K. and T. Ashokkumar. 2017. Plant-mediated biosynthesis of metallic nanoparticles: a Review of literature, factors affecting synthesis, characterization techniques and applications. *Journal of Environmental Chemical Engineering* 5(5): 4866–4883.

Wang, C., Y. J. Kim, P. Singh, R. Mathiyalagan, Y. Jin and D. C. Yang. 2016. Green synthesis of silver nanoparticles by Bacillus methylotrophic, and their antimicrobial activity. *Artificial Cells, Nano-Medicine, and Biotechnology* 44(4): 1127–1132.

Wang, L., C. Hu and L. Shao. 2017. The antimicrobial activity of nanoparticles: Present situation and prospects for the future. *International Journal of Nano-medicine* 12: 1227.

Xu, H., L. Wang, H. Su, L. Gu, T. Han, F. Meng and C. Liu. 2015. Making good use of food wastes: Green synthesis of highly stabilized silver nanoparticles from grape seed extract and their antimicrobial activity *Food Biophysics* 10(1): 12–18.

Yadav, J., S. Kumar, L. Budhwar, A. Yadav and M. Yadav. 2016. Characterization and antibacterial activity of synthesized silver and iron nanoparticles using *Aloe vera*. *Journal of Nanomedicine and Nanotechnology* 7(384): 2.

Yang, N., F. Li, T. Jian, C. Liu, H. Sun, L. Wang and H. Xu. 2017. Biogenic synthesis of silver nanoparticles using ginger (*Zingiber officinale*) extract and their antibacterial properties against aquatic pathogens. *Acta Oceanologica Sinica* 36(12): 95–100.

Zha, J., C. Dong, X. Wang, X. Zhang, X. Xiao and X. Yang. 2017. Green synthesis and characterization of monodisperse gold nanoparticles using *Ginkgo biloba* leaf extract. *Optik* 144: 511–521.
Zikalala, N., K. Matshetshe, S. Parani and O. S. Oluwafemi. 2018. Biosynthesis protocols for colloidal metal oxide nanoparticles. *Nano-Structures and Nano-Objects* 16: 288–299.

13 Tools for Safety Evaluation and Conservation of Medicinal Plants

Shipra Rani Jha and Mohd Mughees

CONTENTS

13.1 INTRODUCTION

India is known for its culture, ancient scripts, Vedas, the invention of zero, and the traditional system of medicine, viz. Ayurveda, Amchi and Unani. Medicinal plants have been used as a source of medicine since ancient times. It is not only in India, but the population of the majority of the developing countries uses the traditional system of medicine for their primary health care. It has been estimated that about three-quarters of the herbal and perfumery products used throughout the world are available in India (Kamboj 2000). The World Health Organization (WHO) estimated that the total herbal medicine market is US $62 billion, and by the year 2050, it is expected to grow to $5 trillion. Ayurveda alone has contributed $813 million annually to the Indian economy. Therefore, the rich genetic diversity of aromatic and medicinal plants is one of the biggest strengths of India and is the first step for the development of all the bio-industries. More than 50% of the best-selling pharmaceuticals are derived from natural products. Medicinal plants contain a large variety of secondary metabolites or important compounds that can be used in the treatment of chronic as well as communicable diseases (Duraipandiyan, Ayyanar and Ignacimuthu 2006).

For most of the world's population, medicinal plants are being used as the most important source of life-saving drugs. India has a rich source of medicinal plants, but rapid population growth and human activities are now destroying the natural environment. Various medicinal plants are now on the stage of extinction. Therefore, some urgent steps for the conservation of those medicinal plants are required. *In vitro* micropropagation requires enormous effort for the production of herbal medicines and is also an important tool for the conservation of their genotypes. Medicinal plants are exploited because of the secondary metabolites produced by them. These metabolites produced

DOI: 10.1201/9781003179139-13

are beneficial to the plants also. These secondary metabolites are generally produced in lesser quantities in plants. Therefore, various *in vitro* cell and organ cultures have been raised or developed to enhance their productivity. This micropropagation technique has received considerable attention from researchers to produce bioactive compounds in eco-friendly conditions.

In the field of herbal medicine discovery, the ordinary methods of isolation, separation, identification and characterization for the desired secondary metabolites suffer from various obstacles like de-replication, lower productivity and difficult separation (Anonymous WHO 1996). However, the innovations of some modern or novel technologies like high throughput screening (HTS) and the current approach in chromatography, spectroscopy and electrophoresis have changed and reconstructed the whole context of pharmacognosy. Secondary metabolites present in the plants can be used as chemical markers, but they may change due to environmental and biological factors like the age of the plant and their geographical distribution, and so on. Secondary metabolites tend to vary with these two factors and, therefore, restrict it only to their qualitative studies. Hence, correct identification of the plants becomes a difficult task for botanists and researchers. Most of the pharmacopeias and regulatory agencies suggest various methods for quality control and standardization like macroscopic and microscopic evaluation and chemo-profiling using chromatographic (TLC and HPTLC) and spectroscopic (MS and LC-MS) techniques 10,11. Techniques like TLC, HPTLC and HPLC used as chemical markers have their own limitations. Since 1990, pharmacognosy has become a molecular science. Molecular markers, especially DNA markers, identify plants at the genomic level and establish new standards in species identification, characterization, and quality control of the herbal medicines (Joshi et al. 2004). In this chapter the reasons for adulteration, further strategies or methods for the identification, authentication and characterization of medicinal plants, and mass multiplication through *in vitro* regeneration to overcome the exploitation of endangered medicinal plants are given.

13.2 ADULTERATION/SUBSTITUTION IN THE MEDICINAL PLANTS

Today, adulteration has become a critical issue in the market for herbal medicine. Therefore, correct identification, authentication and quality control are a preliminary requirement to overcome or minimize the unfair trade of all the traditional herbal medicines. Adulterants are herbs or plants having a similar physiological appearance as that of the genuine plant species, and they mainly replicate each other from the start. Adulteration may be done intentionally for the monetary gain/ profit, or it may be unintentionally due to a lack of awareness. It has taken place in such a way that it became difficult for a layman to distinguish between the genuine plant and the adulterated one. Sometimes the adulterated plant product has no similarities with the genuine plant: for example, the leaves of *Phytolacca*, *Ailanthus* and *Scopalia* are adulterated with the leaves of *Belladonna* and *Cassia*, and so on. Certain medicinal plants are also being adulterated due to their lesser availability just to increase their concentration or weight (Table 13.1). Black pepper is commonly adulterated with papaya seeds which are almost similar in appearance on the basis of their color, shape and size but have a deleterious effect when compared to their medicinal values (Bhalla and Punekar 1975, Tremlova 2001). Turmeric powder has been adulterated or substituted with the rhizome of some related species bearing the coloring pigment curcumin, such as *C. xanthorrhiza* Roxb, *C. Malabarica*, and so on (Sen, Gupta and Dastidar 1974, Tripathi, Khanna and Das 2007).

13.3 CONSEQUENCES OF ADULTERATION IN PLANTS ON HUMAN HEALTH

There is a general belief among the consumers that, besides being cheap and locally available, herbal drugs do not have any side effects. Beyond side effects from the medicinal herb itself, adulteration and lack of awareness about drug interactions have led to some life-threatening adverse conditions.

TABLE 13.1A
Adulterants Found in Some of the Major Traded Medicinal Products

	Adulterants	
Commodity	**Biological materials**	**Chemical/earthy materials**
Nutmeg (*Myristica fragrans*)	Macassar (*Myristica argentea*), Bombay (*M. malabarica*) and *M. otoba*	Pieces of clay for repairing broken nutmeg
Cardamom fruits (*Elettaria cardamomum*)	Orange seeds, unroasted coffee seeds	Small pebbles
Cassia bark (*Cinnamomum cassia*)	Bark of *Cinnamomum japonicum, C. mairei C. burmannii*	-
European dill	Indian dill	Terpenes
Mace (*Myristica fragrans*)	Bombay mace (*Myristica malabarica*), Macassar mace (*M. argentea*)	-
Black pepper berries (*Piper nigrum*)	Dried papaya seed (*Carica papaya*), wild Piper spp. (*P. attenuatum and P. galeatum*), fruits of *Lantana camara* and *Embelia ribes*, seeds of *Mirabilis jalapa*, berries of *Schinus molle*	mineral oil
Clove	Exhausted clove (volatile oil extracted), stem and fruits of clove	Magnesium salt, sand, earth
Allspice powder (*Pimenta dioica*)	Powdered clove stem, berries of *Myrtus Tobasco* and *Lindera benzoin*	-
Turmeric powder (*Curcuma longa*)	Wild *Curcuma* spp. – *C. zedoaria* Rosc or "yellow shotti" syn. *C. xanthorrhiza* Roxb. ('Manjakua') or *C. malabarica*, starch from cheaper source, sawdust	Dye- Metanil Yellow, Orange II lead chromate, chalk powder, yellow soapstone powder

TABLE 13.1B
Lists of Medicinal Plants and Their Adulterant

Genuine plant	Vernacular name	Adulterants
Bacopa monnieri	Brahmi	*Centella asiatica*
Swertia chirayita	Chiraita	*Andrographis paniculata*
Evolvulus alsinoides	Sankhpuspi	*Clitorea ternatea, Ephedra gerardiana, Convolvulus microphyllus*
Aconitum heterophyllum	Ativisha	*Cyperus rotundus*
Plumbago zeylanica	Chitraka	*Baliospermum montanum*
Cassia senna	Senna	*Cassia angustifolia*
Marsdenia tenacissima	Murva	*Rubia cordifolia*
Solanum xanthocarpam	Kateli	*Clerodendron indicum*
Polygonatum cirrhifolium	Meda	*Withania somnifera*
Fumaria parviflora	Parpatta	*Mollungo pentaphylla*
Pluchea lanceolata	Rasna	*Alpinia officinarum, Vanda roxburghii, Geranium wallichianum*
Messua ferrea	Nagakesara	*Callophyllum inophyllum*
Boerhavia diffusa	Punarva	*Trianthema portulacasstrum*
Iris germanica	Pushkarmula	*Saussurea lappa*
Onosma echinoides	Ratanjot	*Clausena pentaphylla, Canscora decussate*

TABLE 13.2
List of Medicinal Plants with Their Side Effects

Name	Common name	Scientific name	Adverse effect
Aconite	monkshood	*Aconitum* species	Abdominal pain, nausea, vomiting, hypotension, paralysis in the respiratory system, death
Broom	Green Broom, Greenweed	*Genista tinctoria*	Uterotonic properties, contraindicated for pregnancy and breast feeding, vomiting
Ephedra		*Ephedra sinica*	Insomnia, hypertension, kidney stones, irregular heart beat
Reserpine		*Rauvolfia serpentina*	Mental depression, increased gastric secretion
Senna		*Cassia senna*	Abdominal pain, liver damage
American ginseng	Ginseng	*Panax quinquefolius*	Hypertension, insomnia, vaginal bleeding, breast tenderness
Aloe vera	Aloe	*Aloe vera*	Carcinogenic, abdominal pain, diarrhea
Afsinthin	wormwood	*Artemisia absinthium*	Neurotoxic
Areca nut	Betel nut	*Areca catechu*	carcinogenic

The efficacy of the herbal product depends upon the use of correct and authenticated plant material. The use of the wrong herb may worsen the condition, or it may lead to death if any toxic adulterants are used or substituted. Herbal formulations can also be dangerously adulterated or contaminated, and these adulterated herbal medicines may unknowingly be used to replace medicines that do have authenticated efficacy. Common side effects of the adulterated medicinal plant are hepatitis, edema, angioedema, hypertension, thrombocytopenia, convulsions and death. There are several plants that have severe adverse effects on human health like the use of *Ephedra* causes cardiovascular diseases, Kava-Kava consumption causes hepatoxic effects, and anticholinergic effect is caused by using *Datura metal* (Cuzzolin et al. 2006) (Table 13.2).

It has been observed that *Ephedra* is a useful medicinal herb in the treatment of respiratory ailments, but overdosing on this herbal drug may lead to liver damage, hypertension, irregular heartbeat, and so on. American ginseng is a herbal tonic and is generally used to control stress and build up immunity besides treating hypertension and diabetes. But this plant has been reported to cause hypertension, insomnia and vaginal bleeding as side effects. Adulteration of *Stephania tetrandra* with the root of *Aristolochia fangchi* had resulted in renal intestinal fibrosis and urothelial carcinoma in some persons (Nortier 2000). This type of misunderstanding regarding herbal medicine may lead to several adverse conditions. With the ever-increasing population number, the need for safer drugs has also become a critical issue which has drawn our attention to the quality and safety of the formulation of traditional Indian medicine.

13.4 VARIATIONS IN THE SECONDARY METABOLITES IN THE SAME SPECIES OF THE PLANT

It has been seen that secondary metabolites vary in the same species of plants. There are certain factors behind these variations of the secondary metabolites. Geographical distribution of plants is one of the important reasons for these variations as it has already been investigated that alkaloid contents in *Aconitum* vary with the altitude. It increases with the increase in altitude (Beigh, Nowchoo and Iqbal 2008). The podophyllotoxin content of *Podophyllum hexandrum* also increases with the increase in altitude (Iqbal, Beigh and Nawchoo 2004). The variations in the alkaloid production with altitude may be due to the adaptation to the species in a particular ecological and climatic niche. *Artemisinin* content in *Artemisia annua* was also varied due to geographical distribution or variations. It has been reported that artemisinin content was 0.01–0.09% in different parts of India (Singh et al. 1988), 0.06–0.1% in Argentina (Liersch et al. 1986), 0.06–0.22% in the USA (Ferreira et al. 1995). Plants behave abnormally or differently under stress conditions or beyond the ecological or climatic niche. The bioactive compounds either get increased or decreased due to the environmental factors. It has also been seen that withafarin content in *Withania somnifera* (Ashwagandha) was affected or varied due to environmental factors (Sangwan et al. 2004). The geographical distribution also tends to variations in vaccine contents in *Adhatoda zelaynica* (Bagchi et al. 2003) and bacoside content in different accession of *Bacopa monnieri* (Mathur et al. 2002).

It has also been postulated in various studies that climatic condition is also an important factor for the variations in the chemical composition of secondary metabolites of the plants. Zobayed, Afreen and Kozai (2005) investigated the effect of temperature stress on secondary metabolites and other important compounds in medicinal plants.

Secondary metabolites concentration in *Hypericum perforatum* (Southwell and Bourke 2001) increased due to the seasonal variations. Marked variations in the active principles present in the medicinal plants are found in different seasons as these secondary metabolites are attributed to environmental factors like rainfall and temperature. In *Calotropis procera*, it has been studied that the optimum concentration of the active ingredients in flowers was obtained in monsoon (Rathore, Gehlot and Shekhawat 2010). The effect of seasonal variations was also observed in *Mentha longifolia*, which showed an increase in the concentration of the active ingredients in a specific season (Ahmad et al. 2011). In *Antirrhinum majus*, marked variations in the content of the four iridoids were shown in the early and late season, and it decreased with the onset of flowering (Hogedal and Molgaard 2000). In *A. nagarum* and *A. elwesii*, alkaloid content in the root was observed at maximum in the month of November, and in leaves, the highest alkaloid concentration was observed in the pre-flowering stage (Sinam and Shantibala 2011). It has also been reported that the concentration of secondary metabolites varied due to the genotypic changes that occurred in the plants. Hypericin and flavonoid contents vary due to the genetic variations (Ghavamaldin et al. 2012).

According to many botanists, bioactive compounds are confined to several specific areas in plants. It has been found that in some plants like species of *Digitalis* and *Mentha*, the highest content of secondary metabolites is present in specific areas like positions of leaves. In *Psoralea corylifolia*, the highest psoralen content was found in seeds than leaves, stem, and root (Ali et al. 2008), but tubers in *Aconitum heterophyllum* had the highest alkaloid content and the least in aerial stem (Beigh, Nowchoo and Iqbal 2008).

13.5 METHODS FOR AUTHENTICATION, IDENTIFICATION AND STANDARDIZATION OF THE MEDICINAL PLANTS

To overcome this baffling situation of adulteration and misconception about the identity of the medicinal plant, various methods have been suggested by most of the regulatory agencies and pharmacopeias for the quality control and standardization of the herbal product. Several markers (taxonomic, chemical, genomic, and proteomic) are used for the identification of various components present in the herbal drug. Common strategies for herb authentication are dependent on its anatomical (Stern, Morris and Judd 1994), molecular (Li et al. 2011), chemical (Kite et al. 2009) and morphological techniques (Carlsward et al. 1997).

13.5.1 Phytochemical Methods for the Identification of Plants

Several methods, such as macroscopic and microscopic evaluation and chemical profiling of the medicinal plants using chromatographic (TLC and HPTLC) and spectroscopic (MS and LC-MS) techniques, are used for authentication of herbal products. HPTLC and HPLC fingerprints are used as valuable tools for the identification of genuine plants from the adulterated ones: HPLC has been used to identify *Cassia* bark (He et al. 2005) and *Psoralea corylifolia* (Zhao et al. 2005). It has also been reported that TLC was used to analyze the chemical profiles of *Tribulus terrestris* (Zhang et al. 2000). After the standardization of various commercial herbal and polyherbal formulations *Madhumehari* Churna (Baidynath), it was found that it contains a mixture of eight herbal antidiabetic medicines *Momordica charantia*, *Emblica officinalis* (fruits), *Syzigium cuminum*, *Gymnema sylvestre* (leaves), *Pterocarpus marsupium* (heartwood). *Trigonella foenum*, *Curcuma longa* (rhizomes) and *Azadirachta indica* (leaves) (Chandel, Pathak and Tailang 2011). It was also observed that "Pancasama Churna" are effective in the treatment of the gastrointestinal disorder (Meena et al. 2010). A traditional herbal formulation, "Dashmularishta", is useful in the treatment of normalization of physiological processes after childbirth or pregnancy. Authentication through chemical profiling is not accurate as it has several limitations, such as the amount of samples, the stability of secondary metabolites and the variable sources of the plant material.

13.5.2 Organoleptic Methods

Herbal drugs can also be evaluated on the basis of their color, odor, taste, shape, touch and texture, and this process is generally known as organoleptic evaluation. This process is based on the involvement of sense organs which include the above parameters and define some distinct characteristic features of the plant material, which may be taken as the first step toward their authentication. The organoleptic evaluation process is the simplest and quickest way to establish the identity of the plant sample. No other preliminary treatment is required for this process. But decisions or evaluations based on the sensory organs vary from person to person depend on the nature of the individual. The following criteria are used for the traditional authentication of *Radix codonopsis:* the root is cylindrical and curved, it has a characteristic odor and is sweet in flavor (British Herbal pharmacopeia 1996, Indian Herbal Pharmacopoeia 2002).

13.5.3 Histological or Microscopic Evaluation

Histological techniques involve some microscopic evaluation to determine the characteristic features of tissue based on their structure and arrangement of the cell, cortex, xylem vessels; for example, thickness and diameter of the exodermis, number of transfusion cells reported to be used as the marker for the identification of *Herba dendrobii* (Li et al. 1995) whereas arrangement of cortex and cork cells have been taken into consideration in identifying *Radix codonopsis* (Namba et al. 1981). Identification through histological approaches is not applicable to modern herbal drugs as closely associated species may have resemblance in their histological characteristics, and, thus, this approach is not so accurate and reliable.

13.5.4 Molecular Methods for the Authentication of the Medicinal Plant

It becomes difficult for botanists to identify a species correctly when its morphological traits are lacking as genetic makeup for each species is unique and is not affected by age and any other environmental as well as physiological factors. Therefore, genetic analysis of medicinal plants also plays an important role in unfolding the disputes regarding their identification and authentication. It has also been investigated that even the plant extracts in the herbal formulation can be authenticated by DNA-based methods (Novak, Gröger and Lukas 2007). Genetic markers can be used in the species characterization and authentication of the medicinal herbs (Shikha and Mishra 2009). These techniques are based on DNA-hybridization and polymerase chain reaction

(PCR), while the hybridization-based methods include Variable Number Tandem Repeat (VNTR), Restriction Fragment Length Polymorphism (RFLP), and Microarray, the PCR-based methods Amplified Fragment Length Polymorphism (AFLP), Randomly Amplified Polymorphic DNA (RAPD), Cleaved Amplified Polymorphic Sequence (CAPS), Amplified Refractory Mutation System (ARMS), Loop-mediated Isothermal Amplification (LAMP) and Sequence Characterized Amplified Region (SCAR).

The term minisatellites was first used by Jeffrey, Wilson and Thein (1985). Minisatellites consist of chromosomal regions having tandem repeat units of a 10–50 base motif. The level of polymorphism is substantial due to the high rate of mutation in minisatellites. It generally results in a unique profile for the different individuals with a high level of polymorphism. These are highly reproducible too. It has similarities in methodological procedures to that of the RFLP and is the major drawback of minisatellites. It has been reported that VNTR analysis was used in the authentication of *Taraxacum officinale* (Steven, Brian and John 2002).

In RFLP, the genome of the plant is digested with a suitable restriction endonuclease enzyme, and in this technique, the similarity in the cleavage pattern can be used to distinguish the plant species from one another. Each detected fragment in RFLP between the individual can be taken as an allele and can be used in the genetic analysis of plant species. Variations in the RFLP were an important technique in genetic fingerprinting, localization of genetic disease genes, genome mapping, and so on. Restriction Fragment Length Polymorphisms, being codominant markers and having high reproducibility, can detect the coupling phase of DNA molecules. If a linked trait is present in a homozygous or heterozygous individual, it can easily be determined by using RFLP. Therefore, RFLPs are very accurate markers in the field of linkage analysis and breeding techniques. It is used in interspecific genetic variations and DNA fingerprinting for genus Capsicum and pepper cultivars (Hosokawa et al. 2000). An attempt has been made to understand the high genetic diversity in *Podophyllum hexandrum* species from Himachal Pradesh, India, differentiation of *Codonopsis* from their adulterants (Fu et al. 1999), authentication of Panax species (Ngan et al. 1999), identification of *Fritillaria pallidiflora* (Wang et al. 2005), essential oil composition in *Atractylodes* plants (Mizukami et al. 1996) and authentication of *Atractylodes* species (Mizukami et al. 2000).

Microarray or chip-based methods can also be useful in the identification and authentication of herbal products. It is known as the collection of several spots of DNA molecules in an array. Each spot of DNA molecules contains picomoles (10^{12}) of probe labeled with fluorescent dyes. They are allowed to hybridize with other DNA or RNA molecules at a suitable experimental condition. It has been reported that DNA microarray was used in the identification of traditional herbal medicine (Carles et al. 2005). It is also used in the authentication of *Dendrobium* species from the formulation of Chinese medicine (Zhang et al. 2003).

The AFLP was developed by Keygene, which is a powerful fingerprinting technique. No prior knowledge of sequencing is required for this technique. Pattern for DNA fingerprinting can be achieved through restriction fragment analysis along with PCR amplification. The whole genomic DNA is digested with restriction endonuclease enzyme having sticky ends. The fragments are then ligated to an adapter sequence and are used for the amplification of DNA fragments obtained after restriction digestion. Amplified PCR products are analyzed on denaturing polyacrylamide gel through autoradiography/ fluorescence methodologies (Vos et al. 1995). AFLP technique is used in the identification of closely related species or individuals and is also useful in gene mapping. Steps involved in AFLPs are (1) restriction of DNA and ligation of oligonucleotide adapters; (2) pre-selective amplification; (3) selective amplification; (4) gel analysis of amplified fragments. It is a technique which detects DNA restriction fragments with the help of PCR amplification. Polyacrylamide gel electrophoresis (PAGE) is used to distinguish between different *Plectranthus* species (Passinho-Soares et al. 2006), *Actaea racemosa* (Zerega et al. 2002), *Rehmannia* (Qi et al. 2008), *Panax japonicus* (Choi et al. 2008), *Panax ginseng* from *Panax quinquefolius* (Ha et al. 2002) and *Cannabis* (Datwyler and Weiblen 2006). The advantage of AFLPs lies in their reproducibility, wide range of applications, high genomic abundance and no prior knowledge of sequencing required for primer construction.

Randomly Amplified Polymorphic DNA was first developed by Weish and McClelland in 1991. It is a PCR-based technology that detects the polymorphism in DNA nucleotide sequence by using a single primer. Each primer approximately directs amplification of several discrete loci in the genome, making the assay useful for efficient screening of nucleotide sequence polymorphism between individuals (Williams et al. 1990). The amplified products are separated or resolved on agarose gel stained with EtBr (ethidium bromide) and viewed under ultraviolet rays. The presence or absence of a band is observed through agarose gel electrophoresis. RAPD is a quick, easy and reliable assay in the identification of medicinal plants. Low quantities of DNA templates (5–50 ng per reaction) are required due to the involvement of PCR. Genetic techniques which involve PCR, randomly amplified polymorphic DNA (RAPD) and sequencing are considered to be more sensitive and reliable for the authentication of the plants (Zhang et al. 2007). RAPD is easy to develop and is one of the most accurate processes for herb authentication, but homoplasy, non-homology, collision, artefactual segregation and lack of reproducibility make it less reliable (Hosokawa et al. 2000). It has been reported that RAPD-based molecular markers were found to be useful in the authentication of different accessions of *Atractylodes* plants (Kohjyouma et al. 1997) and *Piper nigrum* (Khan et al. 2010). RAPD analysis differentiated the dried fruit samples of *Lyciumbarabarum* (Zhang et al. 2001), identified eight types of dried *Coptis*rhizomes (Cheng et al. 1997) and it was also reported that RAPD has been used to authenticate *Ammomum villosum* from its adulterants (Wang et al. 2000), *Panax* species (Shim et al. 2003) and three different species of *Scutellaria* (Hosokawa et al. 2000). RAPD has also proved its importance in the identification of *Cuscuta reflexa* and *Cuscuta chinenesis* (Khan et al. 2010).

Simple sequence repeats (SSRs) are also known as microsatellite-based markers and are helpful in assessing the high-level polymorphism. SSRs are monotonous repetitions of very short (2–6) nucleotide motifs. Microsatellites show variability; therefore, they are suitable for the identification of closely related cultivars (Vosman and Arens 1997) and can distinguish closely related genotypes (Smith and Devey 1994). The microsatellite is a PCR-based technique; hence very small quantity of DNA template (10–100 ng per reaction) is required. Reproducibility of SSRs is high as it uses long PCR primers and is widely used for linkage map construction, paternity testing and DNA fingerprinting, but it is less important in the identification of plant species. The SSR or microsatellite has been used in the evaluation of *Panax ginseng* (Kim et al. 2007), *Schisandra* (Boqian et al. 2009), *Acanthopanax senticosus* (Kim et al. 2007), *Cymbopogon* species (Kumar et al. 2007) and *Dendrobium fimbriatum* (Fan et al. 2009). It has also been reported to be applied in the authentication of *ginseng* (Hon et al. 2003) and used in the characterization of *Gastrodia elata* (Xu et al. 2006). SSR is also useful in studying genomic variations in *Codonopsis lanceolata* (Guo et al. 2006).

The ISSRs are DNA fragments of 100–3000 bp present at an amplifiable distance between two microsatellite repeats. About 10–60 fragments are generated and resolved or separated on gel electrophoresis. This technique uses a single primer for PCR reaction targeting multiple genomic loci for the amplification of ISSRs of different sizes. Inter-Simple Sequence Repeats do not require sequence data for primer construction. It is a PCR-based analytical process that requires only quantities of template DNA (5–50 ng per reaction). Like RAPDs, ISSRs is also a quick and simple process for the identification of plant species, but it can also have reproducibility problems. It has been used in the identification of different accessions of *Cistanche* (Shi et al. 2009), *Salvia* (Song et al. 2010) and *Vitex* (Hu et al. 2007) and *Dendrobium officinale* (Shen et al. 2006).

The CAPS marker technique utilizes the DNA sequences of mapped RFLP markers for the development PCR-based markers. Therefore, these are also known as PCR-RFLP markers (Konieczny and Ausubel 1993). CAPS has been used to distinguish between plants that are homozygous or heterozygous for alleles (Konieczny and Ausubel 1993). When compared to RFLP, the CAPS technique does not involve the technical steps of Southern blot hybridization. CAPS markers are codominant in nature, and polymorphism can be rarely found due to the size of the amplified fragments (300–1800 bp) of the PCR product. It is useful in map-based cloning and the identification of plant species (Weiland and Yu 2003). To increase the possibility of finding DNA polymorphisms, it can combine with SSCP, SCAR, AFLP and RAPD analysis. CAPS markers are preferred for the authentication and identification of medicinal plant species (Konieczny and Ausubei 1993). It is possible to amplify genomic DNA in B. napus using S locus-specific primers (Mohring, Horstmann and Esch 2005). CAPS markers have also been used in the identification of sweet pepper (Capsicum annuum L.) (Jundae et al. 2010). CAPS marker has been used for authentication of *Codonopsis* (Fu et al. 1999), *Alisma* (Li et al. 2007), *Panax* (Diao et al. 2009), *Artemisia* (Lee et al. 2009), *Ephedra* (Guo et al. 2006), *Fritillaria* (Wang et al. 2007) and *Dendrobium* (Zhang et al. 2005).

ARMS is also known as the allele-specific PCR and a variation of PCR in which the primers bind to their target sequence. Oligonucleotides having mismatched 3' end ("mutated") residues will not bind to the "normal" target sequences, and thus no amplification will take place, which causes failure in PCR. ARMS were used for molecular authentication of *Alisma orientale* (Li et al. 2007), Panax species (Zhu et al. 2004) and Curcuma (Ezhu) species (Sasaki et al. 2002) using chloroplast trnK based primer and nuclear 18S rRNA genes. It has also been developed in the identification of gecko (Liu et al. 2001).

LAMP is a simple, rapid, specific and cost-effective nucleic acid amplification method. It uses four different primers specifically designed to recognize six specific regions on the target gene where the amplification and detection of a gene can be completed in a single step by incubating the mixture of samples, primers and DNA polymerase. The reaction process proceeds at a temperature of about 65°C. The target gene can be indicated by the presence of an amplified product due to its high specificity.

SCAR is a PCR-based marker with a pair of specific oligonucleotide primers designed from RAPD fragments. A SCAR marker, generated from polymorphic regions, permits sample authentication based on SCAR size. The conversion of different DNA-based markers like RAPD, RFLP, AFLP, and other molecular marker to SCAR is based on sequence data and improves the reproducibility and reliability of PCR products. SCAR markers have a significant advantage over RAPD markers as it can produce unique PCR products without getting influenced by reaction conditions (Paran et al. 1993, Penner et al. 1993). Conditions under which the performance of PCR gets influenced are the concentration of $MgCl_2$, template DNA, Taq polymerase and thermocycler condition. As it is PCR-dependent technique, a very low amount of template DNA (a nanogram or less) is required to develop this marker. SCAR markers are useful in differentiating a genuine plant from its adulterant as the SCAR primers are able to retrieve a single clear band from the genuine plant sample, therefore, have high detection efficiency. The SCAR marker has been successfully developed in the authentication of *A. japonica* and *A. macrocephala*in whichAjR1, and AmR1 (forward and reverse) were used (Huh et al. 2006). *Pueraria tuberosa* is commonly sold in the Indian market as "vidari" and used in the preparation of "chywanaprash". *Ipomea mauritiana*, *Adeniahondala* and *Cycas circinalis* are common adulterants of *P. tuberosa*. *Ipomea mauritiana* was differentiated from *Pueraria tuberosa*, *Adeniahondala* and *Cycas circinalis* by designing SCAR (IM1F and IM1R) primers producing 323 bp band (Devaiah, Balasubramani and Venkatasubramanian 2011). *Embelia ribes* have also been substituted with *Embelia tsjeriam-cottam*, *Myrsine africana* and *Maesa indica*. The SCAR (ER1 and ER2) primers produce a 594 bp band in *E. ribes* only (Devaiah et al. 2008). Different species of *Phyllanthus*, that is, *Phyllanthus amarus*, *P. debilis* and *P. urinaria*, are morphologically very similar. These are grown in overlapping populations in the same habitat and are collectively known as "Luk-tai-bai". Authentication of these medicinal species was done by PA-B7,

PA-B17, PU-J17 and PU-J19 SCAR primers. SCAR marker (330 bp) was also developed for the authentication of three species of *Echinacea*, *E. angustifolia*, *E. pallida* and *E. purpurea* (Adinolfi et al. 2007). It has also been used for the authentication of *Angelica decursivum*, *Pseucedanum praeruptum* and *Anthriscus sylvestris* (Choo et al. 2009). Development of SCAR markers has been done for the authentication and identification of other medicinal plants like- Zingiber (Chavan et al. 2008), *Panax* (Wang et al. 2001), *Cynanchum* (Moon et al. 2010) and *Artemisia* (Lee et al. 2006). It has also been developed for the authentication of *Bacopa monnieri* L. from their adulterant (Yadav et al. 2012). It is also used for adulterant detection in ground chili, black pepper (Dhanya and Syamkumar 2009, Dhanya, Syamkumar and Sasikumar 2011) and a few endangered plants like *Sinocalycanthus* (Qian et al. 2006), *Morus boninensis* (Tani et al. 2003) and *Paphiopedilum* (Sun et al. 2011). SCAR markers have been used to discriminate important Indian medicinal plants from their close relatives. A SCAR marker has also been developed for the correct identification of *Viola serpens* Wall. ex Ging using an 1135 bp amplicon that was obtained from RAPD analysis of six accessions of *V. serpens* (Jha et al. 2020).

13.6 CONSERVATION OF MEDICINAL PLANTS THROUGH MASS MICROPROPAGATION

With increasing inclination toward the herbal medicines, there is an obligatory demand for large quantity raw materials and the right stage when the secondary metabolites are available in optimum amount for the standardization of herbal formulations. An ever-increasing demand for medicinal plants and their formulations requires their mass propagation or regeneration through *the in vitro* tissue culture technique.

Micropropagation techniques are used for mass-scale multiplication and conservation of an endangered medicinal plant using shoot tip and nodal segments in limited space and within a short period. Conservation of plants having medicinal values is the need of the hour. Due to the increasing demand for medicinal plants, there is a lack of raw materials for the pharmaceutical companies to produce herbal formulations. Therefore, the techniques of plant tissue culture are a great opportunity offered to mankind to grow plants under *in vitro* and aseptic conditions. Micropropagation involves only organized meristems, hence recovery of genetically stable and true to type progenies is allowed. Mass propagation of *Bacopa monnieri* was also reported using leaf explants and nodal segments (Mohapatra and Rath 2005). A miracle medicinal plant of India, *Rauvolfia serpentine* (L.) Benth., was first of all produced from its somatic callus tissue, which grew and fruited normally under *ex-vitro* conditions (Mitra and Chaturvedi 1970). Mass cytodifferentiation is not enough for the regeneration of a whole plant, and there should be differentiation leading to organogenesis. Organogenesis depends on several factors like the type of callus, growth elicitors used for callus induction and the type of explants used. It is possible to initiate organogenesis from the callus cultures by suitable manipulation of plant growth regulators. One hundred percent callogenesis was observed in the explants of *Artemisia absinthium*, independent of cytokinins/auxins ratio (Nin et al. 1996).

The process of formation of embryo-like structure from somatic tissues is known as somatic embryogenesis. Haccius (1978) defined somatic embryogenesis as a non-sexual developmental process producing a bipolar embryo. The first report on plant regeneration via somatic embryogenesis was reported in *Daucus carota* (Steward Mapes and Mears 1958). In addition to being developed from sporophytic cells, somatic embryos have also been developed from the generative cells as in *Datura innoxia* (Guha and Maheshwari 1964). Tissue culture techniques are frequently being used in *in vitro* regeneration and mass multiplication or propagation of various medicinal plants under aseptic conditions (Table 13.3).

TABLE 13.3
List of Important *in vitro* Cultured Medicinal Plants

S. No	Medicinal Plant Species	Explants	Nature of Response	Reference
1	*Tinospora cordifolia*	Nodal segments	Mass propagation	Gururaj et al. (2007)
2	*Swertia chirata*	shoot tip	Shoot culture	Balaraju et al. (2009)
3	*Calastrus paniculetas*	Nodal segment	Mass propagation	Martin et al. (2006)
4	*Zingiber officinali*	Meristem culture	Mass propagation	Bhagyalaxmi et al. (1988)
5	*Tylophora indica*	Nodal segment	Mass propagation	Faisal et al. (2007)
6	*Bacopa monnieri*	Leaf explants	Mass propagation	Mohapatra et al. (2005)
7	*Curcuma domestica*	Shoot apex	Mass propagation	Balachandran et al. (1990)
8	*Picrorhiza kurroa*	Nodal segment	Shoot culture	Sood et al. (2009)
9	*Saussurea lappa*	Shoot tip	Shoot culture	Johnson et al. (1997)
10	*Oroxylum indicum*	Nodal segments	Shoot culture	Dalal et al. (2004)
11	*Glycyrrhiza glabra*	Nodal segments	Axillary bud culture	Vadodaria et al. (2007)
12	*Costus speciosus*	Rhizome	Mass propagation	Chaturvedi et al. (1984)
13	*Ginkgo biloba*	Apical & Nodal segments	Shoot culture	Tommasi and Scaramuzzi (2004)
14	*Cinchona ledgeriana*	Shoot meristem	Mass propagation	Kobilitz et al. (1983)

REFERENCES

Adinolfi, B., Chicca, A., Martinotti, E., Breschi, M.C. and Nieri, P. 2007. Sequence characterized amplified region (SCAR) analysis on DNA from the three medicinal *Echinacea species*. *Fitoterapia* 78(1): 43–45.

Ahmad, I., Ahmad, M.S.A., Ashraf, M., Hussain, M. and Ashraf, M.Y. 2011. Seasonal variation in some medicinal and biochemical ingredients in *Mentha Longifolia* (L.) huds. *Pakistan J. Bot.* 43: 69–77.

Ali, J., Akhtar, N., Sultana, Y., Baboota, S. and Ahmad, S. 2008. Thin-layer chromatographic analysis of psoralen in Babchi (*Psoralea corylifolia*) oil. *Acta Chromatographica* 20: 277–282.

Anonymous. WHO (Geneva) 1996. Quality Control Methods for Medicinal Plant Materials.

Bagchi, G.D., Dwivedi, P.D., Haider, F., Singh, S., Srivastava, S. and Chattopadhyay, S.K. 2003. Seasonal variation in vasicine content in *Adhatoda* species grown under north Indian plain conditions. *J. Med. Aromatic Plant Sci.* 25: 37–40.

Balachandran, S.M., Bhat, S.R. and Chandel, K.P.S. 1990. In vitro clonal multiplication of turmeric (*Curcuma* spp.) and ginger (*Zingiber officinale* Rosc.). *Plant cell reports*, 8(9), 521–524.

Balaraju, K., Paul, A. and Ignacimuthu, S. 2009. Micropropagation of *Swertia chirata* Buch.-Hams. ex Wall.: A critically endangered medicinal herb. *Acta Physiol.Plant.* 31: 487–494.

Beigh, S.Y., Nowchoo, I.A. and Iqbal, M. 2008. Cultivation and conservation of *Aconitum heterophyllum*: a critically endangered medicinal herb of the Northwest Himalayas. *J. Herbs Spi. Med. Plants* 11: 47–56.

Bhagyalakshmi, B. and Singh, N.S. 1988. Meristem culture and micropropagation of a variety of ginger (*Zingiber officinale* Rosc.) with a high yield of oleoresin. *Journal of Hort. Sci.* 63(2): 321–327.

Bhalla, K. and Punekar, B.D. 1975. Incidence and state of adulteration of commonly consumed spices in Bombay city, mustard, black pepper, and asafoetida. *Indian J. Nutri. Dietetics* 12: 216–222.

British Herbal Pharmacopoeia 1996. British Herbal Medicine Association, Exeter.

Boqian, Y., Jing, W., Guopei, C. and Ting, W. 2009. Isolation and characterization of polymorphic microsatellite loci in a traditional Chinese medicinal plant, *Schisandra sphenanthera*. *Conservation Genetics* 10: 615–617.

Carles, M., Cheung, M.K.L., Moganti, S., Dong, T.T.X., Tsim, K.W., Ip, N.Y. and Sucher, N.J. 2005. A DNA Microarray for the authentication of toxic traditional Chinese medicinal plants. *Planta Med.* 71(6): 580–584.

Carlsward, B.S., Stern, W.L., Judd, W.S. and Lucansky, T.W. 1997. Comparative leaf anatomy and systematics in *Dendrobium*, sections *Aporum* and *Rhizobium* (Orchidaceae). *Int. J. Plant Sci.* 158: 332–342.

Chavan, P.D., Warude, K., Joshi, K. and Patwardhan, B. 2008. Development of SCAR (Sequence-characterized amplified region) markers as a complementary tool for identification of ginger from crude drugs and multi-component formulations. *Bio. Appl. Bio.* 50: 61–69.

Chandel, H.S., Pathak, A.K. and Tailang, M. 2011. Standardization of some herbal antidiabetic drugs in polyherbal formulation. *Pharmacognosy Res.* 3(1): 49–56.

Chaturvedi, H. C., Misra, P., and Jain, M. 1984. Proliferation of shoot tips and clonal multiplication of *Costus speciosus* in long-term culture. *Plant Sci. Lett.* 35(1): 67–71.

Cheng, K.T., Chang, H.C., Su, C.H. and Hsu, F.L. 1997. Identification of dried rhizomes of *Coptis* species using random amplified polymorphic DNA. *Botanical Bull. Academia Sinica* 38: 241–244.

Choi, Y.E., Ahn, C., Kim, B.B. and Yoon, E.S. 2008. Development of species specific AFLP-derived SCAR marker for authentication of *Panax japonicus* CA MEYER. *Biol. Pharmaceu. Bull.* 31: 135–138.

Choo, B.K., Moon, B.C., Ji, Y., Kim, B.B., Choi, G., Yoon, T. and Kim, H.K. 2009. Development of SCAR markers for the discrimination of three species of medicinal plants, *Angelica decursiva* (*Peucedanum decursivum*), *Peucedanum praeruptorum* and *Anthriscus sylvestris*, based on the internal transcribed spacer (ITS) sequence and random amplified polymorphic DNA (RAPD). *Biol. Pharmaceu. Bull.* 32(1): 24–30.

Cuzzolin, L., Zaffani, S. and Benoni, G. 2006. Safety implications regarding use of phytomedicines. *European J. Clinical Pharmacol.* 62(1): 37–42.

Dalal, N.V. and Rai, V.R. 2004. In vitro propagation of *Oroxylum indicum* Vent: A medicinally important forest tree. *J. Forest Res.* 9(1): 61–65.

Datwyler, S. and Weiblen, G. 2006. Genetic variation in hemp and marijuana (*Cannabis sativa* L.) according to Amplified Fragment Length Polymorphisms. *J. Forensic Sci.* 51: 371–375.

Devaiah, K., Balasubramani, S.P. and Venkatasubramanian, P. 2011. Development of randomly amplified polymorphic DNA based SCAR marker for identification of *Ipomoea mauritiana* Jacq (Convolvulaceae). *Evidence-Based Compl. Alternative Med.* 2011: 868720

Devaiah, K.M. and Venkatasubramanian, P. 2008. Genetic characterization and authentication of *Embelia ribes* using RAPD-PCR and SCAR marker. *Planta Med.* 74(02): 194–196.

Dhanya, K.S. and Syamkumar, S.B. 2009. Development and application of SCAR marker for the detection of papaya seed adulteration in traded black pepper powder. *Food Biotech.* 23: 97–106.

Dhanya, K.S., Syamkumar, S.S. and Sasikumar, B. 2011. SCAR markers for adulterant detection in ground chilli. *British Food J.* 113: 656–668.

Diao, Y., Lin, X.M., Liao, C.L., Tang, C.Z., Chen, Z.J. and Hu, Z.L. 2009. Authentication of *Panax ginseng* from its adulterants by PCR-RFLP and ARMS. *Planta Med.* 75(05): 557–560.

Duraipandiyan, V., Ayyanar, M. and Ignacimuthu, S. 2006. Antimicrobial activity of some ethnomedicinal plants used by Paliyar tribe from Tamil Nadu, India. *BMC Compl. Alternative Med.* 6: 35–41.

Faisal, M., Ahmad, N. and Anis, M. 2007. An efficient micropropagation system for *Tylophora indica*: an endangered, medicinally important plant. *Plant Biotech. Rep.* 1(3): 155–161.

Fan, W.J., Luo, Y.M., Li, XX, Gu, S., Xie, M.L., He, J., Cai, W.T. and Ding, X.Y. 2009. Development of microsatellite markers in *Dendrobium fimbriatum* hook, an endangered Chinese endemic herb. *Mol. Ecol. Resour.* 9: 373–375.

Ferreira, J.F.S. and Janick, J. 1995. Floral Morphology of *Artemisia annua* with special reference to Trichomes. *International J. Plant Sci.* 156(6).

Fu, R.Z., Wang, J., Zhang, Y.B., Wang, Z.T., But, PPH, Li, N. and Shaw, P.C. 1999. Differentiation of medicinal *Codonopsis* species from adulterants by polymerase chain reaction-restriction fragment length polymorphism. *Planta Med.* 65(7): 648–650.

Ghavamaldin, A., Aptin, R., Khalil, P., Mansour, G. and Mariamalsadat, T. 2012. Study of variation of biochemical components in *Hypericum perforator* L. grown in North of Iran. *J. Med. Plants Res.* 6(3): 366–372.

Gururaj, H.B., Parvatam, G. and Ravishankar G. 2007. Micropropagation of *Tinospora cordifolia* (Willd.) Miers ex Hook. F and Thoms: A multipurpose medicinal plant. *Curr. Sci.* 92: 23–26.

Guo, W.L., Gong, L., Ding, Z.F., Li, YD, Li, F.X., Zhao, S.P. and Liu, B. 2006. Genomic instability in phenotypically normal regenerants of medicinal plant *Codonopsis lanceolata* Benth. et Hook. f., as revealed by ISSR and RAPD markers. *Plant Cell Rep.* 25(9): 896–906.

Guo, Y., Tsuruga, A., Yamaguchi, S., Oba, K., Iwai, K., Sekita, S. and Mizukami, H. 2006. Sequence analysis of chloroplast chlB gene of medicinal *Ephedra* species and its application to authentication of *Ephedra* herb. *Biol. Pharmaceu. Bull.* 29(6): 1207–1211.

Guha, S. and Maheshwari, SC 1964. *In vitro* production of embryos from anthers of *Datura. Nature* (London) 204: 497.

Ha, W.Y., Shaw, P., Liu, J., Yau, F. and Wang, J. 2002. Authentication of *Panax ginseng* and *Panax quinquefolius* using Amplified Fragment Length Polymorphism (AFLP) and Directed Amplification of Minisatellite Region DNA (DAMD). J. Agric. Food Chem. 50: 1871–1875.

Haccius, B. 1978. Questions of unicellular origin of non-zygotic embryos in callus cultures. *Phytomorphology* 28: 74–81.

He, Z.D., Qiao, C.F., Han, Q.B., Cheng, C.L., Xu, H.X., Jiang, R.W., But, P.P.H. and Shaw, P.C. 2005. Authentication and quantitative analysis on the chemical profile of *Cassia* bark (cortex *Cinnamomi*) by High-Pressure Liquid Chromatography. *J. Agric. Food Chem.* 53(7): 2424–2428.

Hogedal, B.D. and Molgaard, P. 2000. HPLC analysis of the seasonal and diurnal variation of iridoids in cultivars of *Antirrhinum majus. Biochem. Syst. Ecol.* 28(10): 949–962.

Hon, C.C., Chow, Y.C., Zeng, F.Y. and Leung, F. 2003. Genetic authentication of ginseng and other traditional Chinese medicine. *Acta Pharmacologica Sinica* 24: 841–846.

Hosokawa, K., Miami, K., Kawahara, K., Nakamura, I. and Shibata T. 2000. Discrimination among three species of medicinal plant *Scutellaria* plants using RAPD markers. *Planta Med.* 66: 270–272.

Hu, Y., Zhang, Q., Xin, H., Qin, L.P., Lu, B.R., Rahman, K. and Zheng, H. 2007. Association between chemical and genetic variation of *Vitex rotundifolia* populations from different locations in China: Its implication for quality control of medicinal plants. *Biomed. Chromatography* 21(9): 967–975.

Huh, M.K. and Bang, K.H. 2006. Identification of *Atractylodes japonica* and *A. macrocephala* by RAPD analysis and SCAR markers. *Silvae Genetica* 55(3): 101–104.

Indian Herbal Pharmacopoeia 2002, Indian Drug Manufacturers Association, Mumbai.

Iqbal, M. and Beigh, S.Y. and Nawchoo I.A. 2004. Variability in morphology and active constituent of *Podophyllum hexandrum*, a Himalayan herb known for its anti-cancer activity. *J. Trop. Med. Plants* 5: 33–36.

Jeffrey, A.J., Wilson, V. and Thein, S. 1985. Hypervariable "minisatellite" regions in human DNA. *Nature* 314: 67–73.

Jha, S.R., Naz, R., Asif, A., Okla, M.K., Soufan, W., Al-Ghamdi, A.A. and Ahmad, A. 2020. Development of an in vitro propagation protocol and a Sequence Characterized Amplified Region (SCAR) Marker of *Viola serpens* Wall. Ex Ging. *Plants* 9(2): 246.

Johnson, T.S., Narayan, S.B. and Narayana, D.B.A. 1997. Rapid in vitro propagation of *Saussurea lappa*, an endangered medicinal plant, through multiple shoot cultures. *In Vitro Cell. Dev. Biol. Plant* 33: 128–130.

Joshi, K., Chavan, P., Warude, D. and Patwardhan, B. 2004. Molecular markers in herbal drug technology. *Curr. Sci.* 87: 159–165.

Jundae, L, Jung-Heon, H., Chul-Geon, A., Won-Phil, L. and Jae-Bok, Y. 2010. A CAPS marker linked to a genic male-sterile gene in the colored sweet pepper, "Paprika" (*Capsicum annuum* L.). *Breed. Sci.* 60: 93–98.

Kamboj, V.P. 2000. Herbal medicine. *Curr. Sci.* 78: 35–39.

Khan, S., Mirza, K.J., Anwar, F. and Abdin, M.Z. 2010. Development of RAPD markers for authentication of *Piper nigrum* (L.). *Environ. We Int. J. Sci. Tech.* 5: 47–56.

Khan, S., Mirzam K.J. and Abdin, M.Z. 2010. Development of RAPD markers for authentication of medicinal plant *Cuscuta reflexa. EurAsian J. BioSci.* 4: 1–7.

Kim, J., Jo, B.H., Lee, K.L., Yoon, E.S., Ryu, G.H. and Chung, K.W. 2007. Identification of new microsatellite markers in *Panax ginseng*. *Molecules Cells* 24: 60–68.

Kim, J. and Chung, K.W. 2007. Isolation of new microsatellite-containing sequences in *Acanthopanax senticosus. J. Plant Biol.* 50: 557–561.

Kite, G.C., Veitch, N.C., Boalch, M.E., Lewis, G.P., Leon, C.J. and Simmonds, M.S.J. 2009. Flavonol tetraglycosides from fruits of *Styphnolobium japonicum* (Leguminosae) and the authentication of *Fructus sophorae* and *Flos sophorae*. *Phytochem.* 70: 785–794.

Koblitz, H., Koblitz, D., Schmauder, H.P. and Gröger, D. 1983. Studies on tissue cultures of the genus *Cinchona* L. alkaloid production in cell suspension cultures. *Plant Cell Rep.* 2(3): 122–125.

Kohjyouma, M., Nakajima, S., Namera, A., Shimizu, R., Mizukami, H. and Kohda, H. 1997. Random amplified polymorphic DNA analysis and variation of essential oil components of *Atractylodes* plants. *Biol. Pharmaceu. Bull.* 20(5): 502–506.

Konieczny, A. and Ausubel, F.M. 1993. A procedure for mapping *Arabidopsis* mutations using co-dominant ecotype-specific PCR-based markers. *Plant J.* 4(2): 403–410.

Kumar, J., Verma, V., Shahi, A., Qazi, G. and Balyan, H. 2007. Development of simple sequence repeat markers in *Cymbopogon* species. *Planta Med.* 73: 262–266.

Lee, J.H., Lee, J.W., Sung, J.S., Bang, K.H. and Moon, S.G. 2009. Molecular authentication of 21 Korean *Artemisia* species (Compositae) by polymerase chain reaction-restriction fragment length polymorphism based on trnL-F region of chloroplast DNA. *Biol. Pharmaceu. Bull.* 32(11): 1912–1916.

Lee, M.Y., Doh, E.J., Park, C.H., Kim, Y.H., Kim, E.S., Ko, B.S. and Oh, S.E. 2006. Development of SCAR marker for discrimination of *Artemisia princeps* and *A. argyi* from other *Artemisia* herbs. *Biol. Pharmaceu. Bull.* 29(4): 629–633.

Li, M., Cao, H., Pui-Hay, B.P. and Shaw, P. 2011. Identification of herbal medicinal materials using DNA barcodes. *J. Syst. Evol.* 49: 271–283.

Li, J. and Xiao, X. 1995. An investigation on medicinal plant resources of *Dendrobium* in Sichuan province. *Chinese J. Chin. Mater. Med.* 20: 7–12.

Li, X., Ding, X., Chu, B., Ding, G., Gu, S., Qian, L., Wang, Y. and Zhou, Q. 2007. Molecular authentication of *Alisma orientale* by PCR-RFLP and ARMS. *Planta Med.* 73(01): 67–70.

Liersch, R., Soicke, H., Stehr, C. and Tüllner, H.U. 1986. Formation of artemisinin in *Artemisia annua* during one vegetation period. *Planta Med.* 52(5): 387–390.

Liu, Z., Wang, Y., Zhou, K., Han, D., Yang, X. and Liu, X. 2001. Authentication of Chinese crude drug, Gecko, by allele-specific diagnostic PCR. *Planta Med.* 67(04), 385–387.

Martin, G., Geetha, S.P., Raja, S.S., Raghu, A.V., Balachandran, I. and Ravindran, P.N. 2006. An efficient micropropagation system for *Celastrus paniculatus* Willd: A vulnerable medicinal plant. *J. Forest Res.* 11(6): 461–465.

Mathur, S., Gupta, M.M., Ram, M., Sharma, S. and Kumar, S. 2002. Herb yield and bacoside-a content of field-grown *Bacopa monnieri* accessions. *J. Herbs. Spi. Med. Plants.* 9: 11–18.

Meena, A.K., Rao, M.M., Panda, P., Kiran, Y.A., Singh, U. and Singh, B. 2010. Standardisation of Ayurvedic polyherbal formulation, *Pancasama* Churna. *Int. J. Pharmacog. Phytochem. Res.* 2(1): 11–14.

Mitra, G.C. and Chaturvedi, H.C. 1970. Fruiting plants from *in vitro* grown leaf tissue of *Rauvolfia serpentina* Benth. *Curr. Sci.* 39: 128–129.

Mizukami, H., Okabe, Y., Kohda, H. and Hiraoka, N. 2000. Identification of the crude drug *Atractylodes* rhizome (Byaku-jutsu) and *Atractylodes lancea* Rhizome (So-jutsu) using chloroplast trnK sequence as a molecular marker. *Biol. Pharmaceu. Bull.* 23: 589–594.

Mizukami, H., Shimizu, R., Kohda, H., Kohjyouma, M., Kawanishi, F. and Hiraoka, N. 1996. Restriction fragment length polymorphisms of medicinal plants and crude drugs. IV. Restriction Fragment Length Polymorphisms of rDNA and variation of essential oil composition in *Atractylodes* plants. *Biol. Pharmaceu. Bull.* 19: 577–580.

Mohapatra, H.P. and Rath, S.P. 2005. In vitro studies of *Bacopa monnieri*: An important medicinal plant with reference to its biochemical variations. *Indian J. Exp. Biology.* 43(4): 373–376.

Mohring, S., Horstmann, V. and Esch, E. 2005. Development of a molecular CAPS marker for the self-incompatibility locus in *Brassica napus* and identification of different S alleles. *Plant Breeding* 124(2): 105–110.

Moon, B.C., Choo, B.K., Cheon, M.S., Yoon, T., Ji, Y., Kim, B.B., Lee, A.Y. and Kim, H.K. 2010. Rapid molecular authentication of three medicinal plant species, *Cynanchum wilfordii*, *Cynanchum auriculatum*, and *Polygonum multiflorum* (*Fallopia multiflorum*), by the development of RAPD-derived SCAR markers and multiplex-PCR. *Plant Biotech. Rep.* 4(1): 1–7.

Ngan, F., Shaw, P., But, P. and Wang, J. 1999. Molecular authentication of *Panax* species. *Phytochem.* 50(5): 787–791.

Namba, T. and Lin, C.C. 1981. Pharmacognostical studies on the crude drugs of *Orchidaceae* from Taiwan (IV) on Chioh-hak. *Japanese J. Pharmacog.* 35: 221–232.

Nin, S., Morosi, E., Schiff, S. and Bennici, A. 1996. Callus cultures of *Artemisia absinthium* L.: Initiation, growth optimization and organogenesis. *Plant Cell Tiss. Org. Cult.* 45(1): 67–72.

Nortier, J.L., Martinez, M.C., Schmeiser, H.H., Arlt, V.M., Bieler, C.A., Petein, M., Depierreux M.F., De Pauw, L., Abramowicz, D., Vereerstraeten, P. and Vanherweghem, J.L. 2000. *Urothelial carcinoma* associated with the use of a Chinese herb (*Aristolochia fangchi*). *New Engl. J. Med.* 342: 1686–1692.

Novak, J., Gröger, S.G. and Lukas, B. 2007. DNA-based authentication of plant extracts. *Food Res. International.* 40(3): 388–392.

Paran, I. and Michelmore, R.W. 1993. Development of reliable PCR-based markers linked to downy mildew resistance genes in lettuce. *Theo. Appl. Gene.* 85: 985–993.

Passinho-Soares, H., Felix, D., Kaplan, M.A., Margis-Pinheiro, M. and Margis, R. 2006. Authentication of medicinal plant botanical identity by Amplified Fragmented Length Polymorphism dominant DNA marker: inferences from the *Plectranthus* Genus. *Planta Med.* 72(10): 929–931.

Penner, G.A., Bush, A., Wise, R., Kim, W., Domier, L., Kasha, K. ... and Fedak, G. 1993. Reproducibility of random amplified polymorphic DNA (RAPD) analysis among laboratories. *Genome Research* 2(4): 341–345.

Rathore, M.S., Gehlot, H.S. and Shekhawat, N.S. 2010. Biotechnological approaches for propagation and prospecting of important medicinal plants from Indian Thar desert. *Int. J. Plant Prod.* 4(1): 1735–8043.

Qi, J., Li, X., Song, J. and Eneji, A. and Ma, X. 2008. Genetic Relationships among *Rehmannia glutinosa* cultivars and varieties. *Planta Med.* 74: 1846–1852.

Qian, Y.Q., Ying-Xiong, Q., Yan-Qi, C., Jian-Xin, Y., Ming-Shui, Z. and Cheng-Xin, F. 2006. Species specific SCAR markers for authentication of *Sinocalycnthus chinensis*. *J. Zhejiang University*-Science A. 7: 868–872.

Sangwan, R.S., Chaurasiya, N.D., Misra, L.N., Lal, P., Uniyal, G.C., Sharma, R., Sangwan, N.S.K., Suri, A., Qazi, G.N. and Tuli, R. 2004. Phytochemical variability in commercial herbal products and preparations of *Withania somnifera* (Ashwagandha). *Curr. Sci.* 86(3): 461–465.

Sasaki, Y., Fushimi, H., Cao, H., Cai, S.Q. and Komatsu, K. 2002. Sequence analysis of Chinese and Japanese *Curcuma* drugs on the 18S rRNA gene and trnK gene and the application of amplification-refractory mutation system analysis for their authentication. *Biol. Pharmaceu. Bull.* 25(12): 1593–1599.

Sen, A.R., Gupta, P.S. and Dastidar, N.G. 1974. Detection of *Curcuma zedoria* and *C. aromatica* in *C. longa* (turmeric) by thin-layer chromatography. *Analyst.* 99: 153–155.

Shen, J., Ding, X.Y., Ding, G., Liu, D.Y., Tang, F. and He, J. 2006. Studies on population difference of *Dendrobium officinale* II establishment and optimization of the method of ISSR fingerprinting marker. *Zhongguo Zhong Yao Za Zhi* (China *J. Chinese Material Med.*) 31: 291–294.

Shi, H., Wang, J., Wang, M., Tu, P. and Li. XB 2009. Identification of *Cistanche* Species by chemical and Inter-simple Sequence Repeat fingerprinting. *Biol. Pharmaceu. Bull.* 32: 142–146.

Shikha, S. and Mishra, N. 2009. Genetic markers – a cutting-edge technology in herbal drug research. *J. Chem. Pharm. Res.* 1: 1–18.

Shim, Y.H., Choi, J.H., Park, C.D., Lim, C.J., Cho, J.H. and Kim, H.J. 2003. Molecular differentiation of *Panax* species by RAPD analysis. *Arch. Pharma. Res.* 26(8): 601–605.

Sinam, Y.M. and Shantibala, D.G.A. 2011. Seasonal variation of bioactive alkaloid content in *Aconitum* spp. from Manipur, India. *Bioscan.* 6(3): 439–442.

Singh, A., Vishwakarma, R.A. and Husain, A. 1988. Evaluation of *Artemisia annua* strains for higher artemisinin production. *Planta Med.* 54(5): 475–476.

Smith, D.N. and Devey, M.E. 1994. Occurrence and inheritance of microsatellites in *Pinus radiata*. Genome, 37: 977–983.

Song, Z., Li, X., Wang, H. and Wang, J. 2010. Genetic diversity and population structure of *Salvia miltiorrhiza* Bge in China revealed by ISSR and SRAP. *Genetica* 138: 241–249.

Sood, H. and Chauhan, H.S. 2009. Development of a low-cost micropropagation technology for an endangered medicinal herb (*Picorhizakurroa*) of North-Western Himalayas. *J. Plant Sci.* 21–31.

Southwell, I.A. and Bourke, CA 2001. Seasonal variation in hypericin content of *Hypericum perforatum* L. (St. John's Wort). *Phytochem.* 56(5): 437–441.

Stern W.L., Morris M.W. and Judd W.S. 1994. Anatomy of the thick leaves in *Dendrobium* section *Rhizobium* (Orchidaceae). *Int. J. Plant Sci.* 155: 716–729.

Steven, R., Brian, K. and John, B. 2002. Genetic variation across VNTR loci in central North American *Taraxacum* surveyed at different spatial scales. *Plant Ecol.* 161: 111–121.

Steward, F.C., Mapes, M.O. and Mears, K. 1958. Growth and organized development of cultured cells. II. Organization in cultures grown from freely suspended cells. *Am. J. Bot.* 45: 705–708.

Sun, Y.W., Liao, Y.J., Hung, Y.S., Chang, J.C. and Sun, J.M. 2011. Development of ITS sequence-based SCAR marker for discrimination of *Paphiopedilum armeniacum*, *Paphiopedilum micranthum*, *Paphiopedilum delenatii* and their hybrids. *Sci. Hort.* 127: 405–410.

Tani, N., Kawahara, T., Yoshimaru, H. and Hoshi, Y. 2003. Development of SCAR marker distinguishing pure seedlings of the endangered species *Morus boninensis* from *M. boninensis* x *M. acidosa* hybrids for conservation in Bonin (*Ogosawara island*). *Conser. Gene.* 4: 605–612.

Tommasi, F. and Scaramuzzi, F. 2004. In vitro propagation of *Ginkgo biloba* by using various bud cultures. *Biol. Plant.* 48(2): 297–300.

Tremlova, B. 2001. Evidence of Spice black pepper adulteration. *Czech J Food Sci.* 1: 235–239.

Tripathi, M., Khanna, S.K. and Das, M. 2007. Surveillance on use of synthetic colours in eatables viz a viz Prevention of Food Adulteration Act of India. *Food Control* 18: 211–219.

Vadodaria, H.K., Samantaray, S. and Maiti, S. 2007. Micropropagation of *Glycyrrhiza glabra* Linn. an important medicinal plant. *J. Cell Tiss. Res.* 7(1): 921–926.

Vos, P., Hogers, R., Bleeker, M., Reijans, M., Van De L.T., Hornes, M., Frijters, A., Pot, J., Peleman, J., Kuiper, M. and Zabeau, M. 1995. AFLP: A new technique for DNA fingerprinting. *Nucl. Acids Res.* 23: 4407–4414.

Vosman, B. and Arens, P. 1997. Molecular characterization of GATA/GACA microsatellite repeats in tomato. *Genome* 40(1): 25–33.

Wang, C.Z., Li, P., Ding, J.Y., Peng, X. and Yuan, C.S. 2007. Simultaneous identification of Bulbus *Fritillaria ecirrhosae* using PCR-RFLP analysis. *Phytomedicine* 14(9): 628–632.

Wang, J., Ha, W.Y., Ngan, F.N., But, P.P.H., Shaw, P.C. 2001. Application of sequence characterized amplified region analysis to authenticate *Panax* species and their adulterants. *Planta Med.* 67: 781–783.

Wang, S., Gao, W., Chen, H. and Xiao, P. 2005. New starches from *Fritillaria* species medicinal plants. *Carbohydrate Polymers.* 61: 111–114.

Wang, P., Huang, F., Zhou, L., Cao, L., Liang, S., Xu, H. and Liu, J. 2000. Analysis of *Amomunvillosum* species and some adulterants of *Zingiberaceae* by RAPD. *Zhongguo Zhong Yao Za Zhi* 23: 71–74.

Weiland, J.J. and Yu, M.H. 2003. A cleaved amplified polymorphic sequence (CAPS) marker associated with root-knot nematode resistance in sugarbeet. *Crop Sci.* 43(5): 1814–1818.

Williams, J.G., Kubelik, A.R., Livak, K.J., Rafalski, J.A. and Tingey, S.V. 1990. DNA polymorphisms amplified by arbitrary primers are useful as genetic markers. *Nucleic Acids Res.* 18(22): 6531–6535.

Xu, Y., Wang, Y., Li, Z., Bao, Z., Zhou, J. and Huang, H. 2006. Characterization of polymorphic microsatellite loci in a traditional Chinese medicinal plant, *Gastrodiaelata. Mol. Ecol. Notes* 6: 316–318.

Yadav, A., Ahmad, J., Chaudhary, A.A. and Ahmad, A. 2012. Development of Sequence Characterized Amplified Region (SCAR) marker for the authentication of *Bacopa monnieri* (L.) Wettst. *European J. Med. Plants* 2: 186–198.

Zerega, N., Mori, S., Lindqvist, C., Zheng, Q. and Motley, T. 2002. Using Amplified Fragment Length Polymorphisms (AFLP) to identify black cohosh (*Actaea racemosa*). *Economic Bot.* 56: 154–164.

Zhao, L., Huang, C., Shan, Z., Xiang, B. and Mei, L. 2005. Fingerprint analysis of *Psoralea corylifolia* L. by HPLC and LC-MS. *J. Chromatography* B. 821: 67–74.

Zhang, J.H., Wang, Z.W., Zhang, X. and Jin, J.Q. and Yang, F.L. 2000. Authentication of *Tribulus terrestris* L. by morphology, microscope and TLC. *Zhong Yao Cai.* 23: 675–677.

Zhang, Y.B., Wang, J., Wang, Z.T. and But, P.P.H. 2003. DNA microarray for identification of the herb of *Dendrobium* species from Chinese medicinal formulations. *Planta Med.* 69(12): 1172–1174.

Zhang, Y.B., Shaw, P.C., Sze, C.W., Wang, Z.T. and Tong, Y.Z. 2007. Molecular authentication of Chinese herbal materials. *Food Drug Anal.* 15: 1–9.

Zhang, K., Leung, H., Yeung, H. and Wong, R. 2001. Differentiation of *Lyciumbarbarum* from its related *Lyciumspecies* using Random Amplified Polymorphic DNA. *Planta Med.* 67: 379–381.

Zhang, Y., But, P.P., Wang, Z. and Shaw, P. 2005. Current approaches for the authentication of medicinal *Dendrobium species* and its products. *Plant Gen. Resou.* 3: 144–148.

Zhu, S., Fushimi, H., Cai, S. and Komatsu, K. 2004. Species identification from Ginseng drugs by multiplex amplification refractory mutation system (MARMS). *Planta Med.* 70(02): 189–192.

Zobayed, S.M.A., Afreen, F. and Kozai, T. 2005. Temperature stress can alter the photosynthetic efficiency and secondary metabolite concentrations in St. John's wort. *Plant Physiol. Biochem.* 43: 977–984.

14 Phytochemical Evaluation and Biotechnological Approach for the Conservation of Some Medicinal Plants

Renu

CONTENTS

DOI: 10.1201/9781003179139-14

14.1 INTRODUCTION

Medicinal plants are moving from hem to the mainstream, where a large number of people are approaching health benefits and remedies free from side effects. Recently, plant-based products for the treatment of different diseases have got considerable attention (Shahid et al. 2013). According to World Health Organization (WHO), about 85% of the population in developing countries depends on medicines derived from the plants for their health care treatment (Pandey and Tripathi 2017). The modern pharmaceutical industry contains about 25% of drugs obtained from plants, and others are prepared synthetically based on prototype compounds derived from plants (Pandey and Tripathi 2017). The raw material for medicines is mostly obtained from the wild. Approximately two-thirds of the 50,000 different medicinal plant species used in preparing medicine are obtained from the wild (Canter, Thomas and Ernst 2005). Due to the increasing demand for medicines, a large number of medicinal plants are in trouble from over-exploitation, habitat destruction, urbanization and the unconstrained collection of medicinal planta from the wild, resulting in the depletion of natural resources of the medicinal plants (Sharma et al. 2010).

Conservation of these threatened medicinal plants is clearly one of the major priorities toward which the technologies must be shifted (Pandey and Tripathi 2017). Techniques such as in vitro regeneration and genetic transformation for genetic enhancement and multiplication of the medicinal plants are some important biotechnological aspects. It can also be utilized for secondary metabolites production using bioreactors (Tripathi and Tripathi 2003). The tissue culture technique is vigorous and has opened a large research area for the conservation of biodiversity. Tissue culture entente has been developed for a large range of endangered, rare and threatened medicinal plants (Sharma et al. 2010). Some important medicinal plants with their medicinal value are given in Table 14.1. This chapter reviews different biotechnological approaches and the application of tissue culture for the in vitro regeneration of medicinal plants from various explants and enhanced production of secondary metabolites.

TABLE 14.1
Some Important Medicinal Plants with Their Medicinal Value

Botanical name	Common name	Family	Medicinal value
Commiphora wightii	Guggul	Burseraceae	Obesity, inflammation, arthritis
Swertia chirata,	Chiretta	Gentianaceae	Cure diabetes, cure anemia, cure diarrhea
Withania somnifera	Ashwagandha	Solanaceae	Boost immunity, reduce blood sugar, reduce stress
Rauvolfia serpentina	Indian snakeroot	Apocynaceae	Treat high blood pressure, insomnia, fever
Tylophora indica	Antamool	Asclepiadaceae	Allergies, congestion, constipation, cough, bloody diarrhea, hemorrhoids
Cassia augustifolia	Seena	Fabaceae	Treat constipation, hemorrhoids and weight loss
Simmondsia chinensis	Jojoba	Simmondsiaceae	Relief of headaches, throat inflammation and in treating wound
Tinospora cardifolia	Giloe	Menispermaceae	Diabetes, high cholesterol, hay fever, lymphoma, rheumatoid arthritis and fever, gonorrhea, syphilis
Asparagus recemosus	Shatavari, Satavar, Shatamull	Liliaceae	Upset stomach, stomach ulcers, diarrhea, bronchitis, tuberculosis and diabetes
Calotropis gigantea.	Crown flower	Apocynaceae	Treat fevers, elephantiasis, nausea and diarrhea

14.2 *COMMIPHORA WIGHTII*

Commiphora wightii, frequently called as guggul, belongs to the family Burseraceae. *Commiphora* consists of about 165 species. Guggul means "protection from diseases"; it has been used from ancient times in the treatment of various diseases such as rheumatoid arthritis, cancer, antiinflammation, wound gastrointestinal problems, and many more. The guggul plant is a shrub or small tree with thin paper bark and branches that are thorny. This plant is slow-growing and reaches up to a height of 3 to 3.5 meters in 7 to 9 years. The guggul plant is distributed in the arid area of northern Africa to central Asia, that is, India, Pakistan and Bangladesh. In India, this plant is found in Karnataka, Gujarat and Rajasthan. The plant is dimorphic. Flowers are small unisexual, and brown to pink in color. The fruit is a drupe (Figure 14.1).

Guggul is an important medicinal plant responsible for the treatment of bone fracture, arthritis, obesity, rheumatism, high cholesterol, blood diuretic, astringent, urinary disorder, neurodegeneration, Parkinson's diseases and aging process (Pareek and Pareek 2012). Guggul is one of the main ingredients for several drugs and medicines and is recommended for the treatment of hay fever and chronic bronchitis and is also used in gout and heart disease. Guggul possesses strong disinfectant properties by increasing white blood cell count. A large number of metabolites has been reported from different parts of plant-like leaves, stem, root and fruit.

14.2.1 Phytochemical Evaluation

The number of steroids, aliphatic esters, amino acids, terpenes, cholesterol and sesamin has been isolated from *Commiphora wightii*. The most important sterol compound is guggulsterone (E and Z form) which is responsible for treating various ailments and disorders and has also been observed by various other workers. Other constituents of guggul are muscanone, myrrhanol, eugenol, mansumbinoic acid, mansumbinoic acid, quercetin, terpenes, steroids lignan, ferulic acid, octadecane-1,2,3,4- tetrol and the essential oil found in different parts (Figure 14.1 D).

14.2.2 Biotechnological Approaches for Improvement and Conservation

Due to this plant's so many therapeutic values, guggul is always in demand. But due to its slow-growing ability, this plant takes about ten years to mature (Kulkarni et al. 2012), and the demand for guggul oleo gum resin is increasing very fast. Vigorous and illegitimate using has made a major of the species in the wild, making the plant vulnerable. Due to the use of guggul in modern medicine, this plant is so scarce because of its extensive use.

14.2.2.1 Biotechnological Approach

Different biotechnological approaches are adopted by the different workers to conserve and for improvement of *Commiphora wightii*.

Micropropagation is an alternative approach for plants that are difficult to generate by conventional methods and for plants whose population decreases due to over-exploitation. This technique can efficiently be used to encounter the growing demand for plant-like *Commiphora wightii*. Micropropagation is used by different workers for the production of valuable compound guggulsterone. In *Commiphora wightii*, different media concentrations were used with a combination of growth hormones for nodal and shoot tip explant establishment. Somatic embryogenesis has successfully produced several plantlets in a short amount of time. It was also used for synthetic seed production. It can also be performed at a large scale as in bioreactors. Somatic embryos were transferred to natural conditions after germinating into plantlets (Table 14.2).

FIGURE 14.1 *Commiphora wightii*: A. A green leafy tree, B. Leafless tree, C. Stem with thin bark, D. Constituents of *Commiphora wightii*.

TABLE 14.2
Different Explants Used for Micropropagation Showing Different Responses

	Explants and their responses	
Explant	**Response**	**References**
Nodal explant	Multiplication of shoot	Yusuf et al. (1999)
Cotyledonary nodes, Microshoot explant	Rooted plantlet produced	Kant et al. (2010)
Shoot, node and leaf explant	Rooted plantlet produced	Singh et al. (2010)
	Responses of explants to different media	
Establishment	**Media used**	**References**
Shoot formation	MS med0 mg/L), thiamineHCL and 0.3%activated charcoal	Barve and Mehta (1993)
Mutiple shoots	BAP(3.0mg/L), kinetin0.5mg/L) and IBA (0.5mg/L)	Soni (2010)
Rooting	White'smedium without any hormone and elevated charcoal concentration	Kant et al. (2010)
	Somatic embryogenesis from different explants	
Explants	**Response**	**References**
Callus	Somatic embryogenesis	Kumar et al. (2003)
Premature fruits	Torpedo stage somatic embryos	Kumar et al. (2004)
Premature fruits	Proliferation of somatic embryo	Kumar et al. (2006)
	secondary metabolite production in different explants	
Explant	**Response**	**References**
Immature fruit somatic embryos	Guggulsterone synthesis in culture	Kumar et al. (2006)
Zygotic embryos, leaf and stem explant	Guggulsterone synthesis in culture	Mathur et al. (2007)
Zygotic embryos	Anthocyanin synthesis through callus culture	Dass et al. (2008)
Zygotic embryos	Guggulsterone synthesis in culture	Dass and Ramawat(2009)
Suspended cells from cell suspension culture	In vitro guggulsterone accumulation	Suhar and Ramawat(2010)

Guggulsterone is a very important secondary metabolite that is found in the oleo gum resin canal and being used in treating a number of diseases since ancient times. This plant is very slow-growing, and the growing demand for this metabolite can't be met. The investigation to increase the production of secondary metabolites through callus culture and bioreactors is still strong. Resin canal formation has been reported using somatic embryogenesis. Guggulsterone has been produced by aseptic resin canals of somatic embryos (Kumar et al. 2006). Guggulsterone production in higher quantity cannot be fulfilled due to its slow-growing habit, for this leaf explant and zygotic embryo-derived callus culture has been used by different workers for high guggulsterone production and other metabolites (Table 14.2).

14.3 *SWERTIA CHIRATA*

Swertia chirata belonging to the family Gentianaceae is generally found in Kashmir to Bhutan, in the temperate Himalayas, and in the Khasi Hills in Meghalaya at the height of 1,200–1,500 m. *Swertia* is a perennial herb with fibrous roots and opposite leaves having cymose inflorescence along with the pentamerous and pedicellate flower, sessile and leaf-like bracts, rotatable calyx and corolla with two nectaries per corolla lobe, either naked or covered by a scale. Fruits are enclosed with persistent calyx and corolla in oblong capsules (Figure 14.2 A). *Swertia chirata* is a local medicinal plant in India. Habitat destruction, human vigorous and illegitimate activities and climate impact have caused a substantial loss of diversity of this precious medicinal plant in the wild. *Swertia chirata* is considered a valuable Indian plant species because of its well-marked active molecule. This plant has now been recorded as being critically endangered in terms of vigorous harvesting and habitat destruction. For the herbal industries, raw materials are collected in quantity from this plant. Due to this, there is a decrease in the population of plants which is leading to its extinction in the wild. The demand for chirata in the national and international market is increasing rapidly. If relevant conservation steps are not taken, a major part of this genus can be lost in the near future.

This herb is a bitter tonic and is used for stomach aches, useful for heart, liver disorders and eyes. It is considered an amazing medicine for nausea and bloating, along with its benefits in protecting the liver against carbon tetrachloride poisoning as it contains amarogentin. Ayurvedic medicine, also named Mahasudarshana churna, contains a number of herbs. *Swertia* is considered the main component. It is also considered effective against malaria and tuberculosis as it contains xanthones. The whole plant is beneficial in curing fevers, bowel regulation, body burning, skin diseases and bronchial asthma. The roots are effective in vomiting.

14.3.1 Phytochemical Evaluation

Chirata consists of a number of chemicals such as xanthones, which are highly effective against malaria and tuberculosis, and amarogentin. It has anticancer properties and promotes the bile flow in the liver. Swertia chirata is cherished in India for its well-studied bioactive compounds, including iridoid and glycosides and xanthone derivatives, which are useful against a wide range of traditional and unconventional illnesses (Samaddar, Jha and Jha 2014). Other than its anti-helminthic, anti-cancer, anti-fungal and antibacterial properties, amarogentin also has gastroprotective properties. Oleanolic acid and swerchirin also exhibit hypoglycemic, hepatoprotective, blood hematopoietic and blood glucose-lowering activity. Swertiamarin acts as a Central Nervous System depressant. Mangiferin, ursolic acid, oleano, syringaresinol, bellidifolin, different polymer of methoxyxanthin, sweroside and other active phytochemical compounds are found in this herb (Singh et al. 2012) (Figure 14.2 B).

14.3.2 Biotechnological Approach for Improvement of *Swertia Chirata*

To maintain the proper supply of this herb, relevant conservational measures are important. For the management of this valuable plant, biotechnological approaches have shown considerable capacity. There are some biotechnological methods such as micropropagation, in vitro propagation, organogenesis, and many more which can save this valuable plant from becoming extinct. These techniques allow for the rapid growth and the long-term preservation of those plants. Valuable germplasm can also be stored using these approaches (Table 14.3). There are other means of conservational plans offered by biotechnology for *Swertia chirata*, such as synthetic seed production through which encapsulated somatic embryos in matrix-like sodium alginate are produced. Synthetic seed production and regeneration of plants in *Swertia chirata* from somatic embryos are reported by Kumar and Chandra (2014).

14.4 *WITHANIA SOMNIFERA*

Withania somnifera is an essential traditional medicinal plant frequently called Ashwagandha, having major demand in the medicinal market both domestically and internationally. *W. somnifera* is distributed in distinct geographical areas such as Jammu and Kashmir, Punjab, Himachal Pradesh, Uttar Pradesh, Madhya Pradesh and some regions of Rajasthan. It is an erect, evergreen shrub, 30–150 cm high, with fleshy roots and simple leaves. Its flowers are inconspicuous, greenish or yellow in axillary, umbellate cymes; its fruit is a berry enclosed in the persistent calyx with yellow seeds. The

(a)

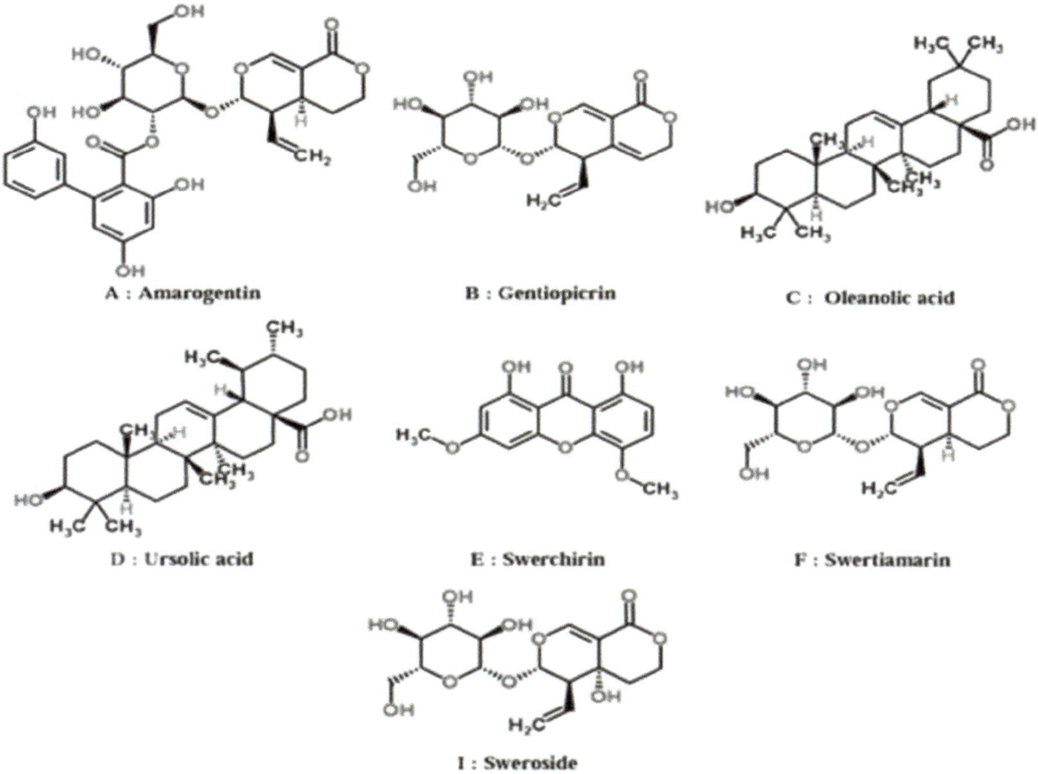

(b)

FIGURE 14.2 A. *Swertia chirata* mature plant with flowers, B. Chemical structures of compounds.
Source: Singh et al. 2012.

roots are a major part of the plant having therapeutical importance (Figure 14.3 A, B). Ashwagandha is one of the most important and best stimulants, having excellent rejuvenating properties, is a tonic for various diseases and the cure for a number of health issues, as well as for increasing energy. It is used as a tranquilizer, has anti-inflammatory properties and acts as a strong immunostimulatory and calming agent (Umadevi et al. 2012). Studies have shown that Ashwagandha increases hemoglobin, stabilizes blood sugar and lowers cholesterol. Ashwagandha is reported to work excellently in

TABLE 14.3
Different Approaches in Different Explant for Improvement of *Swertia chirata*

Explant	Media used	Mode of regeneration	Establishment	References
Root explant	MS medium supplemented with 3 μM 6-benzyladenine + pH of 5.8± 0.1+ solution of NAA (15 ppm)	Micropropagation	Adventitious shoot regeneration	Wawrosch et al. (1999)
Nodal explant	MS medium containing 0.44 μM 6-BAP and 4.65 μM 6-furfurylaminopurine	Micropropagation	Regeneration of shoot	Chaudhuri et al. (2008)
Nodal explant	MS medium supplemented with 4μM BAP and 1.5μM 2ip	Micropropagation	Axillary shoot multiplication	Joshi and Dhaean (2007)
Shoot tip explant	MS medium provided with BAP (1.0 mg/l) along with kinetin (0.1 mg/l), as well as 2% sucrose	Micropropagation	Multiple shoots	Balaraju et al. (2009)
Seeds	MS medium supplemented with 2.22 μM BA, 2.32 μM kinetin as well as 0.54μM NAA	In vitro propagation	In vitro grown shoots	Jha et al. (2011)
Leaf explant	MS medium containing 2.22 μM N-6-benzyladenine as well as 11.6 μM kinetin along with 0.5 μM acetic acid	Organogenesis (direct)	Shoot bud formation	R.K. Chaudhuri and T.B. Jha (2008)

lowering blood pressure and in the treatment of arthritis. A direct spermatogenic effect on the seminiferous tubules of immature rats by applying a testosterone-like effect by extracts of Ashwagandha was concluded in research on this plant (Umadevi et al. 2012). Also, Ashwagandha has an antitumor effect as studies in Chinese Hamsters where ovary cell carcinoma was reported. Additionally, this plant has shown promise in treating mice with lung adenoma caused by urethane (Singh et al. 2011). Ashwagandha is a valuable, long-established medicinal plant with major demand in both the domestic and international markets. Due to its many therapeutic properties, Ashwagandha has been included in the list of the World Health Organization as the most important medicinal plant, and also according to the National Medicinal Plant Board of India, it is concluded as one of the top 32 important medicinal plants because of its major demand in the market. *Withania somnifera* is one of the most important medicinal plants that is cultured to meet the high demand in the pharmaceutical market. Due to the low viability of seeds and long-time storage problems, the global demand for this plant is very difficult to reach through conventional propagation. Rotting seeds and the poor germination of young seedlings are other problems that reduce plant population in the wild.

14.4.1 Phytochemical Evaluation

The presence of a different number of secondary metabolites like alkaloids, sterols, glycosides, glycowithanolides, flavanol, withanolides, Phytoconstituents in *W. somnifera* are responsible for its biological activities. Withanolide A and withaferin A are the two important secondary metabolites. Withanolide A and withaferin A are responsible for the bioactivity of *W. somnifera*. Besides these chemical compounds, 12-deoxy withanstramonolide, withanoside-V ashwagandhanolide, withanone, withanoside-IV, withanolide-B, withanolide-D, pseudo-within 27-hydroxywithanone,

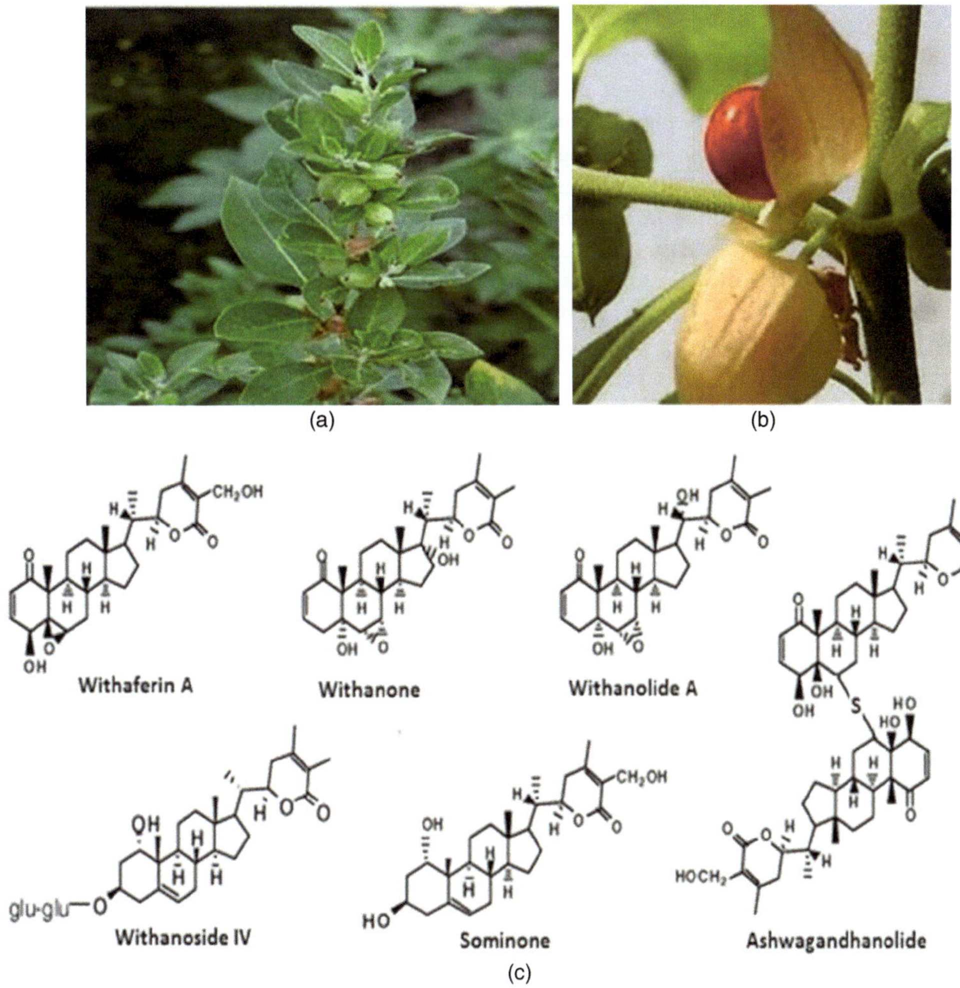

FIGURE 14.3 *W. somnifera*: A. Whole plant; B. Fruits; C. Major chemical composition.

Source: Jain et al. 2012.

somnine, 20-deoxywithanolide A, withastramonolide and somniferous are also some essential compounds present in *W. somnifera* having therapeutical importance (Figure 14.3 C)

14.4.2 Biotechnological Approaches for Improvement and Conservation

The genetic diversity of medicinal plants is becoming endangered at an alarming rate due to overharvesting and the demand for raw materials for the production of medicines. In vitro techniques for propagation of medicinally important plants, conservation of germplasm and the production of therapeutically active compounds for genetic improvement are effective alternative approaches to this problem. Biotechnological approaches, especially plant tissue cultures, are found to be effective for the production of bioactive plant-based secondary metabolites. In order to obtain yield in high concentrations for commercialized utilization to obtain yield in a high concentration of bioactive compounds, attempts have been focused on the energizing of biosynthetic activities using various methods.

Plants are propagated using micropropagation techniques to produce several plantlets in a small space. Micropropagation has been proven a very effective method in the fields of agriculture, horticulture and plant biotechnology. Plant regeneration has been seen by Sen and Sharma (1991) and Teli et al. (1999) using shoot tip and nodal explant in *W. somnifera*. Callus induction from leaf explants along with the MS medium with 2 mg/L Naphthalene acetic acid and 0.5 mg/L kinetin from leaf explants has been reported by Baburaj and Gunasekaran (1995). Meanwhile, Rani and Grover (1999) performed callus cultures from leaf, hypocotyl, and root segments supplemented with MS medium along with 2,4-D (2 mg /L) and kinetin (0.2 mg/ L). The establishment of protocols for fast-growing normal root organ culture and direct rhizogenesis in *W. somnifera* using the MS medium with 0.5 mg/ L IAA and 2 mg/ L IBA has been proposed by Wadegaonkar et al. (2006). Saritha and Naidu (2007) reported flowering of *W. somnifera* in vitro on MS media containing 0.5-4.0 mg/ L kinetin and 0.1 mg/L Indole Acetic Acid, as well as in vitro fruiting on 2.0 mg/L kinetin and 0.1 mg/L Indole Acetic Acid (2007). Also, somatic embryogenesis provides the fastest means for the production of a large number of identical plants in a short period of time. In somatic embryogenesis, generally, a cell that is not involved in the formation of the embryo under natural conditions leads to the formation of somatic embryoids in cultured conditions. These cells are healthy plant tissues, which play an essential role during in vitro propagation by providing an important tool for genetic improvement of therapeutic and commercialized remarkable plants. Somatic embryogenic callus on MS media provided with 3.0 mg/L kinetin from cotyledonary leaf explant has been reported by Rani, Virk and Nagpal (2003). Production and examination of syn seeds in *W. somnifera* using nodal segment as explant has been described by Fatima et al. (2013). Hairy root disease in plants marked by cancerous roots at the site of infection is caused by *Agrobacterium rhizogenes*. These roots are biosynthetic stable, fast-growing and easy to maintain in culture media. Using this technology, a broad range of chemical compounds like pharmaceuticals, therapeutical, nutraceuticals and other useful materials have been produced. For the production of withanolides and other chemical compounds through hairy root culture using *A. rhizogenes* has been widely reported by many researchers.

According to Ray et al. (1996), transformation of hairy roots through the use of *Agrobacterium rhizogenes* resulted in the formation and production of withanolide D in the converted hairy roots (2008). Fungal protein cryptogene encoding genes have been transformed in Ashwagandha through *A. rhizogenes*, and an increment in biomass production was reported by Chaudhuri et al. (2009). Seedlings using the leaf as explant was found more efficient for transformation using *Agrobacterium tumefaciens* by Pandey et al. (2010) than the shoots grown in vitro.

14.5 *RAUVOLFIA SERPENTINA*

Rauvolfia serpentina is frequently called the Indian snakeroot plant and is commonly found in a distinct region of the Indian subcontinent and other Asian countries. It is an erect, evergreen shrub attaining a height of 50 to 80 cm with lanceolate green leaves, unbranched stem and cylindrical roots and cymose inflorescence. Calyx lobes are 1.3–2 mm long. Corolla is an approximately 12-mm long tube, slightly swollen at the location of anthers. Stamens are free, with short filaments. Carpels unite at the base. The fruit is a drupe and spherical shape (Figure 14.4 A).

14.5.1 Phytochemical Evaluation

R. serpentina is a rich source of different varieties of chemical compounds such as alkaloids, ajmaline, ajmalimine, reserpine, deserpidine, indobidine, reserpine, rescinnamine, suaveline, rescinnamidine, serpentine and yohimbine, serpentinine and surpaline (Kaur and Amrinder 2017). The main alkaloid of *R. serpentina* is reserpine (Kaur and Amrinder 2017) (Figure 14.4 B). Due to presence of various important secondary metabolites of high medicinal importance this plant is

(a)

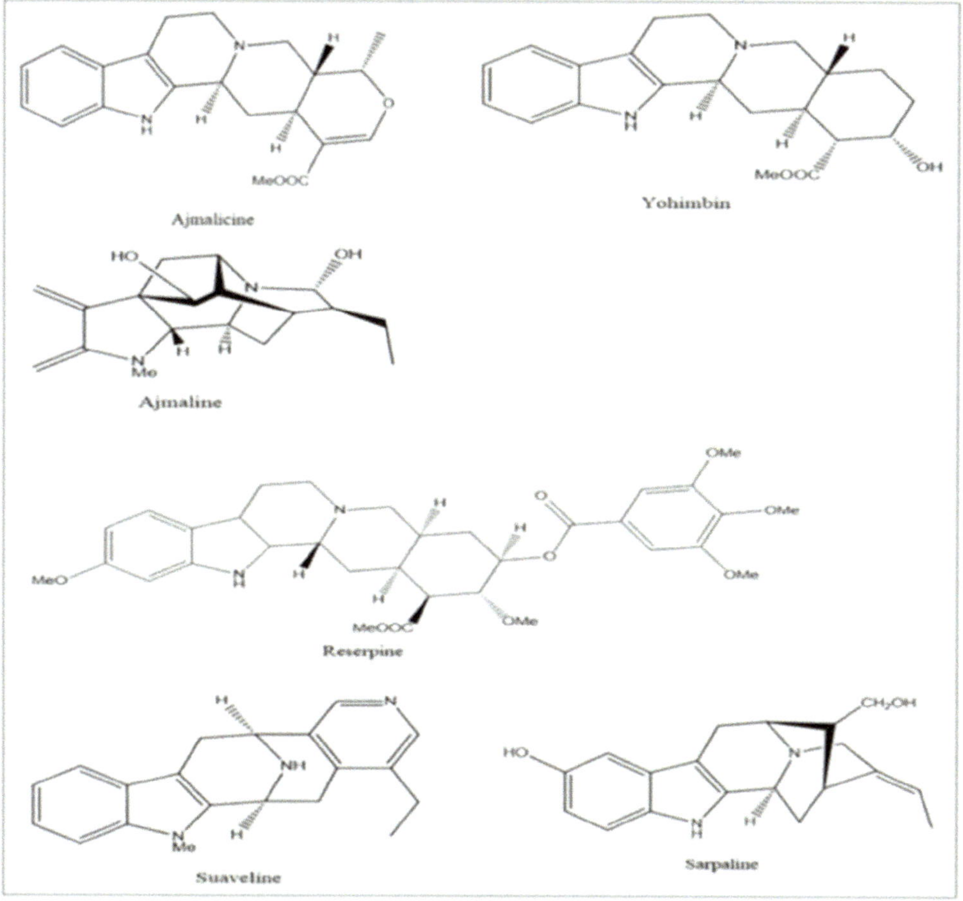

(b)

FIGURE 14.4 *Rauvolfia serpentina*: A. Full-grown plant with flower, B. Major chemical compounds.

Source: Kaur and Amrinder 2017.

more valuable and increases its demand in local and international markets. Roots are mainly utilized in various disease treatment and homeopathic medicine preparations. Roots of *Rauvolfia serpentina* are used for fever treatment, wound healing, infections and psychogenic disorders and are also found to have anti-venom properties. Roots are chiefly used as slumberous and used to reduce blood pressure. Roots are also used as tranquilizers and against nervous disorders like over-exultation, anxiety, psychogenic disorders, insanity, sleeplessness and epilepsy. Roots of *Rauvolfia* extracts are also efficient for intestinal problems and diarrhea. Leaves are often utilized for cataract removal of the eyes.

14.5.2 Biotechnological Approaches for Improvement and Conservation

As this plant has such great medicinal importance it is always in high demand. This plant can be propagated through seeds and root cuttings. However, the conventional breeding practice in the plant faces many problems such as poor or no viability of seeds and germination. In vitro biotechnological approaches like regeneration, direct and indirect, through callus development, embryogenesis, syn seeds production and secondary metabolite estimation gives an alternative pathway for the preservation and support of this plant as given in Table 14.4. Direct regeneration is the regeneration of required plant parts through explants without any callus formation. Different growth regulators used during regeneration which promote root and shoot formation and proliferation are used and organogenesis is achieved. Various combinations of plant growth regulators have been used by several workers for direct regeneration. Akram and Ilahi (1986) used BAP and NAA in combinations for root bud differentiation. Shoot multiplication has been observed by Goel et al. (2009). Senapati, Lahere and Tiwary (2014) reported multiple shoot formation. Also, BAP in combination with IBA or IAA were used by Jain et al. (2003) for regeneration of shoots and proliferation of axillary bud. Mondal, Silva and Ghosh (2011) used different combinations of growth regulators such as IBA in combination of kinetin and NAA for shoot regeneration and in vitro flowering in *Rauvolfia*.

Indirect regeneration is the regeneration of required plants and plant parts through explants which involves the presence of callus phase. Using different combinations of plant growth regulators, explants are converted into callus, which is further treated to develop the required plant part or organ. Auxins like 2, 4-D in combination with cytokinins like BAP and kinetin are the most efficient growth regulators that are efficient to produce callus.. Somatic embryogenesis is the induction of embryos from somatic cells, and can be induced directly or indirectly with or without

TABLE 14.4
Different Responses to Different Media Combination for Improvement of *Rauvolfia serpentina*

Establishment	Media with Different Combination	References
Formation of callus	5 mg/l 2, 4-D along with 2 mg/l NAA.	Gantait et al. (2017).
	0.5 mg/l BAP and 0.2 mg/l NAA along with 0.5 mg/l BAP as well as 2 mg/l NAA	Ahmad et al. (2002) Rani et al. (2014)
	0.5 IAA and 0.5 NAA	Yahya et al. (2007).
	10 mg/l IBA as well as 10 mg/l NAA	
Formation of shoots	Callus along with kinetin along with NAA + BAP	Rani et al. (2014) and Kaur 2018
	Callus and 3 mg/l 2,4-D, as well as 10% coconut milk as an additive 4 μM and 2.27 μM TDZ	Aryal and Joshi (2009) Pandey et al. (2007) and Rohela et al. (2013)
Formation of roots	IBA concentration vary	Pandey et al. (2007)

the involvement of callus. Components of culture media and a different supply of plant growth regulators play an essential role in somatic embryogenesis. Direct somatic embryogenesis using root explants were observed by Sudha and Seeni (1996) using shoot tip and single node explant. Uikey et al. (2016) observed indirect somatic embryogenesis through callus induction. For somatic embryos to convert into plantlets, a variety of combinations of growth regulators such as auxin and cytokinin were used. Zafar et al. (2019) used GA3 (0.25 mg/l) in the culture media for induction of the embryos.

Other means of conservation of *Rauvolfia serpentina* are offered through biotechnological approaches such as synthetic seed production through encapsulation of growing regions of the plant, which itself can't produce any part of the plant in natural growing conditions; however, under artificial conditions using an encapsulation media, these can efficiently perform like true seeds which can be used for various purposes either for propagation or for the genetic improvement. Explants for synseeds production in the *Rauvolfia* are generally nodal segments and shoot tips, and the gelling agent used in synseeds development is sodium alginate and calcium chloride. Different researchers used different concentrations of sodium alginate and calcium chloride in synseeds production, such as Faisal et al. (2012, Faisal, Alatar and Hegazy 2013), used 100 mM calcium chloride, and Gantait et al. (2017) used about 75 mM calcium chloride for the encapsulation matrix. These synthetic seeds were then used for regeneration purposes in high auxin and cytokinin concentration, as observed by Faisal et al. (2012, Faisal, Alatar and Hegazy 2013) and germplasm storage and for genetic improvement.

14.6 *TYLOPHORA INDICA*

Tylophora indica, belonging to the family Asclepiadaceae, is frequently known as Indian Ipecse or Antmool. It is a perennial and climbing herb dispersed all around the southern and eastern parts of Indian forests, making a dense patch in the humid environment and also found in hilly places and in some parts of the Himalayas. *Tylophora indica* is indigenous to India. The plant grows in areas with lesser rainfall. Leaves are ovate. The flowers are small with greenish purple in color. The fruit is a follicle (Figure 14.5 A).

14.6.1 Phytochemical Evaluation

Many alkaloids have been reported in *Tylophora* plants, such as Septicine, Tylophorine, Tylophorinidine, Tylophoricine, flavonoids, resins, tannin and wax. Tylophorine is the main the major component of *Tylophora*. Tylophorine, phenanthroindolizidine and alkaloids have high inflammatory responses. Not long ago, some rare alkaloids as in tyloindicines A to J, 6-desdimethyltylophorine, desmethyltylophorinine, 5-hydroxy-O-methyltylophorinidine, anhydroustylophorinine, anhydrous-dehydrotylophorinine, 14-hydroxyiso tylocrebrine and 4,6-disdimethylisotylo-crebrine were also reported. Some non-alkaloidal compounds, such as octaosanyl, stigmasterol, tyloindane, cetyl-alcohol, octacosanoate, coutchone, pigments, glucose, potassium chloride and quercetin, are also found in *Tylophora indica* (Figure 14.5 B). Historically, for various treatments of diseases, *Tylophora* has been used such as asthma, inflammation, infection and also as a sedative as it has anti-inflammatory, antiasthmatic, anti oxidant and tranquilizing properties. In some regions of the Karnataka, the leaves and shoots of this plant are used for the treatment of jaundice. It has also been used for the treatment of inflammation, various allergies, asthma, colds, diarrhea and arthritis. It is an excellent blood purifier and is often used for the same. The leaves and roots are also used in hydrophobia. The leaves are used for treating tumors as they have antitumor properties. For the preservation of food, its roots are highly beneficial. Ipecacuanha is considered one of the best medicines.

(a)

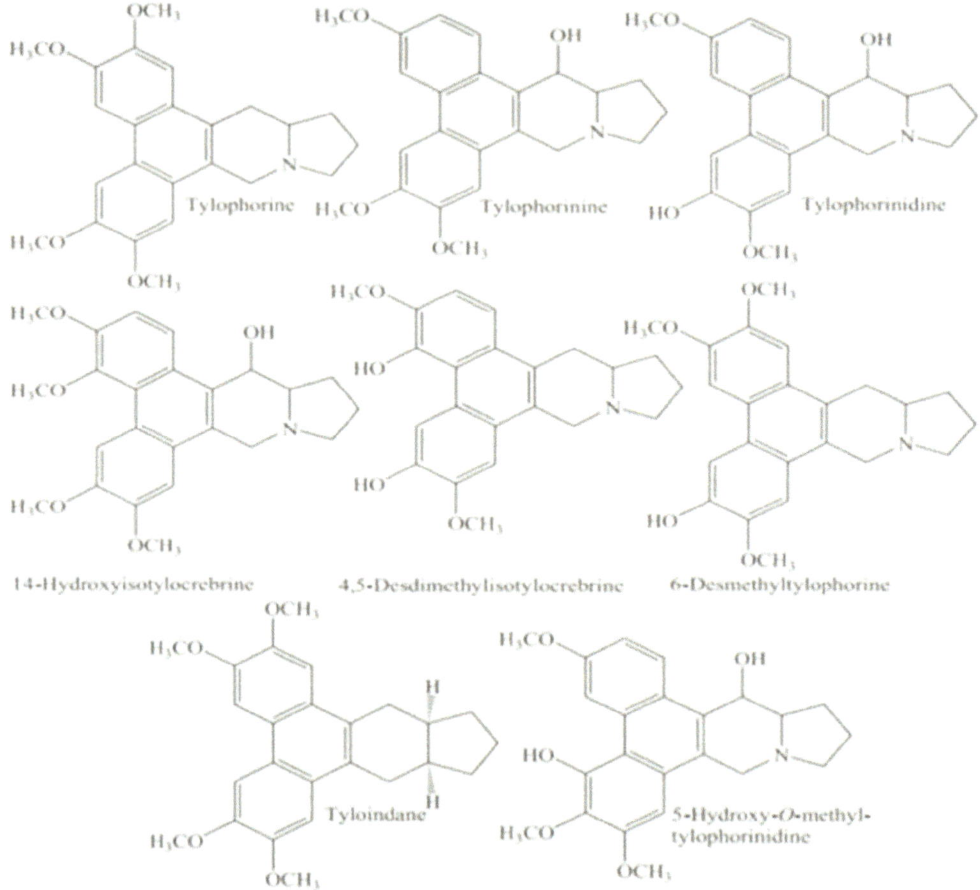

(b)

FIGURE 14.5 *Tylophora indica*: A. Plant with flowers; B. Structures of the alkaloids isolated from different parts of *T. indica*.

Source: Nazar et al. 2020.

14.6.2 Biotechnological Approaches for Improvement and Conservation

As discussed above, *Tylophora* has been utilized for the treatment of asthma since old times. Due to its effectiveness in treating asthma, *Tylophora* is in high demand in the national and international medicinal market. Due to the presence of alkaloids, *Tylophora indica* is highly effective in respiratory disorders. Plants raised from the seeds face various issues such as poor seedlings and poor germination, which is not good for commercial purpose. In addition, overharvesting of roots has become a very serious problem as it is a major source of various drugs making this plant prone to being endangered. Therefore, it is important to take essential steps for the conservation of this plant. The biotechnological approach has proven to be one of the best solutions to developing large-scale production of clones in a very short time (Table 14.5). Micropropagation is one of the frequently utilized plant tissue culture techniques for mass production and enhancing the rate of multiplication. Biotechnological techniques are now used extensively for not only clonal production but also for the conservation of endangered and threatened plant species. There are many such successful

TABLE 14.5
Some Biotechnological Approaches in *Tylophora indica* for Conservation and Improvement

Explant	Method	Culture Condition	References
Mature leaves	Somatic embryogenesis	Continuous light. CWFT. 40 lmol m^{-2} s^{-1} 25± 2 C	Jayanthi and Mandal (2001)
Young shoots	Synced	16-h PP. CWFL. 50 lmol $m^{-2}s^{-1}$. 24± 2 C. 60 % RH	Faisal and Anis (2007)
Shoot tip, leaf and nodal explants	Callus induced organogenesis	No conditions reported	Patel and Patel (2012)
Leaf explant	Adventitive rooting	Constant darkness. 25± 1C	Rashmi et al. (2012)
Response in different culture condition using various explants			
Explant and culture media	**Response**		**References**
Mature leaf + MS medium 0.5μM TDZ+1.5μm 2,4 D	Friable embryogenic callus (FEC) formation		Chandrasekhar et al. (2006)
Root +MS media+ BA or 2iP	Nodular meristem induced. 42% of explants develop into shoot buds directly in presence of 10.72μMBA 39% NMS dedifferentiated into FEC in presence of BA		Chaudhuri et al. (2004)
Leaf +MS media+ 9.04microM 2,4-D + 0.05μM kinetin. After subculture, transfer to MS media containing 9.84μM 2ip.	Embryogenic callus formation which mature in later transfer medium highly.		Jayanthi et al. (2001)
Nodal explant + MS media+ BA 2.5μM + NAA 0.5μM along with different concentration of ascorbic acid	Multiple shoot induction along with callus embryogenesis.		Anis et al. (2009)
Invitro green root system + MS media+ 5μM BA+ 0.10μM NAA	Callus Embryogenesis with mean embryo culture 18.07 Induced		Sahai et al. (2010)

approaches that have been reported by different workers in *Tylophora indica*. The synseed production or artificial seed development is an approach that has been successfully reported in *Tylophora indica*. Faisal and Anis (2007) presented their findings on ex vitro nodal segments explant for artificial seed formation in Tylophora indica. They found a maximum conversion frequency of 91% in MS media containing 2.5uM BA and 0.5µM NAA using 3% (w/v) sodium alginate and 100 mM $CaCl_2.2H_2O$. They were able to successfully preserve fake seeds at 4 degrees Celsius for 1 to 8 weeks, with subsequent conversion into plantlets.

Devendra et al. (2011) investigated the generation of artificial seeds utilizing somatic embryos produced from in vitro grown leaves. Somatic embryos were encapsulated in 12 MS media containing 50mM and 3% NA, resulting in bead formation with survival and germination frequencies of 22.4% and a plant conversion rate of 6.8%: 4 degrees Celsius was preferable to room temperature 22 degrees Celsius for the retailer. When held for a maximum of 60 days, however, the conversion rate dropped from 80% to 15% after storage survival.

14.7 *CASSIA AUGUSTIFOLIA*

Cassia augustifolia, belonging to the family Caesalpinaceae, popularly known as seena, is an annual herb highly cultivated in the southern parts of India. This plant, on a large scale, is used both in ancient and in modern systems of medicines. This plant has oval short-stalked compound leaves with large and brilliant-colored flowers with flat seeds (Figure 14.6 A). *Cassia augustifolia* is utilized extensively in the wild through therapeutic companies and local people for medicinal purposes. For commercial purposes, this plant is propagated by seed, but poor viability and low germination rate are very big threats to its survival. For large-scale production, it is now very important to develop methods for biotechnological approaches.

14.7.1 Phytochemical Evaluation

Senna contains dianthrone glycosides (1.5% to 3%). Sennoside A and B are crystalline glucosides reported from the leaves and pods. Sennosides C and D are some minor sennosides, which have been identified by some workers. All these sennosides reported from *Cassia angustifolia* were found to have a laxative effect. The plant was also found to have anthraquinones, Estragole, rhein and chrysophanol in a small amount. Flavonols present include isorhamnetin and kaempferol. Other compounds in senna include tetrahydropyran, chrysophanic acid, Estragole, 2-Propyl-tetrahydropyran-3-ol, 2-hydroxybenzoic acid, resin and essential oil. Sennosides are known for their laxative effect. Senna also contains mannitol, Caryophyllene and phytol (Al-Marzoqi, Hadi and Hameed 2016). Senna is used for the treatment of stomach problems, mostly in Eastern and Western countries. Glycosides, 1-(1,5-dimethyl-4- hexenyl)-4-methyl, 2,5-dimethyl-4-hydroxy-3(2H)- furanone, y-Tocopherol (Al-Marzoqi, Hadi and Hameed 2016) and tinnevellin are also present (Figure 14.6 B). Seena has a laxative property which has a tendency to loosen and relax muscles and boost hair texture, give relief from headaches and epilepsy. It improves the hair texture by making hair thicker and more smooth. The leaves and pods of this plant have the property of laxatives in excess. It is an important medicine for stomach-related problems by detoxifying the body. *Cassia augustifolia* is also used in the treatment of heart disease, jaundice, herpes, headache, hair loss, pimples and for relaxation of the muscle. *Cassia* has also been reported to have anti-inflammatory, antibiotic antioxidant and anticancer activities. Tetraacetyl-d-xylonic nitrile found in *Cassia* shows antiviral property.

14.7.2 Biotechnological Approaches for Improvement and Conservation

Biotechnology approaches, including rapid micropropagation, are used for the conservation of this threatened species. Rapid clonal multiplication of this plant is one of the strategies, which is

(a)

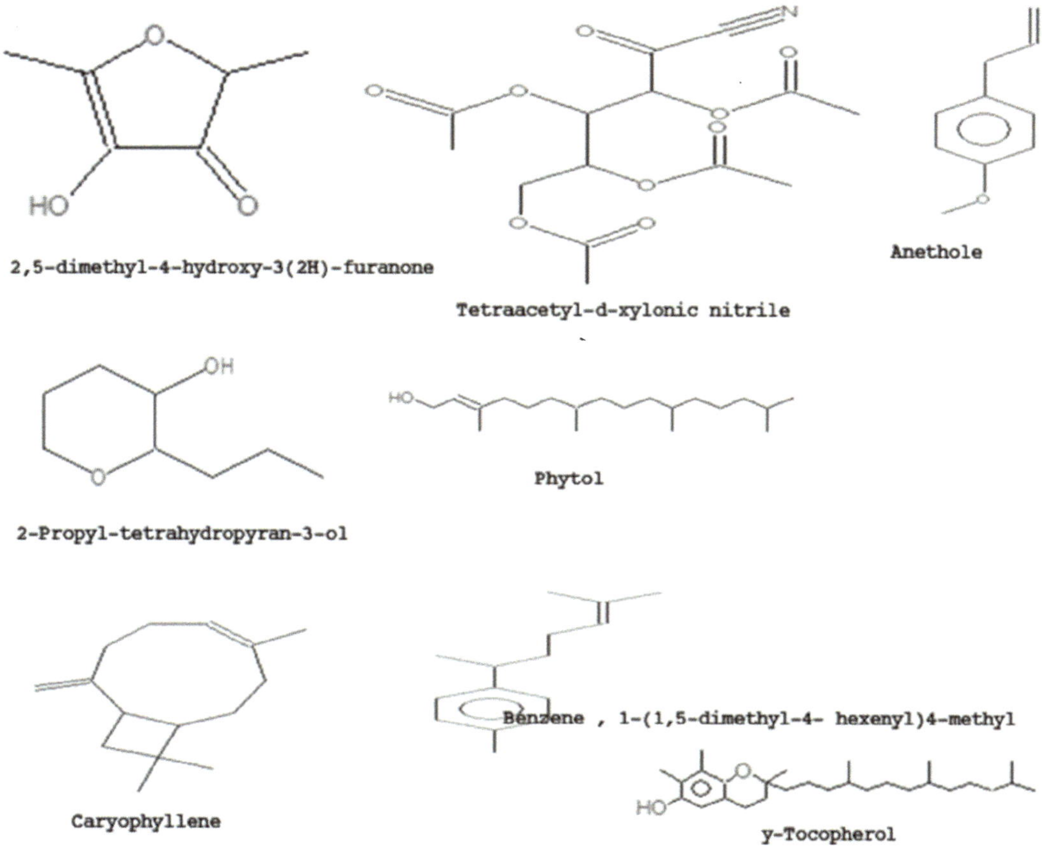

(b)

FIGURE 14.6 *Cassia augustifolia*: A. Plant with flowers, B. Some major phytochemical components of *Cassia angistifoila*.

not possible through the conventional procedure. Not only this species but many more threatened and endangered but important plant species can be multiplied rapidly, and large-scale production is now possible by different biotechnological methods. There are very few reports available using the biotechnological approach in *Cassia augustifolia*. Biotechnological approaches have been adopted by different workers for micropropagation purposes through different explants such as node segment by Siddiqui and Anis (2007), root as explant by Parveen and Shahzad (2011) and petiole explant by Siddique, Anis and Aref (2010). Siddiqui and Anis (2007) used nodal explants to achieve shoot multiplication and plant regeneration on MS media along with cytokinins, 6-benzyladenine and thidiazuron for shoot multiplication and MS medium along with indole-3-butyric acid and activated charcoal for root multiplication. Similarly, Siddique, Anis and Aref (2010), via direct organogenesis, induced shoot regeneration on MS media supplied with 2,4-dichlorophenoxy acetic acid and thidiazuron was done taking petiole as explant. Shoot number and length both increased with auxin and cytokinin subculturing. Parveen and Shahzad (2011) also work on micropropagation in *Cassia augustifolia* using root explants. They cultured root explant on Murashige and Skoog medium provided with 6-benzyladenine and kinetin along with thidiazuron.

14.8 *SIMMONDSIA CHINENSIS*

Simmondsia chinensis is commonly known as jojoba and belongs to the family Simmondsiaceae found in California, USA and Mexico. Simmondsia chinensis is a tall and dense plant with oval-shaped and oppositely placed leaves. The flowers are small, with 5 to 6 sepals and no petals, and are greenish-yellow in color. The plant is dioecious. The fruit is a three-angled-shaped capsule. The mature seed contains an oil known as liquid wax in a large amount. The seed is oval in shape (Figure 14.7A).

14.8.1 Phytochemical Evaluation

The uniqueness of the jojoba plant stem is the presence of liquid wax (jojoba oil), which consists of approximately 58% of seed weight. The major constituents in *Simmondsia chinensis* are the wax esters, eicosenyl eicosenoate, docosenyl eicosenoate, eicosenyl docosenoate and tetracosenyl eiosenoate, eicosenyl octadecenoate. Aside from oil, jojoba meal also has various potential uses. In *Simmondsia chinensis*, the majority of proteins are albumins and globulins. Also, in different concentrations, some ether extracts, the pinnitol alpha-D-galactosides, simmondsin, crude fiber,

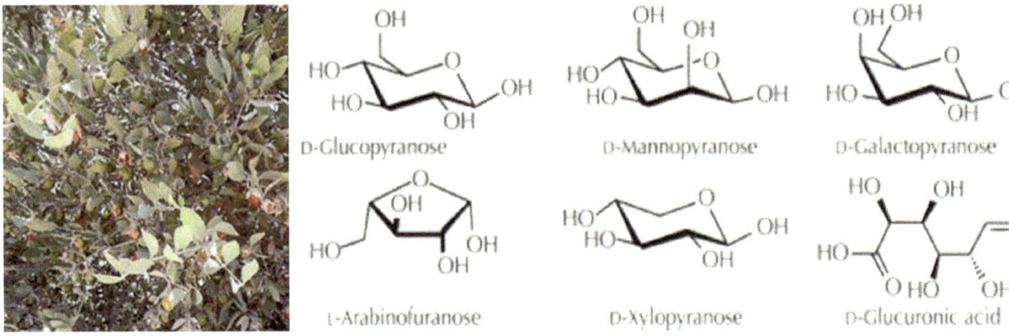

(a) (b)

FIGURE 14.7 A. *Simmondsia chinensis* B. some chemical compounds of *Simmondsia chinensis*.

Source: McKeon 2016.

1-oleoyl-3-lysophosphatidylcholine ash, 1,2-dioleoyl-3-phosphatidylcholine, nitrogen-free extract, 5-alpha-D-galactopyranosyl-D-pinitol, trypsin inhibitor factors, 3-O-methyl-D-chiro-inositol and 2-alpha-D-galactopyranosyl-D-pinitol and simmondsin ferulates, were also isolated from jojoba (Figure 14.7B). In the jojoba plant, leaves are found to possess potential medicinal properties. Jojoba extracts derived from hulls have been seen to have magnificent antibacterial properties against some bacteria such as *E. coli*, *Bacillus cereus*, *Salmonella typhimurium* by Wagdy and Taha (2012). The high effectivity of jojoba against bacteria and fungi has also been reported by Abu-Salem and Ibrahim (2014). Jojoba seeds are also found to have appetite-suppressing effects, with the elimination of simmonds in jojoba seed by treatment with NaOH at 20°C. Jojoba oil was also found to help to maintain blood cholesterol levels by some workers. Jojoba oil, for a long time, has been used in cosmetic products. Because jojoba contains a number of oxidative stress (Kahremany et al. 2019), it was reported to boost skin hydration by reducing water depletion by Meyer et al. (2008). El Mogy (2005) also proposed that jojoba oil, along with salicylic acid, is very good for rejuvenating the skin.

14.8.2 Biotechnological Approaches for Improvement and Conservation

As we discussed above, *Simmondsia chinensis* has varying medicinal importance and is used by a number of therapeutic, pharmaceutics and cosmetic industries due to its health benefits. Due to jojoba having antibacterial and skin improvement properties, this plant is extensively in demand and is the major reason for its depletion in the wild, as it is not able to multiply itself rapidly to meet the growing demand for *Simmondsia chinensis* in the market nationally and internationally. Jojoba is found to be rich in liquid wax. Because of its importance in the medicinal field, a number of researchers are now focusing on the conservation and in vitro production of this plant to meet the growing demand. Kumar, Singh and Mangal (2009) regenerated plantlets on MS medium provided with 6-Benzylaminopurine and Naphthalene acetic acid with the use of node as explant. Singh, Reddy and Patolia (2008) also used nodal segments and achieved bud initiation on MS medium supplemented with 6-benzylaminopurine and adenine. Eed and Burgoyne (2015) chose a nodal segment with two full or half leaves and shoot tip explants, grew it on MS as a basal medium in addition to Gamborg's vitamins (B5), and improved MS medium provided with various concentrations of plant growth regulators for shoot initiation. They then rooted the nodal segment in MS media supplemented with Gamborg's vitamins (B5) and Indole-3-butyric acid Chaturvedi and Meena (1989) have seen the shoot induction using the axillary bud of a nodal explant when allowed to grow in an improved Schenk and Hildebrandt's medium along with the different concentrations of 6-benzylaminopurine and indole-3-acetic acid. Rost and Hinchee (1980) reported the initiation of callus on MS medium along with sucrose and different concentrations of plant growth regulators such as Indole Acetic Acid and cytokinin like kinetin using stem, shoot tip and cotyledon as explant. In order to significantly induce numerous shoots, Agrawal et al. (2002) removed nodal segments and grew them on MS media enriched with N6-benzyladenine.

14.9 *TINOSPORA CARDIFOLIA*

Tinospora cordifolia belongs to the Menispermaceae family and is indigenous to the Indian subcontinent. It is a climbing shrub with a number of branches. Roots are arial. Stems are long, fleshy and climbing in nature. The leaves of this plant are simple, ex-stipulate, petiolate and heart-shaped. Inflorosence is recemose type. Flowers are unisex, greenish-yellow in color, with clustered male flowers and solitary female flowers. Sepals and petals are six in number in a series of two. Fruits are fleshy, orange to red in color, with thick stalk with a terminal scar. Due to the presence of curved seeds, this family is frequently called as moonseed family (Figure 14.8).

FIGURE 14.8 Morphology of different parts of *T. cardifolia*: A. Stem, B. Leaf, C. fruit, D. flower, E. aerial root.

14.9.1 Phytochemical Evaluation

A number of phytochemicals have been reported in *T. cardifolia* till now. Alkaloids such as Berberine, Choline, Tembetarine, Magnoflorine, Tinosporin, Palmetine, Isocolumbin, Aporphine alkaloids, Jatrorrhizine, Tetrahydropalmatine (Mittal, Sharma and Batra 2014) have anticancerous, anti-diabetes, immunomodulatory roles in various disease. Various steroids like β-sitosterol, δ-sitosterol, 20 β-hydroxyecdysone, Ecdysterone, Makisterone A and Giloinsterol showed activity of induced cell cycle arrest in G2/M phase and apoptosis through c-Myc suppression. Some diterpenes such as Furanolactone, Clerodane derivatives, 16-diepoxy-cleroda-13, Tinosporon, Tinosporides, Jateorine, Columbin (Mittal, Sharma and Batra 2014) are also found in *Tinospora*. Aliphatic compounds having antiinflamatory properties showed protection against 6-hydroxydopamine-induced parkinsonisms in rats. These compounds are Octacosanol, Heptacosanol and Nonacosan-15-one Dichloromethane (Mittal, Sharma and Batra 2014). *T. cardifolia* has been used as a revitalizing herb and for other medicinal purposes for far too long. *T. cardifolia* has been shown to activate macrophages and other immune cells, promoting the animal's immune potential by releasing interleukins and TNF. Antibacterial and antioxidant properties have been discovered in *T. cardifolia*. The plant's leaf extracts contain a variety of phytochemically active chemicals, including methanol, chloroform and ethyl acetate extracts, which have been shown to have potent antibiotic activity in bacteria. *T. cardifolia* could also be used as a drug vector for degenerative diseases caused by free radicals,

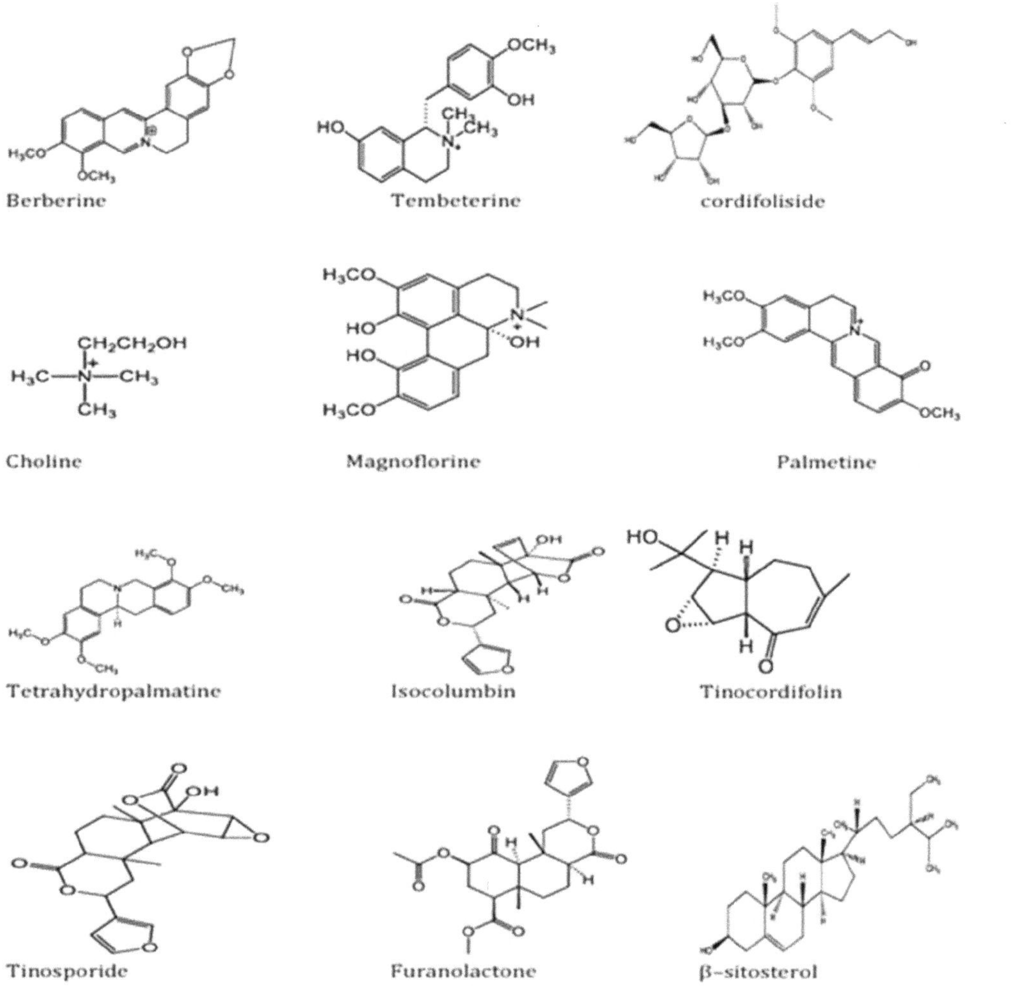

FIGURE 14.9 Chemical constituents present in *T. cardifolia*. .

according to the findings. *T. cardifolia* has also been reported to be effective in the treatment of jaundice, and it is used in the treatment of snakebite and scorpion bites when combined with other drugs. To observe a positive effect on HIV patients, a polyherbal *T. cardifolia* formulation was created. *T. cardifolia* has also been shown to help with Parkinson's disease and bronchitis, as well as promote survival and increase body resilience. It has also been shown to help the immune system fight COVID 19 (Modi et al. 2020) (Figure 14.9).

14.9.2 Biotechnological Approaches for Improvement and Conservation

Because *Tinospora cardifolia* has been found to have enormous medicinal properties, this plant has been over-utilized by therapeutic companies and local people for conventional medication. This growing demand creates a shortage of raw materials in the market. Therefore, there is a need to conserve medicinally valuable plants. A better and an effective approach, that is, the biotechnological approach, is the best option to conserve this valuable medicinal plant (Table 14.6). Micropropagation using different explants is generally used for mass multiplication and conservation of an endangered and medicinally important plant species in a short time period. Propagation through tissue culture

TABLE 14.6
Showing Some Biotechnological Approaches Done for Improvement of *T. cardifolia* Using Different Explants

Explant and culture media	Nature of response	Establishment	References
Nodal explant mixed in MS medium and woody plant medium (WPM) containing 2.32 mM kinetin along with 8.87 mM BA	*In vitro* clonal propagation	Rooted plantlets	Raghu et al. (2006)
Leaf explant in MS medium supplemented with kinetin as well as BAP	Micropropagation	Callus induction	Bhalerao et al. (2013)
Stem, leaf and nodal explant in MS medium containing 1.5 mg/l kinetin along with 6-benzylaminopurine and naphthalene acetic acid	Micropropagation	Callus induction and organogenesis	Singh et al. (2009)
Nodal explant + MS medium containing BA alone or in combination with NAA or IAA with their concentrations vary	Micropropagation	Shoot proliferation	Sivakumar et al. (2014)
Nodal explant	Mass propagation	Shoot induction	Gururaj et al. (2007)

provides clonal mass production of medicinally important and endangered species. There is a need to conserve plants with medicinal values.

14.10 *ASPARAGUS RECEMOSUS*

Asparagus racemosus, belonging to the Liliaceae family, is also known as Shatavari. It is a climbing shrub with enormous branching stems that grow naturally in the tropical and sub-tropical region of Nepal, with long roots and thick branching stems and small scaly or needle-like leaves commonly called cladodes. The inflorescence is racemose type with small white-colored flowers. The fruit is a berry with globular 3-lobed. The seeds are hard (Figure 14.10A).

14.10.1 Phytochemical Evaluation

The main active compounds of *A. recemosus* are steroidal saponins, shatavarin, sarsasapogenin, recemosol and asparagimine (Figure 14.10B). *Asparagus recemosus* has been used in Ayurveda for a long time. Roots are generally utilized in diarrhea and dysentery. When roots are boiled with some oil, this mixture is used for different skin diseases. Roots are also found to be effective in ayurvedic treatment for women's fertility-related problems when roots are boiled in milk. *Asparagus recemosus* is found to be proven a fantastic medicine in the therapy of women's infertility, miscarriage, various menopausal-related problems, acidity, bronchial infection and stomach problems. The paste of this plant is used for joint stiffness. The whole plant is utilized for diarrhea, diabetes and various brain diseases. The roots are also beneficial as a tonic used a lot for digestive and female reproductive organs.

14.10.2 Biotechnological Approaches for Improvement and Conservation

The demand for *A. recemosus* has increased due to the global acceptance of its medicinal value. But, the production of *Asparagus racemosus* depends solely on the seeds. Due to the increasing demand of the raw material for this plant in therapeutic companies, the number of this plant species is decreasing at an alarming rate. Pharmaceutical companies and local tribes are majorly responsible

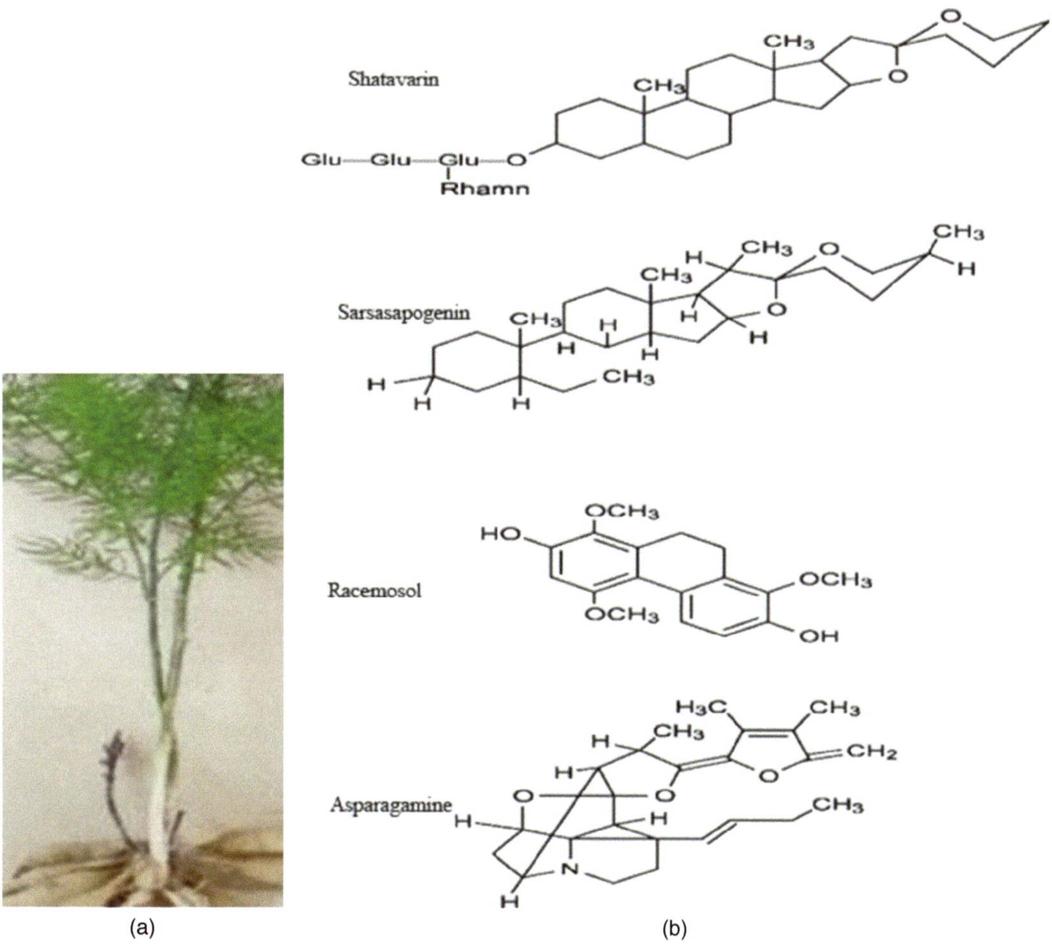

FIGURE 14.10 A. Tuber of *A. recemosus*; B. Active compounds of *A. recemosus*.

for this plant being endangered, threatened and overexploited. All these factors, along with habitat destruction, have resulted in the shrinkage of *the Asparagus* population. Therefore, there is an urgent need to adopt some solution to protect this plant from being threatened and over-exploitation. Many Biotechnological approaches have been taken into consideration for its conservation by a number of workers. The biotechnological approach can help to sustain the numbers of this important plant in the wild by adopting various plant tissue culture techniques like micropropagation, organogenesis, regeneration and many more, as shown in Table 14.7. Through biotechnological approaches, large-scale production can be achieved, which is not possible through conventional methods.

14.11 *CALOTROPIS GIGANTEA*

Calotropis gigantea, belonging to the family Asclepiadaceae (Apocynaceae), is an important medicinal plant. *Calotropis* produce milky sap hence commonly called giant milkweed. It is native to India, China and Malaysia and distributed in other countries like Algeria, Egypt, Eritrea, Ethiopia, Gambia, Myanmar, Nepal, Niger, Pakistan, Saudi Arabia, Senegal, sierra Leone, Somalia, Sudan, Syrian Arab Republic, Tanzania, Thailand, Uganda, Bolivia, Brazil, Chile, Colombia, Cuba, Dominica, Venezuela and the Virgin Islands. *Calotropis gigantea* is a branched shrub with an erect

TABLE 14.7
Biotechnological Approaches Performed by Different Scientists for Improvement of *A. recemosus*

Explant and culture media	Response	References
Shoot apex and nodal segment + MS medium containing 6-Benzyl Adenine with 8.9 μM concentration + Naphthalene Acetic Acid in 0.27 μM concentration at 25 degree C and forward rooting elongated shoots + MS medium along with IBA (7.35 μM) then transfer to MS medium with 15% coconut milk	Micropropagation	Manisha Sharan , Chitra Nene and Madhuri Sharon (2011)
Callus and shoot induction + MS media supplemented with different concentrations of NAA (a-naphthaleneacetic acid) and BAP(6-benzylaminopurine)	Callus culture	Krishna Kumar Pant and Sanu Devi Joshi (2009)
Nodal explant in MS medium supplemented with MS basal salts, sucrose, myo-inositol, agar along with a-naphthaleneacetic acid and 6- benzylaminopurine and as well as kinetin in varying concentration	Plant Regeneration	Krishna Kumar Pant, Sanu Devi Joshi (2018)
Callus culture + MS medium provided with 1.0 mg/L NAA along with 2, 4-D as well as 0.5 mg/L BAP	Saponin production	Pise and Rudra (2011)
Nodal segment in MS medium containing 3.69 μM 2-isopentyl adenine and sucrose. Then shoot clusters cultured on half-strength MS medium with 1.61 μM Naphthalene acetic acid along with 0.46 μM kinetin, adenine sulfate, malt extract, 198.25 μM phloroglucinol and sucrose	In vitro propagation	Nishritha Bopana, Sanjay Saxena (2007)
Shoot explant + MS Media along with varying concentrations of IAA and BAP	Regeneration of shoot bud and clonal propagation	Kar and Sen 1985

and cylindrical stem, which contains milky latex, often sessile leaves. The inflorescence is an umbellate cymose type with white to purplish flowers; five calyx lobes, and five corollas with valvate aestivation. Five stamens have short membranous anthers. The presence of pollinium is one of the characteristic features of this family. There are two distinct carpels with two styles. Ovules are present in many numbers. The fruit is simple and fleshy (Figure 14.11).

14.11.1 Phytochemical Evaluation

Phytochemical elements such as cardenolides, glycosides, terpenoids, phenol, sugars, flavonoids, alkaloids and saponins, as well as glycosides, calotropin, uscharin, calotoxin and calactin, have been identified in leaves by various researchers (Al-Snafi 2015). *Calotropis gigantea* has been shown to contain polyphenols, triterpene glycosides, flavonoids, steroids, tannins, coumarins, anthraquinones, saponins, cardiac glycosides, sterols and alkaloids. Benzoyllineolone, benzoylisolineolone, alotropterpenyl ester, calotropursenyl acetate and calotropfriedelenyl acetate are also found in the root. -amyrin, -amyrin, taraxasterol and its four J-isomer, taraxasteryl isovalerate, taraxasteryl acetate, -sitosterol and quercetin-3-rutinoside have also been found in the root and root bark. Calotropursenyl acetate and calotropfriedelenyl acetate, as well as akundarol isovalerate, mundarol isovalerate and quercetin -3- rutinoside, are all new norditerpenyl esters. *Calotropis gigantea* is a very important medicinal plant and has been used since ancient times for different disease treatments. Every part of this plant has a medicinal value, from leaves to the roots, bark, flowers and even latex.

FIGURE 14.11 *Calotropis gigantea.*

In Ayurveda, root bark and leaves of this plant have been used in dysentery cutaneous affections. *Calotropis gigantea* is found to have antifungal, antibacterial and insecticidal properties. For a long time, this plant has been in use to treat common flu, joint pain, cold, cramps, inflammation, diarrhea, constipation and nausea. The latex obtained from this plant also has many therapeutic implications, such as in sprain, pain, pimples and making poisnous arrow. The bark is used for toothache, inflammation, parasitic infection, cramps and joint pain. The leaves are applied for burn injury, snake bite, inflammation and body pain. The milk is useful in leukoderma, rheumatism, fever, cold, cough and leprosy.

14.11.2 Biotechnological Approaches for Improvement and Conservation

Due to the increasing therapeutical and pharmaceutical implications of *Calotropis gigantea*, this plant species is continuously being overexploited by local tribes and also by different medicinal companies. In 2012 this plant was included in the red list. This plant conventionally propagates via seeds, and the growth of this plant species is slow, which cannot meet the rising demand for *Calotropis* in the market. Propagation by seeds faces many problems in the natural environment; therefore, there is a need to conserve and propagate this plant through biotechnological approaches and plant tissue culture such as micropropagation and regeneration for large-scale production (see Table 14.8).

14.12 CONCLUSION

A global shift of preference for herbal medicines is continuously increasing the consumption of medicinal plants. The therapeutic potential of the medicinal plant has led to their depletion in the wild due to habitat loss and over-exploitation. At this rate, these plants will become extinct soon. There is a need to adopt some better techniques for conservation as well as mass production. The present tissue culture techniques have shown significant progress in the synthesis and regulation of pharmaceutically important secondary metabolites. Some new techniques, experimental tools and a modern biotechnological approach can be proven to save these valuable medicinal plants in the wild.

TABLE 14.8
Different Explant Showing Different Responses for Conservation and Improvement of *Calotropis gigantea*

Explant with culture media	Type of response	Establishment	References
Immature embryos + Murashige and Skoog supplemented with 1 mgl^{-1} 2,4-D as well as 0.1 mgl^{-1} of NAA	Callus culture	Shoot and root induction	Roy and Deepesh (1990)
Hypocotyl segment supplement with 10μM BAP along with IAA	Micropropagation	Shoot multiplication	Amuthapriya T and P Ravichandran (2014)
Mature embryo in MS medium supplemented with + 2 mg/l 2,4-D + combination of 1 mg/l BA (benzyl amino purine) and 2 mg/l NAA (naphthalene acetic acid) + 1 mg/l IBA	Indirect organogenesis	Callus formation and rooting	Abbasi (2017)
Stem with node + MS media supplemented with 2mg/l BAP+0.1mg/l NAA, while 1mg/l IBA	Micropropagation	Induction of shoot and root	Silva et al. (2017)
Hypocotyl segment + MS medium along with 2.69 μM α-naphthaleneacetic acid as well as 4.44 μM 6-benzyl-aminopurine	Cell suspension	Shoots induction	Roy et al. (2000)

REFERENCES

Abbasi, H. 2017. Indirect organogenesis in milkweed (*Calotropis procera*) from mature zygotic embryo explants. *Biotechnol. J. Biotech. Computational Biol. Bionanotech* 98(3).

Abu-Salem, F. and Ibrahim, H.M. 2014. Antimicrobial activity and phytochemicals screening of jojoba (*Simmondsia chinensis*) root extracts and latex. *Int. J. Biol. Biomol. Agric. Food Biotechnol. Eng.* 8(5): 516–522.

Agrawal, V., Prakash, S. and Gupta, S.C. 2002. Effective protocol for in vitro shoot production through nodal explants of *Simmondsia chinensis*. *Biol. Plant.* 45(3): 449–453.

Ahmad, S., Amin, M.N., Azad, M.A.K. and Mosaddik, M.A. 2002. Micropropagation and plant regeneration of *Rauvolfia serpentina* by tissue culture technique. *Pakistan J. Biol. Sci.* 5(1): 75–79.

Akram, M., Ilahi, I. and Mirza, M.A. 1986. *Acclimatization of in vitro raised plants of* Rauwolfia serpentina Benth. *VIth Int Congr* Plant Tissue Cell Cult. Minneapolis, MN, Abstr: 29.

Al-Marzoqi, A.H., Hadi, M.Y. and Hameed, I.H. 2016. Determination of metabolites products by *Cassia angustifolia* and evaluate antimicrobial activity. *J. Pharmacog. Phytother.* 8(2): 25–48.

Al-Snafi, A.E. 2015. The constituents and pharmacological properties of *Calotropis procera*-An Overview. *Int. J. Pharmacy Rev. Res.* 5(3): 259–275.

Amuthapriya, T. and Ravichandran, P. 2014. Rapid in vitro shoot multiplication through hypocotyl segment culture in *Calotropis gigantea* R. BR. *Int. J. Bioassays* 3(11): 147–153.

Anis, M., Hussain, M.K., Faisal, M.O.H.D., Shahzad, A. , Ahmad, N., Siddique, I. and Khan, H. 2009. In vitro approaches for plant regeneration and conservation of some potential medicinal plants. Advances in Plant Biotechnology and its Application, IK Int. Publ.: New Delhi, India: 407–419.

Aryal, S. and Joshi, S.D. 2009. Callus Induction and Plant Regeneration in *Rauvolfia Serpentina* (L.) Benth Ex. Kurz. *J. of Nat. His.and Mus.* 24: 82–88.

Baburaj, S. and Gunasekaran, K. 1995. In vitro differentiation of shoots from leaf callus cultures of *Withania somnifera* (L.) Dunal. *J. Indian Bot. Soc* 74(1–4): 323–324.

Balaraju, K., Agastian, P. and Ignacimuthu, S. 2009. Micropropagation of *Swertia chirata* Buch.-Hams. ex Wall.: A critically endangered medicinal herb. *Acta Physiol. Plant* 31(3): 487–494.

Barve, D.M. and Mehta, A.R. 1993. Clonal propagation of mature elite trees of *Commiphora wightii*. Plant cell, tissue and organ *cult.* 35(3): 237–244.

Bhalerao, B.M., Vishwakarma, K.S. and Maheshwari, V.L. 2013. *Tinospora cordifolia* (Willd.) Miers ex Hook. f. and Thoms.-plant tissue culture and comparative chemo-profiling study as a function of different supporting trees. *IJNPR* 4(4): 380–386.

Bopana, N. and Saxena, S. 2007. *Asparagus racemosus* – Ethnopharmacological evaluation and conservation needs. *J. Ethnopharmacol.* 110(1): 1–15.

Canter, P.H., Thomas, H. and Ernst, E. 2005. Bringing medicinal plants into cultivation: opportunities and challenges for biotechnology. *Trends Biotechnol.* 23(4): 180–185.

Chandrasekhar, T., Hussain, T.M., Gopal, G.R. and Rao, J.S. 2006. Somatic embryogenesis of *Tylophora indica* (Burm. f.) Merril., an important medicinal plant. *Int. J. Appl. Sci. Eng.* 4(1): 33–40.

Chaturvedi, H.C. and Sharma, M. 1989. In vitro production of cloned plants of Jojoba (*Simmondsia chinensis* (Link) Schneider) through shoot proliferation in long-term culture. *Plant sci.* 63(2): 199–207.

Chaudhuri, K., Das, S., Bandyopadhyay, M., Zalar, A., Kollmann, A., Jha, S. and Tepfer, D. 2009. Transgenic mimicry of pathogen attack stimulates growth and secondary metabolite accumulation. Transgenic Res. 18: 121–134.

Dass, S. and Ramawat, K.G. 2009. Studies on somatic cell variability in *Commiphora wightii* (Arnott.) Bhandari for guggulsterone production. *Nat. Prod. Radiance* 8(5): 532–536.

Dass, S., Tanwar, Y.S. and Ramawat, K.G. 2008. *Commiphora wightii* callus cultures: A new source of anthocyanin. *J. Herb. Med. Toxicol.* 2: 1720.

Devendra, B.N., Srinivas, N. and Naik, G.R. 2011. Direct somatic embryogenesis and synthetic seed production from *Tylophora indica* (Burm. f.) Merrill an endangered, medicinally important plant. *Int. J. of Botany* 7(3): 216–222.

Eed, A.M. and Burgoyne, A.H. 2015. Tissue culture of *Simmondsia chinensis* (Link) Schneider. *Banats J. Biotechnol.* 6(11): 45.

El Mogy, N.S. 2005. *US Patent No. 6,846,499.* US Patent and Trademark Office, Washington, DC.

Faisal, M., Alatar, A.A., Ahmad, N., Anis, M. and Hegazy, A.K. 2012. Assessment of genetic fidelity in *Rauvolfia serpentina* plantlets grown from synthetic (encapsulated) seeds following in vitro storage at 4 C. *Molecules*, 17(5): 5050–5061.

Faisal, M., Alatar, A.A. and Hegazy, A.K. 2013. Molecular and biochemical characterization in *Rauvolfia tetraphylla* plantlets grown from synthetic seeds following in vitro cold storage. *Appl. Biochem. Biotechnol.* 169(2): 408–417.

Faisal, M. and Anis, M. 2007. Regeneration of plants from alginate-encapsulated shoots of *Tylophora indica* (Burm. f.) Merrill, an endangered medicinal plant. *J. Hortic. Sci. Biotechnol.* 82(3): 351–354.

Fatima, N., Ahmad, N., Anis, M. and Ahmad, I. 2013. An improved in vitro encapsulation protocol, biochemical analysis and genetic integrity using DNA based molecular markers in regenerated plants of *Withania somnifera* L. *Ind. Crops Prod.* 50: 468–477.

Gantait, S.S., Dutta, K. and Majumder, J. 2017. In vitro regeneration and conservation of an endangered medicinal plant sarpagandha (*Rauvolfia serpentina L.*). *J. Horticul. Sci.* 12(1): 71–77.

Goel, M.K., Mehrotra, S., Kukreja, A.K., Shanker, K. and Khanuja, S.P.S. 2009. In vitro propagation of *Rauwolfia serpentina* using liquid medium, assessment of genetic fidelity of micropropagated plants, and simultaneous quantitation of reserpine, ajmaline, and ajmalicine. In Protocols for *In Vitro Cult. Sec. Metabolite Anal. Ar. and Med. Plants* (pp. 17–33). Humana Press, Totowa, NJ.

Gururaj, H.B., Giridhar, P. and Ravishankar, G.A. 2007. Micropropagation of *Tinospora cordifolia* (Willd.) Miers ex Hook. f and Thoms – A multipurpose medicinal plant. *Curr. Sci.* 92(1): 23–26.

Jain, R., Kachhwaha, S. and Kothari, S.L. 2012. Phytochemistry, pharmacology, and biotechnology of *Withania somnifera* and *Withania coagulans*: A review. *J. Med. Plant Res.* 6(41): 5388–5399.

Jain, V., Singh, D., Saraf, S. and Saraf, S. 2003. In vitro micropropagation of *Rauwolfia serpentine* through multiple shoot generation. *Anc. Sci. Life* 23(1): 44.

Jayanthi, M. and Mandal, P. K. 2001. Plant regeneration through somatic embryogenesis and RAPD analysis of regenerated plants in *Tylophora indica* (Burm. f. Merrill.). *In Vitro Cell Dev. Biol. Plant.* 37(5): 576–580.

Jha, T.B., Dafadar, A. and Chaudhuri, R.K. 2011. Somatic embryogenesis in *Swertia chirata* Buch.-Ham. ex Wall. – A multipotent medicinal herb. *Asian J. Biotechnol.* 3: 186–193.

Joshi, P. and Dhawan, V. 2007. Assessment of genetic fidelity of micropropagated *Swertia chirayita* plantlets by ISSR marker assay. *Biol. Plant.* 51(1): 22–26.

Kahremany, S., Babaev, I., Gvirtz, R., Ogen-Stern, N., Azoulay-Ginsburg, S., Senderowitz, H. ... and Gruzman, A. 2019. Nrf2 activation by SK-119 attenuates oxidative stress, UVB, and LPS-induced damage. *Skin Pharmacol. Physiol.* 32(4): 173–181.

Kant, T., Prajapati, S. and Parmar, A.K. 2010. Efficient micropropagation from cotyledonary node cultures of *Commiphora wightii* (Arn.) Bhandari, an endangered medicinally important desert plant. *J. Plant Dev.* 17: 37–48.

Kar, D.K. and Sen, S. 1985. Propagation of *Asparagus racemosus* through tissue culture. *Plant Cell. Tissue Organ Cult.* 5(1): 89–95.

Kaur, A. 2017. Pharmacobotanical and pharmacological evaluation of ayurvedic crude drug: *Rauwolfia serpentina* (Apocynaceae). *Int. J. Green Pharm.* (IJGP), 11(04): s686–s693.

Kaur, A. 2018. Pharmacobotanical and pharmacological evaluation of ayurvedic crude drug: *Rauwolfia serpentina* (Apocynaceae). *Int. J. Green Pharm.* (IJGP), 11(04): s686–s689.

Kulhari, A., Sheorayan, A., Kalia, S., Chaudhury, A. and Kalia, R.K. 2012. Problems, progress, and future prospects of improvement of *Commiphora wightii* (Arn.) Bhandari, an endangered herbal magic, through modern biotechnological tools: a review. *Genet. Resour. Crop Evol.* 59(6): 1223–1254.

Kumar, S., Mathur, M., Jain, A.K. and Ramawat, K.G. 2006. Somatic embryo proliferation in *Commiphora wightii* and evidence for guggulsterone production in culture. *Ind. J.l of Biotech.* 5: 217–222.

Kumar, S., Singh, N. and Mangal, M. 2009. Micropropagation of *Simmondsia chinensis* through enhanced axillary branching from nodal segments. *J. Plant Biol.* 36: 75–81.

Kumar, S., Sonie, K.C. and Ramawat, K.G. 2004. Development of resin canals during somatic embryogenesis in callus cultures of *Commiphora wightii*. *Indian J. Biotech.* 3: 267–270.

Kumar, S., Suri, S.S., Sonie, K.C. and Ramawat, K.G. 2003. Establishment of embryonic cultures and somatic embryogenesis in callus culture of guggul-*Commiphora wightii* (Arnott.) Bhandari. *Indian J. Exp. Biol.* 41: 69–77.

Kumar, V. and Chandra, S. 2014. High frequency somatic embryogenesis and synthetic seed production of the endangered species *Swertia chirayita*. *Biologia*. 69(2): 186–192.

Mathur, M., Jain, A.K., Dass, S. and Ramawat, K.G. 2007. Optimization of guggulsterone production in callus cultures of *Commiphora wightii* (Arnott.) Bhandari. *Indian J. Biotech.* 6: 525–531.

McKeon, T.A. 2016. Emerging industrial oil crops. In *Industrial Oil Crops* (pp. 275–341). AOCS Press.

Meyer, J., Marshall, B., Gacula Jr, M. and Rheins, L. 2008. Evaluation of additive effects of hydrolyzed Jojoba (*Simmondsia chinensis*) esters and glycerol: A preliminary study. *J. cosmet. Dermatol.* 7(4): 268–274.

Mittal, J., Sharma, M. M. and Batra, A. 2014. *Tinospora cordifolia*: a multipurpose medicinal plant. *J. Med. Plants* 2(2): 32–47.

Modi, B., Kumari Shah, K., Shrestha, J., Shrestha, P., Basnet, A., Tiwari, I. and Prasad Aryal, S. 2020. Morphology, biological activity, chemical composition, and medicinal value of *tinospora cordifolia* (willd.) miers. *Adv. J. Chem.* Section B: 36–54.

Mondal, S., Silva, J.T. and Ghosh, P.D. 2011. In vitro flowering *in Rauvolfia serpentina* (L.) Benth. ex. Kurz. *Int. J. Plant Dev. Biol.* 5(1): 75–77.

Murthy, H.N., Dijkstra, C., Anthony, P., White, D.A., Davey, M.R., Power, J.B. ... and Paek, K.Y. 2008. Establishment of *Withania somnifera* hairy root cultures for the production of withanolide A. *J. Integr. lant boil.* 50(8): 975–981.

Nazar, S., Hussain, M.A., Khan, A., Muhammad, G. and Bukhari, S.N.A. 2020. Alkaloid-rich plant *Tylophora indica*; Current trends in isolation strategies, chemical profiling and medicinal applications. *Arab. J. Chem.* 13(8): 6348–6365.

Pandey, A.K. and Tripathi, Y.C. 2017. Biotechnological approach for medicinal plant conservation and enhanced production of secondary metabolites. *Int. J. Pharmacogn.* 4(11): 352–362.

Pandey, V., Misra, P., Chaturvedi, P., Mishra, M.K., Trivedi, P.K. and Tuli, R. (2010). *Agrobacterium* tumefaciens-mediated transformation of Withania somnifera (L.) Dunal: an important medicinal plant. *Plant Cell rep.* 29(2): 133–141.

Pandey, V.P., Kudakasseril, J., Cherian, E. and Patani, G. 2007. Comparison of two methods for in vitro propagation of *Rauwolfia serpentina* from nodal explants. Indian Drugs-Bombay 44(7): 514.

Pant, K.K. and Joshi, S.D. 2009. In vitro multiplication of wild Nepalese *Asparagus racemosus* through shoots and shoot induced callus cultures. *Bot. Res. Intl.* 2(2): 88–93.

Pant, K.K. and Joshi, S.D. 2018. In vitro bud induction from the nodes of *Asparagus recemosus* wild and their germination behaviours. *Int. J. Tech. Res. Sci.* 3: 47–50.

Pareek, A. and Pareek, L.K. 2012. *Commiphora wightii* (Guggal), an endangered medicinal plant of rajasthan needs attention of biotechnologists for its conservation. *Res. J. Pharm. Biol. Chem. Sci.* 3(1): 83–89.

Parveen, S. and Shahzad, A. 2011. A micropropagation protocol for *Cassia angustifolia* Vahl. from root explants. *Acta Physiol. Plant.* 33(3): 789–796.

Patel, A. and Patel, I.C. 2012. Micropropagation of medicinally important climber: *Tylophora indica* Merrill. future demand. *Life Sci. Leafl.* 2012: 311–316.

Pise, M., Rudra, J., Bundle, S., Begde, D. and Upadhyay, A. 2011. Shatavari production from in vitro cultures of *Asparagus racemosus* Wild. *J. Med. Plant Res.* 5(4): 507–513.

Raghu, A.V., Geetha, S.P., Martin, G., Balachandran, I. and Ravindran, P.N. 2006. In vitro clonal propagation through mature nodes of *Tinospora cordifolia* (Willd.) Hook. F. and Thoms.: An important ayurvedic medicinal plant. *In Vitro Cell. Dev. Biol. Plant.* 42(6): 584–588.

Rani, G., Virk, G.S. and Nagpal, A. 2003. Callus induction and plantlet regeneration in *withania somnifera* (L.) Dunal. *In Vitro Cell. Dev. Biol. Plant* 39: 468–474.

Rani, A., Kumar, M. and Kumar, S. 2014. Micropropagation of serpgandha (*Rauvolfia serpentina* L. Benth): An endangered medicinal plant. *Appl. Biol. Res.* 16(1): 114–118.

Rani, G. and Grover, I.S. 1999. In vitro callus induction and regeneration studies in *Withania somnifera. Plant Cell Tissue Organ Cult.* 57(1): 23–27.

Rashmi, M.P., Vinaya, M., Vedamurthy, A.B. and Nayeem, A. 2012. Effectiveness of auxins in inducing in vitro adventitious root formation in Tylophora indica (Burm. F.) Merrill. *J. Cell Tissue Res.* 12(3): 3357–3360.

Ray, S., Ghosh, B., Sen, S. and Jha, S. 1996. Withanolide production by root cultures of *Withania somnifera* transformed with Agrobacterium rhizogenes. *Planta Med.* 62(06): 571–573.

Rochelle, G.K., Bella, P., Kota, S., Abbagani, S., Chithakari, R. and Reuben, T.C. 2013. In vitro plantlet regeneration from leaf and stem calluses of *Rauwolfia tetraphylla (R. canescens)* and confirmation of genetic fidelity of plantlets using the ISSR-PCR method. *J. Herbs Spices Med. Plants* 19(1): 66–75.

Rost, T.L. and Hinchee, M.A. 1980. Preliminary report of the production of callus, organogenesis and regeneration of Jojoba (*Simmondsia chinensis, Link, Schneid.*) in tissue culture. *Int. J. Hortic. Sci.* 55(3): 299–305.

Roy, A.T. and De, D.N. 1990. Tissue culture and plant regeneration from immature embryo explants of *Calotropis gigantea* (L.) R. Br. *Plant Cell Tissue Organ Cult.* 20(3): 229–233.

Roy, A.T., Koutoulis, A. and De, D.N. 2000. Cell suspension culture and plant regeneration in the latex-producing plant, *Calotropis gigantea* (Linn.) R. Br. *Plant cell Tissue Organ Cult.* 63(1): 15–22.

Sahai, A., Shahzad, A. and Sharma, S. 2010. Histology of organogenesis and somatic embryogenesis in excised root cultures of an endangered species *Tylophora indica* (Asclepiadaceae). *Aust. J. Bot.* 58(3): 198–205.

Samaddar, T., Jha, S. and Jha, T.B. 2014. Indian *Swertia* from Eastern Himalaya: Strategies of conservation and biotechnological improvements. *The Gentianaceae-Volume 1: Characterization and Ecology* 1: 279–301.

Saritha, K.V. and Naidu, C.V. 2007. In vitro flowering of *Withania somnifera* Dunal – An important antitumor medicinal plant. *Plant Sci.* 172: 847–851.

Sen, J. and Sharma, A.K. 1991. Micropropagation of *Withania somnifera* from germinating seeds and shoot tips. *Plant Cell Tissue Organ Cult.* 26(2): 71–73.

Senapati, S.K., Lahere, N. and Tiwary, B.N. 2014. Improved in vitro clonal propagation of *Rauwolfia serpentina* L. Benth–An endangered medicinal plant. *Plant Biosystems* 148(5): 885–888.

Shahid, M., Dar, F.K., Ismaeel, A.Y., Al-Mahmood, A., Sindi, K.A., Malik, A. and Khan, H.M. 2013. Plant natural products as a potential source of antimicrobial agents: an overview and a glimpse on recent developments. *Recent Trends in Biotechnology and Therapeutic Applications of Medicinal Plants*, 93–107.

Sharan, M., Nene, C. and Sharon, M. 2011. Regeneration of *Asparagus racemosus* by shoot apex and nodal explants. *Asian. J. Plant Sci. Res.* 1: 49–56.

Sharma, S., Rathi, N., Kamal, B., Pundir, D., Kaur, B. and Arya, S. 2010. Conservation of biodiversity of highly important medicinal plants of India through tissue culture technology-a review. *Ag. Biol. J. North America* 1(5): 827–833.

Siddique, I. and Anis, M. 2007. In vitro shoot multiplication and plantlet regeneration from nodal explants of *Cassia angustifolia* (Vahl.): a medicinal plant. *Acta Physiol. Plant.* 29(3): 233–238.

Siddique, I., Anis, M. and Aref, I.M. 2010. In vitro adventitious shoot regeneration via indirect organogenesis from petiole explants of *Cassia angustifolia Vahl.* – a potential medicinal plant. *Appl. Biochem. Biotechnol.* 162(7): 2067–2074.

Silva, L.N.S., Perea, P.C.D., Senanayake, S.G.J.N. and Dahanayake, N. 2017. Development of an appropriate protocol for micropropagation of *Calotropis gigantea* (L.)(Wara). *Trop. Agric. Res. Ext.* 20(3–4).
Singh, A., Reddy, M.P. and Patolia, J.S. 2008. An improved protocol for micropropagation of elite genotypes of *Simmondsia chinensis (Link) Schneider*. *Biol. Plant.* 52(3): 538–542.
Singh, A., Sah, S.K., Pradhan, A., Rajbahak, S. and Maharajan, N. 2009. In vitro study of *Tinospora cordifolia (Willd.) Miers* (Menispermaceae). *Botanica Orientalis*: *J. Plant Sci.* 6: 103–105.
Singh, N., Bhalla, M., de Jager, P. and Gilca, M. 2011. An overview ashwagandha: A Rasayana (rejuvenator) of Ayurveda. *Afr. J. Tradit. Complement. Altern. Med.* 8(5S): 208–213.
Singh, N., Garg, A., Yadav, K. and Kumari, S. 2010. Influence of growth regulators on the explants of *Commiphora Mukul* (Hook. ex Stocks) Engl. under in vitro conditions. *Researcher* 2(7): 41–48.
Singh, R. L., Singh, P. and Agarwal, A. 2012. Chemical constituents and bio-pharmacological activities of *Swertia chirata*: A review. *NPAIJ* 8(6): 238–247.
Sivakumar, V., Rajan, M.D., Sadiq, A.M. and Jayanthi, M. 2014. In vitro micropropagation of *Tinospora cordifolia (Willd.)* Miers ex Hook. F. and Thoms-An important medicinal plant. *J. Pharmacogn. Phytochem.* 3(2): 208–213.
Soni, V. 2010. Efficacy of in vitro tissue culture versus stem cuttings for propagation of *Commiphora wightii* in Rajasthan, India. *Conservation Evidence*, 7: 91–93.
Sudha, G.G. and Seeni, S. 1996. In vitro propagation of *Rauwolfia micrantha*, a rare medicinal plant. *Plant Cell Tissue Organ Culture* 44(3): 243–248.
Suthar, S. and Ramawat, K.G. 2010. Growth retardants stimulate guggulsterone production in the presence of fungal elicitor in fed-batch cultures of Commiphora wightii. Plant. Biotech. Rep. 4(1): 9–13.
Teli, N.P., Patil, N.M., Pathak, H.M., Bhalsing, S.R. and Maheshwari, V.L. 1999. *Withania somnifera* (Ashwagandha): Regeneration through meristem culture. *J. Plant Biochem. Biotechnol.* 8(2): 109–111.
Tripathi, L. and Tripathi, J.N. 2003. Role of biotechnology in medicinal plants. *Trop. J. Pharm. Res.* 2(2): 243–253.
Uikey, D.S., Tripathi, M.K., Tiwari, G., Patel, R.P. and Ahuja, A. 2016. Embryogenic cell suspension culture induction and plantlet regeneration of *Rauvolfia serpentina* (L.) Benth.: influence of different plant growth regulator concentrations and combinations. *Med. Plants – Int. J. Phytomed. Relat. Ind.* 8(2): 153–162.
Umadevi, M., Rajeswari, R., Rahale, C.S., Selvavenkadesh, S., Pushpa, R., Kumar, K S. and Bhowmik, D. 2012. Traditional and medicinal uses of *Withania somnifera*. *J. Pharm. Innov.* 1(9, Part A): 102.
Wadegaonkar, P.A., Bhagwat, K.A. and Rai, M.K. 2006. Direct rhizogenesis and establishment of fast growing normal root organ culture of *Withania somnifera* Dunal. *Plant Cell Tiss. Org. Cult.* 84(2): 223–225.
Wagdy, S.M. and Taha, F.S. 2012. Primary assessment of the biological activity of jojoba hull extracts. *Life Sci. J.* 9(2): 244–253.
Wawrosch, C., Maskay, N. and Kopp, B. 1999. Micropropagation of the threatened Nepalese medicinal plant *Swertia chirata* Buch. – Ham. ex Wall. *Plant Cell Rep.* 18(12): 997–1001.
Yahya, A.F., Hyun, J., Lee, J. and Jung, M. 2007. Effect of explant types, auxin concentration and light condition on in vitro root production and alkaloid content of *Rauvolfia serpentina* (L.) Benth. ex Kurz. *J. Korean For. Soc.* 96(2): 178.
Yusuf, A., Rathore, T.S. and Shekhawat, N.S. 1999. Micropropagation of *Commiphora wightii* (Arnott) Bhandari–A threatened medicinal plant of semi-arid region. *Indian J. Plant Genet. Res.* 12(3): 371–375.
Zafar, N., Mujib, A., Ali, M., Tonk, D., Gulzar, B., Malik, M. ... and Mamgain, J. 2019. Genome size analysis of field-grown and tissue culture regenerated *Rauvolfia serpentina* (L) by flow cytometry: Histology and scanning electron microscopic study for in vitro morphogenesis. *Ind. Crops Prod.* 128: 545–555.

Index

Note: Page numbers in *italics* indicate figures and in **bold** indicate tables on the corresponding pages.